Programming the Dynamic Analysis of Structures

Programming the Dynamic Analysis of Structures

P Bhatt

CRC Press
Taylor & Francis Group
Boca Raton London New York

CRC Press is an imprint of the
Taylor & Francis Group, an **informa** business
A SPON PRESS BOOK

CRC Press
Taylor & Francis Group
6000 Broken Sound Parkway NW, Suite 300
Boca Raton, FL 33487-2742

First issued in paperback 2019

Publisher's Note:
This book has been prepared from camera-ready-copy supplied by the author.

ISBN-13: 978-0-419-15610-9 (hbk)
ISBN-13: 978-0-367-86349-4 (pbk)

British Library Cataloguing in Publication Data
A catalogue record for this book is available from the British Library

Library of Congress Cataloging in Publication data
A catalog record for this book has been requested

Visit the Taylor & Francis Web site at
http://www.taylorandfrancis.com

and the CRC Press Web site at
http://www.crcpress.com

Dedicated with affection to the younger generation

(Arun, Ranjana, Ramendra, Sashidhar, Sudeep, Sujaatha, Sumeeta, and Vinod).

CONTENTS

PREFACE

The subject of structural dynamics is an important branch of structural engineering. Its importance has increased with the need for more and more flexible structures such as tall buildings, long span bridges, cable roofs, etc., many founded in earthquake prone areas of the world. Many books covering either the theoretical aspects of structural dynamics or the mathematical aspects of eigen-value problems or the finite element method or listing of finite element programs are available, but rarely in the same book. The calculations involved in structural dynamics are by their very nature numerically intensive and the use of a digital computer is almost obligatory, even for the solution of simple problems. The existing books do give some simple programs, almost as an afterthought, but rarely in any detail.

The object of this book is to provide, in a single comprehensive volume, both the theory and the associated computer programs. Only methods amenable to automatic computation are included and all hand calculation methods are omitted. Full details of the theory of solution methods for linear eigen-value problems, finite element and finite strip methods are given. All concepts are explained in detail and illustrated by numerous fully worked out numerical examples. Corresponding elementary programs follow many of the methods. These programs have two main aims. Firstly to teach the reader the steps needed to convert theory to computer programs and secondly to provide programs which the reader can use to check their working of numerical examples. These programs are finally translated to 'full blown' programs in FORTRAN-90 for the eigen-value solution of two-dimensional rigid-jointed frames, plane grids, elastic plates, with the dynamic stiffness matrices established using finite element and finite strip methods. Programs are also provided for the solution by direct integration of differential equations. It is hoped that the book will be welcomed both by students studying courses in structural dynamics and also by practicing engineers.

As in any such undertaking, many people have helped and I express my sincere thanks to all. To my colleague Mr Robert Watson for some of the diagrams, to Mrs Tessa Bryden for enthusiastic secretarial assistance, to the late Ernie Hinton for enthusiastic review of the initial proposal and subsequent unstinting encouragement and provision of his programs, to Sheila, Arun and Ranjana for constantly asking me 'When is the book going to finish?', to the many editors at Spon Press for mild pressure and for being very understanding. Finally when deciding on the 'balance' between various topics of the book, to Ovid for the advice 'Medio tutissimus ibis' (you will go most safely in the middle).

P. Bhatt
2nd October (Mahatma Gandhi's birthday) 2001, Glasgow

CHAPTER 1

SINGLE DEGREE OF FREEDOM SYSTEM - I

1.1 INTRODUCTION

Structures in practice are subjected to a variety of forces, both static and dynamic. Static forces, such as gravity forces, remain constant with time. Structures are also subjected to dynamic forces, which vary with respect to time. Some of these forces can act over a long period of time while others act over a relatively short period. Typical of forces which act over a long period of time are vehicular loading on a bridge and wind loads on buildings. On the other hand, forces due to an earthquake or an explosion act over a fairly short period of time. In the vast majority of design situations, the dynamic forces affect the serviceability limit state. For example, vibrations could cause discomfort to users of a structure such as a building or a bridge. In some cases vibrations could lead to the malfunctioning of delicate apparatus. In extreme cases, dynamic forces could affect the ultimate limit state by causing collapse due to violent shaking during an earthquake or cause fatigue failure of joints and components. As will be shown later, dynamic properties of a structure are governed mainly by the mass and stiffness of the structure. Many design advances of recent years, such as the use of higher strength materials, the use of welding or friction grip bolts in steel structures, the wide spread use of prestressed concrete, the emergence of flexible structures such as long span bridges and tall buildings, have all made structures more sensitive to dynamic forces. Many of the changes in construction practice have also reduced the inherent damping present in structures, making them more susceptible to vibration. It is for these reasons that the study of structural dynamics has assumed great importance. The nature of the dynamic forces that act on a structure vary widely. Some, such as the force due to a rotating machine, can be described almost completely as a function of time both in magnitude and direction. Such forces are deterministic forces. On the other hand, forces due to wind can only be described in terms of statistical properties, such as mean and standard deviation. Such forces are called stationary random forces. Forces due to an earthquake are even more complicated. Each earthquake is almost unique. Earthquake forces cannot be described even in statistical terms. Such forces are called non-stationary random forces. In this book, only analysis of structures subjected to deterministic force is considered.

1.2 INERTIAL FORCE

According to Newton's second law of motion, the resultant force acting on a particle and the corresponding acceleration are related by the equation

Resultant force = Mass x Acceleration

If a particle is in static equilibrium, then the acceleration is equal to zero and, hence, the resultant force should also be equal to zero. It should be remembered that since both force and acceleration are vector quantities, the relationship is a vector equation.

In contrast to a problem in statics, while considering equilibrium in a dynamic problem, acceleration is not equal to zero. Newton's second law can be restated as

Resultant force on the mass - Mass x Acceleration = 0

This shows that a dynamic problem can be treated as an equivalent static problem by including, in addition to external forces acting on the mass, an additional force equal to - (mass x acceleration). This additional force is called inertial force. This way of looking at equilibrium under dynamic situation is called D'Alembert's principle.

1.3 DAMPING FORCE

When a structure is vibrating, it moves relative to the surrounding medium, such as air as in the case of most structures or water as in the case of structures such as oil rigs. The surrounding medium resists motion of the structure and causes additional forces to act on the structure. These forces generated by the relative motion with respect to the surrounding medium are called damping forces. It should be appreciated that damping forces can also arise due to relative motion between parts of the structure such as at a bolted joint in a steel structure or across cracks in a reinforced concrete structure.

1.4 SINGLE DEGREE OF FREEDOM SYSTEM

Consider the simple 'structure' shown in Fig. 1.1. The mass M is attached to the support through a 'weightless' spring of stiffness K and a dashpot simulating damping normally present in all structures. It is assumed for simplicity, that the mass can move only horizontally. In other words, the structure has only one degree of freedom of movement.

Figure 1.2 shows some real structures modelled for mathematical purposes as single degree freedom systems. Fig. 1.2a shows a simply supported beam of span L and flexural rigidity EI, supporting a concentrated mass at midspan. The 'spring' stiffness K is the force required to cause unit displacement at the midspan of a simply supported beam. Therefore $K = 48\ EI/L^3$.

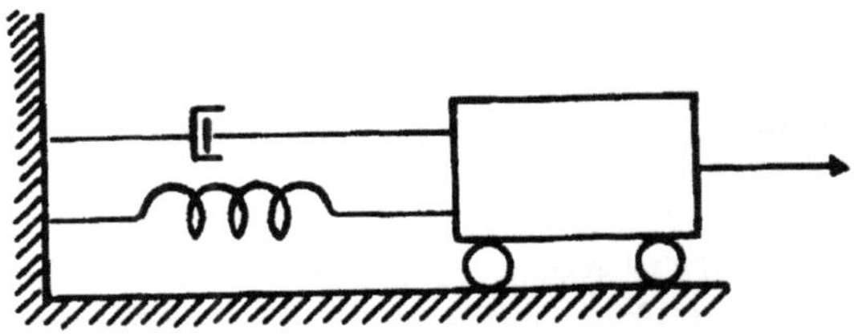

Fig. 1.1 Mass-spring-damper System

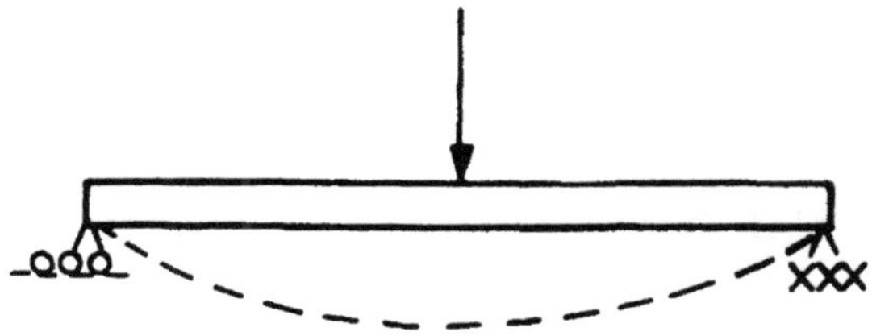

Fig. 1.2a A simply supported beam

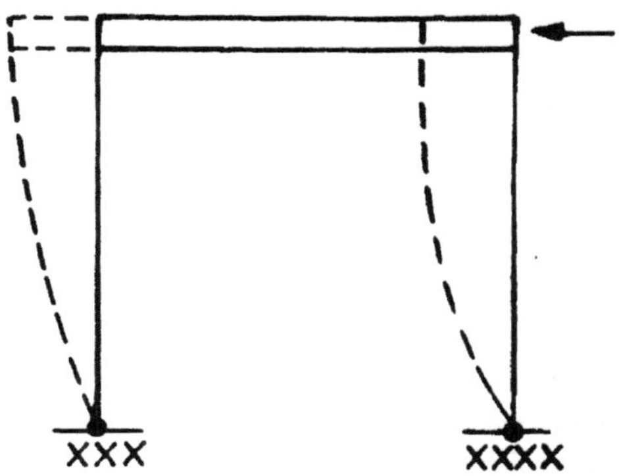

Fig. 1.2b Single bay portal frame

Figure 1.2b is a single bay portal. The height of the columns is H and their flexural rigidity is EI. It is assumed that the beam is sufficiently rigid to prevent rotation of the columns at the top so that the motion of the mass can be described by the sway displacement at the top of columns. In this case, the 'spring' stiffness K is the force required to cause unit sway displacement of the two legs of the portal frame.

Fixed feet columns, $K = 2 \text{ legs} \times \{12\, EI/H^3\}$.

Pinned feet column, $K = 2 \text{ legs} \times \{3\, EI/H^3\}$.

The assumption that the spring is 'weightless' is made purely to simplify the problem at this early stage of discussion. In reality, the columns and beams, which contribute to spring stiffness, are not weightless.

In Fig. 1.1, the damping present in the structure is shown by a dashpot. It should be appreciated that in real structures, in general, there are no identifiable dashpots causing damping. A dashpot is only a simple way of modelling the presence of damping in structures.

The rest of this chapter is devoted to the study of single degree of freedom systems. The study of SDOF systems is important because it brings out many of the important properties affecting the behaviour of structures subjected to dynamic loads. Very often in practice, an SDOF system is the simplest idealisation used for the preliminary study of quite complex structures. In addition, as will be shown in Chapter 3, multi-degree freedom systems can be 'reduced' to a series of SDOF systems, thus facilitating the study of complex systems.

1.5 MATHEMATICAL STUDY OF THE SDOF SYSTEM

The differential equation governing the behaviour of the system shown in Fig. 1.1 is established quite simply by using the concept of dynamic equilibrium. Consider the forces acting on the free body shown in Fig. 1.3.

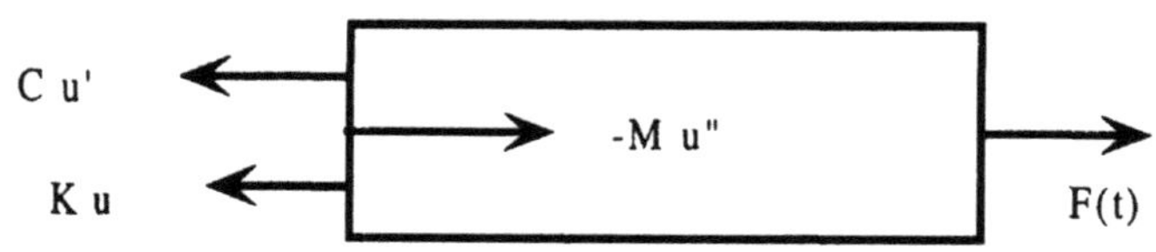

Fig. 1.3 Forces on the free body

The forces acting on the mass are the spring force, damping force, inertial force and external force. It should be remembered that the damping force always opposes motion. The dynamic equilibrium requires that the sum of the forces is equal to zero. Assuming that the positive direction of motion is to the right, let the displacement of the mass be u(t). The forces acting on the mass are

i. Spring force acting on the mass = K u(t)

ii. Inertial force = -M $\dfrac{d^2u}{dt^2}$

Note that the inertial force acts in the positive direction.

iii. Damping force

It is conventional to assume that damping force is proportional to the velocity of the mass. Therefore

Damping force = C $\dfrac{du}{dt}$

C = Coefficient of viscous damping.
iv. External force = $F(t)$, which is a function of time, t.
Summing up the forces to zero for dynamic equilibrium, we have

$$M\frac{d^2u}{dt^2} + C\frac{du}{dt} + Ku = F(t) \tag{1.1}$$

For convenience in writing mathematical expressions, let

$$K/M = \omega^2,\ C/M = 2\beta \tag{1.2}$$

The differential equation (1.1) can therefore be expressed as

$$\frac{d^2u}{dt^2} + 2\beta\frac{du}{dt} + \omega^2 u = \frac{F(t)}{M} \tag{1.3}$$

1.6 INFLUENCE OF GRAVITATIONAL FORCES

Consider the system shown in Fig. 1.4. The system is identical in all respects to that shown in Fig. 1.1, except that the motion is vertical. Under static conditions, the displacement of the mass is equal to the extension Δ of the spring due to the weight W. Therefore, Δ is equal to W/ K. Under dynamic conditions, when considering the forces acting on the mass, in addition to the forces shown in Fig. 1.3, we have to include the weight, W = Mg, g = acceleration due to gravity. The equilibrium equation is therefore given by

$$M\frac{d^2(u+\Delta)}{dt^2} + C\frac{d(u+\Delta)}{dt} + K(u+\Delta) = W + F(t) \tag{1.4}$$

In the above equation, the displacement u is measured from the static position. Therefore, (u + Δ) is the total displacement. Since Δ is a constant and $K\Delta = W$, the above equation simplifies to equation (1.1). This shows that equation 1.1 is applicable to cases where acceleration due to gravity is to be included provided that the dynamic displacement is measured from the static position of rest as the origin.

1.7 SOLUTION OF THE DIFFERENTIAL EQUATION

The differential equation (1.1) is an ordinary differential equation with constant coefficients. The solution is obtained as the sum of a complementary solution and the particular integral.

Complementary solution is the solution to the equation when F(t) is equal to zero. Therefore, complementary function is the solution of the equation

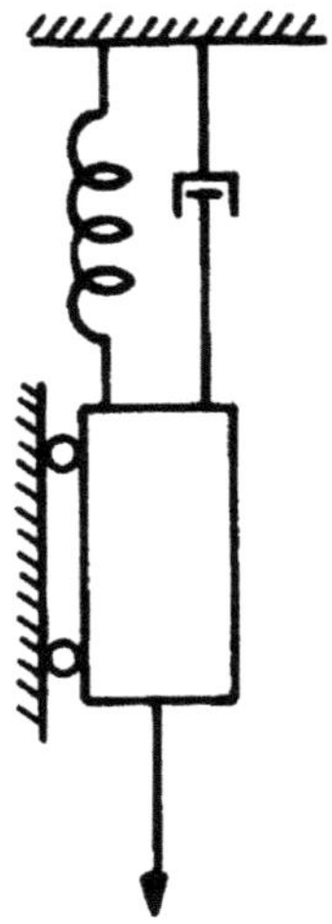

Fig. 1.4 Mass-spring-damper system

$$\frac{d^2u}{dt^2} + 2\beta\frac{du}{dt} + \omega^2 u = \frac{F(t)}{M} \tag{1.5}$$

The solution to $u(t)$ in the above equation is obtained by assuming that

$$u(t) = A\,e^{\alpha t} \tag{1.6}$$

where A is an arbitrary constant and α is yet to be determined.
Differentiating $u(t)$ with respect to time t, we have

$$u'(t) = \alpha A e^{\alpha t} = \alpha\,u(t) \tag{1.7}$$

$$u''(t) = \alpha^2 A e^{\alpha t} = \alpha^2 u(t) \tag{1.8}$$

Substituting equations (1.7) and (1.8) in equation (1.5), and simplifying

$$(\alpha^2 + 2\,\alpha\beta + \omega^2)\,u(t) = 0 \tag{1.9}$$

Since u(t) is not equal to zero, the terms inside the bracket must be zero

$$\alpha^2 + 2\,\alpha\beta + \omega^2 = 0 \tag{1.10}$$

This is called the characteristic equation of the differential equation. Since the characteristic equation is a quadratic equation, the roots are given by

$$\alpha_1,\,\alpha_2 = -\,\beta \pm \sqrt{\{\beta^2 - \omega^2\}} \tag{1.11}$$

The complementary equation is thus given by

$$u(t) = A_1\,e^{\alpha_1 t} + A_2\,e^{\alpha_2 t} \tag{1.12}$$

The constants A_1 and A_2 are determined from the initial conditions prescribing the displacement and velocity at t = 0.

The particular integral is the solution of the equation, when the force F(t) is present but without any reference to the initial conditions.

Because the complementary function part of the solution depends on the initial boundary conditions and exists even when the external force F(t) is zero, the complementary function is called the natural motion solution. Similarly, because the particular integral part of the solution depends on the external force, this solution is called the forced motion solution.

Depending on the presence of damping and external force, the solution to the differential equation is obtained for two distinct cases as follows.

i. Free vibration: Free vibration refers to the case when the external force is equal to zero and the motion results when the system is disturbed from its state of rest and allowed to vibrate. If damping is equal to zero, then such a motion is called undamped free vibration. On the other hand, if damping is present, then the resulting motion is called damped free vibration.

ii. Forced vibration: When vibration takes place due to an external vibratory force acting on the system, then the resulting motion is described as forced vibration. As in the case of free vibration, depending on the presence or absence of damping, one can have damped forced vibration or undamped forced vibration respectively.

1.8 SOLUTION TO THE FREE VIBRATION PROBLEM

The solution to $u(t)$ is given by the equations (1.11) and (1.12). Depending on the sign of $(\omega^2 - \beta^2)$ in equation (1.11), three possible cases arise, as follows.

i. $\beta < \omega$. This is called an underdamped case for reasons to be explained in the next section.

Setting $\omega_d = \sqrt{(\omega^2 - \beta^2)}$, then from equation (1.11),

$\alpha_1, \alpha_2 = -\beta \pm i\,\omega_d,\ i = \sqrt{-1}$

The roots α_1, α_2 are complex. Substituting for α_1, α_2 in equation (1.12) u(t) is given by

$$u(t) = A_1\, e^{(-\beta + i\,\omega_d)t} + A_2\, e^{(-\beta - i\omega_d)t} \tag{1.13}$$

where A_1 and A_2 are conjugate complex constants to be determined from boundary conditions.

Noting that

$$\cos \omega_d t = 0.5(e^{i\,\omega_d t} + e^{-i\omega_d t})$$

$$\sin \omega_d t = -0.5i(e^{i\,\omega_d t} - e^{-i\omega_d t})$$

$$u(t) = e^{-\beta t}(B_1 \cos\omega_d t + B_2 \sin\omega_d t) \tag{1.14}$$

where B_1 and B_2 are real constants of integration to be determined so as to satisfy the initial conditions.

ii. $\beta = \omega$. This is called the critically damped case. In this case there are two real repeated roots $\alpha_1 = \alpha_2 = -\beta$. The solution is given by

$$u(t) = e^{-\beta t}(B_1 + B_2\, t) \tag{1.15}$$

where B_1 and B_2 are integration constants.

iii. $\beta > \omega$. This is called the over damped case. In this case there are two real roots. The solution is given by

$$u(t) = A_1\, e^{(-\beta + \omega_d)t} + A_2\, e^{(-\beta - \omega_d)t} \tag{1.16}$$

The hyperbolic sine and cosine functions are related to the exponential functions by

$$\cosh\omega_d t = 0.5\{e^{\omega_d t} + e^{-\omega_d t}\}$$
$$\sinh\omega_d t = 0.5\{e^{\omega_d t} - e^{-\omega_d t}\}$$

$$u(t) = e^{-\beta t}\{B_1 \cosh\omega_d t + B_2 \sinh\omega_d t\} \tag{1.17}$$

The solutions to the above three cases are discussed in more detail in the next section.

1.9 UNDAMPED FREE VIBRATION

The displacement $u(t)$ is given by equation (1.14). The constants B_1 and B_2 are determined from initial conditions. If the system is disturbed from its initial stationary position by giving an initial displacement of u_o and an initial velocity of u_o' at t = 0, then $u_o = B_1$ and $u_o' = B_2\,\omega$.

$$u(t) = u_o \cos\omega t + \{\frac{u'_0}{\omega}\}\sin\omega t \tag{1.18}$$

The above expression can be expressed more elegantly as follows. Let

$$u_0 = R\cos\theta\ ,\ \frac{u'_0}{\omega} = R\sin\theta,$$
$$R = \sqrt{\{u_0^2 + (\frac{u'_0}{\omega})^2}\ ,\ \theta = \tan^{-1}(\frac{u'_0}{\omega u_0}) \tag{1.19}$$

$$u(t) = R(\cos\omega t\ \cos\theta + \sin\omega t\ \sin\theta) = R\cos(\omega t - \theta) \tag{1.20}$$

This indicates that the motion is described by a cosine curve with an amplitude equal to R and a time lag of $t_o = \theta/\omega$.

It is perhaps worth noting that the frequency of vibration is independent of the amplitude R. This is of course true only if the amplitude is not large enough to invalidate the basic assumptions involved in the derivation of the equations of motion.

Since the trigonometric cosine and sine functions are periodic functions with a period of 2π, the motion is periodic with a period T. The system vibrates with a period of T or a frequency f. The three quantities ω, f and T are related as follows

$$\text{circular frequency } \omega = \sqrt{\frac{K}{M}}\ \omega \text{ radians/seconds} \tag{1.21a}$$

$$\text{frequency } f = \frac{1}{T} = \frac{\omega}{2\pi} \text{ Hertz} \tag{1.21b}$$

$$\text{period } T = \frac{1}{f} \text{ seconds} \tag{1.21c}$$

1.9.1 Units Used in Dynamic Analysis

In vibration problems, both mass and force units occur. It is therefore important that a consistent system of units is used. It is suggested that the following consistent system be adopted. By definition, a force of 1 N acting on a mass of 1 kg produces an acceleration of 1 m/sec^2. Therefore, consistency of units is satisfied when length is expressed in metres, force in Newtons, mass in kilograms and time in seconds.

1.9.2 Example

A simply supported beam of 3 m span supports a load at midspan of 100 kN. Calculate the natural frequency of vibration. It is given that the second moment of area I of the beam is equal to 11710 cm^4, Young's modulus E = 210 kN/mm^2.
Solution: Express all parameters in terms of units specified in Section 1.9.1.
$I = 11710\ cm^4 = 11710 \times 10^{-8}\ m^4$, $E = 210\ kN/mm^2 = 210 \times 10^3 \times 10^6\ N/m^2$

$$EI = 24.591 \times 10^6\ Nm^2,\ L = span = 3\ m$$

As the beam is a simply supported and carries a mass at the midspan,

$$K = 48\ EI/L^3 = 48 \times [24.591 \times 10^6]/3^3 = 43.717 \times 10^6\ N/m$$

Load W acting at midspan = 100 kN = 100×10^3 N
Assuming that the gravitational constant $g \approx 10$ m/sec/sec, the corresponding mass M = W/g. Therefore

$$M = 100 \times 10^3 / g = 10 \times 10^3\ kg$$

$$\omega = \sqrt{\{K/M\}} = 66.11\ \text{radians/second}$$

$$f = \frac{\omega}{2\pi} = 10.52\ \text{Hertz},\ T = \frac{1}{f} = 0.095\ \text{seconds}$$

Figure 1.5 shows a plot of variation of *u(t)* with t. Calculations were made on the assumption that $u_0' = 0$. As can be seen, the motion is a simple harmonic and repeats itself indefinitely.

1.10 DAMPED FREE VIBRATION

Damped free vibrations are similar to undamped free vibrations considered in Section 1.9, except that the damping constant C is not equal to zero. We have to consider three separate cases as follows.

1.10.1 Underdamped System (β < ω)

In this case, the solution is given by equation (1.14). Assuming that the system is given at time t = 0, $u(t) = u_o$ and $u'(t) = u_o'$ then

$$u_o = B_1 \text{ and } u_o' = -\beta B_1 + B_2 \omega_d$$

$$B_1 = u_o,\ B_2 = \{u_o' + \beta u_o\} / \omega_d$$

$$u(t) = e^{-\beta t}\,[u_o \cos\omega_d t + \{(u_o' + \beta\, u_o)/\omega_d\}\sin\omega_d t] \tag{1.22}$$

Using steps similar to those used in deriving equation (1.19)

$$u(t) = e^{-\beta t}\,R\cos(\omega_d t - \theta) = e^{-\beta t}\,R\cos(\omega_d t - \theta) \tag{1.23}$$

$$R^2 = u_o^2 + \{(u_o' + \beta\, u_o)/\omega_d\}^2,\ \tan\theta = \{u_o' + \beta\, u_o\}/(u_o\,\omega_d)$$

As can be seen, $cos(\omega_d t - \theta)$ is periodic with a period $T_d = \omega_d/(2\pi)$ but the presence of $e^{-\beta t}$ term damps out the vibration. As the value of C is increased, the vibrations are damped out at a greater rate per cycle.

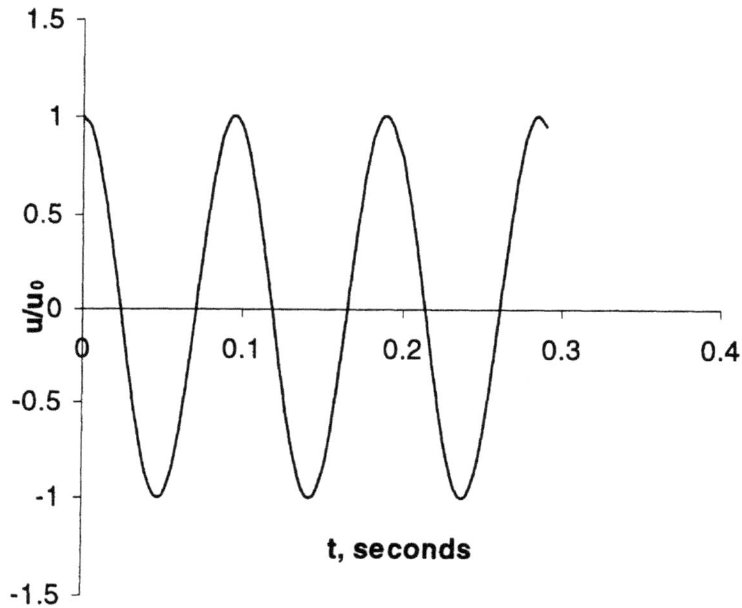

Fig. 1.5 Undamped free vibration

1.10.2 Critically damped case ($\beta = \omega$)

The solution to u(t) is given by equation (1.15). Assuming that $u(t) = u_o$ and $u'(t) = u_o'$ at t = 0,

$$B_1 = u_o,\ B_2 = \beta\, u_o + u_o'$$

$$u(t) = e^{-\beta t}\,\{u_o(1 + \beta\, t) + u_o'\,t\} \tag{1.24}$$

As can be seen, vibratory motion is completely damped out. The smallest amount of damping constant C required to damp out all vibration is called critical damping C_{cr}.

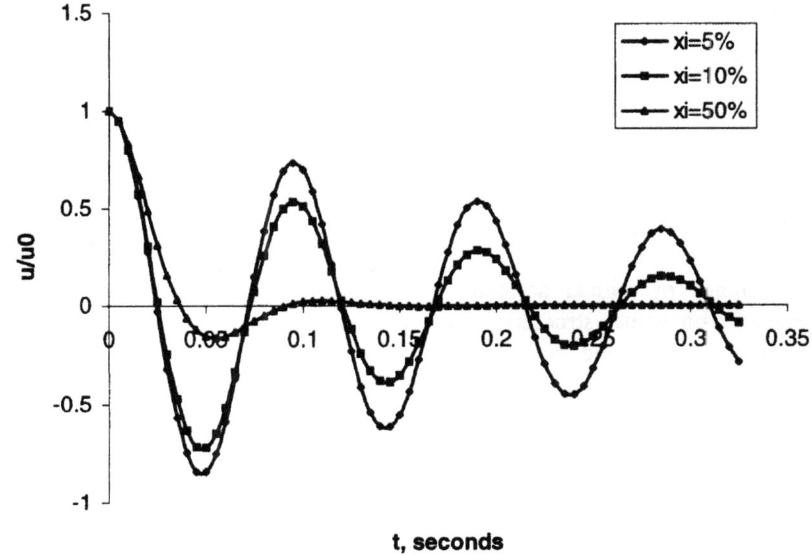

Fig. 1.6 Damped free vibrations

1.11 CRITICAL DAMPING

Critical damping C_{cr} can be expressed in terms of M and K as follows. At critical damping $\beta = \omega$, but $\beta = C_{cr}/(2M)$ and $\omega = \sqrt{(K/M)}$. Therefore

$$C_{cr} = 2\sqrt{(KM)} \tag{1.25}$$

Using the expression for C_{cr}, the expression for ω_d in the case of an under damped system is given by

$$\omega_d = \sqrt{(\omega^2 - \beta^2)} = \omega\sqrt{\{1 - (\beta/\omega)^2\}}$$

Substituting for $\beta = C/(2M)$, $\omega = \sqrt{(K/M)}$, $C_{cr} = 2\sqrt{(KM)}$

$$(\beta/\omega)^2 = C^2/(4KM) = (C/C_{cr})^2$$

Letting $C/C_{cr} = \xi$, damping ratio and $\beta = \xi\,\omega$

$$\omega_d = \omega\sqrt{\{1 - \xi^2\}} \tag{1.26}$$

Fig. 1.6 shows a plot of the variation of $u(t)$ with t. Calculations were made for the same simply supported beam considered in Section 1.9.2 assuming that $u_0' = 0$ and using damping ratio ξ of 5%, 10% and 50%.

1.12 DAMPING IN STRUCTURES

Damping in structures has various origins. The simplest is friction due to relative movement between parts of structures, for example at bolted joints in steel structures and at cracked surfaces in the case of concrete structures. This type of frictional damping is called Coulomb damping. It is independent of velocity. On the other hand when a body moves in a fluid, such as air or water, then the resistance is normally proportional to the square of velocity. In the interest of simplicity, damping force is often assumed to be proportional to velocity. This type of damping is called linear 'Viscous damping'. In addition to the above external causes, there is also energy dissipated due to internal friction caused by slipping and sliding of particles at internal planes during deformation. The phenomenon of damping in structures is thus very complex involving many causes. It is almost impossible to determine what proportion of damping can be assigned to a particular aspect. Therefore in practice the value of damping present is assumed as a percentage of critical damping. Some experimental data obtained from measurements on actual structures, such as buildings and bridges, exist. These values provide reasonable guidance for practical calculation. Table 1.1 shows typical values of C/C_{cr} met in practice. As can be seen, in practice the damping present is such that $C/C_{cr} < 15\%$. If $C/C_{cr} = 0.15$, then $\omega_d = 0.99\ \omega$. Therefore it can be concluded that the effect of a small amount of damping is mainly to damp out the vibrations but leaving the frequency of vibration practically unaltered from the undamped value.

Table 1.1 Damping ratios in practical structures

	$C/C_{cr}\%$	
Type and condition of structure	Working stress	Near yield
Welded steel, prestressed concrete, well reinforced concrete	2-3	5-7
Reinforced concrete with considerable cracking	3-5	7-10
Bolted steel, wood structures with nailed or bolted joints	5-7	10-15

In practical calculations it is safe to ignore damping, because it has the effect of reducing the stresses under dynamic loading. As will be shown in Chapter 2, in the case of sudden dynamic loading, such as wind gusts or earthquake disturbance, there is generally insufficient time for damping to have any significant effect. However, it is desirable to include it in the case of continuous dynamic loading.

1.13 OVERDAMPED SYSTEM

As indicated in Section 1.12, the amount of damping present in practical situations is very small. The case of an overdamped system where ($\beta > \omega$) does not commonly occur in problems of structural engineering interest. The general expression for displacement is given by equation 1.17.

Assuming that the system is given at time t = 0, $u(t) = u_o$ and $u'(t) = u_o'$ then

$$u_o = B_1 \text{ and } u_o' = -\beta B_1 + B_2 \omega_d$$

$$B_1 = u_o,\ B_2 = \{u_o' + \beta u_o\}/\omega_d$$

$$u(t) = e^{-\beta t}\,[u_o \cosh\omega_d t + \{(u_o' + \beta u_o)/\omega_d\}\sinh\omega_d t] \tag{1.27}$$

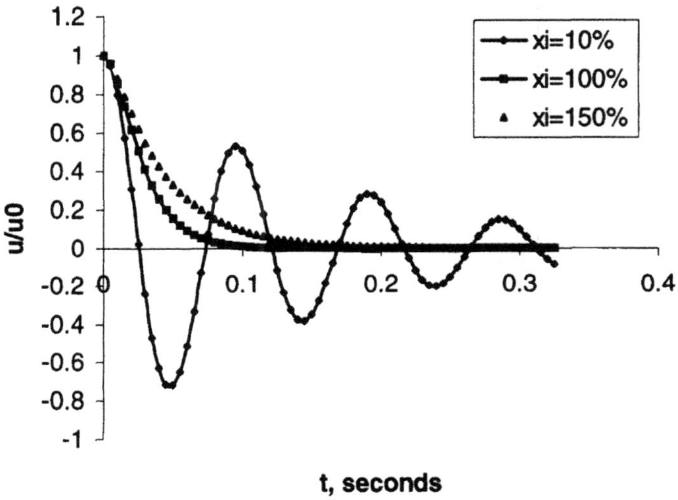

Fig. 1.7 Underdamped, critically damped and overdamped systems

It is interesting to note that from equation (1.27), that although both cosh ω_d t and sinh ω_d t, tend to infinity as t tends to infinity, but because of the presence of the $e^{-\beta t}$ term, vibrations are damped out. Fig.1.7 shows a plot of displacement of the simply supported beam considered before for three cases of damping of viz. underdamped with $\xi = 0.10$, critically damped with $\xi = 1.0$ and an overdamped system with $\xi = 1.5$. It is worth noting that the displacements of an overdamped system are larger than that of the critically damped system, although in both cases all vibratory motion is completely suppressed and displacements fade out exponentially.

1.14 MEASUREMENT OF DAMPING

In the case of underdamped systems, which are of practical interest, a simple measure of damping is the ratio of displacements at times one cycle apart. If T_d is the period of vibration, using equation (1.14), the ratio between displacements T_d apart can be calculated as follows

$$u(t) = e^{-\beta t}\{B_1 \cos\omega_d t + B_2 \sin\omega_d t\}$$

$$u(t + T_d) = e^{-\beta(t + T_d)}\{B_1 \cos\omega_d(t + T_d) + B_2 \sin\omega_d(t + T_d)\}$$

$$T_d = 2\pi / \omega_d,\ \omega_d = \omega\sqrt{\{1 - (C/C_{cr})^2\}}$$
$$\cos\omega_d(t + T_d) = \cos(\omega_d t + 2\pi) = \cos\omega_d t$$
$$\sin\omega_d(t + T_d) = \sin(\omega_d t + 2\pi) = \sin\omega_d t$$
$$u(t + T_d) = e^{-\beta(t + T_d)}\{B_1 \cos\omega_d t + B_2 \sin\omega_d t\}$$

Therefore the ratio of displacements T_d apart is given by

$$u(t)/u(t + T_d) = e^{\beta Td}$$
$$\beta T_d = \log_e [u(t)/u(t + T_d)]$$

The term $\log_e [u(t)/u(t+T_d)]$ is called log decrement δ.

As

$$\beta = \xi\,\omega,\ \omega_d = \omega\sqrt{\{1 - \xi^2\}},\ T_d = 2\pi / \omega_d$$
$$\delta = \beta\, T_d = [2\pi\,\xi]/\sqrt{\{1 - \xi^2\}}$$

Since the damping ratio ξ is generally less than 0.10,

$$\delta \approx 2\,\pi,\ \xi \approx \delta/(2\pi) \quad (1.28)$$

Note that if $\xi = 0.10$, then $\delta = 0.63$ and the ratio of displacements at times T_d apart is equal to $e^{\beta Td} = 1.88$. Since $1/1.88 \approx 0.5$, as a rough rule of thumb, 10% critical damping reduces the amplitude by 50% per cycle. From the point of view of calculating the damping ratio from an actual displacement versus time plot, if the ratio of displacements n cycles apart is used, then

$$u(t)/u(t + nT_d) = e^{\beta n\, Td}$$
$$n\,\beta T_d = \log_e [u(t)/u(t + nT_d)$$

$$n\,\delta = \log_e [u(t)/u(t + nT_d)] \quad (1.29)$$

This allows a damping ratio to be calculated to a better accuracy than if the ratio of displacements just T_d apart are used.

In general, using accelerometers, it is easier to measure acceleration at a point in a structure rather than the corresponding displacement, because measurement of displacement requires a datum. Assuming that the system is given at time t = 0, u(t) = u_o and u'(t) = 0 then, using equation (1.23),

$$u(t) = e^{-\beta t} R \cos(\omega_d t - \theta)$$
$$R^2 = u_o^2 + \{\beta\, u_o\}\ \omega_d\}^2$$
$$\tan\theta = \{u_o' + \beta\, u_o\}/(u_o\, \omega_d)$$

Differentiating displacement twice with respect to t, acceleration is given by

$$U''(t) = e^{-\beta t} R\{(\beta^2 - \omega_d^2) \cos\omega_d(t - t_o) + 2\,\beta\,\omega_d \sin\omega_d(t - t_o)\}$$

Using the same steps as used in deriving equation (1.22) in Section 1.9, the acceleration is given by

$$u''(t) = \omega^2 e^{-\beta t} R \cos(\omega_d t - \theta + \varphi)$$
$$R^2 = u_o^2 + \{\beta\, u_o\}/\omega_d\}^2$$
$$\tan\varphi = (-2\beta\,\omega_d)/\omega^2$$

This shows that plots of $u(t)$ and $u(t)''/\omega^2$, will be identical except for a phase shift of φ. This means that acceleration–time plot rather than displacement–time plot can be used to calculate damping present in real structures.

1.15 SUMMARY OF FREE VIBRATION ANALYSIS

The results of free vibration response can be summarised as follows:
i. With zero damping, motion is purely simple harmonic.
ii. When damping is below critical damping, the motion is still periodic but there is amplitude decay.
iii. As damping is increased up to critical damping all vibratory motion is completely damped out.
iv. The factors, which affect the free vibration response are mass M, stiffness K and damping C of the system.
v. Damping present in practical situations is in less than 10% of critical damping. The effect of light damping is to cause amplitude decay leaving the frequency of vibration practically unaltered.
vi. The frequency of vibration in the case of lightly damped system is given by

$$\omega \approx \sqrt{(K/M)},\ f \approx \omega/(2\pi),\ T \approx 2\pi/\omega.$$

vii. Logarithmic decrement δ is given by

$$\delta = \log_e [(u(t)/u(t + T_d)] \approx 2\pi\ (C/C_{cr}).$$

1.16 SYSTEMS SUBJECTED TO HARMONIC EXCITATION

In the previous sections, motion in the absence of external forces was considered. As an introduction to the study of SDOF systems subjected to external force and also to bring out the important concept of resonance, the underdamped system shown in Fig. 1.1 subjected to an external sinusoidal force $F = F_o \sin\Omega t$ will be studied. It is interesting to mention in passing that because a general periodic force can be expressed as a Fourier series in terms of sine and cosine functions, the results obtained in this section are of more general interest than might appear at first glance.

The differential equation to be solved is

$$M\frac{d^2u}{dt^2} + C\frac{du}{dt} + Ku = F(t) = F_0 \sin\Omega t \tag{1.30}$$

The solution to the differential equation (1.30) is the sum of complementary solution (also called natural motion solution or starting transient) and particular integral (also called forced motion solution or steady state solution). In the case of an underdamped system, the complementary solution is given by

Complementary Function = $e^{-\beta t}\{B_1 \cos\omega_d t + B_2 \sin\omega_d t\}$

The particular integral is assumed to be given by

Particular Integral = $\{D_1 \cos\Omega t + D_2 \sin\Omega t\}$

where D_1 and D_2 are constants.

Substituting the P.I. in equation (1.30)

$$-\Omega^2 (D_1 \cos\Omega t + D_2 \sin\Omega t) + 2\beta\Omega(-D_1 \sin\Omega t + D_2 \cos\Omega t)$$
$$+ (D_1 \cos\Omega t + D_2 \sin\Omega t) = (F/M) \sin\Omega t$$

Equating the coefficients of terms in sinΩt and cosΩt we have

$$-2\beta\,\Omega\,D_1 + (\omega^2 - \Omega^2)\,D_2 = F_o/M$$
$$2\,\beta\Omega\,D_2 + (\omega^2 - \Omega^2)\,D_1 = 0$$

Adopting the notation $r = \Omega/\omega$ and since $\beta/\omega = C/C_{cr} = \xi$ and $K/M = \omega^2$, D_1 and D_2 can be expressed as follows

$$D_1 = -\frac{F_o}{K}\frac{2r\xi}{R^2}, \quad D_2 = \frac{F_o}{K}\frac{(1 - r^2)}{R^2}, \; R^2 = (1-r^2)^2 + (2r\xi)^2$$

$$PI = \frac{F_o}{KR^2}\left[(1-r^2)\sin\Omega t - 2r\xi\cos\Omega t\right]$$

This can be simplified further by setting

$$(1 - r^2) = R\cos\theta,\ 2r\xi = R\sin\theta,\ \tan\theta = (2r\xi)/(1 - r^2)$$
$$PI = (F_o/KR)\ si\ (\Omega t - \theta)$$

$$u(t) = e^{-\beta t}\{B_1\cos\omega_d t + B_2\sin\omega_d t\} + (F_o/KR)\sin(\Omega t - \theta) \qquad (1.31)$$

If starting from rest, then

$$B_1 = (Fo/KR)\sin\theta$$
$$B_2 = (Fo/KR)\,[(\beta/\omega_d)\sin\theta - (\Omega/\omega_d)\cos\theta]$$

The complementary solution represents damped free vibration. In the presence of damping, the effect of this is quickly damped out. Remembering that damping equal to only 10% critical damping reduces the amplitude by 50% per cycle, the presence of the $e^{-\beta t}$ term in the complementary solution ensures rapid decay of this part of the solution. This is why the complementary solution is also called starting transient. Once the effect of the starting transient has disappeared, we are left with the particular integral part of the solution. This is why the particular integral is also called steady state solution. Therefore the steady state motion is given by

$$u(t) = (F_o/KR)\sin(\Omega t - \theta) \qquad (1.32)$$

Since the applied force $F = Fo\sin\Omega t$, it is clear that the frequency of vibration is the same as the frequency of the applied force except that the motion lags behind the applied force by $t_o = \theta/\Omega$. It should be noted that since

$$\tan\theta = (2r\xi)\,/\,(1-r^2),\ t_0 = 0 \text{ if } \xi = 0$$

Therefore damping present in the system causes the lag between the displacement and the applied force.

1.17 DYNAMIC MAGNIFICATION FACTOR

The steady state response is given by

$$u(t) = Fo/(KR)]\sin\Omega(t - t_o)$$
$$u(t)_{max} = (Fo/KR)$$

Since Fo/K = maximum static deflection Δ_{st}, the maximum dynamic deflection can be expressed as

$$u(t)_{max} = \Delta_{st}/R$$

Therefore the maximum dynamic deflection is 1/R maximum static deflection. 1/R is called dynamic magnification factor (DMF).

$$DMF = \frac{1}{\sqrt{[(1-r^2)^2 + (2r\xi)^2]}} \tag{1.33}$$

where $r = \Omega/\omega$, the ratio of applied to undamped natural frequency of the system.

Evidently DMF is a function of frequency ratio r and damping ratio ξ. Fig. 1.8, shows a plot of DMF versus r for various values of ξ.

For a given value of damping ratio, DMF is a maximum when

$$d(DMF)/dr = 0$$

Carrying out the differentiation and simplifying

$$1 - r^2 - 2\xi^2 = 0,\ r = \sqrt{(1 - 2\xi^2)}$$

For lightly damped system (i.e. $\xi = < 0.10$), the maximum DMF occurs when r is almost equal to 1 and the maximum DMF is equal to $0.5/\xi$. If $\xi = C/C_{cr} = 0.10$, the maximum dynamic displacement is five times the corresponding maximum static displacement.

Figure 1.8 shows that for values of r less than about 0.5, DMF remains fairly near unity. This corresponds to a quasi-static situation and dynamic effects can be safely ignored and only static analysis carried out.

Similarly if r is greater than about 1.5, the dynamic magnification is less than unity. This is the case where the structure is being isolated from the effects of vibration. The structure will show little response to forces pulsating at frequencies above the resonant frequency. The region where it is important to consider dynamic is $0.5 < r < 1.5$.

1.17.1 Response Near Resonance

In the previous section, it was shown that if the frequency of applied force is the same as the frequency of the system, then large displacements can result. In the case of undamped system, the solution to the differential equation is given by

$$u'' + \omega^2 u = (1/M)Fo \sin\Omega t$$

The solution is given by

$$u(t) = \{B_1 \cos\omega t + B_2 \sin\omega t\} + \{D_1 \cos\Omega t + D_2 \sin\Omega t\}$$

If $\omega = \Omega$, then because of the repeated nature of the Complementary Function and Particular Integral part of the solution, the solution is given by

$$u(t) = \{B_1 \cos\omega t + B_2 \sin\omega t\} + t\{D_1 \cos\omega t + D_2 \sin\omega t\}$$

In order to satisfy the differential equation, the values of constants D_1 and D_2 become

$$D_2 = 0,\ D_1 = -F_0/(2M\omega)$$

Therefore

$$u(t) = \{B_1 \cos\omega t + B_2 \sin\omega t\} - F_0/(2M\omega)\, t \cos\Omega t$$

Clearly, because of the presence of the term t, the displacements can become very large. However it also takes time to build up a large amplitude. Hence it is a safe procedure to accelerate a machine through a resonant frequency so long as the normal working frequency is well above the resonant frequency.

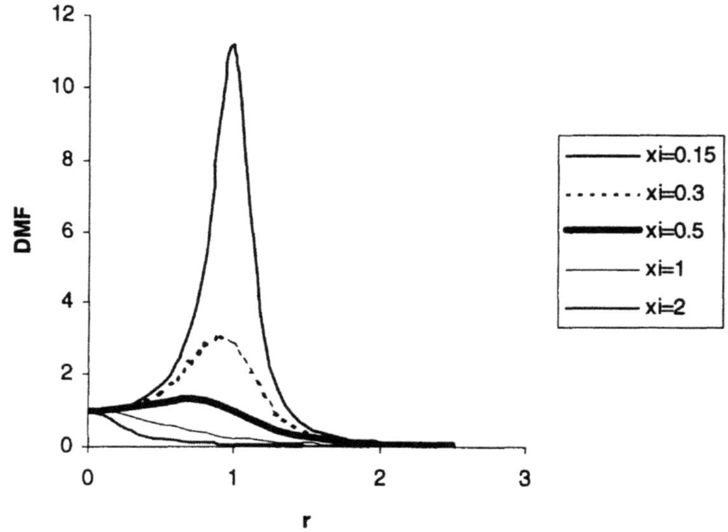

Fig. 1.8 Dynamic magnification factor versus frequency ratio

1.18 RESPONSE TO BASE EXCITATION

In Section 1.17, the SDOF system subjected to an external force was studied. As an introduction to the study of systems subjected to forces arising from the acceleration of the foundation such as that due to seismic disturbance, the SDOF system shown in Fig. 1.1 subjected to foundation movement will be studied. Let the displacement of the base be u_b and the displacement of the mass be u_m. The forces acting on the mass are

a. Inertial force which depends purely on the acceleration of the mass = $-M\,u''_m$

b. Damping force which depends on the relative velocity of the mass with respect to the base = $-C(u'_m - u'_b)$

c. Spring force which depends on the extension of the spring = $-K(u_m - u_b)$

Using D'Alembert's principle, the equation of equilibrium is given by

$$M\,u''_m + C\,(u'_m - u'_b) + K\,(u_m - u_b) = 0 \tag{1.34}$$

Adding $-Mu''_b$ to both sides of the equation

$$M\,(u''_m - u''_b) + C\,(u'_m - u'_b) + K\,(u_m - u_b) = -M\,u''_b$$

Setting $u = u_m - u_b$, the differential equation can be written as

$$Mu'' + Cu' + Ku = -Mu''_b \tag{1.35}$$

where u is the relative displacement of the mass with respect to the foundation. As can be seen, equation (1.35) is identical to equation (1.1) except that *F(t)* has been replaced by $-Mu_b''$. In other words, the analysis of systems subjected to acceleration of the base is similar to the analysis of systems subjected to external force.

CHAPTER 2

SINGLE DEGREE OF FREEDOM SYSTEM - II

2.1 INTRODUCTION

In Chapter 1, the response of a single degree of freedom (SDOF) system to free and forced vibration under a harmonic force was investigated. It was shown that for lightly damped systems, the frequency is dependent mainly on the mass and stiffness of the system. It was also shown that if the frequency of the applied force is nearly equal to the natural frequency of the system then the system resonates resulting in very large displacements. In this chapter the response of the SDOF system to general loads will be investigated and methods both analytical and numerical will be described for solving the differential equation viz.

$$\frac{d^2u}{dt^2} + 2\beta\frac{du}{dt} + \omega^2 u = \frac{F(t)}{M} \tag{2.1}$$

$\omega^2 = K/M,\ \xi = C/C_{cr},\ \omega_d = \omega\sqrt{(1-\xi^2)},\ \beta = \xi\,\omega$

2.2 LAPLACE TRANSFORM METHOD

The differential equation (2.1) is best solved using the Laplace transform method especially because the forcing function F(t) is, in general, discontinuous and the constants of integration are determined on the basis of initial conditions. As some readers might not be familiar with the method, the procedure is set out in some detail along with some simple examples.

It is perhaps worth reminding the reader that the Laplace transform method is similar to Macaulay's Method (also called Singularity Functions Method) used in the solution of beam deflection problems when the lateral load on the beam is discontinuous.

The Laplace transform $\bar{\Phi}(s)$ of a function $\Phi(t)$ is defined by

$$\bar{\Phi}(s) = \int_0^\infty \Phi(t)\, e^{-st}\, dt \tag{2.2}$$

where s is complex.

2.2.1 Laplace Transform: Examples

The basic step of calculating the Laplace transform is illustrated by a few simple examples.

Example 1: Calculate the Laplace transform of $\Phi(t) = e^{at}$.
Using equation (2.2),

$$\bar{\Phi}(s) = \int_0^\infty \Phi(t)\, e^{-st}\, dt = \int_0^\infty e^{at}\, e^{-st}\, dt = \int_0^\infty e^{-(s-a)t}\, dt$$

$$= -\frac{1}{(s-a)} e^{-(s-a)t} \Big|_0^\infty = \frac{1}{(s-a)}$$

The last step is obtained by noting that e^{-t} tends to zero as t tends to infinity.
As a corollary, if $\Phi = (e^{at} - e^{bt})$, then from the result in Example 1,

$$\bar{\Phi}(s) = \frac{1}{(s-a)} - \frac{1}{(s-b)} = \frac{(a-b)}{(s-a)(s-b)}$$

Example 2: Calculate the Laplace transform of $\Phi(t) = t$.
Using equation (2.2),

$$\bar{\Phi}(s) = \int_0^\infty \Phi(t)\, e^{-st}\, dt = \int_0^\infty t\, e^{-st}\, dt$$

Integrating by parts,

$$\bar{\Phi}(s) = e^{-st}\left[-\frac{t}{s} - \frac{1}{s^2}\right]\Big|_0^\infty \frac{1}{s^2}$$

2.3 INVERSE LAPLACE TRANSFORM

Examples in Section 2.2.1, showed how to calculate the Laplace transform for a given function. In the solution of differential equations using this method, it is necessary to calculate the original function, if only the transform of the function is given. Unfortunately this 'Inverse' process is not straightforward. Fortunately, with the help of ready-made tables, which tabulate for a large number of well-known functions corresponding to transforms, the reverse process can be accomplished.

2.3.1 Inverse Laplace Transform: Examples

A simple example to illustrate the inverse process.

Example 1: Calculate the function Φ(t), given that the Laplace transform is

$$\bar{\Phi}(s) = \frac{s}{(s-a)(s-b)}$$

Using the concept of partial fractions, the transform can be written as

$$\bar{\Phi}(s) = \frac{s}{(s-a)(s-b)} = \frac{1}{(a-b)}\left[\frac{a}{(s-a)} - \frac{b}{(s-b)}\right]$$

Using the result from Example 1 in Section 2.2.1, the function corresponding to each of the partial fractions can be determined. Therefore

$$\Phi(t) = \frac{1}{(a-b)}\left[a\,e^{at} - be^{bt}\right]$$

2.4 LAPLACE TRANSFORM OF DERIVATIVES

The Laplace transforms of derivatives of Φ(t) are obtained as follows.

i. First derivative of Φ (t):

The Laplace transform of the first derivative of Φ(t) is, by definition, given by

$$\bar{\Phi}'(s) = \int_0^\infty \frac{d\Phi(t)}{dt}\, e^{-st}\, dt$$

Integrating by parts

$$\bar{\Phi}'(s) = e^{-st}\,\Phi(t)\,\Big|_0^\infty + s\int_0^\infty \Phi(t)\, e^{-st}\, dt$$

The integral on the right-hand side is equal to $\bar{\Phi}$ and since e^{-st} tends to zero as t approaches infinity

$$\bar{\Phi}'(s) = -\Phi(0) + s\,\bar{\Phi}(s) \tag{2.3}$$

ii Second derivative of Φ (t):

The Laplace transform of the second derivative of Φ(t) is, by definition, given by

$$\bar{\Phi}''(s) = \int_0^\infty \frac{d^2\Phi(t)}{dt^2}\, e^{-st}\, dt$$

Integrating by parts once

$$\bar{\Phi}''(s) = \Phi'(t)\,e^{-st}\,\Big|_0^\infty + s\int_0^\infty \frac{d\Phi(t)}{dt}\, e^{-st}\, dt$$

The integral on the right-hand side is evidently $\bar{\Phi}'(s)$. Substituting for $\bar{\Phi}'(s)$ and in addition noting that since e^{-st} tends to zero as t approaches infinity

$$\bar{\Phi}''(s) = -\Phi'(0) - s\,\Phi(0) + s^2\,\bar{\Phi}(s) \tag{2.4}$$

2.5 SOLUTION OF DIFFERENTIAL EQUATION

In order to solve the differential equation (2.1), by the Laplace transform method, we take the Laplace transform of both sides of the above equation. Substituting for the Laplace transform of *u'(t)* and *u''(t)* in terms of the Laplace transform of *u(t)* and its initial conditions *u(0)* and *u'(0)*, we have

$$\{-u'(0) - s\,u(0) + s^2\,\bar{u}(s)\} + 2\beta\,\{-u(0) + s\,\bar{u}(s)\} + \omega^2\,\bar{u}(s) = \frac{\bar{F}(s)}{M}$$

Simplifying

$$\{s^2 + 2\beta\,s + \omega^2\}\bar{u}(s) = u'(o) + (2\beta + s)\,u(0) + \frac{\bar{F}(s)}{M} \tag{2.5}$$

Factorising, $\{s^2 + 2\beta\,s + \omega^2\}$ as (s-a) (s-b), where,

$$a = -\beta + j\omega_d\,,\quad b = -\beta - j\omega_d\,, j = \sqrt{-1}\,, \omega_d = \sqrt{(\omega^2 - \beta^2)}$$

The Laplace transform of u becomes

$$\bar{u}(s) = 'u'(o)\,\bar{g}(s) + (2\beta + s)\,u(0)\,\bar{g}(s) + \frac{1}{M}\bar{F}(s)\,\bar{g}(s) \tag{2.6}$$

$$\bar{g}(s) = \frac{1}{(s-a)(s-b)}$$

If the system is undamped, then $\beta = 0$ and

$$\bar{g}(s) = \frac{1}{(s^2 - \omega^2)}$$

The solution to u(t) is obtained by taking the inverse transform of both sides of equation (2.6).

2.6 SOME USEFUL RESULTS

Application of Laplace transform for the solution of practical problems is facilitated by the introduction of some special functions and some important 'theorems'. These are discussed in this section.

2.6.1 Unit Step Function

In studying the response of SDOF systems to general loading, the unit step function (also called Heaviside step function) is useful for defining discontinuous loading. As shown in Fig. 2.1, it is defined as follows

$$H(t - \tau) = 0, \text{ if } t < \tau \text{ and } H(t - \tau) = 1, \text{ if } t \geq \tau \tag{2.7}$$

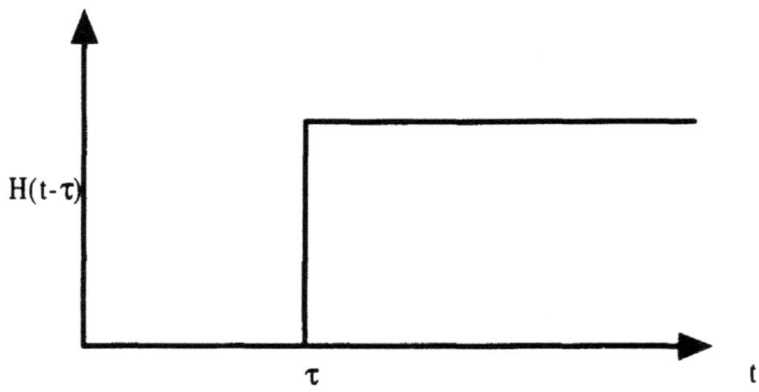

Fig. 2.1 Unit step function

Unit step function is similar to uniformly distributed load in the case of beam problems, where the uniformly distributed load does not start from the origin. Substituting for $\Phi(t) = H(t - \tau)$, the Laplace transform of unit step function is given by

$$\bar{\Phi} = \int_0^\infty H(t-\tau)e^{-st}\,dt = \int_\tau^\infty e^{-st}\,dt = -\frac{1}{s}e^{-st}\Big|_\tau^\infty = \frac{1}{s}e^{-s\tau} \tag{2.8}$$

Note that because $H(t - \tau) = 0$, if $t < \tau$, the lower limit of integration changes from 0 to τ.

Note that if an impulse lasts only over the period $t = t_1$ to $t = t_2$, then the corresponding $\Phi(t)$ is given by

$$\Phi(t) = H(t - t_1) - H(t - t_2)$$

The corresponding Laplace transform is given by

$$\bar{\Phi} = \int_0^\infty \{H(t-t_1) - H(t-t_2)\}e^{-st}\,dt = \frac{1}{s}\{e^{-st_1} - e^{-st_2}\} \tag{2.9}$$

2.6.2 Shift Theorems

There are two useful theorems that facilitate the evaluation of Laplace transforms and their respective inverses.

2.6.2.1 Shift Theorem 1

If $\Phi(t) = \Psi(t)\, e^{at}$, then

$$\bar{\Phi}(s) = \int_0^\infty \Phi(t)\, e^{-st}\, dt = \int_0^\infty [\Psi(t)\, e^{at}]\, e^{-st}\, dt = \int_0^\infty \Psi(t)\, e^{-(s-a)t}\, dt$$

If $\bar{\Psi}(s) = \int_0^\infty \Psi(t)\, e^{-st}\, dt$, then comparing the two expressions, one can see that

$$\bar{\Phi}(s) = \bar{\Psi}(s-a)$$

The theorem can be stated as follows

$$\Phi(t) = \Psi(t)\, e^{at},\ \bar{\Phi}(s) = \bar{\Psi}(s-a) \tag{2.10}$$

As a simple application of Shift Theorem 1, let $\Psi = t^2$. Using the definition for Laplace transform and also successively integrating by parts

$$\bar{\Psi}(s) = \int_0^\infty t^2\, e^{-st}\, dt = -\frac{1}{s} t^2\, e^{-st} \Big|_0^\infty + \frac{2}{s}\int_0^\infty t\, e^{-st}\, dt$$

$$= \frac{2}{s}\Big[-\frac{1}{s} t\, e^{-st} \Big|_0^\infty \Big] + \frac{2}{s^2}\int_0^\infty e^{-st}\, dt$$

$$= \frac{2}{s^2}\Big[\int_0^\infty e^{-st}\, dt \Big] = \frac{2}{s^2}\Big[-\frac{1}{s} e^{-st} \Big|_0^\infty \Big] = \frac{2}{s^3}$$

Therefore if $\Psi(t) = t^2$, then $\bar{\Psi}(s) = \dfrac{2}{s^3}$. From the Shift Theorem 1, the Laplace transform of $t^2 e^{-at}$ is equal to $\dfrac{2}{(s-a)^3}$.

2.6.2.2 Shift Theorem 2

If $\Phi(t) = \Psi(t - a)\, H(t - a)$, then

$$\bar{\Phi}(s) = \int_0^\infty \Phi(t)\, e^{-st}\, dt = \int_0^\infty \Psi(t-a)\, H(t-a)\, e^{-st}\, dt$$

$$= \int_a^\infty \Psi(t-a)\, e^{-st}\, dt$$

Note that in the last integral the lower limit changes from 0 to a. This is because $H(t - a) = 0$ for $t < a$.

$$\bar{\Phi}(s) = \int_0^\infty \Psi(t-a)\, e^{-st}\, dt$$

Substituting $(t - a) = u$, then

$$\bar{\Phi}(s) = e^{-as} \int_0^{\infty} \Psi(u) \quad e^{-su} \ du = e^{-as} \ \bar{\Psi}(u)$$

Therefore, the second shift can be stated as follows. If $\Phi(t)$ is a given function, then the Laplace transform of the product of $\Phi(t - a)\, H(t - a)$ is equal to $\bar{\Phi} \ e^{-as}$. For example, the Laplace transform of t^2 is equal to $\frac{2}{s^3}$. Therefore the Laplace transform of $(t-a)^2 \ H(t-a)$ is given by $\frac{2}{s^3} \ e^{-as}$

2.6.3 Convolution Theorem

Convolution theorem is useful for evaluating the **inverse transform** of the product of the transforms of individual functions. This is particularly important in obtaining the particular integral of ordinary differential equations. The theorem says that if $\Phi(t)$ and $\Psi(t)$ are two functions whose Laplace transforms are, respectively, $\bar{\Phi}$ and $\bar{\Psi}$, then the **inverse transform** of the product $\bar{\Phi} \ \bar{\Psi}$ is given by

$$\int_0^t \Phi(\tau) \ \Psi(t - \tau)\, d\tau = \int_0^t \Phi(t - \tau) \ \Psi(\tau)\, d\tau$$

The proof of this theorem is quite simple. By definition

$$\bar{\Psi} = \int_0^{\infty} \Psi \ e^{-st} dt \qquad \therefore \ \bar{\Phi}\bar{\Psi} = \bar{\Phi} \int_0^{\infty} \Psi \ e^{-st} \ dt$$

Note that Ψ is a function of t but $\bar{\Phi}$ is a function of s. Taking $\bar{\Phi}$ which is a function of s inside the integral sign

$$\bar{\Phi} \ \bar{\Psi} = \int_0^{\infty} \Psi \ [\bar{\Phi} \ e^{-st} \ ] dt \tag{2.11}$$

In order to avoid confusion with t later on, we can, in the above integral, change the variable t to u without making any difference. Therefore

$$\bar{\Phi} \ \bar{\Psi} = \int_0^{\infty} \Psi \ [\bar{\Phi} \ e^{-su} \ ] \ du$$

However from the second shift theorem, $\bar{\Phi} \ e^{-su}$ is equal to the Laplace transform of $\Phi(t - u)\, H(t - u)$. Therefore

$$\bar{\Phi} \ e^{-su} = \int_0^{\infty} \Phi(t - u) \ H(t - u) \ e^{-st} \ dt$$

Substituting the above integral into the original expression for $\bar{\Phi} \ \bar{\Psi}$, we have

$$\bar{\Phi} \ \bar{\Psi} = \int_0^{\infty} \Psi \ [\int_0^{\infty} \ \Phi(t - u) \ H(t - u) \ e^{-st} \ dt \ ] \ du$$

Reorganizing the above expression as follows (which is permissible in this case), we get

$$\bar{\Phi}\,\bar{\Psi} = \int_0^\infty [\int_0^\infty \Phi(t-u)\, H(t-u)\, \Psi\, du]\; e^{-st}\, dt$$

We notice now that the variable in the inside integral is u If t is the variable, then

$H(t-u) = 0$, if $t < u$ and $H(t-u) = 1$, if $t \geq u$ (2.12a)

However if u is treated as the variable, then

$H(t-u) = 0$, if $u > t$ and $H(t-u) = 1$, if $u \leq t$ (2.12b)

Since in the inside integral, u is the variable, using equation (2.12b), the upper limit changes from infinity to t because $H(t - u)$ is zero for all values of $u > t$.

Introducing this change, the expression for the product $\bar{\Phi}\,\bar{\Psi}$ can be written as

$$\bar{\Phi}\,\bar{\Psi} = \int_0^\infty [\int_0^t \Phi(t-u)\, \Psi\, du]\; e^{-st}\, dt$$

This shows that the right-hand side is nothing but the Laplace transform of inner integral inside square brackets. Therefore we can say that the **inverse** Laplace transform of $\bar{\Phi}\,\bar{\Psi}$ is equal to

$$\int_0^t \Phi(t-u)\; \Psi\; du$$

In a similar way we can show that the inverse Laplace transform of $\bar{\Phi}\,\bar{\Psi}$ is also equal to

$$\int_0^t \Phi\; \Psi(t-u)\; du$$

This result is known as convolution integral theorem and will be used in later sections to derive the particular integral part of the solution to the differential equation.

2.7 SUMMARY OF SOME RESULTS

As already remarked, using a table of functions and their transforms, knowing $\bar{\Phi}(s)$, one can determine the corresponding Φ(t). Table 2.1 gives some standard results.

The following results (see Example 1 in Section 2.2.1) will be useful in applying Laplace transforms to the solution of the differential equation.

If $a = -\beta + j\,\omega_d$, $b = -\beta - j\,\omega_d$, $(a - b) = 2\,j\,\omega_d$, then if

(i) $\bar{\Phi}(s) = \dfrac{1}{(s-a)(s-b)}$,

$$\Phi(t) = e^{-\beta t}\,\frac{[e^{j\omega_d t} - e^{-j\omega_d t}]}{2j\omega_d} = \frac{e^{-\beta t}}{\omega_d}\sin\omega_d t$$

If $\beta = 0$ for an undamped system, then

$$\Phi(t) = \frac{1}{\omega}\sin\omega t$$

(ii) $\bar{\Phi}(s) = \dfrac{s}{(s-a)(s-b)}$

$$\Phi(t)=e^{-\beta t}\frac{[(-\beta+j\,\omega_d)e^{j\omega_d t}-(-\beta-j\,\omega_d)e^{-j\omega_d t}]}{2j\omega_d}$$

$$\Phi(t)=e^{-\beta t}[\frac{e^{j\omega_d t}+e^{-j\omega_d t}}{2} - \frac{\beta}{\omega_d}\frac{e^{j\omega_d t}-e^{-j\omega_d t}}{2j}] = e^{-\beta t}[\cos\,\omega_d t - \frac{\beta}{\omega_d}\sin\,\omega_d t]$$

If $\beta = 0$ for an undamped system, then

$$\Phi(t) = \cos\omega\, t$$

(iii) Shift Theorem 1: $\Phi(t) = \Psi(t)\; e^{at}$, $\bar{\Phi}(s) = \bar{\Psi}(s-a)$

(iv) Shift Theorem 2: The Laplace transform of $\Phi(t - a)\, H(t - a) = \bar{\Phi}\; e^{-as}$

(v) Convolution Theorem: The **inverse transform** of the product $\bar{\Phi}\;\bar{\Psi}$ is given by $\int_0^t \Phi(\tau)\,\Psi(t-\tau)d\tau$ or $\int_0^t \Phi(t-\tau)\,\Psi(\tau)d\tau$

Table 2.1 Laplace transforms

Function	Transform
$\frac{1}{(a-b)}[e^{at} - e^{bt}]$	$\frac{1}{(s-a)(s-b)}$
$\frac{1}{(a-b)}[ae^{at} - be^{bt}]$	$\frac{s}{(s-a)(s-b)}$
$H(t - \tau)$	$\frac{e^{-\tau s}}{s}$
$F(t - \tau).H(t - \tau)$	$e^{-\tau s}\,\bar{F}(s)$
$\int_0^t \Phi(\tau)\, g(t-\tau)d\tau$	$\bar{\Phi}(s)\;\bar{g}(s)$
$\frac{t^n}{n!}$	$\frac{1}{s^{(n+1)}}$

2.8 GENERAL SOLUTION OF DUHAMEL INTEGRAL

Using equation (2.6) and the results of Sections 2.7, the solution to the differential equation (2.1) can be written as

$$u(t)=e^{-\beta t}\{\cos\omega_d t+\frac{\beta}{\omega_d}\sin\omega_d t\}\ u(0)+e^{-\beta t}\{\ \frac{1}{\omega_d}\sin\omega_d t\}u'(0)$$

$$+ \text{ Inverse transform of } [\frac{1}{M}\bar{F}(s)\ \bar{g}(s)]$$

where

$$\bar{g}(s)=\frac{1}{(s-a)(s-b)},\quad a=-\beta+j\omega_d\ ,\ b=-\beta\ -j\omega_d$$

$$g(t)=\ e^{-\beta t}\{\ \frac{1}{\omega_d}\sin\omega_d t\},\quad g(t-\tau)=\ e^{-\beta(t-t)}\ \frac{1}{\omega_d}\sin\omega_d(t-\tau)$$

From the Convolution theorem (Section 2.6.3), the **inverse** Laplace transform of $\bar{\Phi}(s)\ \bar{g}(s)$ is $\int_0^t \Phi(\tau)\,g(t-\tau)\,d\tau$. Therefore the inverse Laplace transform of $\frac{1}{M}\bar{F}(s)\,\bar{g}(s)$ is

$$\frac{1}{M}\int_0^t F(\tau)\,g(t-\tau)\,d\tau\ =\ \frac{1}{M\,\omega_d}\int_0^t F(\tau)\ \ e^{-\beta(t-\tau)}\sin\omega_d(t-\tau)\ \ d\tau$$

The solution to the differential equation (2.1) is

$$u(t)=e^{-\beta t}\{\cos\omega_d t+\frac{\beta}{\omega_d}\sin\omega_d t\}\ u(0)+e^{-\beta t}\{\ \frac{1}{\omega_d}\sin\omega_d t\}u'(0)$$

$$+\ \frac{1}{M\,\omega_d}\int_0^t F(\tau)\ \ e^{-\beta(t-\tau)}\sin\omega_d(t-\tau)\ \ d\tau \tag{2.13}$$

In many cases it is simpler to operate on the above integral as it is. However, using the relationship for compound angles,

$$\sin\omega_d(t-\tau)=\sin\omega_d t\ \cos\omega_d\tau-\cos\omega_d t\ \sin\omega_d\tau$$

$$\frac{1}{M}\int_0^t F(\tau)\ \ g(\tau-t)\ \ d\tau=\ \frac{1}{M\ \ \omega_d}\int_0^t F(\tau)\ \ e^{-\beta(t-\tau)}\ \ \sin\ \omega_d(t\ -\tau)\ \ d\tau$$

$$=\ \frac{e^{-\beta t}}{M\ \ \omega_d}\int_0^t F(\tau)\ \ e^{\beta\tau}\ \ [\sin\ \omega_d t\ \ \cos\omega_d\tau\ \ -\ \ \cos\ \omega_d t\ \ \sin\omega_d\tau\ ]d\tau$$

$$=\ \frac{e^{-\beta t}}{M\ \ \omega_d}[\sin\omega_d\ t\int_0^t F(\tau)\,e^{\beta\tau}\ \ \cos\omega_d\ \tau\,d\tau\ -\ \cos\omega_d\ t\int_0^t F(\tau)\ \ e^{\beta\tau}\ \ \sin\omega_d\tau\ d\tau] \tag{2.14}$$

The above integrals are commonly known as Duhamel integrals. This representation is generally convenient for numerical evaluation of the integral.

2.9 SOLUTION OF DUHAMEL INTEGRAL: SPECIFIC EXAMPLES

Three simple cases of load input will be considered. The object is to investigate how ratio of the duration of the load to the period of vibration of the system affects the maximum response.

2.9.1 Suddenly Applied Load

Figure 2.2 shows the relationship between force and time. In this case $F(t) = P$.

$$\frac{1}{M}\int_0^t F(\tau)\,g(t-\tau)\,d\tau \;=\; \frac{P}{M\,\omega_d}\int_0^t e^{-\beta(t-\tau)}\sin\omega_d(t-\tau)\;d\tau$$

The evaluation of the integral is simplified by substituting $t - \tau = x$.
Then $d\tau = -dx$ and the limits in terms of x become, when $\tau = t$, $x = 0$ and when $\tau = 0$, $x = t$.

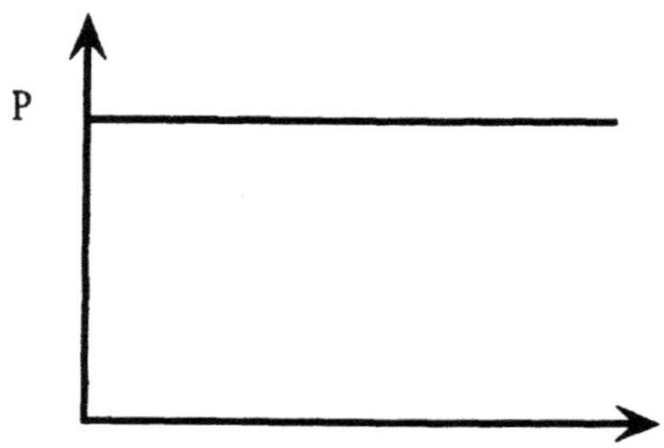

Fig. 2.2 Suddenly applied load

$$\frac{1}{M}\int_0^t F(\tau)\,g(t-\tau)\,d\tau \;=\; \frac{P}{M\,\omega_d}\int_0^x e^{-\beta x}\sin\omega_d x\;dx \qquad (2.15)$$

Using the standard integral

$$\int_0^t e^{ax}\sin bx\;dx \;=\; \frac{1}{(a^2+b^2)}[e^{at}\;(a\sin bt\;-\;b\;\cos bt)+\,b]$$

where $a = -\beta$, $b = \omega_d$, $a^2 + b^2 = \beta^2 + \omega_d^2 = \omega^2$
Substituting for the integrals in equation (2.15)

$$\frac{1}{M}\int_0^t F(\tau)\,g(t-\tau)\,d\tau \;=\; \frac{P}{M\omega_d\omega^2}[e^{-\beta t}\{-\beta\;\sin\omega_d t\;-\;\omega_d\cos\omega_d t\}\;+\;\omega_d]$$

$$=\;\frac{P}{M\omega^2}\;[e^{-\beta t}\{-\frac{\beta}{\omega_d}\sin\omega_d t\;-\;\cos\omega_d t\}\;+\;1]$$

Substituting $K = M\omega^2$ and static displacement $\Delta = P/K$

$$u = e^{-\beta t}[\;(\frac{\beta}{\omega_d}\sin\omega_d t\;+\;\cos\omega_d t)\;u(0)\;+\;\frac{\sin\omega_d t}{\omega_d}\;u'(0)\,]$$

$$+\;\Delta[1\;-\;e^{-\beta t}\{\frac{\beta}{\omega_d}\sin\omega_d t\;+\;\cos\omega_d t\}]$$

If the system starts from rest, the $u(0) = u'(0) = 0$.

$$\therefore\ u = \Delta[1 - e^{-\beta t}\{\frac{\beta}{\omega_d}\sin\omega_d t + \cos\omega_d t\}] \tag{2.16a}$$

If damping is ignored, then $\beta = 0$ and $\omega_d = \omega$. Therefore the response of an undamped system to a suddenly applied load is given by

$$u = \Delta[1 - \cos\omega t] \tag{2.16b}$$

The dynamic magnification factor is given by the ratio of dynamic displacement u to static displacement Δ. Fig. 2.3 shows a plot of the dynamic magnification factor for undamped motion and for damped motion with 10% of critical damping and ω = 66.11, the same as the one used for the simply supported beam in Section 1.9.2. Note that in the absence of damping, the dynamic magnification factor is 2. It is a well-known result that if the load is suddenly applied, then the resulting maximum response is twice the value that would result if the same load is applied 'gradually' i.e. when the acceleration is very small. Note that even a small amount of damping reduces considerably the effect of a suddenly applied load. In this example, the dynamic magnification factor in the damped case is only 1.73.

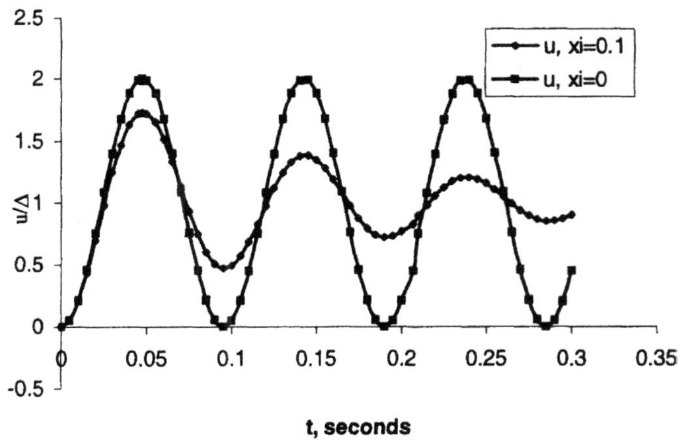

Fig. 2.3 Response to suddenly applied load

2.9.2 Rectangular Impulse Loading

Figure 2.4 shows the force–time relationship. In contrast to the previous case shown in Fig. 2.2, where once the load was applied it remained constant, in this case the load acts only for a short time. The force variation is given by

$$F(t) = P,\ t \le t_1$$
$$= 0,\ t > t_1$$

Using the definition of step function,

$$F(t) = P - H(t - t_1)\,P$$

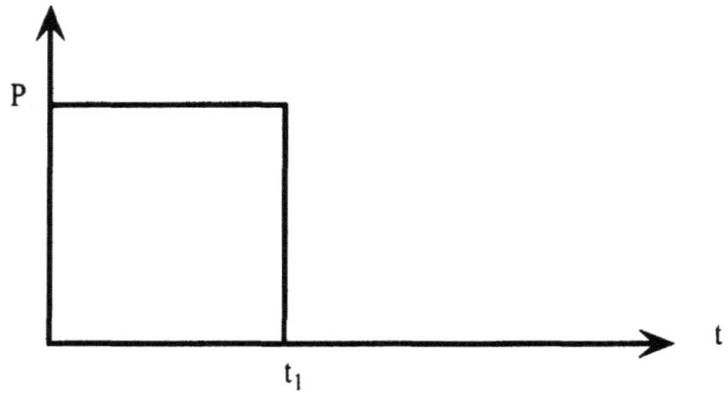

Fig. 2.4 Rectangular impulse

$$\frac{1}{M}\int_0^t F(\tau)\; g(\tau - t)\; d\tau = \frac{P}{M\;\omega_d}\int_0^t [1 - H(\tau - t_1)]\; e^{-\beta(t-\tau)}\;\sin\,\omega_d(t-\tau)\; d\tau$$

$$= \frac{P}{M\;\omega_d}\int_0^t e^{-\beta(t-\tau)}\;\sin\omega_d(t-\tau)\;\;d\tau \;-\; \frac{P}{M\;\omega_d} H(t\; - \;t_1)\int_0^t e^{-\beta(t-\tau)}\;\sin\omega_d(t-\tau)\;\;d\tau$$

As in the case of suddenly applied load, the evaluation of the integral is simplified by substituting $t - \tau = x$, $d\tau = -dx$ and the limits for the first integral in terms of x become, when $\tau = t$, $x = 0$ and when $\tau = 0$, $x = t$. However because $H(t - t_1)$ is zero if $t < t_1$, then the limits for the second integral become, when $\tau = t$, $x = 0$ and when $\tau = t_1$, $x = t - t_1$. Therefore, in terms of the variable x

$$\frac{1}{M}\int_0^t F(\tau)\,g(t-\tau)\,d\tau \;=\; \frac{P}{M\;\omega_d}[\int_0^t \;e^{-\beta\,x}\;\sin\omega_d x\;\;dx$$

$$- H(t - t_1)\int_0^{(t\;-\;-t_1)} e^{-\beta\;x}\;\sin\omega_d x\;dx]$$

Using equation (2.16a), the above integral can be written as

$$u \;=\; \Delta[1 \;-\; e^{-\beta t}\{\frac{\beta}{\omega_d}\;\sin\omega_d t \;+\; \cos\omega_d t\}]$$

$$-\Delta H(t-t_1)[1-e^{-\beta(t-t_1)}\{\frac{\beta}{\omega_d}\sin\omega_d(t-t_1) + \cos\omega_d(t-t_1)\}] \tag{2.17a}$$

If damping is zero, then

$$u = \Delta[1 - \cos\omega t - H(t\; - \;t_1)\{1 - \cos\omega(t\; - \;t_1)\}] \tag{2.17b}$$

Figure 2.5 shows a plot of the dynamic magnification factor (DMF) for damped motion with *10%* of critical damping. Note that other things being equal, the DMF depends on the ratio of the duration of the impulse to the natural period of the system. As the duration of the impulse increases, the DMF also increases. As opposed to the static situation, in the dynamic case, it is not just the magnitude of the load but also the ratio of the duration of the load to the period of vibration of the structure which affects the dynamic magnification factor. If the duration of the load is too short, then its effectiveness is greatly reduced because of the inertia in the system. For long duration, the DMF approaches that of a suddenly applied load. When the duration is approximately half the period of the system, the maximum magnification is reached. For further increase of the duration of the impulse, the DMF remains almost a constant.

2.9.3 Triangular Impulse Loading

Figure 2.6 shows the type of loading considered. This type of loading approximates loading due to a blast.

$$F(t) = P(1 - t/t_1),\ t \leq t_1$$
$$= 0,\ t > t_1$$

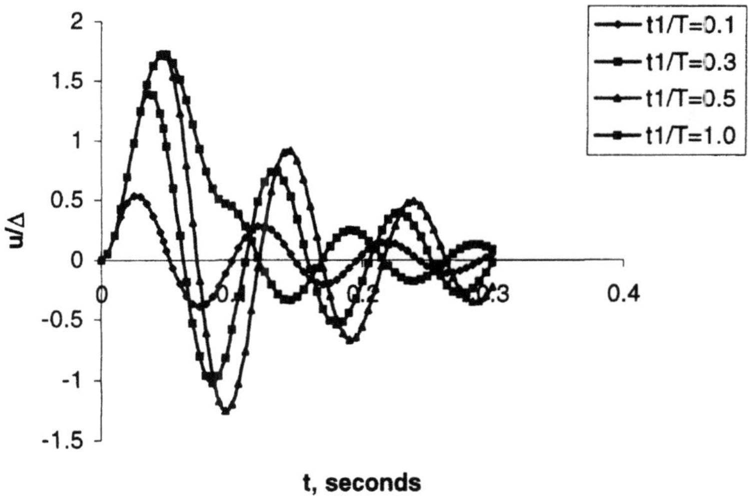

Fig. 2.5 Response to rectangular impulse load of different durations

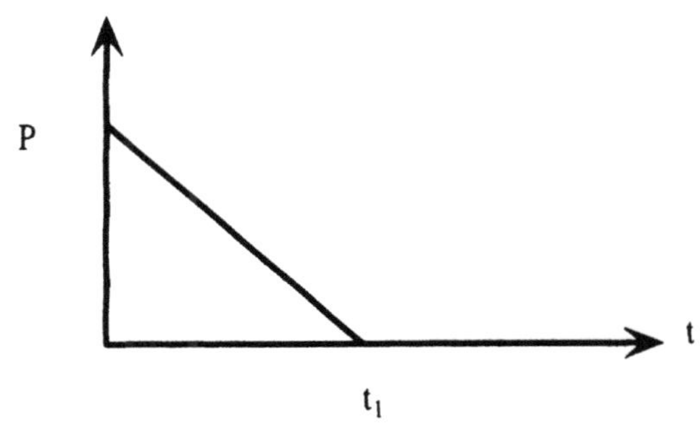

Fig. 2.6 Triangular impulse

As shown in Fig. 2.7, $P(1 - t/t_1)$ becomes negative for $t > t_1$ and in order to cancel this effect, we need another triangular load starting at $t = t_1$. The second loading is given by $(P/t_1)(t - t_1) = P(t/t_1 - 1)$. So the load function $F(t)$ can be represented by

$$F(t) = P(1 - t/t_1) + H(t - t_1)\,P(t/t_1 - 1) = P(1 - t/t_1)\{1 - H(t - t_1)\}$$

Note this approach is similar to what is done using Macaulay's method when dealing with triangular loading in beam problems.

$$\frac{1}{M}\int_0^t F(\tau)\, g(t-\tau) = \frac{P}{M\,\omega_d}\int_0^t (1-\frac{\tau}{t_1})\{1 - H(\tau - t_1)\}[\, e^{-\beta(t-\tau)}\{\sin\omega_d(t-\tau)\}]\,d\tau$$

Note that $H(t - t_1) = 0$ for $t < t_1$. Therefore the above integral can be written as

$$\frac{1}{M}\int_0^t F(\tau)\, g(t-\tau) = \frac{P}{M\,\omega_d}\int_0^t [(1-\frac{\tau}{t_1})[\, e^{-\beta(t-\tau)}\{\sin\omega_d(t-\tau)\}]\,d\tau$$

$$- H(t - t_1)\frac{P}{M\omega_d}\int_{t_1}^t [\,(1 - \frac{\tau}{t_1})\,e^{-\beta(t-\tau)}\sin\omega_d(t-\tau)\,]\,d\tau$$

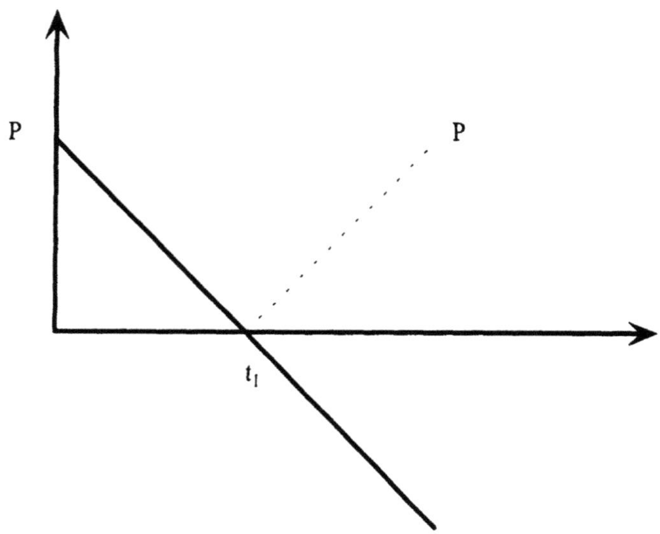

Fig. 2.7 Two shifted triangular impulses

Integration of the above can be carried out as in Section 2.9.2

$$\frac{1}{M}\int_0^t F(\tau)\,g(t-\tau) = \frac{P}{M\omega_d}\int_0^t [\{1 - \frac{(t-x)}{t_1}\}\, e^{-\beta x}\sin\omega_d x\,]\,dx$$

$$- H(t - t_1)\frac{P}{M\omega_d}\int_0^{t-t_1} [\,\{1 - \frac{(t-x)}{t_1}\}\,e^{-\beta x}\sin\omega_d x]\,dx$$

$$= \frac{P}{M\,\omega_d t_1}[\{(t_1 - t)\int_0^t e^{-\beta x}\sin\omega_d x\,dx + \int_0^t x e^{-\beta x}\sin\omega_d x\,dx\}$$

$$- H(t - t_1)\{(t_1 - t) \int_0^{t-t_1} e^{-\beta x} \sin \omega_d x \, dx + \int_0^{t-t_1} x e^{-\beta x} \sin \omega_d x \, dx\}]$$

Using the standard integrals

$$\int_0^t e^{ax} \sin bx \, dx = \frac{1}{(a^2 + b^2)}[e^{at} (a \sin bt - b \cos bt) + b]$$

$$\int_0^t x e^{ax} \sin bx \, dx = \frac{t\, e^{at}}{(a^2 + b^2)}[a \sin bt - b \cos bt]$$

$$+ \frac{1}{(a^2 + b^2)^2}[e^{at}\{(b^2 - a^2)\sin bt + 2a\, b \cos bt\} - 2ab]$$

Substituting $a = -\beta$, $b = \omega_d$, $a^2 + b^2 = \omega^2$, $b^2 - a^2 = \omega_d^2 - \beta^2$
As $\beta = \xi\,\omega$, where ξ = damping ratio, $b^2 - a^2 = \omega^2(1 - 2\,\xi^2)$
and $2ab = -2\,\omega^2\,\xi\sqrt{(1 - \xi^2)}$

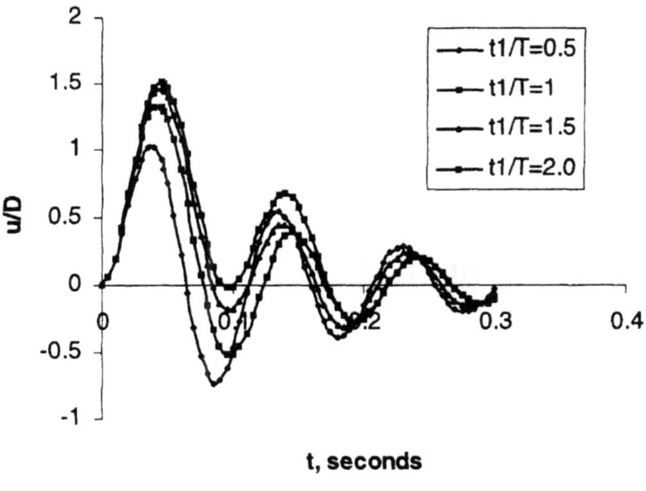

Fig. 2.8 Response to a triangular impulse

$$\int_0^t e^{ax} \sin bx \, dx = \frac{1}{\omega^2}\omega_d [1 - e^{-\beta t}\{\frac{\beta}{\omega_d}\sin \omega_d t + \cos \omega_d t\}]$$

$$\int_0^t x e^{ax} \sin bx \, dx = -\frac{1}{\omega^2}\omega_d t[e^{-\beta t}\{\frac{\beta}{\omega_d}\sin \omega_d t + \cos \omega_d t\}]$$

$$+ \frac{1}{\omega^2}[2\xi\sqrt{(1-\xi^2)} + e^{-\beta t}\{(1 - 2\xi^2)\sin \omega_d t - 2\xi\sqrt{(1-\xi^2)} \cos \omega_d t\}]$$

Setting $M\,\omega^2 = K$, $P/K = \Delta$ and assuming that the system starts from rest

$$\frac{u}{\Delta}=(1-\frac{t}{t_1})[1-e^{-\beta t}\{\frac{\beta}{\omega_d}\sin\omega_d t+\cos\omega_d t\}]-(\frac{t}{t_1})e^{-\beta t}\{\frac{\beta}{\omega_d}\sin\omega_d t+\cos\omega_d t\}$$

$$+\ \frac{1}{\omega_d t_1}[2\xi\sqrt{(1-\xi^2)}+e^{-\beta t}\{(1-2\xi^2)\sin\omega_d t-2\xi\sqrt{(1-\xi^2)}\cos\omega_d t\}]$$

$$-\ H(t-t_1)[\ (1-\frac{t}{t_1})[1-e^{-\beta(t-t_1)}\{\frac{\beta}{\omega_d}\sin\omega_d(t-t_1)+\cos\omega_d(t-t_1)\}]$$

$$+\ (1-\frac{t}{t_1})e^{-\beta(t-t_1)}\{\frac{\beta}{\omega_d}\sin\omega_d(t-t_1)+\cos\omega_d(t-t_1)\}$$

$$+\frac{1}{\omega_d t_1}[2\xi\sqrt{(1-\xi^2)}+e^{-\beta(t-t_1)}\{(1-2\xi^2)\sin\omega_d(t-t_1)-2\xi\sqrt{(1-\xi^2)}\cos\omega_d(t-t_1)\}]]$$

The above equations can be further simplified as

$$\frac{u}{\Delta}=(1-\frac{t}{t_1})-e^{-\beta t}\{\frac{\beta}{\omega_d}\sin\omega_d t+\cos\omega_d t\}$$

$$+\ \frac{1}{\omega_d t_1}[2\xi\sqrt{(1-\xi^2)}+e^{-\beta t}\{(1-2\xi^2)\sin\omega_d t-2\xi\sqrt{(1-\xi^2)}\cos\omega_d t\}]$$

$$-\ H(t-t_1)[\ (1-\frac{t}{t_1})$$

$$+\frac{1}{\omega_d t_1}[2\xi\sqrt{(1-\xi^2)}+e^{-\beta(t-t_1)}\{(1-2\xi^2)\sin\omega_d(t-t_1)-2\xi\sqrt{(1-\xi^2)}\cos\omega_d(t-$$

(2.18a)

If damping is ignored, then

$$\frac{u}{\Delta}=(1-\frac{t}{t_1})-\cos\omega t+\frac{1}{\omega t_1}\sin\omega t-H(t-t_1)[(1-\frac{t}{t_1})+\frac{1}{\omega t_1}\sin\omega(t-t_1)]$$

(2.18b)

Figure 2.8 shows the DMF for damping ratio equal to 0.10 and different durations of a triangular pulse. As can be seen, the longer the duration of the pulse, the larger the dynamic response and as the duration gets longer and longer, the dynamic magnification approaches that of a suddenly applied load.

2.9.4 Exponentially Decaying Force

A mathematically useful continuous force variation with time as opposed to a triangular pulse is an exponentially decaying force given by

$$F = F_0\, e^{-\alpha t}$$

where α is a constant.

Figure 2.9 shows the force distribution for three values of α. A large value of α gives rise to a very peaky force while a small value of α gives rise to a gently decaying force.

Assuming that the system starts from rest, then

$$u = \frac{1}{M}\int_0^t F(\tau)\, g(t-\tau)\, d\tau \;=\; \frac{1}{M\,\omega_d}\int_0^t F(\tau)\; e^{-\beta(t-\tau)}\,\sin\omega_d(t-\tau)\; d\tau$$

Substituting for F, we have

$$u = \frac{F_0}{M\,\omega_d}\int_0^t e^{-\alpha\tau}\; e^{-\beta(t-\tau)}\sin\omega_d(t-\tau)\; d\tau$$

The integration is carried out by substituting $t - \tau = x$ or $\tau = (t - x)$. The integral becomes

$$u = \frac{F_0\, e^{-\alpha t}}{M\,\omega_d}\int_0^t e^{(\alpha-\beta)\,x}\sin\omega_d x\; dx$$

$$= \frac{F_0}{M\,\{(\alpha-\beta)^2+\omega_d^{\,2}\}}[e^{-\alpha t} - e^{-\beta t}\{\cos\omega_d t - \frac{(\alpha-\beta)}{\omega_d}\sin\omega_d t\}]$$

If $\Delta = F_0/K$, $K = M\omega^2$,

$$\frac{u}{\Delta} = \frac{\omega^2}{\{(\alpha-\beta)^2+\omega_d^{\,2}\}}[e^{-\alpha t} - e^{-\beta t}\{\cos\omega_d t - \frac{(\alpha-\beta)}{\omega_d}\sin\omega_d t\}]$$

Figure 2.10 shows a plot of u/Δ versus t for various values of α.

2.10 SHOCK SPECTRA

In previous sections, the response was calculated for typical impulses. For design purposes, the interest lies mainly in the maximum displacement. A plot of the period of the system is generally known as displacement response spectra for impulsive loads or shock spectra. Because of the fact that damping present in actual structures is generally small and, in any case, ignoring damping is conservative, shock spectra is normally calculated for an undamped system.

2.10.1 Rectangular Impulse

The displacement-time relationship for an undamped system for the rectangular impulsive loading shown in Fig. 2.4 is given by equation (2.17b)

$$u = \Delta[1 - \cos\omega t],\;\; t \le t_1$$

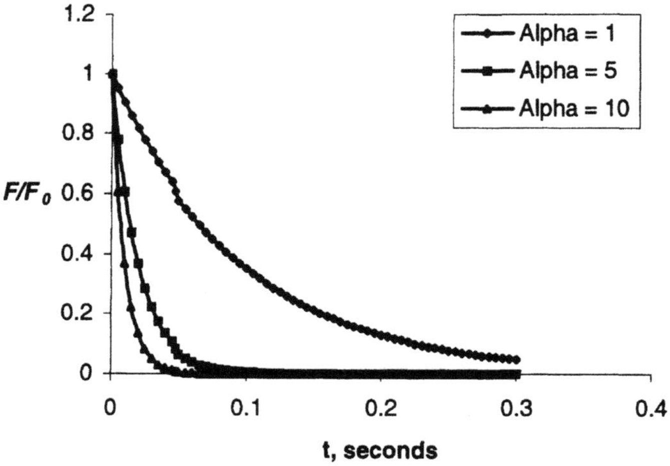

Fig. 2.9 Variation of force with Alpha

u/Δ is a maximum of 2 when $\omega t = \pi$. Substituting for $\omega = 2\pi/T$, u/Δ = when $t_1/T \geq 2$.

$$u = \Delta[\cos\omega(t - t_1) - \cos\omega t], \quad t \geq t_1$$

u/Δ is a maximum when $\sin\omega(t - t_1) = \sin\omega t$. Solving the transcendental equation for t in terms of t_1, the DMF can be determined when $t_1/T > 1$. Fig. 2.11 shows shock spectra for the rectangular impulsive loading.

2.10.2 Triangular Impulse

The displacement-time relationship for an undamped system for the triangular impulsive loading shown in Fig. 2.6 is given by equation (2.18b)

$$\frac{u}{\Delta} = (1 - \frac{t}{t_1}) - \cos\omega t + \frac{1}{\omega t_1}\sin\omega t \quad, t \leq t_1$$

$$\frac{u}{\Delta} = -\cos\omega t + \frac{1}{\omega t_1}\sin\omega t + \frac{1}{\omega t_1}\sin\omega(t - t_1), \; t \geq t_1$$

Proceeding as in Section 2.10.1, the maximum DMF can be determined. Fig. 2.12 shows the shock spectra for triangular impulse load.

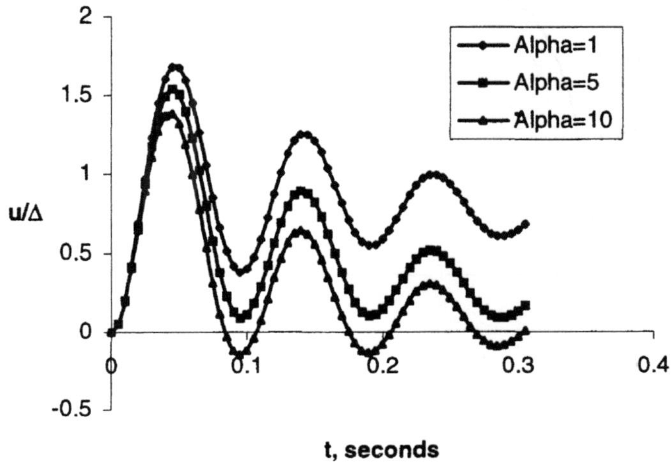

Fig. 2.10 Response to exponential force

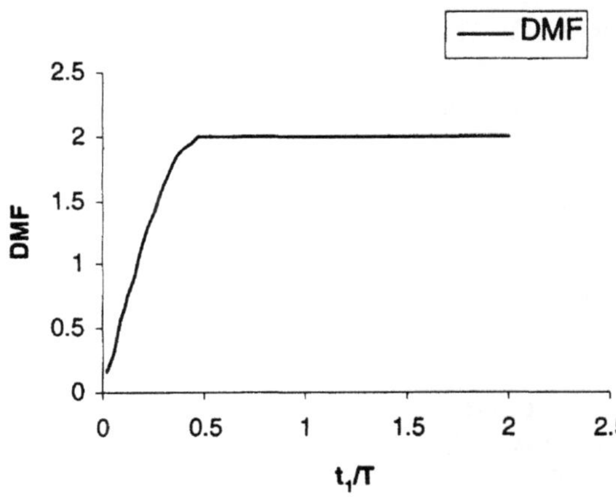

Fig. 2.11 Shock spectra for rectangular impulse load

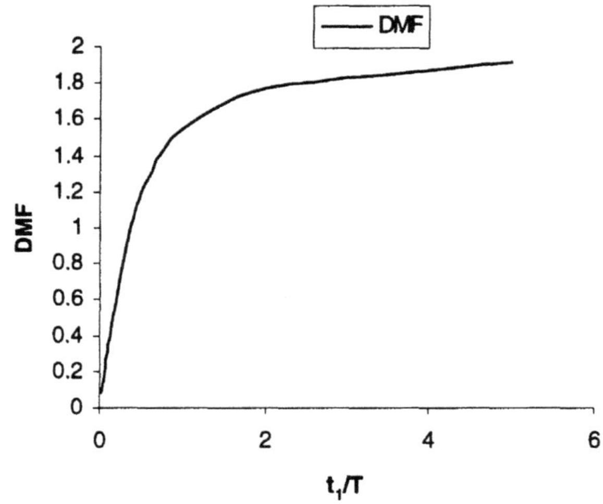

Fig. 2.12 Shock spectra for triangular impulse load

2.11 NUMERICAL EVALUATION OF DUHAMEL INTEGRAL

In Section 2.9, the Duhamel integral was evaluated analytically for some impulsive force variations. This becomes difficult if the load function F(t) is complex. In such a case one has to resort to numerical integration techniques. If the SDOF system is damped, and the system starts from rest, then the displacement u at time t is given by equation (2.14). The determination of u(t) boils down to the evaluation of two integrals, which can be evaluated in a step-by-step fashion as follows. In general we can divide the region of integration into N steps such that $t_N = t$ and $t_0 = 0$

$$\int_0^t F(\tau)\ e^{\beta\tau} \cos\omega_d\ \tau\ d\tau \ =\ \sum_{i=1}^{N} \int_{t_{i-1}}^{t_i} F(\tau)\ e^{\beta\tau} \cos\omega_d\ \tau\ d\tau$$

$$\int_0^t F(\tau)\ e^{\beta\tau} \sin\omega_d\ \tau\ d\tau \ =\ \sum_{i=1}^{N} \int_{t_{i-1}}^{t_i} F(\tau)\ e^{\beta\tau} \sin\omega_d\ \tau\ d\tau$$

and the value of displacement is evaluated at intervals of time each t_i, i = 1 to N.

The evaluation of the integral in any segment $t_{i-1} \le t \le t_i$ is facilitated by assuming that the variation of the forcing function in the region of interest is linear as shown in Fig. 2.13. Therefore if $F(\tau) = C_3 + C_4\,\tau$

Where

$$C_3 = \frac{(F_{i-1}\, t_i - F_i\, t_{i-1})}{(t_i - t_{i-1})}\ , \ C_4 = \frac{(F_i - F_{i-1})}{(t_i - t_{i-1})}$$

The task of evaluating the Duhamel integral boils down to evaluating the four integrals

$$\int_{t_{i-1}}^{t_i} F(\tau)\ e^{\beta\,\tau} \cos\omega_d\, \tau\ d\tau = C_3 \int_{t_{i-1}}^{t_i} e^{\beta\,\tau} \cos\omega_d\, \tau\ d\tau + C_4 \int_{t_{i-1}}^{t_i} \tau\ e^{\beta\,\tau} \cos\omega_d\, \tau\ d\tau$$

$$\int_{t_{i-1}}^{t_i} F(\tau)\ e^{\beta\,\tau} \sin\omega_d\, \tau\ d\tau = C_3 \int_{t_{i-1}}^{t_i} e^{\beta\,\tau} \sin\omega_d\, \tau\ d\tau + C_4 \int_{t_{i-1}}^{t_i} \tau\ e^{\beta\,\tau} \sin\omega_d\, \tau\ d\tau$$

The integrals involved are well-known standard integrals.

$$\int e^{ax} \cos bx\ dx = \frac{e^{ax}}{(a^2+b^2)}[a \cos bx + b \sin bx] = I_1(x)$$

$$\int e^{ax} \sin bx\ dx = \frac{e^{ax}}{(a^2+b^2)}[a \sin bx - b \cos bx] = I_2(x)$$

$$\int x e^{ax} \cos bx\ dx = \frac{x e^{ax}}{(a^2+b^2)}[a \cos bx + b \sin bx]$$

$$- \frac{e^{ax}}{(a^2+b^2)^2}[(a^2-b^2)\cos bx + 2ab \sin bx]$$

$$= x I_1(x) - \frac{e^{ax}}{(a^2+b^2)^2}[C_1 \cos bx + C_2 \sin bx] = I_4(x)$$

$$\int x e^{ax} \sin bx\ dx = \frac{x\ e^{ax}}{(a^2+b^2)}[a \sin bx - b \cos bx]$$

$$- \frac{e^{ax}}{(a^2+b^2)^2}[(a^2-b^2)\sin bx - 2ab \cos bx]$$

$$= x I_2(x) - \frac{e^{ax}}{(a^2+b^2)^2}[C_1 \sin bx - C_2 \cos bx] = I_3(x)$$

where $a = \beta$, $b = \omega_d$ and $a^2 + b^2 = \omega^2$, $C_1 = a^2 - b^2$, $C_2 = 2ab$

$$\int_{t_{i-1}}^{t_i} F(\tau)\ e^{\beta\,\tau} \cos\omega_d\, \tau\ d\tau = C_3 I_1(\tau) + C_4\, I_4(\tau)\Big|_{t_{i-1}}^{t_i}$$

$$\int_{t_{i-1}}^{t_i} F(\tau)\ e^{\beta\,\tau} \sin\omega_d\, \tau\ d\tau = C_3 I_2(\tau) + C_4\, I_3(\tau)\Big|_{t_{i-1}}^{t_i}$$

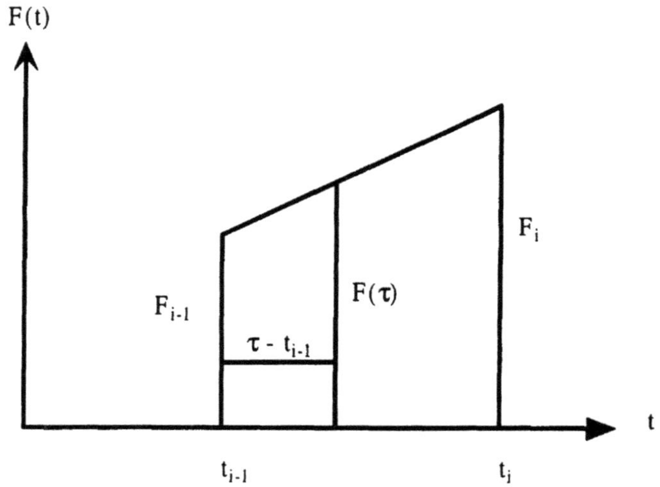

Fig. 2.13 Linear variation of force

The velocity and acceleration of the mass can be obtained by differentiating u(t) with respect to t. When differentiating u with respect to t, the important result of differentiation of an integral containing a parameter has to be followed. In this specific case this becomes

$$\frac{d}{dt}\left[\int_{\tau=0}^{\tau=t} f(\tau)\,d\tau\right] = f(t)$$

Therefore

$$\frac{d}{dt}\left[\int_0^t F(\tau)e^{\beta\tau}\cos\omega_d\tau\;\;d\tau\right] = F(t)\;e^{\beta t}\;\cos\omega_d t$$

$$\frac{d}{dt}\left[\int_0^t F(\tau)e^{\beta\tau}\sin\omega_d\tau\;\;d\tau\right] = F(t)\;e^{\beta t}\;\sin\omega_d t$$

Using the above result

$$\frac{du}{dt} = \frac{-\beta e^{-\beta t}}{M\;\omega_d}\left[\sin\omega_d t\int_0^t F(\tau)e^{\beta\tau}\cos\omega_d\tau\;\;d\tau - \cos\omega_d t\int_0^t F(\tau)\;e^{\beta\tau}\;\sin\omega_d\tau\;\;d\tau\right]$$

$$+\frac{e^{-\beta t}}{M}\left[\cos\omega_d t\int_0^t F(\tau)e^{\beta\tau}\cos\omega_d\tau\;\;d\tau + \sin\omega_d t\int_0^t F(\tau)\;e^{\beta\tau}\;\sin\omega_d\tau\;\;d\tau\right]$$

$$\frac{d^2u}{dt^2} = \frac{e^{-\beta t}}{M\;\omega_d}\left[A\int_0^t F(\tau)e^{\beta\tau}\cos\omega_d\tau\,d\tau - B\int_0^t F(\tau)\;e^{\beta\tau}\;\sin\omega_d\tau\,d\tau\right] + \frac{F(t)}{M}$$

$$A = \{(\beta^2 - \omega_d^2) \sin \omega_d t - 2\beta \omega_d \cos \omega_d t\}$$
$$B = \{(\beta^2 - \omega_d^2) \cos \omega_d t + 2\beta \omega_d \sin \omega_d t\}$$

2.12 PROGRAM DUHAMEL

Program DUHAMEL uses the theory developed above to compute the Duhamel integral. The data for the variation of the forcing function with time is stored in two arrays FORCEINPUT(NDPTS) and TIMEINPUT(NDPTS), where NDPTS is the number of data points defining the force variation. The computations are carried on to a maximum prescribed time limit TMAX at a time increment not exceeding DTIME. The following variable names are used in the program.

MASS = M, STIFF = K, DRATIO = damping as a ratio of critical damping,
OMEGA = ω, OMEGAD = ω_d, BETA = DRATIO x OMEGA

The first step is to calculate the force at intervals not greater than DTIME. The interval between linear segments of specified variation of force with time input is subdivided in to intervals of DTIME. If it cannot be divided into an equal number of subdivisions each equal to DTIME, then the last segment will be less than DTIME. The interpolated force–time relationship is stored in two arrays FORCE and TIME. The computations follow closely the theory developed above.

```
!*******************************************************************
          PROGRAM DUHAMEL
!*******************************************************************
!   THIS PROGRAM USES DUHAMEL INTEGRAL TO CALCULATE THE
!   RESPONSE OF A SINGLE DEGREE OF FREEDOM ELASTIC SPRING-
!  MASS-DAMPER SYSTEM TO EITHER AN APPLIED FORCE OR A BASE
!  DISPLACEMENT
! *******************************************************************
!************VARIABLE DECLARATION*****************************
          IMPLICIT NONE
          INTEGER::NDPTS,I,IP1,NDPTSP1, NMAX,INCR,II
          REAL::MASS,STIFF,DRATIO,TMAX,DTIME,OMEGA,OMEGA2,&
          &OMEGA4,OMEGAD,BETA,C1,C2,C3,C4,C5,C6,C7,C8,C9,AT,BT,&
          &TI,TIM1,T0,T1,F0,F1,Y,T,VEL,ACLR,PERIOD,SD1,CD1
          REAL,ALLOCATABLE::FORCEINPUT(:),TIMEINPUT(:),FORCE(:),&
          &TIME(:)
!*******FILE MANAGEMENT*****************************************
         CHARACTER(LEN=40)::FILEIN, FILEOUT
         PRINT*,'NAME OF INPUT FILE'
         READ*,FILEIN
         PRINT*,'NAME OF OUTPUT FILE'
         READ*,FILEOUT
         OPEN(UNIT=1, FILE =FILEIN,STATUS='OLD',ACTION='READ')
         OPEN(UNIT=2, FILE =FILEOUT,STATUS='REPLACE',ACTION='WRITE')
!*******************************************************************
!***READ AND PRINT BASIC DATA INPUT**************************
          READ(1,*)MASS, STIFF, DRATIO
!***CALCULATE BASIC PARAMETERS******************************
```

```
            OMEGA2 = STIFF/MASS
            OMEGA4 = OMEGA2*OMEGA2
            OMEGA = SQRT(OMEGA2)
            OMEGAD = OMEGA*SQRT(1-DRATIO*DRATIO)
            BETA = DRATIO*OMEGA
            PERIOD= 6.283185307/OMEGA
            WRITE(2,'(A,E14.6)')'MASS = ',MASS
            WRITE(2,'(A,E14.6)')'STIFFNESS = ',STIFF
            WRITE(2,'(A,E14.6)')'DAMPING RATIO = ',DRATIO
            WRITE(2,'(A,E14.6)')'OMEGA = ',OMEGA
            WRITE(2,'(A,E14.6)')'OMEGAD = ',OMEGAD
            WRITE(2,'(A,E14.6)')'PERIOD = ',PERIOD
!***DATA ABOUT FORCE-TIME RELATIONSHIP********************
           READ(1,*)NDPTS
           WRITE(2,'(A,I4)')'NUMBER OF DATA POINTS DEFINING &
           &FORCE-TIME VARIATION =  ',NDPTS
           READ(1,*)TMAX
           WRITE(2,'(A,F10.4)')'MAXIMUM TIME UP TO WHICH &
           &RESPONSE IS REQUIRED = ',TMAX
           READ(1,*)DTIME
           WRITE(2,'(A,F10.4)')'TIME INCREMENT =',DTIME
           NMAX = NINT(TMAX/DTIME) + 1
           WRITE(2,*)'FORCE-TIME RELATIONSHIP'
           ALLOCATE(FORCEINPUT(NDPTS),TIMEINPUT(NDPTS),&
           &FORCE(NMAX),TIME(NMAX))
!*** READ AND WRITE INPUT DATA FOR FORCE-TIME RELATIONSHIP
! ****ASSUMED****************PIECEWISE LINEAR****************
           WRITE(2,*)'TIME        FORCE'
            DO  I = 1,NDPTS
            READ(1,*)FORCEINPUT(I),TIMEINPUT(I)
            WRITE(2,*)TIMEINPUT(I),FORCEINPUT(I)
            END DO
!****INTERPOLATE THE FORCE-TIME INPUT TO OBTAIN FORCE AND
! **TIME SPACED AT INTERVALS NOT GREATER THAN DTIME******
            FORCE(1)=FORCEINPUT(1)
            TIME(1)=TIMEINPUT(1)
            T=0
            II=1
            DO I=2,NMAX
            T=T+DTIME
            IF(T<= TIMEINPUT(II+1)) THEN
            FORCE(I)=FORCEINPUT(II)+(FORCEINPUT(II+1)-&
            &FORCEINPUT(II))*(T-TIMEINPUT(II))/(TIMEINPUT(II+1)&
            &-TIMEINPUT(II))
            TIME(I)=T
            ELSE IF(T>TIME(II+1) .AND. T< TIMEINPUT(NDPTS))THEN
            FORCE(I)=FORCEINPUT(II+1)
            TIME(I)= TIMEINPUT(II+1)
            T=TIME(I)
            II=II+1
            ELSE IF(T>TIMEINPUT(NDPTS))THEN
            EXIT
            END IF
            END DO
            WRITE(2,*)'INTERPOLATED FORCE-TIME RELATIONSHIP'
```

```
          WRITE(2,*)'TIME        FORCE'
          DO I= 1,NMAX
          WRITE(2,*)TIME(I),FORCE(I)
          END DO
!***START OF DUHAMEL INTEGRATION*****************************
! *** CYCLE ON THE TIME INCREMENTS UNTIL MAXIMUM TIME ******
!*AND CALCULATE  DISPLACEMENT, VELOCITY AND ACCELARATION.*
        C1 = BETA*BETA - OMEGAD*OMEGAD
        C2 = 2.0*BETA*OMEGAD
        NMAX=NMAX-1
        AT = 0
        BT = 0
        WRITE(2,*)'TIME        DISPLACEMENT        VELOCITY&
        &    ACCLERATION'
         INCREMENT: DO  I = 1,NMAX
         IP1= I  + 1
         T0 = TIME(I)
         T1 = TIME(IP1)
         F0 = FORCE(I)
         F1 = FORCE(IP1)
         C3 = (F0*T1 - F1*T0)/(T1-T0)
         C4 = (F1-F0)/(T1-T0)
         C5 = C3*(DUHA1(T1) - DUHA1(T0))
         C6 = C4*(DUHA4(T1) - DUHA4(T0))
         C7 = C3*(DUHA2(T1) - DUHA2(T0))
         C8 = C4*(DUHA3(T1) - DUHA3(T0))
         AT = AT + C5 + C6
         BT = BT + C7 + C8
         SD1= SIN(OMEGAD*T1)
         CD1=COS(OMEGAD*T1)
          C9 = EXP(-BETA*T1)/(MASS*OMEGAD)
          Y =  C9*(SD1*AT – CD1*BT)
          VEL=C9*((OMEGAD*CD1-BETA*SD1)*AT +& &(OMEGAD*SD1+BETA*CD1)*BT)
          ACLR=-C9*((C2*CD1-C1*SD1)*AT + &
          &(C2*SD1+C1*CD1)*BT)+F1/MASS
          WRITE(2,'(F10.4,3(3X,E14.6))')T1,Y,VEL,ACLR
          END DO  INCREMENT
!******DEFINE FUNCTION STATEMENTS NEEDED FOR LATER
!*CALCULATIONS*****************************************
         CONTAINS
         FUNCTION DUHA1(TAU)
         REAL::DUHA1,TAU
         DUHA1 = EXP(BETA*TAU)*(BETA*COS(OMEGAD*TAU) +&
         & OMEGAD*SIN(OMEGAD*TAU))/OMEGA2
         END FUNCTION
         FUNCTION DUHA2(TAU)
         REAL::DUHA2,TAU
         DUHA2 = EXP(BETA*TAU)*(BETA*SIN(OMEGAD*TAU)  -&
         & OMEGAD*COS(OMEGAD*TAU))/OMEGA2
         END FUNCTION
         FUNCTION DUHA3(TAU)
         REAL::DUHA3, TAU
         DUHA3=TAU*DUHA2(TAU)-EXP(BETA*TAU)*(C1&
         &*SIN(OMEGAD*TAU)- C2*COS(OMEGAD*TAU))/OMEGA4
         END FUNCTION
```

```
FUNCTION DUHA4(TAU)
REAL::DUHA4,TAU
DUHA4 = TAU*DUHA1(TAU)-EXP(BETA*TAU)*(C1&
&*COS(OMEGAD*TAU)+ C2*SIN(OMEGAD*TAU))/OMEGA4
END FUNCTION
END PROGRAM DUHAMEL
```

2.12.1 Use of Program DUHAMEL: Example

The program DUHAMEL was used to calculate the response of a mass-spring-damper system to a trapezoidal impulse as shown in Fig. 2.14. The properties assumed were

$K = 43.717 \times 10^6$ N/m, $M = 10 \times 10^3$ kg, $C/Ccr = 5\%$.

The natural circular frequency of the system $\omega = \sqrt{(K/M)} = 66.11$ radians/second and period $T = 2\pi/\omega = 0.095$ seconds.

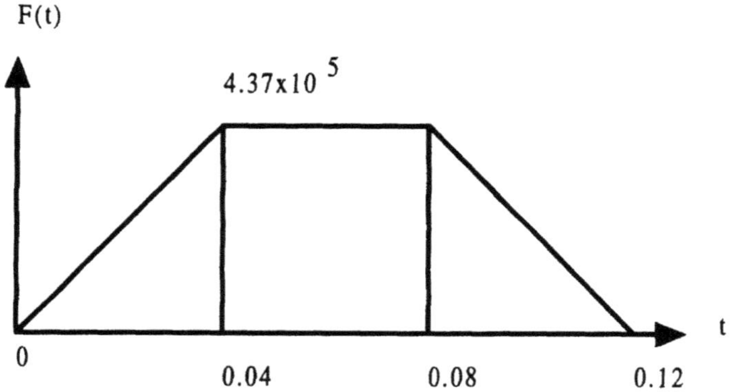

Fig. 2.14 Assumed relationship between force and time

Calculations were done for a time interval of 0.005 seconds and the response was required up to 0.3 seconds. Fig 2.15 shows the results obtained.

2.13 EARTHQUAKE RESPONSE SPECTRA

It was shown in Section 1.18 that an SDOF system subjected to acceleration of the base could be considered as an equivalent system subjected to external force given by negative of the product $Mu''_b(t)$, where $u_b(t)$ is the base displacement. Substituting for $F(t) = - Mu''_b(t)$, in equation 2.13

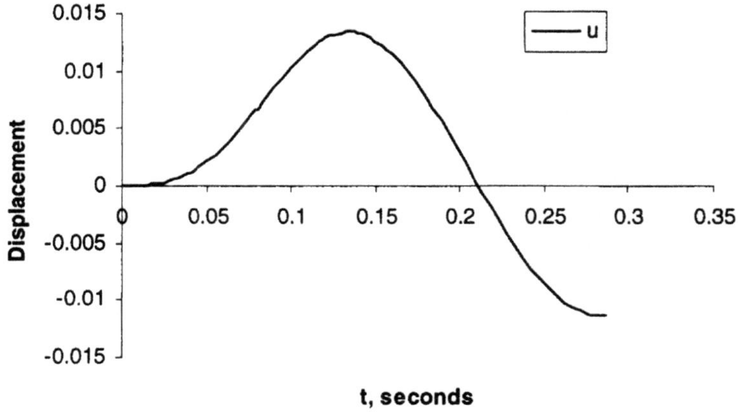

Fig. 2.15 Displacement response

$$u(t) = e^{-\beta t}\{\cos\omega_d t + \frac{\beta}{\omega_d}\sin\omega_d t\}\, u(0) + e^{-\beta t}\{\frac{1}{\omega_d}\sin\omega_d t\}\, u'(0)$$

$$-\frac{1}{\omega_d}\int_0^t \frac{d^2u_b(\tau)}{d\tau^2}\, e^{-\beta(t-\tau)}\sin\omega_d(t-\tau)\, d\tau \qquad (2.15)$$

where u(t) = relative displacement of the mass with respect to the base.
Assuming that the system starts from rest

$$u(t) = -\frac{1}{\omega_d}\int_0^t \frac{d^2u_b(\tau)}{d\tau^2}\, e^{-\beta(t-\tau)}\sin\omega_d(t-\tau)\, d\tau$$

The velocity and acceleration are obtained by differentiating u(t) with respect to t. These are given by

$$\frac{du(t)}{dt} = -\int_0^t \frac{d^2u_b(\tau)}{d\tau^2}\, e^{-\beta(t-\tau)}\cos\omega_d(t-\tau)\, d\tau$$

$$+\frac{\beta}{\omega_d}\int_0^t \frac{d^2u_b(\tau)}{d\tau^2}\, e^{-\beta(t-\tau)}\sin\omega_d(t-\tau)\, d\tau$$

$$\frac{d^2u(t)}{dt^2} = [\omega_d - \frac{\beta^2}{\omega_d}]\int_0^t \frac{d^2u_b(\tau)}{d\tau^2}\, e^{-\beta(t-\tau)}\sin\omega_d(t-\tau)\, d\tau$$

$$+ 2\beta\int_0^t \frac{d^2u_b(\tau)}{d\tau^2}\, e^{-\beta(t-\tau)}\cos\omega_d(t-\tau)\, d\tau - \frac{d^2u_b(t)}{dt^2}$$

$$\frac{d^2\{u(t) + u_b(t)\}}{dt^2} = [\omega_d - \frac{\beta^2}{\omega_d}]\int_0^t \frac{d^2u_b(\tau)}{d\tau^2}\, e^{-\beta(t-\tau)}\sin\omega_d(t-\tau)\, d\tau$$

$$+ 2\beta\int_0^t \frac{d^2u_b(\tau)}{d\tau^2}\, e^{-\beta(t-\tau)}\cos\omega_d(t-\tau)\, d\tau$$

If the system is undamped, then $\beta = 0$ and $\omega_d = \omega$. Therefore

$$u(t) = -\frac{1}{\omega}\int_0^t \frac{d^2u_b(\tau)}{d\tau^2} \sin\omega(t-\tau)\ d\tau$$

$$\frac{du(t)}{dt} = -\int_0^t \frac{d^2u_b(\tau)}{d\tau^2} \cos\omega(t-\tau)\ d\tau$$

$$\frac{d^2\{u(t) + u_b(t)\}}{dt^2} = \omega\int_0^t \frac{d^2u_b(\tau)}{d\tau^2} \sin\omega(t-\tau)\ d\tau$$

The absolute acceleration of the mass is related to relative displacement by

$$\frac{d^2\{u(t) + u_b(t)\}}{dt^2} = -\omega^2 u(t)$$

It is assumed that for low values of damping, the above relationship holds good. If the dynamic magnifications for relative displacement and absolute acceleration are given, respectively, by, S_d and S_a, then

$$S_a \approx \omega^2\ S_d$$

If in the expression for velocity, the cosine function is replaced by sine function, and if the dynamic magnification for relative velocity is S_v, then

$$S_v \approx \omega S_d$$

The two relationships

$$\omega\ S_d \approx S_v\ ,\ \frac{1}{\omega} S_a = S_v$$

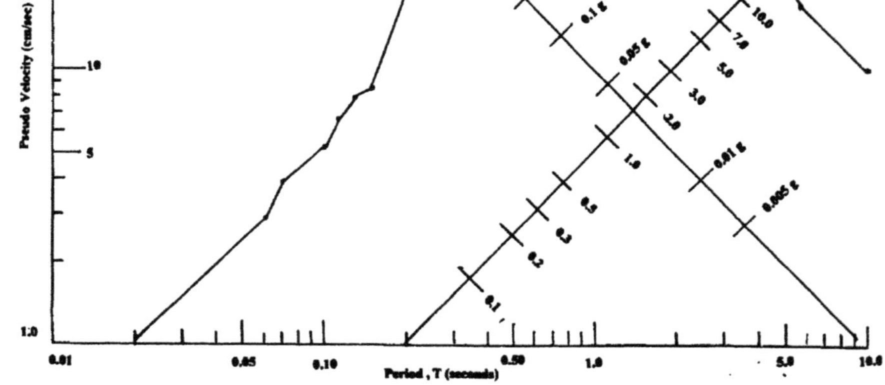

Fig. 2.16 Tripartite earthquake response spectra

can both be plotted on the same 'Tripartite' graph as follows. If along the horizontal axis is plotted the logarithm of period T = 2π/ω and along the vertical axis the logarithm of S_v, then

$$\log S_v = \log T + \log(\frac{S_a}{2\pi})$$

$$\log S_v = -\log T + \log(2\pi S_d)$$

This shows that for constant values of S_a, the relationship between $\log S_v$ and $\log T$ is a line inclined at 45^0 to logT axis. Similarly for constant values of S_d, the relationship between $\log S_v$ and $\log T$ is a line inclined at 135^0 to logT axis. Fig. 2.16 shows one such tripartite graph.

CHAPTER 3

INTRODUCTION TO MULTIPLE DEGREE OF FREEDOM SYSTEMS

3.1 INTRODUCTION

In the previous two chapters, the behaviour of a single degree of freedom (SDOF) system was considered. A typical example was a single-storey building with the mass concentrated at the roof level and capable of only a sway displacement as shown in Fig. 1.2b. In practice, systems with multiple degree of freedom (MDOF) such as the three-storey 'shear' building shown in Fig. 3.1 are quite common.

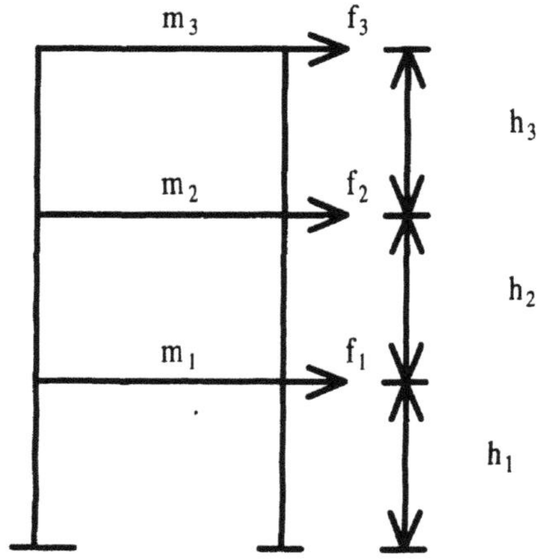

Fig. 3.1 A three-storey shear frame

In order to establish the stiffness, force and mass relationship, it is assumed for the sake of simplicity, that all the mass is concentrated at the storey level and only sway displacement is permitted at each storey. Let the shear stiffness of ith storey be k_i and the mass at the ith level be m_i, where

$$k_i = 2 \text{ columns } [12\ EI/h^3]_i$$

Let the external dynamic force acting at the ith storey level be $f_i(t)$ and the corresponding displacement $x_i(t)$.

Since damping present is generally small, to simplify discussions, let damping be ignored. Using D'Alembert's principle, the dynamic equilibrium equation of mass at each storey can be established. On each mass m_i, the forces acting are the external force f_i, the inertial force equal to $-m_i\, x''_i$ and the 'spring' forces due to the relative drift of the columns which restrain motion. As shown in Fig. 3.2, the three equilibrium equations, one for each floor, are

$$m_1 x''_1 + k_1 x_1 - k_2(x_2 - x_1) = f_1$$
$$m_2 x''_2 + k_2(x_2 - x_1) - k_3(x_3 - x_2) = f_2$$
$$m_3 x''_3 + k_3(x_3 . x_2) = f_3$$

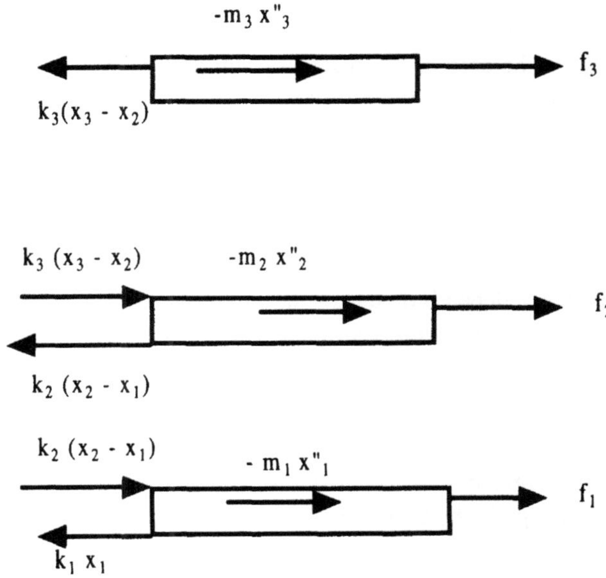

Fig. 3.2 Dynamic equilibrium of the masses

Expressing these equations in matrix form

$$\begin{bmatrix} m_1 & 0 & 0 \\ 0 & m_2 & 0 \\ 0 & 0 & m_3 \end{bmatrix} \begin{bmatrix} x''_1 \\ x''_2 \\ x''_3 \end{bmatrix} + \begin{bmatrix} k_1+k_2 & -k_2 & 0 \\ -k_2 & k_2+k_3 & -k_3 \\ 0 & -k_3 & k_3 \end{bmatrix} \begin{bmatrix} x_1 \\ x_2 \\ x_3 \end{bmatrix} = \begin{bmatrix} f_1 \\ f_2 \\ f_3 \end{bmatrix}$$

Or setting

$$M = \begin{bmatrix} m_1 & 0 & 0 \\ 0 & m_2 & 0 \\ 0 & 0 & m_3 \end{bmatrix}, \quad X'' = \begin{bmatrix} x''_1 \\ x''_2 \\ x''_3 \end{bmatrix}$$

$$K = \begin{bmatrix} k_1+k_2 & -k_2 & 0 \\ -k_2 & k_2+k_3 & -k_3 \\ 0 & -k_3 & k_3 \end{bmatrix}, \quad X = \begin{bmatrix} x_1 \\ x_2 \\ x_3 \end{bmatrix}, \quad F = \begin{bmatrix} f_1 \\ f_2 \\ f_3 \end{bmatrix}$$

The equilibrium equations can be expressed in matrix form as

$$M X'' + K X = F \tag{3.1}$$

It has to be appreciated that X, X'' and F are all functions of time, t. This is similar to the equilibrium equation (1.1), in the case of SDOF system. The only difference between equations (1.1) and (3.1) is that quantities such as M and K are matrices while X'', X and F are vectors rather than single quantities. The matrices M and K are called mass and stiffness matrices. The vectors X'', X and F are called acceleration, displacement and force vectors, respectively. It is worth noting that both mass and stiffness matrices are symmetric.

If the force vector F is equal to zero, then the system represents undamped free vibration, similar to the case of the SDOF system discussed in Chapter 1. In the case of undamped free vibration, if the system is simply disturbed from its equilibrium position and allowed to vibrate, the resulting motion is simple harmonic. Therefore if, say, x is the displacement at time t, we can represent

$$x(t) = \bar{x}\cos(\omega t + \theta)$$

Where $\bar{x}$ = amplitude of displacement, ω = circular frequency of vibration and θ = phase difference which depends on the displacement and velocity at time t = 0.

Differentiating x(t) twice with respect to time enables the relationship between acceleration and displacement to be established.

$$x''(t) = -\omega^2\ \bar{x}\cos(\omega t + \theta) = -\omega^2 x(t)$$

$$x''(t) = -\omega^2 x(t)$$

If this result is substituted into the matrix equation, then we have the equation governing free undamped vibration of the MDOF system.

$$KX = \omega^2 MX \tag{3.2}$$

In the case of the SDOF system, for non-zero values of X, the above equation is satisfied when $\omega^2 = K/M$. In the case of MDOF systems, as both K and M are matrices, the interpretation of the equation $K X = \omega^2 M X$ will have to be different. The matrix equation (3.2) is known as an eigen-value or characteristic value problem. The vector X is known as an eigen-vector or characteristic vector and ω^2 is known as eigen-value or characteristic value.

3.2 EIGEN-VALUE PROBLEM

Before discussing the solution of the eigen-value problem and the many useful general properties associated with it, it is useful to consider a simple and direct approach for the solution of the eigen-value problem. The approach is not best suited for the solution of a large-scale eigen-value problem with many hundreds (or thousands) of variables. Methods suited for the solution of practical eigen-value problems will be presented in Chapter 5. The objective for the present is to consolidate and clarify important concepts in the solution of the eigen-value problem.

3.2.1 Determinants

The idea of a determinant is best introduced through simple examples. The determinant of a matrix A is generally indicated by the symbol $|A|$.

i. The determinant of a scalar is simply, the scalar itself.

$$A = [a_{11}] \, , \; |A| \; = \; a_{11}$$

ii. A is a 2 x 2 matrix.

$$A = \begin{bmatrix} a_{11} & a_{12} \\ a_{21} & a_{22} \end{bmatrix}$$

$$|A| \; = \; (-1)^{(1+1)} a_{11} [a_{22}] \; + (-1)^{(1+2)} a_{12} [a_{21}] = a_{11}\, a_{22} \; - a_{12}\; a_{21}$$

iii. A is a 3 x 3 matrix.

$$A = \begin{bmatrix} a_{11} & a_{12} & a_{13} \\ a_{21} & a_{22} & a_{23} \\ a_{31} & a_{32} & a_{33} \end{bmatrix}$$

$$|A| = (-1)^{(1+1)} a_{11} \begin{vmatrix} a_{22} & a_{23} \\ a_{32} & a_{33} \end{vmatrix} + (-1)^{(1+2)} a_{12} \begin{vmatrix} a_{21} & a_{23} \\ a_{31} & a_{33} \end{vmatrix}$$

$$+ (-1)^{(1+3)} a_{13} \begin{vmatrix} a_{21} & a_{22} \\ a_{31} & a_{32} \end{vmatrix}$$

As can be seen a_{11} is multiplied by the determinant of the 2 x 2 matrix obtained by ignoring from A the elements of the first row and first column. Similarly a_{12} is multiplied by the determinant of the 2 x 2 matrix obtained by ignoring from A the elements of the first row and second column. Lastly a_{13} is multiplied by the determinant of the 2 x 2 matrix obtained by ignoring from A the elements of the first row and third column.

iv. A is an N x N matrix: The determinate of an N x N matrix is obtained by the general formula

$$|A| = \sum_{k=1}^{N} (-1)^{(k+1)} \; a_{1k} \, |A_{1k}| \tag{3.3}$$

where A_{1k} is the *(N-1) x (N-1)* matrix obtained by ignoring the first row and k^{th} column elements of the *N x N* matrix *A*.

In fact it is not necessary to use the first row for calculating the determinant. If the ith row is used, then

$$|A| = \sum_{j=1}^{N} (-1)^{(i+j)} \; a_{ij} \, |A_{ij}| \tag{3.4a}$$

where A_{ij} is the *(N-1) x (N-1)* matrix obtained by ignoring the ith row and j^{th} column elements of the *N x N* matrix *A*.

Similarly, if j^{th} column is used, then

$$|A| = \sum_{i=1}^{N} (-1)^{(i+j)} \; a_{ji} \, |A_{ji}| \tag{3.4b}$$

It is perhaps worth stating that the equations (3.4a, b) are purely of theoretical significance. They are not used in practical calculation except in the case of upper and lower triangular matrices. The reason for this is that there are other more efficient methods for achieving the same result. If a matrix has a zero determinant, it is called a singular matrix and such a matrix cannot be inverted.

3.2.1.1 Upper and Lower Triangular Matrices

Using the equation (3.4a), the determinant of an upper or a lower triangular matrix is simply the continued product of the diagonals.

i. If L is a lower triangular matrix, then expanding by rows, $|L|$ is given by

$$L=\begin{bmatrix} l_{11} & 0 & 0 & 0 \\ l_{21} & l_{22} & 0 & 0 \\ l_{31} & l_{32} & l_{33} & 0 \\ l_{41} & l_{42} & l_{43} & l_{44} \end{bmatrix}$$

$$|L| = l_{11}\begin{bmatrix} l_{22} & 0 & 0 \\ l_{32} & l_{33} & 0 \\ l_{42} & l_{43} & l_{44} \end{bmatrix} = l_{11}\, l_{22}\begin{vmatrix} l_{33} & 0 \\ l_{43} & l_{44} \end{vmatrix} = l_{11}\, l_{22}\, l_{33}\,|l_{44}| = l_{11}\, l_{22}\, l_{33}\, l_{44}$$

ii. If U is an upper triangular matrix, then expanding by rows, $|U|$ is given by

$$U=\begin{bmatrix} u_{11} & u_{12} & u_{13} & u_{14} \\ 0 & u_{22} & u_{23} & u_{24} \\ 0 & 0 & u_{33} & u_{34} \\ 0 & 0 & 0 & u_{44} \end{bmatrix}$$

$$|U| = u_{11}\begin{bmatrix} u_{22} & u_{23} & u_{24} \\ 0 & u_{33} & u_{34} \\ 0 & 0 & u_{44} \end{bmatrix} = u_{11}\, u_{22}\begin{vmatrix} u_{33} & u_{34} \\ 0 & u_{44} \end{vmatrix} = u_{11}\, u_{22}\, u_{33}\,|u_{44}| = u_{11}\, u_{22}\, u_{33}\, u_{44}$$

3.2.1.2 Calculation of Determinants: Example

Calculate the determinant of the following 3 x 3 matrix.

$$A=\begin{bmatrix} 1 & 3 & 5 \\ 7 & 11 & 13 \\ 19 & 23 & 29 \end{bmatrix}$$

$$|A| = (-1)^{(1+1)}\,1\begin{vmatrix} 11 & 13 \\ 23 & 29 \end{vmatrix} + (-1)^{(1+2)}\,3\begin{vmatrix} 7 & 13 \\ 19 & 29 \end{vmatrix} + (-1)^{(1+3)}\,5\begin{vmatrix} 7 & 11 \\ 19 & 23 \end{vmatrix}$$

$$|A| = 1(11\times29 - 23\times13) - 3\,(7\times29 - 19\times13) + 5(7\times23 - 19\times11)$$
$$= 1(20)\ \ -3(-44) + 5(-48) = -88$$

3.2.2 Cramer's Rule

An important rule in the solution of homogeneous linear equations defined by the matrix relationship $AX = 0$, says that a non-trivial solution exists only if the determinant of A is zero. This is known as Cramer's rule and is an important result in the solution of linear eigen-value problems.

3.2.3 Positive Definiteness

An N x N matrix is said to be positive definite when sub matrices of size 1 x 1, 2 x 2, 3 x 3...up to and including N x N matrices have determinant greater than zero. If any one of the sub matrices has a zero determinant, then it is said to be positive semi-definite. If the determinant of the matrix is zero, then it is called a singular matrix. It is perhaps worth recalling that in the case of all stable structures, the stiffness matrix is real and positive definite. However, if the structure can undergo rigid body movements, then the stiffness matrix will be semi-definite. An example of such a structure is a structure in space not fixed to a foundation.

3.2.3.1 Positive Definiteness: Examples

Example 1: Verify that the following stiffness matrix for the four-storey shear frame shown in Fig. 3.3 is positive definite.

$$K = \begin{bmatrix} 14 & -6 & 0 & 0 \\ -6 & 10 & -4 & 0 \\ 0 & -4 & 8 & -4 \\ 0 & 0 & -4 & 4 \end{bmatrix}$$

The determinant is given by

$$|K| = 14\begin{vmatrix} 10 & -4 & 0 \\ -4 & 8 & -4 \\ 0 & -4 & 4 \end{vmatrix} + (-1)^{(1+2)}(-6)\begin{vmatrix} -6 & -4 & 0 \\ 0 & 8 & -4 \\ 0 & -4 & 4 \end{vmatrix}$$

$$= 14[10\{8\times4-(-4)^2\} + (-1)^{(1+2)}(-4)\{-4\times4-0\}]$$
$$+\ 6[(-6)\{8\times4-(-4)^2\}+(-1)^{(1+2)}(-4)\{0\}] = 768$$

The sub matrices are

$$K_1=[14], K_2=\begin{bmatrix}14 & -6\\ -6 & 10\end{bmatrix}, K_3=\begin{bmatrix}14 & -6 & 0\\ -6 & 10 & -4\\ 0 & -4 & 8\end{bmatrix}, K_4=\begin{bmatrix}14 & -6 & 0 & 0\\ -6 & 10 & -4 & 0\\ 0 & -4 & 8 & -4\\ 0 & 0 & -4 & 4\end{bmatrix}$$

The determinants of the sub matrices are 14, $[14 \times 10 - (-6)^2] = 104$, $[14\{10 \times 8 - (-4)^2\} + 6\{-6 \times 8 - 0\}] = 608$ for the 1 x 1, 2 x 2 and 3 x 3 matrices respectively.

Thus all the relevant matrices are symmetric and have positive determinant. Thus the stiffness matrix is positive definite.

<u>Example 2:</u> Verify that the following stiffness matrix for the three-storey shear frame shown in Fig. 3.1, but is not fixed to any base and hence capable of rigid body motion, is positive semi-definite.

$$K=\begin{bmatrix}8 & -8 & 0 & 0\\ -8 & 14 & -6 & 0\\ 0 & -6 & 10 & -4\\ 0 & 0 & -4 & 4\end{bmatrix}$$

$$|K| = 8\begin{vmatrix}14 & -6 & 0\\ -6 & 10 & -4\\ 0 & -4 & 4\end{vmatrix} + (-1)^{(1+2)}(-8)\begin{vmatrix}-8 & -6 & 0\\ 0 & 10 & -4\\ 0 & -4 & 4\end{vmatrix}$$

$$= 8[14\{10\times 4-(-4)^2\} + (-1)^{(1+2)}(-6)\{-6\times 4-0\}]$$
$$+ \; 8[(-8)\{10\times 4-(-4)^2\}+(-1)^{(1+2)}(-6)\{0\}] = 0$$

The sub matrices are

$$K_1=[8], K_2=\begin{bmatrix}8 & -8\\ -8 & 14\end{bmatrix}, K_3=\begin{bmatrix}8 & -8 & 0\\ -8 & 14 & -6\\ 0 & -6 & 10\end{bmatrix}, K_4=\begin{bmatrix}8 & -8 & 0 & 0\\ -8 & 14 & -6 & 0\\ 0 & -6 & 10 & -4\\ 0 & 0 & -4 & 4\end{bmatrix}$$

The determinants of the sub matrices are 8, $(8 \times 14 - (-8)^2) = 48$, $8\{14 \times 10 - 6)^2\} + 8\{-8 \times 10 - 0\} = 192$ for the 1 x 1, 2 x 2 and 3 x 3 matrices respectively. Thus all the relevant matrices are symmetric and have positive determinant except the 4 x 4 matrix, which has a zero determinant. Thus the stiffness matrix is positive semi-definite. The structural stiffness matrix of structures capable of rigid body motion will always be positive semi-definite.

3.2.4 Eigen-values of [K - ω² M]: Examples

The following examples illustrate the use of Cramer's rule to calculate the eigen-values of a few simple examples.

<u>Example 1:</u> For the three-storey shear frame shown in Fig. 3.1, the stiffness and mass matrices are given by

$$K = \begin{bmatrix} 14 & -6 & 0 \\ -6 & 10 & -4 \\ 0 & -4 & 4 \end{bmatrix}, \quad M = \begin{bmatrix} 4 & 0 & 0 \\ 0 & 4 & 0 \\ 0 & 0 & 2 \end{bmatrix}$$

Calculate the eigen-values using Cramer's rule.

The matrix [K - ω^2 M] is given by

$$K - \omega^2 M = \begin{bmatrix} (14 - \omega^2 4) & -6 & 0 \\ -6 & (10 - \omega^2 4) & -4 \\ 0 & -4 & (4 - \omega^2 2) \end{bmatrix}$$

The determinant of the 3 x 3 matrix is given by

$$\left|K - \omega^2 M\right| = (-1)^{(1+1)}(14 - \omega^2 4)\begin{vmatrix} (10 - \omega^2 4) & -4 \\ -4 & (4 - \omega^2 2) \end{vmatrix}$$

$$+ (-1)^{(1+2)}(-6)\begin{vmatrix} -6 & -4 \\ 0 & (4 - \omega^2 2) \end{vmatrix} + (-1)^{(1+3)}(0)\begin{vmatrix} -6 & (10 - \omega^2 4) \\ 0 & -4 \end{vmatrix}$$

$$\left|K - \omega^2 M\right| = (14 - \omega^2 4)\{(10 - \omega^2 4)(4 - \omega^2 2) - (-4)^2\} - (-6)\{(-6)(4 - \omega^2 2)\}$$

$$= (14 - \omega^2 4)\{8\omega^4 - 36\omega^2 + 24\} - 36\{4 - \omega^2 2\}$$

$$= -32\omega^6 + 256\omega^4 - 528\,\omega^2 + 192$$

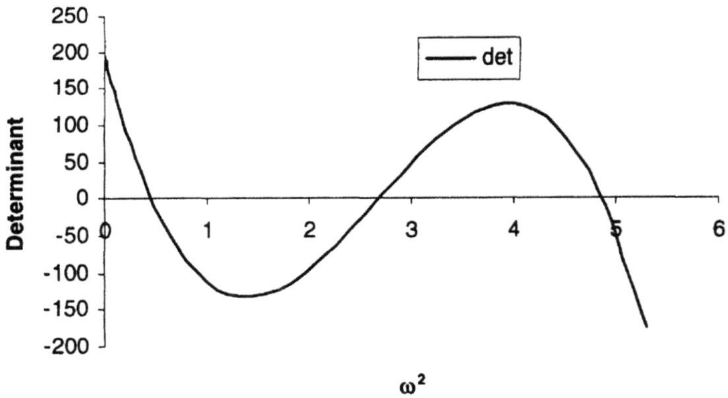

Fig. 3.3 Variation of the determinant with ω^2

The determinant is a cubic equation in ω^2. Fig. 3.3 shows a plot of the determinant versus ω^2. The three values of ω^2 for which the determinant is zero are

$$\omega^2 = 0.46056, 2.681993 \text{ and } 4.857448.$$

The structure thus has three eigen-values. These are called the natural frequencies of the system.

Example 2: As a more complicated example, consider the four-storey shear frame shown in Fig. 3.4.

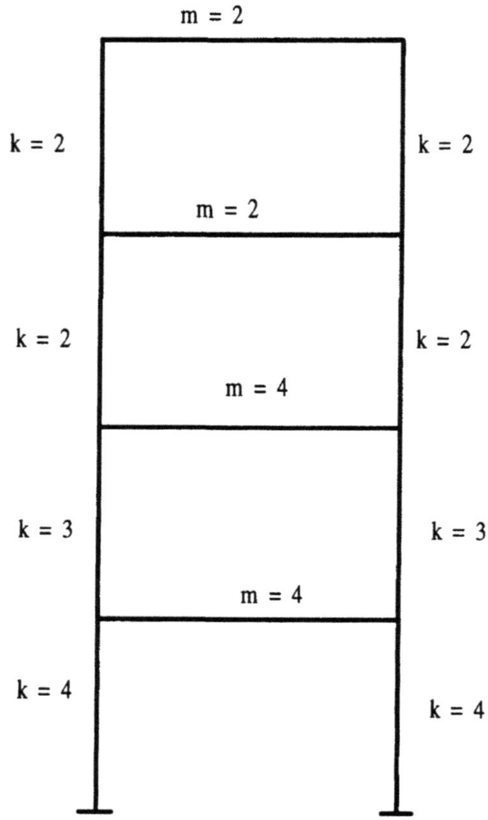

Fig. 3.4 Four-storey shear frame

The stiffness and mass matrices are given by

$$K = \begin{bmatrix} 14 & -6 & 0 & 0 \\ -6 & 10 & -4 & 0 \\ 0 & -4 & 8 & -4 \\ 0 & 0 & -4 & 4 \end{bmatrix}, \quad M = \begin{bmatrix} 4 & 0 & 0 & 0 \\ 0 & 4 & 0 & 0 \\ 0 & 0 & 2 & 0 \\ 0 & 0 & 0 & 2 \end{bmatrix}$$

$$K - \omega^2 M = \begin{bmatrix} (14-\omega^2 4) & -6 & 0 & 0 \\ -6 & (10-\omega^2 4) & -4 & 0 \\ 0 & -4 & (8-\omega^2 2) & -4 \\ 0 & 0 & -4 & (4-\omega^2 2) \end{bmatrix}$$

$$\left|K-\omega^2 M\right| = (-1)^{(1+1)}(14-\omega^2 4)\begin{bmatrix}(10-\omega^2 4) & -4 & 0\\ -4 & (8-\omega^2 2) & -4\\ 0 & -4 & (4-\omega^2 2)\end{bmatrix}$$

$$+\ (-1)^{(1+2)}(-6)\begin{bmatrix}-6 & -4 & 0\\ 0 & (8-\omega^2 2) & -4\\ 0 & -4 & (4-\omega^2 2)\end{bmatrix}$$

$$\left|K-\omega^2 M\right| = (14-4\omega^2)[(10-4\omega^2)\{(8-2\omega^2)(4-2\omega^2)-16\}$$

$$+(-1)(-4)\{-4(4-2\omega^2)\}+(-1)(-6)[(-6)\{(8-2\omega^2)(4-2\omega^2)-16\}]$$

$$= 64\omega^8 - 768\omega^6 + 2848\omega^4 - 3328\omega^2 + 768$$

The determinant is a quartic equation in ω^2. Fig. 3.5 shows a plot of the determinant versus ω^2. The four values of ω^2 for which the determinant is zero are

$$\omega^2 = 0.303146,\ 1.591674,\ 4.241005 \text{ and } 5.864176$$

The structure thus has four eigen-values which are the natural frequencies of the system.

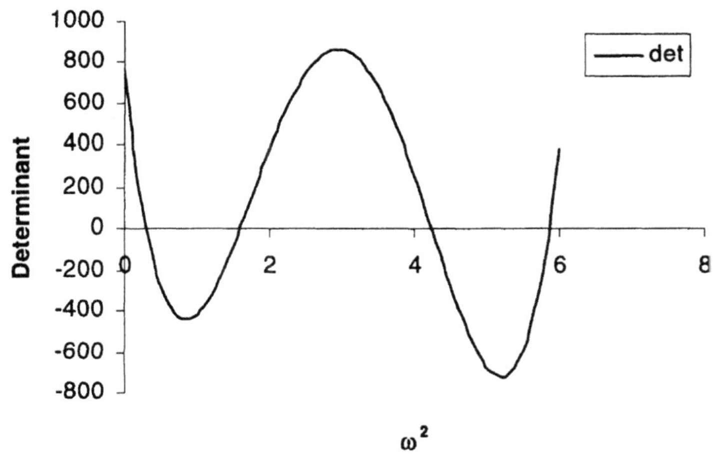

Fig. 3.5 Variation of the determinant for the four-storey shear frame

3.2.5 EIGEN-VECTORS OF [K - ω^2 M]

Having obtained the eigen-values of a matrix, the next step is to calculate the eigen-vectors **corresponding to each** eigen-value. Equation (3.1) governs the undamped free vibration of a MDOF system. There are no external forces acting on the system. The displacement of the structure results from giving an initial displacement and velocity at a particular storey level and letting the structure vibrate. The displacements at any level will be dependent on the value of the initial displacement, which is perfectly arbitrary. Therefore, all one can do is to determine the relative rather than absolute displacements.

3.2.5.1 Eigen-vectors: Examples

Example 1: Using the following data for the three-storey shear frame discussed in Section 3.1.3, determine the eigen-vectors corresponding to the three eigen-values.

$$K - \omega^2 M = \begin{bmatrix} (14 - \omega^2 4) & -6 & 0 \\ -6 & (10 - \omega^2 4) & -4 \\ 0 & -4 & (4 - \omega^2 2) \end{bmatrix}$$

$\omega^2 = 0.46056$, 2.681993 and 4.857448.

i. Eigen-vector for $\omega^2 = 0.46056$

$$K - \omega^2 M = \begin{bmatrix} 12.15776 & -6 & 0 \\ -6 & 8.15776 & -4 \\ 0 & -4 & 3.07888 \end{bmatrix}$$

Fixing arbitrarily, the value of $x_1 = 1.0$, then
From row 1: $12.15776\, x_1 - 6\, x_2 = 0.$, $x_2 = 2.02629$
From row 3: $-4\, x_2 + 3.07888\, x_3 = 0$, $x_3 = 2.63251$
Therefore the eigen-vector corresponding to $\omega^2 = 0.46056$ is

$$X = \begin{bmatrix} 1.0 \\ 2.02629 \\ 2.63251 \end{bmatrix}$$

It is quite common, instead of fixing the value of one variable such as x_1 to a specific value of say unity, but to normalise the vector X such that

$$X^T M X = 1$$

The reason for doing this is to keep the values of displacement from being either too large or too small. This is called M normalization. This is just one way of normalising. There are other ways too. For example some people normalise such that $X^T X = 1$. This is called Euclidian normalisation. It is perhaps worth stating that whichever method of 'normalising' is used, it does not affect the ratio between displacements.

Using M normalisation, in this example

$$X^T M X = \begin{bmatrix} 1.0 & 2.02629 & 2.63251 \end{bmatrix} \begin{bmatrix} 4 & 0 & 0 \\ 0 & 4 & 0 \\ 0 & 0 & 2 \end{bmatrix} \begin{bmatrix} 1.0 \\ 2.02629 \\ 2.63251 \end{bmatrix} = 34.28365$$

Therefore the M normalised value of X is given by

$$X = \frac{1}{\sqrt{34.28365}}\begin{bmatrix} 1.0 \\ 2.02629 \\ 2.63251 \end{bmatrix} = \begin{bmatrix} 0.170788 \\ 0.346066 \\ 0.449600 \end{bmatrix}$$

ii. <u>Eigen-vector for $\omega^2 = 2.68199$</u>: Proceeding as in the case of the first eigen-value,

$$K - \omega^2 M = \begin{bmatrix} 3.27204 & -6 & 0 \\ -6 & -0.72796 & -4 \\ 0 & -4 & -1.36398 \end{bmatrix}$$

Fixing arbitrarily, the value of $x_1 = 1.0$, then
From row 1: $3.27204\ x_1 - 6\ x_2 = 0$, $x_2 = 0.54534$
From row 3: $-4\ x_2 - 1.36398\ x_3 = 0$, $x_3 = -1.59926$
Therefore the eigen-vector corresponding to $\omega^2 = 2.68199$ is

$$X = \begin{bmatrix} 1.0 \\ 0.54534 \\ -1.59926 \end{bmatrix}$$

$$X^T M X = \begin{bmatrix} 1.0 & 0.54534 & -1.59926 \end{bmatrix}\begin{bmatrix} 4 & 0 & 0 \\ 0 & 4 & 0 \\ 0 & 0 & 2 \end{bmatrix}\begin{bmatrix} 1.0 \\ 0.54534 \\ -1.59926 \end{bmatrix} = 10.30485$$

Therefore the M normalised value of X is given by

$$X = \frac{1}{\sqrt{10.30485}}\begin{bmatrix} 1.0 \\ 0.54534 \\ -1.59926 \end{bmatrix} = \begin{bmatrix} 0.31152 \\ 0.16988 \\ -0.49819 \end{bmatrix}$$

iii. <u>Eigen-vector for $\omega^2 = 4.857448$</u>

$$K - \omega^2 M = \begin{bmatrix} -5.42979 & -6 & 0 \\ -6 & -9.42979 & -4 \\ 0 & -4 & -5.714896 \end{bmatrix}$$

Fixing arbitrarily, the value of $x_1 = 1.0$, then
From row 1: $-5.42979\ x_1 - 6\ x_2 = 0$, $x_2 = -0.90497$
From row 3: $-4\ x_2 - 5.714896\ x_3 = 0$, $x_3 = 0.633408$
Therefore the eigen-vector corresponding to $\omega^2 = 4.857448$ is

$$X = \begin{bmatrix} 1.0 \\ -0.90497 \\ 0.633408 \end{bmatrix}$$

$$X^T M X = \begin{bmatrix} 1.0 & -0.90497 & 0.633408 \end{bmatrix} \begin{bmatrix} 4 & 0 & 0 \\ 0 & 4 & 0 \\ 0 & 0 & 2 \end{bmatrix} \begin{bmatrix} 1.0 \\ -0.90497 \\ 0.63340 \end{bmatrix} = 8.07861$$

Therefore the M normalised value of X is given by

$$X = \frac{1}{\sqrt{8.078261}} \begin{bmatrix} 1.0 \\ -0.90497 \\ 0.633408 \end{bmatrix} = \begin{bmatrix} 0.35184 \\ -0.31840 \\ 0.22286 \end{bmatrix}$$

Figure 3.6 shows the three mode shapes. Essentially each storey deforms in a shearing mode. In the first mode all the three storeys deform in the same direction. In the second mode the top storey deforms in the opposite direction to the bottom two storeys. In the last mode, the deformation of the three storeys alternate.

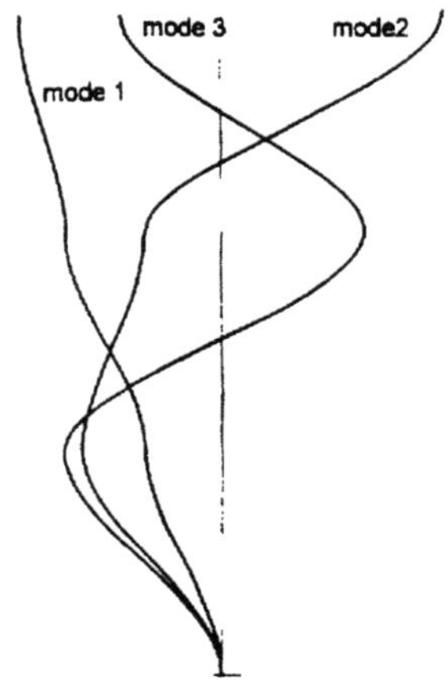

Fig. 3.6 Mode shapes of the three-storey frame

Example 2: Using the following data, determine the eigen-vectors for the four-storey shear frame discussed in Section 3.2.3.

$$K - \omega^2 M = \begin{bmatrix} (14-\omega^2 4) & -6 & 0 & 0 \\ -6 & (10-\omega^2 4) & -4 & 0 \\ 0 & -4 & (8-\omega^2 2) & -4 \\ 0 & 0 & -4 & (4-\omega^2 2) \end{bmatrix}$$

$\omega^2 = 0.303146$, 1.591674, 4.241004 and 5.864176.

i. Eigen-vector for $\omega^2 = 0.303146$

$$K - \omega^2 M = \begin{bmatrix} 12.78742 & -6 & 0 & 0 \\ -6 & 8.78742 & -4 & 0 \\ 0 & -4 & 7.39371 & -4 \\ 0 & 0 & -4 & 3.39371 \end{bmatrix}$$

Fixing arbitrarily, the value of $x_1 = 1.0$, then
Row 1: $12.78742\ x_1 - 6\ x_2 = 0,\ x_2 = 2.13124$
Row 2: $-6\ x_1 + 8.78742\ x_2 - 4\ x_3 = 0,\ x_3 = 3.18202$
Row 3: $-4\ x_2 + 7.39371\ x_3 - 4\ x_4 = 0,\ x_4 = 3.75049$

$$X = \begin{bmatrix} 1.0 \\ 2.13124 \\ 3.18202 \\ 3.75049 \end{bmatrix}$$

$$X^T M X = [1.0 \quad 2.13124 \quad 3.18202 \quad 3.75049] \begin{bmatrix} 4 & 0 & 0 & 0 \\ 0 & 4 & 0 & 0 \\ 0 & 0 & 2 & 0 \\ 0 & 0 & 0 & 2 \end{bmatrix} \begin{bmatrix} 1.0 \\ 2.13124 \\ 3.18202 \\ 3.75049 \end{bmatrix} = 70.5514$$

$$X = \frac{1}{\sqrt{70.55140}} \begin{bmatrix} 1.0 \\ 2.13124 \\ 3.18202 \\ 3.75049 \end{bmatrix} = \begin{bmatrix} 0.11906 \\ 0.25373 \\ 0.37884 \\ 0.44651 \end{bmatrix}$$

ii. Eigen-vector for $\omega^2 = 1.591674$

$$K - \omega^2 M = \begin{bmatrix} 7.63330 & -6 & 0 & 0 \\ -6 & 3.63330 & -4 & 0 \\ 0 & -4 & 4.81665 & -4 \\ 0 & 0 & -4 & 0.81665 \end{bmatrix}$$

Fixing arbitrarily, the value of $x_1 = 1.0$, then
Row 1: $7.63330\ x_1 - 6\ x_2 = 0,\ x_2 = 1.27222$
Row 2: $-6\ x_1 + 3.63330\ x_2 - 4\ x_3 = 0,\ x_3 = -0.34441$
Row 3: $-4\ x_2 + 4.81665\ x_3 - 4\ x_4 = 0,\ x_4 = -1.68695$

$$X = \begin{bmatrix} 1.0 \\ 1.27222 \\ -0.34441 \\ -1.68695 \end{bmatrix}$$

$$X^T M X = \begin{bmatrix} 1.0 & 1.27222 & -0.34441 & -1.68695 \end{bmatrix} \begin{bmatrix} 4 & 0 & 0 & 0 \\ 0 & 4 & 0 & 0 \\ 0 & 0 & 2 & 0 \\ 0 & 0 & 0 & 2 \end{bmatrix} \begin{bmatrix} 1.0 \\ 1.27222 \\ -0.34441 \\ -1.68695 \end{bmatrix}$$

$$= 16.403$$

$$X = \frac{1}{\sqrt{16.40301}} \begin{bmatrix} 1.0 \\ 1.27222 \\ -0.34441 \\ -1.68695 \end{bmatrix} = \begin{bmatrix} 0.24691 \\ 0.31412 \\ -0.08504 \\ -0.41652 \end{bmatrix}$$

iii. Eigen-vector for $\omega^2 = 4.241005$

$$K - \omega^2 M = \begin{bmatrix} -2.96402 & -6 & 0 & 0 \\ -6 & -6.96402 & -4 & 0 \\ 0 & -4 & -0.48201 & -4 \\ 0 & 0 & -4 & -4.48201 \end{bmatrix}$$

Fixing arbitrarily, the value of $x_1 = 1.0$, then
Row 1: $-2.96402\ x_1 - 6\ x_2 = 0$, $x_2 = -0.49400$
Row 2: $-6\ x_1 - 6.96402\ x_2 - 4\ x_3 = 0$, $x_3 = -0.63994$
Row 3: $-4\ x_2 - 0.48201\ x_3 - 4\ x_4 = 0$, $x_4 = 0.57111$

$$X = \begin{bmatrix} 1.0 \\ -0.49400 \\ -0.63994 \\ 0.57111 \end{bmatrix}$$

$$X^T M X = \begin{bmatrix} 1.0 & -0.49400 & -0.63994 & 0.57111 \end{bmatrix} \begin{bmatrix} 4 & 0 & 0 & 0 \\ 0 & 4 & 0 & 0 \\ 0 & 0 & 2 & 0 \\ 0 & 0 & 0 & 2 \end{bmatrix} \begin{bmatrix} 1.0 \\ -0.49400 \\ -0.63994 \\ 0.57111 \end{bmatrix}$$

$$= 6.448$$

$$X = \frac{1}{\sqrt{6.447546}} \begin{bmatrix} 1.0 \\ -0.49400 \\ -0.63994 \\ 0.57111 \end{bmatrix} = \begin{bmatrix} 0.39383 \\ -0.194551 \\ -0.25202 \\ 0.22492 \end{bmatrix}$$

iv. Eigen-vector for $\omega^2 = 5.864176$

$$K - \omega^2 M = \begin{bmatrix} -9.45670 & -6 & 0 & 0 \\ -6 & -13.45670 & -4 & 0 \\ 0 & -4 & -3.72835 & -4 \\ 0 & 0 & -4 & -7.72835 \end{bmatrix}$$

Fixing arbitrarily, the value of $x_1 = 1.0$, then

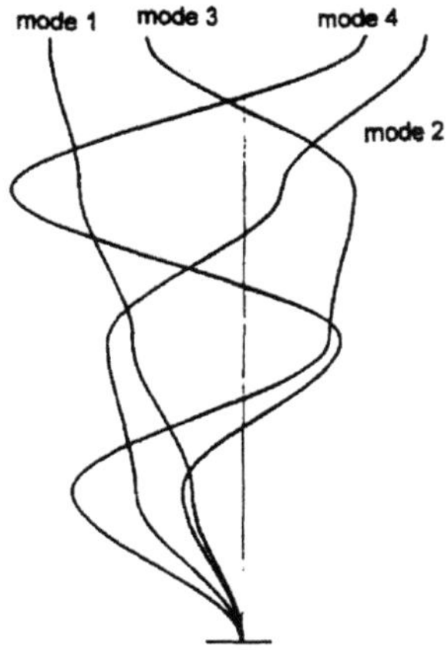

Fig. 3.7 Mode shapes of the four-storey frame

Row 1: -9.45670 x_1 – 6 x_2 = 0, x_2 = -1.57612
Row 2: -6 x_1 – 13.45670 x_2 – 4 x_3 = 0, x_3 = 3.80233
Row 3: -4 x_2 – 3.72835 x_3 – 4 x_4 = 0, x_4 = -1.967997

$$X = \begin{bmatrix} 1.0 \\ -1.57612 \\ 3.80233 \\ -1.967997 \end{bmatrix}$$

$$X^T M X = \begin{bmatrix} 1.0 & -1.57612 & 3.80233 & -1.967997 \end{bmatrix} \begin{bmatrix} 4 & 0 & 0 & 0 \\ 0 & 4 & 0 & 0 \\ 0 & 0 & 2 & 0 \\ 0 & 0 & 0 & 2 \end{bmatrix} \begin{bmatrix} 1.0 \\ -1.57612 \\ 3.80233 \\ -1.967997 \end{bmatrix}$$

$$= 50.598098$$

$$X = \frac{1}{\sqrt{50.598098}} \begin{bmatrix} 1.0 \\ -1.57612 \\ 3.80233 \\ -1.967997 \end{bmatrix} = \begin{bmatrix} 0.140583 \\ -0.221575 \\ 0.534544 \\ -0.27667 \end{bmatrix}$$

Figure. 3.7 shows the four mode shapes. Each storey deforms in a shearing mode.

3.3 GENERAL AND SPECIAL EIGEN-VALUE PROBLEMS

In the previous sections specific examples were studied for which the eigen-value equation was equation (3.2)

$$K X = \omega^2 M X \tag{3.2}$$

This type of matrix relationship is called a general eigen-value problem. Mathematicians have devoted a great deal of attention to another type of eigen-value problem called a special eigen-value problem, given by

$$A\,Y = \lambda\,Y \tag{3.5}$$

where Y = eigen-vector and λ = eigen-value.

The reason for mentioning the special eigen-value problem is that some solution methods convert the general problem to special problem before attempting a solution.

3.3.1 Conversion from General to Special Eigen-value Problem

Provided K is non-singular, theoretically the conversion is done easily by the equation

$$K^{1}\,M X = (1/\omega^2)\,X$$

However, this destroys the bandedness and symmetry inherent in matrices K and M and in practical solution procedures, and it is necessary to ensure that both symmetry and bandedness are preserved in order to minimize storage requirements for the matrices involved. As a simple demonstration that even though K and M are symmetric, $K^{1}\,M$ is not symmetric, consider the following matrices

$$K = \begin{bmatrix} 2 & -1 \\ -1 & 2 \end{bmatrix},\ K^{-1} = (\frac{1}{3})\begin{bmatrix} 2 & 1 \\ 1 & 2 \end{bmatrix},\ M = \begin{bmatrix} 1 & 0 \\ 0 & 2 \end{bmatrix},\ K^{-1}\,M = \frac{1}{3}\begin{bmatrix} 2 & 2 \\ 1 & 4 \end{bmatrix}$$

Bandedness and symmetry can be preserved by the following procedure, which is commonly employed if conversion has to be carried out.

Let the stiffness matrix K of size N x N be symmetric and positive definite. In this case K can be expressed as

$$K = U^T U \tag{3.6}$$

where U is an upper triangular matrix.

This is called Cholesky decomposition and will be explained fully in Chapter 4. Substituting for $K = U^T U$, equation (3.2) becomes,

$$U^T\,U X = \omega^2\,M X \tag{3.7}$$

Pre-multiplying equation (3.7) by U^T

$$U X = \omega^2\,U^T\,M X = U^T\,M\,[U^1\,U]\,X \tag{3.8}$$

In the above manipulation, it should be noted that the terms inside the square brackets is equal to a unit matrix.

Setting

$$U X = Y,\ A = U^T\,M\,U^1 \text{ and } \omega^2 = \lambda^{-1} \tag{3.9}$$

the general eigen-value problem $K X = \omega^2\,M X$ is converted to the special eigen-value problem $A\,Y = \lambda\,Y$

It can be shown that if K and M are symmetric, then A is also symmetric. In practical computations, calculation of U^1 and explicit computations of A are avoided as will be explained in a later section. The following example demonstrates the above procedure.

Example: Given K and M, calculate U.

$$K = \begin{bmatrix} 2 & -1 \\ -1 & 2 \end{bmatrix}, \ M = \begin{bmatrix} 1 & 0 \\ 0 & 2 \end{bmatrix}$$

$$U = \begin{bmatrix} u_{11} & u_{12} \\ 0 & u_{22} \end{bmatrix}, U^T U = \begin{bmatrix} u_{11}^2 & u_{11} u_{12} \\ u_{11} u_{12} & u_{12}^2 u_{22} \end{bmatrix} = K = \begin{bmatrix} 2 & -1 \\ -1 & 2 \end{bmatrix}$$

$$u_{11} = \sqrt{2} \ , \ u_{12} = -\frac{1}{\sqrt{2}} \ , \ u_{22} = \frac{\sqrt{3}}{\sqrt{2}}$$

$$U = \frac{1}{\sqrt{2}} \begin{bmatrix} 2 & -1 \\ 0 & \sqrt{3} \end{bmatrix} , \ U^{-1} = \frac{1}{\sqrt{6}} \begin{bmatrix} \sqrt{3} & 1 \\ 0 & 2 \end{bmatrix}$$

$$A = U^{-T} \ M \ U^{-1} = \frac{1}{6} \begin{bmatrix} 3 & \sqrt{3} \\ \sqrt{3} & 9 \end{bmatrix}$$

As can be seen, A is symmetric. It can be shown that this procedure preserves not only symmetry but also bandedness.

3.4 PROPERTIES OF EIGEN-VALUES AND EIGEN-VECTORS

There are some very important properties of eigen-values and eigen-vectors, which are used in the solution of both general and special eigen-value problems. It is assumed that the matrices K, M and A are real and symmetric and, in addition, K and A are assumed to be positive definite.

3.4.1 Characteristic Equation

As shown by examples in Section 3.1.4, the determinant of matrices $[K - \omega^2 M]$ or $[A - \lambda I]$ of size N x N results in a polynomial of Nth degree in ω^2 or λ. The determinant of $[A - \lambda I]$, for example, can be expressed as an Nth degree polynomial

$$\lambda^N + C_{N-1}{}^{N-1} + \ldots + C_1 \lambda + C_0 = 0$$

where C_i, i = 1 to N are constants.

The above equation can be factorised as

$$(\lambda - \lambda_1)(\lambda - \lambda_2)\ldots(\lambda - \lambda_N) = 0$$

This is called the characteristic equation and it indicates that there are N roots for the equation, each of which is an eigen-value of A. It should be noted that it is not necessary for all the eigen-values to be distinct. Associated with each eigen-value is an eigen-vector. If A is positive definite then all eigen-values are real and greater than zero. However if A is semi-definite, then all eigen-values are real but some, which correspond to rigid body displacement, are zero.

3.4.1.1 Normalisation of Eigen-vectors

It can be easily verified that for the problem defined by $AY_i = \lambda_i Y_i$, that if Y_i is a solution, then αY_i is also a solution, where α is a non-zero constant because

$$A(\alpha Y_i) = \lambda_i (\alpha Y_i)$$

This shows that eigen-vector is only defined to within a multiple of itself. This arises from the fact that since the determinant of $[A - \lambda_i I]$ is zero, the matrix $[A - \lambda_i I]$ is singular. Therefore, it is not possible to solve for Y_i uniquely. All that can be done is to determine the ratio of the components of the vector. In practice, to prevent the occurrence of large/small numbers, it is common to fix the length of the vector Y_i such that $Y_i^T Y_i = 1$. This is called Euclidian norm.

In the case of eigen-vectors corresponding to the problem defined by equation (3.2), the eigen-vector X is M normalised such that $X^T M X = 1$

3.4.2 Orthogonal Property of Eigen-vectors of a Symmetric Matrix

Consider two eigen-pairs given by

$$K X_i = \omega_i^2 M X_i \qquad \text{(i)}$$

$$K X_j = \omega_j^2 M X_j \qquad \text{(ii)}$$

Pre-multiplying equation (i) by X_j^T

$$X_j^T K X_i = \omega_i^{\,2} X_j^T M X_i \qquad \text{(iii)}$$

Taking the transform of equation (iii) and remembering that $[AB]^T = B^T A^T$,

$$X_i^T K^T X_j = \omega_i^{\,2} X_i^T M^T X_j$$

Since K and M are symmetric, $K^T = K$ and $M^T = M$, therefore

$$X_i^T K X_j = \omega_i^2 X_i^T M X_j \qquad \text{(iv)}$$

Pre-multiplying equation (ii) by X_i^T

$$X_i^T K X_j = \omega_j^{\,2} X_i^T M X_j \qquad \text{(v)}$$

Subtracting equation (iv) from (v)

$$(\omega_j^{\,2} - \omega_i^2) X_i^T M X_j = 0$$

$$\text{If } \omega_j^{\,2} \neq \omega_i^{\,2},\ X_i^T M X_j = 0 \qquad (3.10)$$

This is called the orthogonal property of eigen-vectors of a symmetric matrix.
In the case when $\omega_i^{\,2} = \omega_j^{\,2}$, then it is not necessary for $X_i^T M X_j$ to be equal to zero. However it is always possible to find vectors X_i and X_j, which are orthogonal. But these pairs of vectors are not unique.

Example 1: Check that the eigen-vectors determined in Section 3.2.5 for the three-storey shear frame shown in Fig. 3.1 are orthogonal. The eigen-vectors are

$$X_1 = \begin{bmatrix} 0.170788 \\ 0.346066 \\ 0.449600 \end{bmatrix},\ X_2 = \begin{bmatrix} 0.31152 \\ 0.16988 \\ -0.49819 \end{bmatrix},\ X_3 = \begin{bmatrix} 0.35184 \\ -0.31840 \\ 0.22286 \end{bmatrix},\ M = \begin{bmatrix} 4 & 0 & 0 \\ 0 & 4 & 0 \\ 0 & 0 & 2 \end{bmatrix}$$

$$X_1^T M X_2 = [0.170788 \quad 0.346066 \quad 0.44960] \begin{bmatrix} 4 & 0 & 0 \\ 0 & 4 & 0 \\ 0 & 0 & 2 \end{bmatrix} \begin{bmatrix} 0.31152 \\ 0.16988 \\ -0.49819 \end{bmatrix} = -3.003 \times 10^{-8}$$

$$X_1^T M X_3 = [0.170788 \quad 0.346066 \quad 0.44960] \begin{bmatrix} 4 & 0 & 0 \\ 0 & 4 & 0 \\ 0 & 0 & 2 \end{bmatrix} \begin{bmatrix} 0.35184 \\ -0.31840 \\ 0.22286 \end{bmatrix} = 9.76 \times 10^{-8}$$

$$X_3^T M X_2 = [0.35184 \quad -0.31840 \quad 0.22286] \begin{bmatrix} 4 & 0 & 0 \\ 0 & 4 & 0 \\ 0 & 0 & 2 \end{bmatrix} \begin{bmatrix} 0.31152 \\ 0.16988 \\ -0.49819 \end{bmatrix} = -2.725 \times 10^{-6}$$

As can be seen, the eigen-vectors do satisfy the orthogonality condition to within permissible round-off errors.

<u>Example 2</u>: Check that the eigen-vectors determined in Section 3.2.5 for the four-storey shear frame shown in Fig. 3.4 are orthogonal.

$$M = \begin{bmatrix} 4 & 0 & 0 & 0 \\ 0 & 4 & 0 & 0 \\ 0 & 0 & 2 & 0 \\ 0 & 0 & 0 & 2 \end{bmatrix}, \; X_1 = \begin{bmatrix} 0.11906 \\ 0.25373 \\ 0.37883 \\ 0.44651 \end{bmatrix},$$

$$X_2 = \begin{bmatrix} 0.24691 \\ 0.31412 \\ -0.08504 \\ -0.41652 \end{bmatrix}, \; X_3 = \begin{bmatrix} 0.39383 \\ -0.19455 \\ -0.25202 \\ 0.22492 \end{bmatrix}, \; X_4 = \begin{bmatrix} 0.140583 \\ -0.221575 \\ 0.534544 \\ -0.27667 \end{bmatrix}$$

$$X_1^T M X_2 = [0.1191 \quad 0.2537 \quad 0.3788 \quad 0.4465] \begin{bmatrix} 4 & 0 & 0 & 0 \\ 0 & 4 & 0 & 0 \\ 0 & 0 & 2 & 0 \\ 0 & 0 & 0 & 2 \end{bmatrix} \begin{bmatrix} 0.2469 \\ 0.3141 \\ -0.0850 \\ -0.4165 \end{bmatrix} \approx 0$$

$$X_1^T M X_3 = [0.1191 \quad 0.2537 \quad 0.3788 \quad 0.4465] \begin{bmatrix} 4 & 0 & 0 & 0 \\ 0 & 4 & 0 & 0 \\ 0 & 0 & 2 & 0 \\ 0 & 0 & 0 & 2 \end{bmatrix} \begin{bmatrix} 0.3938 \\ -0.1946 \\ -0.2520 \\ 0.2249 \end{bmatrix} \approx 0$$

$$X_1^T M X_4 = [0.1191 \quad 0.2537 \quad 0.3788 \quad 0.4465] \begin{bmatrix} 4 & 0 & 0 & 0 \\ 0 & 4 & 0 & 0 \\ 0 & 0 & 2 & 0 \\ 0 & 0 & 0 & 2 \end{bmatrix} \begin{bmatrix} 0.1406 \\ -0.2216 \\ 0.5345 \\ -0.2767 \end{bmatrix} \approx 0$$

$$X_3^T M X_2 = \begin{bmatrix} 0.3938 & -0.1946 & -0.2520 & 0.2249 \end{bmatrix} \begin{bmatrix} 4 & 0 & 0 & 0 \\ 0 & 4 & 0 & 0 \\ 0 & 0 & 2 & 0 \\ 0 & 0 & 0 & 2 \end{bmatrix} \begin{bmatrix} 0.2469 \\ 0.3141 \\ -0.0850 \\ -0.4165 \end{bmatrix} \approx 0$$

$$X_4^T M X_2 = \begin{bmatrix} 0.1406 & -0.2216 & 0.5345 & -0.2767 \end{bmatrix} \begin{bmatrix} 4 & 0 & 0 & 0 \\ 0 & 4 & 0 & 0 \\ 0 & 0 & 2 & 0 \\ 0 & 0 & 0 & 2 \end{bmatrix} \begin{bmatrix} 0.2469 \\ 0.3141 \\ -0.0850 \\ -0.4165 \end{bmatrix} \approx 0$$

$$X_4^T M X_3 = \begin{bmatrix} 0.1406 & -0.2216 & 0.5345 & -0.2767 \end{bmatrix} \begin{bmatrix} 4 & 0 & 0 & 0 \\ 0 & 4 & 0 & 0 \\ 0 & 0 & 2 & 0 \\ 0 & 0 & 0 & 2 \end{bmatrix} \begin{bmatrix} 0.3938 \\ -0.1946 \\ -0.2520 \\ 0.2249 \end{bmatrix} \approx 0$$

Once again it is clear that the eigen-vectors are orthogonal to within the limits of round-off errors.

3.4.3 Arbitrary Vector in Terms of Eigen-vectors

One important consequence of the orthogonal property of eigen-vectors is that an arbitrary vector can be expressed in terms of eigen-vectors.

Let Y be an arbitrary vector of size N which is expressed as a linear combination of eigen-vectors X_i, I = 1, N.

$$Y = C_1 X_1 + C_2 X_2 ... + C_N X_N \tag{3.11}$$

Where C_i s are constants and it is assumed that the eigen-vectors X_i are M-normalised, i.e. $X_i^T M X_i = 1$.

Pre-multiplying equation (3.11) by M,

$$M Y = C_1 M X_1 + C_2 M X_2 ... + C_N M X_N \tag{3.12}$$

Multiplying equation (3.12) by say X_i^T, then

$$X_i^T M Y = C_1 X_i^T M X_1 + C_2 X_i^T M X_2 ... + C_N X_i^T M X_N$$

But because of the orthogonal property given by equation (3.10),

$$X_i^T M X_j = 0, \text{ if } i \neq j. \text{ and } X_i^T M X_j = 1, \text{ if } i = j.$$

$$\therefore \; C_i = X_i^T M Y \tag{3.13}$$

Example 1: Using the eigen-vectors calculated in Section 3.2.5 for the three-storey shear frame shown in Fig. 3.1, express the arbitrary vector Y in terms of the eigen-vectors.

$$Y^T = [1 \quad 2 \quad 3]$$

It is given that

$$X_1 = \begin{bmatrix} 0.170788 \\ 0.346066 \\ 0.449600 \end{bmatrix}, \; X_2 = \begin{bmatrix} 0.31152 \\ 0.16988 \\ -0.49819 \end{bmatrix}, \; X_3 = \begin{bmatrix} 0.35184 \\ -0.31840 \\ 0.22286 \end{bmatrix}, \; M = \begin{bmatrix} 4 & 0 & 0 \\ 0 & 4 & 0 \\ 0 & 0 & 2 \end{bmatrix}$$

$$C_1 = X_1^T M\,Y = \begin{bmatrix}0.17079 & 0.34607 & 0.44960\end{bmatrix}\begin{bmatrix}4 & 0 & 0\\0 & 4 & 0\\0 & 0 & 2\end{bmatrix}\begin{bmatrix}1\\2\\3\end{bmatrix} = 6.14928$$

$$C_2 = X_2^T M\,Y = \begin{bmatrix}0.31152 & 0.16988 & -0.49819\end{bmatrix}\begin{bmatrix}4 & 0 & 0\\0 & 4 & 0\\0 & 0 & 2\end{bmatrix}\begin{bmatrix}1\\2\\3\end{bmatrix} = -0.38402$$

$$C_3 = X_3^T M\,Y = \begin{bmatrix}0.35184 & -0.31840 & 0.22286\end{bmatrix}\begin{bmatrix}4 & 0 & 0\\0 & 4 & 0\\0 & 0 & 2\end{bmatrix}\begin{bmatrix}1\\2\\3\end{bmatrix} = 0.19732$$

Example 2: Using the eigen-vectors calculated in Section 3.2.5 for the four-storey shear frame shown in Fig. 3.4, express the arbitrary vector Y in terms of the eigen-vectors.

$$Y^T = [1 \quad -2 \quad 3 \quad -4]$$

$$M = \begin{bmatrix}4 & 0 & 0 & 0\\0 & 4 & 0 & 0\\0 & 0 & 2 & 0\\0 & 0 & 0 & 2\end{bmatrix},\; X_1 = \begin{bmatrix}0.11906\\0.25373\\0.37883\\0.44651\end{bmatrix},\; X_2 = \begin{bmatrix}0.24691\\0.31412\\-0.08504\\-0.41652\end{bmatrix},\; X_3 = \begin{bmatrix}0.39383\\-0.19455\\-0.25202\\0.22492\end{bmatrix},$$

$$X_4 = \begin{bmatrix}0.140583\\-0.221575\\0.534544\\-0.27667\end{bmatrix}$$

$$C_1 = X_1^T MY = \begin{bmatrix}0.11906 & 0.25373 & 0.37883 & 0.44651\end{bmatrix}\begin{bmatrix}4 & 0 & 0 & 0\\0 & 4 & 0 & 0\\0 & 0 & 2 & 0\\0 & 0 & 0 & 2\end{bmatrix}\begin{bmatrix}1\\-2\\3\\-3\end{bmatrix}$$

$$= -1.95968$$

$$C_2 = X_2^T MY = \begin{bmatrix}0.24691 & 0.31412 & -0.08504 & -0.41652\end{bmatrix}\begin{bmatrix}4 & 0 & 0 & 0\\0 & 4 & 0 & 0\\0 & 0 & 2 & 0\\0 & 0 & 0 & 2\end{bmatrix}\begin{bmatrix}1\\-2\\3\\-3\end{bmatrix}$$

$$= 0.46356$$

$$C_3 = X_3^T MY = \begin{bmatrix}0.39383 & -0.19455 & -0.25202 & 0.22492\end{bmatrix}\begin{bmatrix}4 & 0 & 0 & 0\\0 & 4 & 0 & 0\\0 & 0 & 2 & 0\\0 & 0 & 0 & 2\end{bmatrix}\begin{bmatrix}1\\-2\\3\\-3\end{bmatrix}$$

$$= 0.27008$$

$$C_4 = X_4^T MY = [0.140583 \quad -0.221575 \quad 0.534544 \quad -0.27667]\begin{bmatrix} 4 & 0 & 0 & 0 \\ 0 & 4 & 0 & 0 \\ 0 & 0 & 2 & 0 \\ 0 & 0 & 0 & 2 \end{bmatrix}\begin{bmatrix} 1 \\ -2 \\ 3 \\ -3 \end{bmatrix}$$

$$= 7.202216$$

3.5 MODE SUPERPOSITION METHOD

The fact that any arbitrary vector can be expressed in terms of the eigen-vectors can be used to develop a very powerful procedure for solving a damped or an undamped forced vibration problem of MDOF systems.

3.5.1 Undamped Forced Response

Equation (3.1), is the equilibrium equation expressed in matrix form for the forced vibration of an undamped MDOF system.

$$M X'' + K X = F \tag{3.1}$$

Where, X and X'' as well as the force vector F are functions of time, t.

Let X be expressed in terms of the eigen-vectors Φ_i of the system.

$$X(t) = \sum_1^N C_i(t)\, \Phi_i \tag{3.14}$$

where the coefficients $C_i(t)$ are functions of time and Φ_i is the ith eigen-vector. Substituting for X and its derivatives in terms of Φ and C, equation (3.1) is converted to one in terms of C and its derivative.

$$\sum_1^N C_i'' M\, \Phi_i \; + \sum_1^N C_i\, K\, \Phi_i = F \tag{3.15}$$

These now represent N simultaneous linear differential equations in terms of C. If equation (3.15) is pre-multiplied by Φj^T, then

$$\sum_1^N C_i'' \Phi_j^T\, M\, \Phi_i + \sum_1^N C_i \Phi_j^T\, K\, \Phi_i = \Phi_j^T\, F \tag{3.16}$$

But because of the orthogonality property of eigen-vectors expressed by equation (3.10), and the fact that the eigen-vectors are M normalised

$$\Phi_j^T\, M\, \Phi_i = 0, \; if\; i \neq j\;, \Phi_i^T\, M\, \Phi_i = 1 \tag{3.17}$$

In addition, the eigen-vectors satisfy the relationship

$$K\, \Phi_i = \omega_i^2\, M\, \Phi_i \tag{3.18}$$

Therefore, equation (3.16) simplifies to

$$C_i'' + \omega_i^2\; C_i = \Phi_i^T\; F \tag{3.19}$$

Depending on the forcing function $F(t)$, this differential equation can be solved by the Duhamel integral approach discussed in Sections (2.8) and (2.11).

3.5.2 Undamped Forced Response: Example

Example 1: Use mode superposition method to analyse the three-storey shear building shown in Fig. 3.1 and analysed in Section 3.2.5. The frame is subjected to the following forcing function, which represents a suddenly applied load of P at first storey level.

$$F(t) = \begin{bmatrix} f_1(t) \\ f_2(t) \\ f_3(t) \end{bmatrix} = \begin{bmatrix} P \\ 0 \\ 0 \end{bmatrix}$$

From the previous analysis in Section 3.2.5, the eigen-values are $\omega^2 = 0.46056$, 2.68199 and 4.857448. The eigen-vectors are

$$\Phi_1 = \begin{bmatrix} 0.170788 \\ 0.346066 \\ 0.449600 \end{bmatrix}, \quad \Phi_2 = \begin{bmatrix} 0.31152 \\ 0.16988 \\ -0.49819 \end{bmatrix}, \quad \Phi_3 = \begin{bmatrix} 0.35184 \\ -0.31840 \\ 0.22286 \end{bmatrix}$$

Using equation (3.18), the force corresponding to each mode is given by

$$\Phi_1^T F = [0.170788 \quad 0.346066 \quad 0.449600] \begin{bmatrix} P \\ 0 \\ 0 \end{bmatrix} = 0.170788 P$$

$$\Phi_2^T F = [0.31152 \quad 0.16988 \quad -0.49819] \begin{bmatrix} P \\ 0 \\ 0 \end{bmatrix} = 0.31152 P$$

$$\Phi_3^T F = [0.35184 \quad -0.31840 \quad 0.22286] \begin{bmatrix} P \\ 0 \\ 0 \end{bmatrix} = 0.35184 P$$

From equation (3.19), the three differential equations are

$$C_1'' + 0.46056 C_1 = 0.170788 P \tag{3.20a}$$

$$C_2'' + 2.68199 C_2 = 0.31152 P \tag{3.20b}$$

$$C_3'' + 4.857448 C_3 = 0.35184 P \tag{3.20c}$$

It was shown in Section 2.9.1, that for an undamped system subjected to a suddenly applied force W the differential equation is

$$M u''(t) + K u = W \tag{3.21}$$

Assuming that the system starts from rest, the response is given by

$$u = \frac{W}{M \omega^2} [1 - \cos \omega t] \tag{3.22}$$

The maximum value of u is given by

$$u_{Max} = \frac{2W}{M\,\omega^2} \tag{3.23}$$

In the present case the solution to equations (3.20 a-c) is obtained by substituting the appropriate values of *M, K,* ω and the force *W* in equation (3.21). The maximum value is twice the static value as shown in Section 2.9.1.

$$u = C_1\,, M = 1,\, W = 0.170788\,P,\ \omega^2 = 0.46056,\, \omega = 0.67865$$

$$\therefore C_1 = 0.3708\,P\,[1 - \cos 0.67865t],\, C_{1\,MAX} = 0.7416$$

$$u = C_2\,, M = 1,\, W = 0.31152\,P,\ \omega^2 = 2.68199,\, \omega = 1.63767$$

$$\therefore C_2 = 0.1162\,P\,[1 - \cos 1.63767t]\,,\, C_{2\,MAX} = 0.2324$$

$$u = C_3\,, M = 1,\, P = 0.35184W,\ \omega^2 = 4.857448\,,\, \omega = 2.20396$$

$$\therefore C_3 = 0.0724\,P\,[1 - \cos 2.20396t]\,,\, C_{3\,MAX} = 0.1448$$

Having calculated the solutions for C_i, i= 1,3, the solution to $x_i(t)$ can be calculated from equation (3.14). The final solution to X is given by

$$x_1(t) = 0.170788\,C_1 + 0.31152\,C_2 + 0.35184\,C_3$$

$$x_2(t) = 0.346066\,C_1 + 0.16988\,C_2 - 0.31840\,C_3$$

$$x_3(t) = 0.449600\,C_1 - 0.49819\,C_2 + 0.22286\,C_3$$

Fig 3.8 shows a plot of the floor displacements as a function of time, t.

3.5.3 Approximation Using Square Root of the Sum of the Squares

Although in the previous section, the response of the structure as a function of time was calculated from a design point of view, what is of interest is simply the maximum response. If one can estimate to reasonable accuracy the maximum response without undertaking the onerous task of tracing the time history of the response, considerable reduction in calculation can be done. Unfortunately, because different values of ω_i are associated with each C_i, they do not reach their maximum values at the same time. Calculation of $C_{j\,MAX}$ can be done from a knowledge of the DMF using the response spectra plots such as the one shown in Fig. 2.11. If the maximum value of x_i is calculated by substituting the maximum values of C_j for j = 1,2 and 3, it can result in a significant over estimation of the maximum displacement, as shown in Table 3.1.

A second compromise approach to estimating the maximum values of x_i is to use what is called as Square Root of the Sum of the Squares (SRSS) approach. This is given by

$$x_{1\,MAX} \approx \sqrt{\{(0.170788\,C_{1\,MAX})^2 + (0.31152\,C_{2\,MAX})^2 + (0.35184\,C_{3\,MAX})^2\}} = 0.1545$$

$$x_{2\,MAX} \approx \sqrt{\{(0.346066\,C_{1\,MAX})^2 + (0.16988\,C_{2\,MAX})^2 + (-0.31840\,C_{3\,MAX})^2\}} = 0.2638$$

$$x_{3\ MAX} \approx \sqrt{\{(0.449600\ C_{1\ MAX})^2 + (-0.49819\ C_{2\ MAX})^2 + (0.22286\ C_{3\ MAX})^2\}} = 0.3544$$

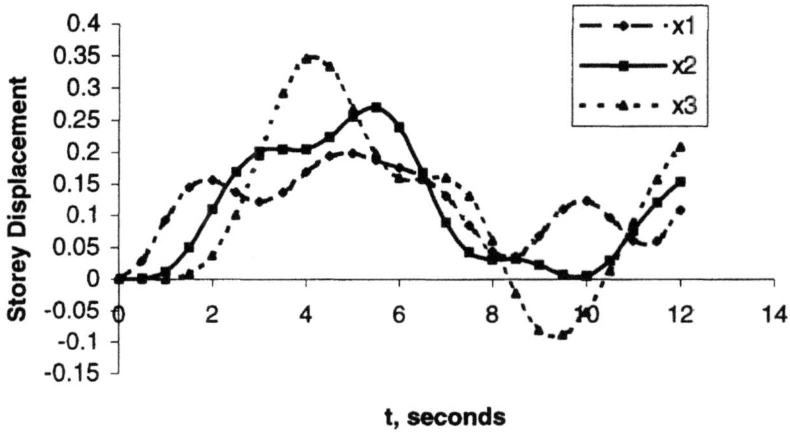

Fig. 3.8 Storey displacements (undamped case)

Table 3.1 Maximum displacements

	Exact maximum	Using $C_{i\ MAX}$	SRSS
x_1	0.1986	0.2500	0.1545
x_2	0.2708	0.3423	0.2638
x_3	0.3489	0.4815	0.3544

The SRSS values can underestimate or overestimate the maximum values. In spite of this shortcoming, this method is very often used in practice.

It is worth noting that the static displacements are given by solving the matrix equation $K\,X = F$, where

$$\begin{bmatrix} 14 & -6 & 0 \\ -6 & 10 & -4 \\ 0 & -4 & 4 \end{bmatrix} \begin{bmatrix} x_1 \\ x_2 \\ x_3 \end{bmatrix} = \begin{bmatrix} P \\ 0 \\ 0 \end{bmatrix}$$

The solution is $x_1 = x_2 = x_3 = 0.125$ P. As is to be expected, this means that there is no relative drift in the top two storeys.

Note that with time, if there is damping, then the terms involving t disappear and the solution converges to the static values. It is worth noting that the maximum value of x_3 in the undamped forced motion is approximately equal to 0.35 P. In other words, the magnification factor is equal to 0.35 P/(0.125 P) = 2.8. In the case of SDOF system, it was shown in Section 2.9.1, that the maximum DMF is equal to

2.0. In the case of MDOF systems, one can have DMFs larger than two. It is also worth pointing out that there is no periodicity in the displacements.

Figure 3.9 shows the inter-storey drift, which is an indication of shear force in the columns.

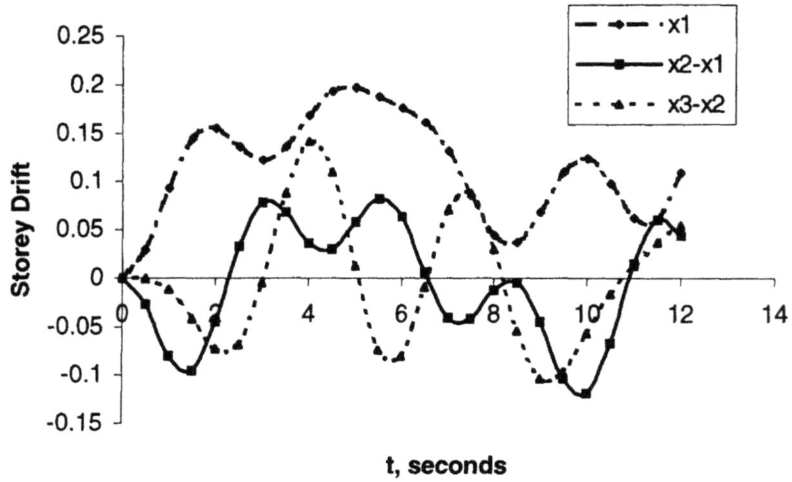

Fig. 3.9 Inter storey drift (undamped motion)

3.5.4 Damped Forced Response

In the previous section, undamped response was analysed. It was shown in Chapter 1 that in the case of damped forced motion of an SDOF system, the differential equation is given by equation (1.3).

In the case of damped force vibration of an MDOF system, a similar equation is assumed to be given by

$$M X'' + C X' + K X = F \tag{3.24}$$

where C is the damping matrix.

Unfortunately there is in practice not enough information to specify the coefficients of the damping matrix, C. It is usually assumed that

$$C = a M + b K \tag{3.25}$$

where a and b are constants. This is known as proportional or Rayleigh damping. The manner in which the constants a and b affect the damping ratio can be understood by considering an SDOF, for which the critical damping $C_{cr} = 2\sqrt{(KM)}$ and $\omega = \sqrt{(K/M)}$. Therefore

$$\frac{C}{C_{cr}} = a \frac{M}{2\sqrt{(K M)}} + b \frac{K}{2\sqrt{(K M)}} = \frac{1}{2}[\frac{a}{\omega} + b\omega]$$

This shows that the first part due to the parameter *a* has the effect of having lower damping for higher frequency while the second part due to the parameter *b* has the exact opposite effect.

In other words the damping matrix *C* can be manipulated by adjusting just two constants to fit the desired damping for any two frequencies. Unfortunately, in an MDOF system, the structure has many more than just two frequencies. In practice the constants *a* and *b* are calculated such that the required values of damping are obtained for the first two or any two desired frequencies and this is assumed to hold good for higher frequencies too.

Following the steps in Section 3.5.1, substituting equations (3.12) and (3.25) in equation (3.24), the differential equation becomes

$$\sum_1^N C_i'' M\ \Phi_i\ + \{a\sum_1^N C_i' M\ \Phi_i + b\sum_1^N C_i' K\ \Phi_i\} + \sum_1^N C_i\ K\ \Phi_i = F \qquad (3.26)$$

Pre-multiplying equation (3.26) by Φ_j^T

$$\sum_1^N C_i''\Phi_j^T M\Phi_i + \{a\sum_1^N C_i'\Phi_j^T M\Phi_i + b\sum_1^N C_i'\Phi_j^T K\Phi_i\} + \sum_1^N C_i\Phi_j^T K\Phi_i = \Phi_j^T F \quad (3.27)$$

Using equations (3.17) and (3.18), equation (3.27) transforms to

$$C_i''\ + 2\ \beta_i\ C_i'\ + \omega_i^2\ C_i = \Phi_i^T\ F \qquad (3.28)$$

where $2\beta_i\ = (a\ + b\,\omega_i^2)$

Thus, as in the case of, an undamped system, Rayleigh damping allows the mode superposition method to be applied to damped forced system analysis as well resulting in the N simultaneous differential equations being reduced to N differential equations in single unknowns.

3.5.5 Damped Forced Response: Example

Use the mode superposition method to analyse the three-storey shear building shown in Fig. 3.1 and analysed in Section 3.5.2 for undamped forced motion. The eigen-values are $\omega^2 = 0.46056$, 2.68199 and 4.857448.

The first step is to calculate the parameters of Rayleigh damping. Assuming damping ratio $\xi = 5\%$ for the first two frequencies.

i. First frequency: $\omega^2 = 0.46056$ or $\omega\ = 0.678646$
   0.05 = a 0.736762 + b 0.339323 (i)
ii. Second frequency: $\omega^2 = 2.68199$ or $\omega\ = 1.637678$
   0.05 = a 0.305310 + b 0.818839 (ii)

Solving a = 0.047982, b = 0.043172.

Using these values, the value of damping ratio for the third frequency is $\xi = 0.059$.

The values of $\beta = (a + b\omega^2\}/2$ for the three frequencies are 0.034, 0.082 and 0.1288. The damped frequency $\omega_d\ = \sqrt{(\omega^2\ - \beta^2)} = 0.677794$, 1.635624 and 2.200195.

It was shown in Section 2.9.1 that the damped forced vibration solution in the case of suddenly applied load for the differential equation assuming starting from rest is given by

$$u'' + 2\beta u' + \omega^2 u = W/M \qquad (3.29)$$

$$u = \frac{W}{M\omega^2}[1 - e^{-\beta t}\{\frac{\beta}{\omega_d}\sin \omega_d t + \cos \omega_d t\}] \qquad (3.30)$$

In the present case, the solution to equation (3.28) is obtained by substituting the appropriate values of M, K, ω and the force W in equation (3.30)

$$u = C_1, M = 1, W = 0.170788\,P,\ \omega^2 = 0.46056$$

$$\beta = 0.034, \omega_d = 0.677794$$

$$C_1 = 0.3708\ P\ [1 - e^{-0.034t}\{0.05\ \sin 0.677794t + \cos\ 0.677794t\}]$$

$$u = C_2, M = 1, W = 0.31152P,\ \omega^2 = 2.68199$$

$$\beta = 0.082, \omega_d = 1.635624$$

$$C_2 = 0.1162\ P[1 - e^{-0.082t}\{0.05\ \sin 1.635624t + \cos\ 1.635624t\}]$$

$$u = C_3, M = 1, W = 0.35184P,\ \omega^2 = 4.857448$$

$$\beta = 0.1288, \omega_d = 2.200195$$

$$C_3 = 0.0724\ P[1 - e^{-0.1288t}\{0.0586\ \sin 2.200195t + \cos\ 2.200195t\}]$$

The final solution to X is given by

$$X(t) = \sum_1^N C_i(t)\ \Phi_i$$

$$x_1(t) = 0.170788\ C_1 + 0.31152\ C_2 + 0.35184\ C_3$$

$$x_2(t) = 0.346066\ C_1 + 0.16988\ C_2 - 0.31840\ C_3$$

$$x_3(t) = 0.449600\ C_1 - 0.49819\ C_2 + 0.22286\ C_3$$

Figure 3.10 shows the variation of floor displacement and Fig. 3.11 shows the inter-storey drift with time.

As in the case of undamped response, the maximum displacements can be calculated. Because of the low damping present, it is reasonable to assume that the maximum DMF is equal to two in this case also. Table 3.2 shows the comparison between various methods of estimating the maximum response.

Table 3.2 Maximum displacements

	Exact maximum	Using C_{jMAX}	SRSS
x_1	0.1824	0.2500	0.1545
x_2	0.2350	0.3423	0.2638
x_3	0.3009	0.4815	0.3544

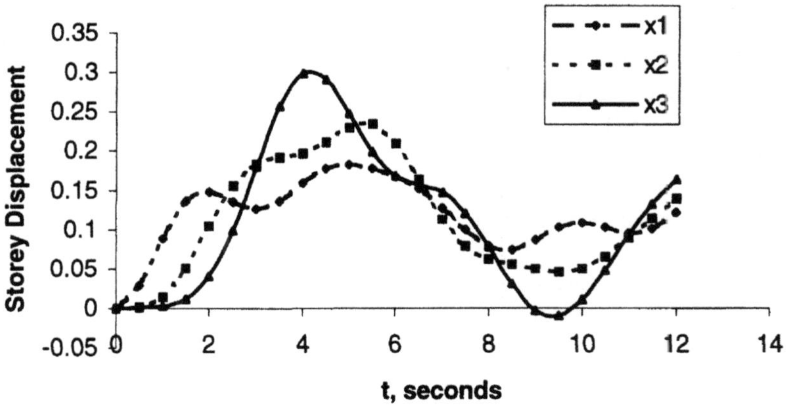

Fig 3.10 Storey displacement (damped response)

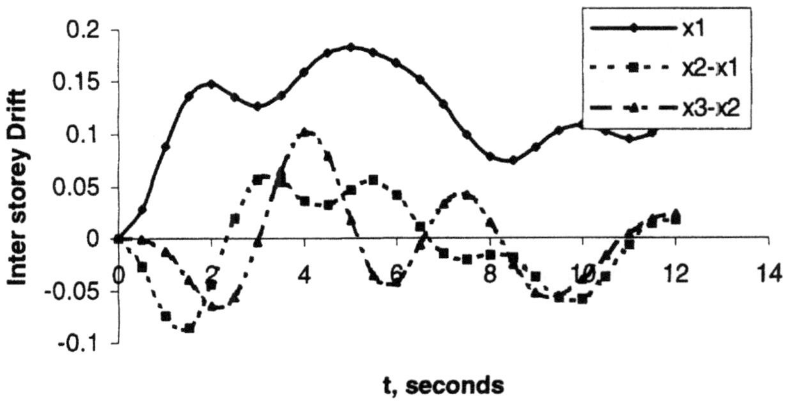

Fig 3.11 Inter Storey drift (damped response)

3.6 RESPONSE TO BASE ACCELERATION

In sections 3.5.1 and 3.5.5, response of a multi-storey frame subjected to prescribed force inputs was discussed. If, instead of a force input, the frame is subjected to an acceleration of the base, as would happen in an earthquake, then the problem can be analysed in a similar way to force input. As was shown in Section 1.18, the acceleration of the masses is governed by absolute displacement but the elastic forces represented by $K X$ are governed by displacements measured relative to the base.

Considering the three-storey shear frame shown in Fig. 3.1, the equilibrium of the masses at the three storeys are given by

$$m_1\,(x''_1 + x''_{base}) + k_1 x_1 - k_2(x_2 - x_1) = 0$$
$$m_2\,(x''_1 + x''_{base}) + k_2(x_2 - x_1) - k_3(x_3 - x_2) = 0$$
$$m_1\,(x''_1 + x''_{base}) + k_3(x_3 - x_2) = 0$$

Expressing these equations in matrix form

$$\begin{bmatrix} m_1 & 0 & 0 \\ 0 & m_2 & 0 \\ 0 & 0 & m_3 \end{bmatrix}\begin{bmatrix} x''_1 \\ x''_2 \\ x''_3 \end{bmatrix} + \begin{bmatrix} k_1+k_2 & -k_2 & 0 \\ -k_2 & k_2+k_3 & -k_3 \\ 0 & -k_3 & k_3 \end{bmatrix}\begin{bmatrix} x_1 \\ x_2 \\ x_3 \end{bmatrix} = -\,x''_{base}\begin{bmatrix} m_1 \\ m_2 \\ m_3 \end{bmatrix}$$

In other words, at each storey level, the product of the storey mass and the acceleration of the base acts as the force at the storey level. The matrix form of the above equation is therefore

$$M\ X'' + K\ X = F \tag{3.31}$$

$$F = -\,x''_{base}\begin{bmatrix} m_1 & 0 & 0 \\ 0 & m_2 & 0 \\ 0 & 0 & m_3 \end{bmatrix}\begin{bmatrix} 1 \\ 1 \\ 1 \end{bmatrix} \tag{3.32}$$

The above equation can be written in matrix form as

$$F = -\,x''_{base}\, M\, I \tag{3.33}$$

where I is a unit vector.

When using the mode superposition method to solve the problem of a structure subjected to base acceleration, the load for the jth mode is given by the $\phi_j^T F$. In earthquake analysis, the term earthquake participation factor is used. For the jth mode, the earthquake participation factor Γ_j is defined as

$$\Gamma_j = \phi_j^T\ M\ I \tag{3.34}$$

Using this factor, the force corresponding to j[th] mode is given by

$$\phi_j^T\ F = -\,x''_{Base}\ \Gamma_j \tag{3.35}$$

The term Effective Mass M^j_{Eff} for the jth mode is defined as

$$M^j_{Eff} = \Gamma_j^2 \tag{3.36}$$

The sum of M_{Eff}^{j} for all the modes can be shown to be equal to the total mass on the structure.

3.6.1 Earthquake Participation Factors: Example

Calculation of earthquake participation factors and effective masses are illustrated by two simple examples.

Example 1: Using the eigen-vectors determined in Section 3.2.5 for the three-storey shear frame shown in fig. 3.1, calculate the earthquake participation factors and the effective masses for all the three modes. It is give that

$$\phi_1 = \begin{bmatrix} 0.170788 \\ 0.346066 \\ 0.449600 \end{bmatrix} \quad \phi_2 = \begin{bmatrix} 0.31152 \\ 0.16988 \\ -0.49819 \end{bmatrix}, \quad \phi_3 = \begin{bmatrix} 0.35184 \\ -0.31840 \\ 0.22286 \end{bmatrix}$$

$$M = \begin{bmatrix} 4 & 0 & 0 \\ 0 & 4 & 0 \\ 0 & 0 & 2 \end{bmatrix}, \quad M_{Total} = 4+4+2=10$$

$$\Gamma_1 = \phi_1^T M\, I = \begin{bmatrix} 0.170788 & 0.346066 & 0.44960 \end{bmatrix} \begin{bmatrix} 4 & 0 & 0 \\ 0 & 4 & 0 \\ 0 & 0 & 2 \end{bmatrix} \begin{bmatrix} 1 \\ 1 \\ 1 \end{bmatrix} = 2.967$$

$$M_{Eff}^{1} = \Gamma_1^2 = 8.80, \quad \frac{M_{Eff}^{1}}{M_{Total}} = 88.0\%$$

$$\Gamma_2 = \phi_2^T M\, I = \begin{bmatrix} 0.31152 & 0.16988 & -0.49819 \end{bmatrix} \begin{bmatrix} 4 & 0 & 0 \\ 0 & 4 & 0 \\ 0 & 0 & 2 \end{bmatrix} \begin{bmatrix} 1 \\ 1 \\ 1 \end{bmatrix} = 0.929$$

$$M_{Eff}^{2} = \Gamma_2^2 = 0.863, \quad \frac{M_{Eff}^{2}}{M_{Total}} = 8.6\%$$

$$\Gamma_3 = \phi_3^T M\, I = \begin{bmatrix} 0.35184 & -0.31840 & 0.22286 \end{bmatrix} \begin{bmatrix} 4 & 0 & 0 \\ 0 & 4 & 0 \\ 0 & 0 & 2 \end{bmatrix} \begin{bmatrix} 1 \\ 1 \\ 1 \end{bmatrix} = 0.5795$$

$$M_{Eff}^{3} = \Gamma_3^2 = 0.335, \quad \frac{M_{Eff}^{3}}{M_{Total}} = 3.4\%$$

Example 2: Using the eigen-vectors determined in Section 3.2.5 for the three-storey shear frame shown in Fig. 3.4, calculate the earthquake participation factors and the effective masses for all the three modes. It is given that

$M_{total} = 4 + 4 + 2 + 2 = 12$

$$M = \begin{bmatrix} 4 & 0 & 0 & 0 \\ 0 & 4 & 0 & 0 \\ 0 & 0 & 2 & 0 \\ 0 & 0 & 0 & 2 \end{bmatrix}, \phi_1 = \begin{bmatrix} 0.11906 \\ 0.25373 \\ 0.37883 \\ 0.44651 \end{bmatrix}, \quad \phi_2 = \begin{bmatrix} 0.24691 \\ 0.31412 \\ -0.08504 \\ -0.41652 \end{bmatrix}, \phi_3 = \begin{bmatrix} 0.39383 \\ -0.19455 \\ -0.25202 \\ 0.22492 \end{bmatrix},$$

$$\phi_4 = \begin{bmatrix} 0.140583 \\ -0.221575 \\ 0.534544 \\ -0.27667 \end{bmatrix}$$

$$\Gamma_1 = \phi_1^T MI = \begin{bmatrix} 0.11906 & 0.25373 & 0.37883 & 0.44651 \end{bmatrix} \begin{bmatrix} 4 & 0 & 0 & 0 \\ 0 & 4 & 0 & 0 \\ 0 & 0 & 2 & 0 \\ 0 & 0 & 0 & 2 \end{bmatrix} \begin{bmatrix} 1 \\ 1 \\ 1 \\ 1 \end{bmatrix}$$

$$\Gamma_1 = 3.142, \; M^1_{Eff} = \Gamma_1^2 = 9.871, \frac{M^1_{Eff}}{M_{Total}} = 82.2\%$$

$$\Gamma_2 = \phi_2^T MI = \begin{bmatrix} 0.2469 & 0.3141 & -0.0850 & -0.4165 \end{bmatrix} \begin{bmatrix} 4 & 0 & 0 & 0 \\ 0 & 4 & 0 & 0 \\ 0 & 0 & 2 & 0 \\ 0 & 0 & 0 & 2 \end{bmatrix} \begin{bmatrix} 1 \\ 1 \\ 1 \\ 1 \end{bmatrix}$$

$$\Gamma_2 = 1.2501, \; M^2_{Eff} = \Gamma_2^2 = 1.563, \frac{M^2_{Eff}}{M_{Total}} = 13.0\%$$

$$\Gamma_3 = \phi_3^T MI = \begin{bmatrix} 0.3938 & -0.1946 & -0.2520 & 0.2249 \end{bmatrix} \begin{bmatrix} 4 & 0 & 0 & 0 \\ 0 & 4 & 0 & 0 \\ 0 & 0 & 2 & 0 \\ 0 & 0 & 0 & 2 \end{bmatrix} \begin{bmatrix} 1 \\ 1 \\ 1 \\ 1 \end{bmatrix}$$

$$\Gamma_3 = 0.743, \; M^3_{Eff} = \Gamma_3^2 = 0.552, \frac{M^3_{Eff}}{M_{Total}} = 4.6\%$$

$$\Gamma_4 = \phi_4^T MI = \begin{bmatrix} 0.1406 & -0.2216 & 0.5345 & -0.2767 \end{bmatrix} \begin{bmatrix} 4 & 0 & 0 & 0 \\ 0 & 4 & 0 & 0 \\ 0 & 0 & 2 & 0 \\ 0 & 0 & 0 & 2 \end{bmatrix} \begin{bmatrix} 1 \\ 1 \\ 1 \\ 1 \end{bmatrix}$$

$$\Gamma_4 = 0.192, \; M^4_{Eff} = \Gamma_4^2 = 0.037, \frac{M^1_{Eff}}{M_{Total}} = 0.31\%$$

It is clear from the values of the participation factors and effective mass, their values decrease as the mode number increases. The practical significance of this fact is that, in general, it is not necessary to include all the modes in a calculation. Only a few significant modes need to be included in order to obtain accurate enough results for practical problems. It is therefore not necessary to determine all the eigen-values and the corresponding eigen-vectors. This results in significant savings in practical computations. In other words

$$X(t) \approx \sum_{1}^{p} C_i(t)\, \Phi_i \ , \ p << N$$

In Chapter 6, attention will therefore be focussed on methods that enable one to determine only the lower few eigen-pairs. For any given acceleration of the base, the problem can be solved following the steps used in the previous section.

CHAPTER 4

MATRIX ROUTINES

4.1 INTRODUCTION

In this chapter, some important routines involving matrix manipulation will be described. These routines are used in later chapters. Since most readers will be familiar with basic matrix algebra, the discussions are intentionally kept quite short.

4.2 MATRIX MANIPULATION

Unlike FORTRAN 77, many of the basic matrix routines can be performed most conveniently by using FORTRAN 90 intrinsic procedures. It is assumed the matrices involved A, B, C, D etc have been declared as arrays.

- Addition and Subtraction: If A, B, C and D are all of same size, i.e. have identical number of rows and columns, then operations of matrix addition and subtraction are achieved by the commands
  C = A + B
  D = A – B

- Multiplication: Two matrices A and B can be multiplied to obtain the matrix C = AB, if and only if the number of columns in A is equal to the number of rows in B. The two matrices A and B are said to be conformable in the order AB. Thus if the matrix A is of order ($m \times \ell$) and B is of order ($\ell \times n$), then C is of order ($m \times n$). The ij^{th} element of matrix C is $c_{ij} = \sum a_{ik}\ b_{kj}$, $k = 1$ to ℓ. If two matrices A and B are conformable in the order AB, then the product C = AB is achieved by the command
  C = MATMUL(A,B)

- Transpose: If $D = C^T$, this is achieved by the command
  D = TRANSPOSE(C)

- If $E = B^T\ C\ B$ is permissible, then this is calculated by the command
  E = MATMUL(TRANSPOSE(B),MATMUL(C,B))
- Initialising a matrix: All elements of A can be set to zero by the command
  A = 0

These commands greatly simplify the process of writing programs.

4.3 CHOLESKY FACTORISATION

In the structural dynamics computations, it is necessary to factorise the stiffness matrix K, which is a square symmetric positive definite matrix, as

$$K = U^T U$$

where, U is an upper triangular matrix and U^T is transform of U. The elements of U can be obtained by direct multiplication.

$$K = \begin{bmatrix} k_{11} & k_{12} & k_{13} & k_{14} \\ & k_{22} & k_{23} & k_{24} \\ \textit{symmetric} & & k_{33} & k_{34} \\ & & & k_{44} \end{bmatrix}$$

$$K = U^T U = \begin{bmatrix} u_{11} & 0 & 0 & 0 \\ u_{12} & u_{22} & 0 & 0 \\ u_{13} & u_{23} & u_{33} & 0 \\ u_{14} & u_{24} & u_{34} & u_{44} \end{bmatrix} \begin{bmatrix} u_{11} & u_{12} & u_{13} & u_{14} \\ 0 & u_{22} & u_{23} & u_{24} \\ 0 & 0 & u_{33} & u_{34} \\ 0 & 0 & 0 & u_{44} \end{bmatrix}$$

Multiplying through and equating the terms,

$$k_{11} = u_{11}^2 \,,\; u_{11}\, u_{12} = k_{12} \,,\; u_{11}\, u_{13} = k_{13} \,,\; u_{11}\, u_{14} = a_{14}$$

$$\therefore \quad u_{11} = \sqrt{k_{11}} \quad ,\; u_{1j} = \frac{k_{1j}}{u_{11}},\; j = 2,4$$

$$k_{22} = u_{12}^2 + u_{22}^2 \,,\; k_{23} = u_{12}\, u_{13} + u_{22}\, u_{23} \quad ,\; k_{24} = u_{12}\, u_{14} + u_{22}\, u_{24}$$

$$\therefore \quad u_{22} = \sqrt{(k_{22} - u_{12}^2)} \,,\; u_{2j} = \frac{k_{2j} - u_{12}\, u_{1j}}{u_{22}} \,,\; j = 3,4$$

$$k_{33} = u_{13}^2 + u_{23}^2 + u_{33}^2 \,,\; k_{34} = u_{13}\, u_{14} + u_{23}\, u_{24} + u_{33}\, u_{34}$$

$$\therefore \quad u_{33} = \sqrt{(k_{33} - u_{13}^2 - u_{23}^2\,)} \,,\; u_{3j} = \frac{k_{34} - \sum_{k=1}^{2} u_{k3}\, u_{kj}}{u_{33}},\; j = 4$$

$$k_{44} = u_{14}^2 + u_{24}^2 + u_{34}^2 + u_{44}^2$$

$$u_{44} = \sqrt{(k_{44} - u_{14}^2 - u_{24}^2 - u_{34}^2)}$$

Clearly, the general formula will be

$$u_{ii} = \sqrt{(k_{ii} - \sum_{k=1}^{i-1} u_{ki}^2\,)} \quad ,\; i \ge 2$$

$$u_{ij} = \frac{k_{ij} - \sum_{k=1}^{i-1} u_{ki}\, u_{kj}}{u_{ii}},\; j > i, \;\; \textit{and} \;\; u_{ij} = 0 \,,\; j < i$$

It is possible to operate such that the matrix U replaces the upper triangular part of K, thus avoiding the need to store both K and U. This approach is useful if the stiffness matrix K is not required for further computations.

In the above, computations were performed row by row. It is possible to perform the calculations column-wise using the following set of equations.

$$u_{11} = \sqrt{k_{11}}$$

$$u_{12} = \frac{k_{12}}{u_{11}}, \quad u_{22} = \sqrt{(k_{22} - u_{12}^2)}$$

$$u_{13} = \frac{k_{13}}{u_{11}}, \quad u_{23} = \frac{(k_{23} - u_{12}\,u_{13})}{u_{22}}, \quad u_{33} = \sqrt{(k_{33} - u_{113}^2 - u_{23}^2)}$$

$$u_{14} = \frac{k_{14}}{u_{11}}, \quad u_{24} = \frac{(k_{24} - u_{12}\,u_{14})}{u_{22}}, \quad u_{34} = \frac{(k_{34} - u_{13}\,u_{14} - u_{23}\,u_{24})}{u_{33}}$$

$$u_{44} = \sqrt{(k_{44} - u_{14}^2 - u_{24}^2 - u_{34}^2)}$$

The factorisation is useful for two separate purposes.

a. For the solution of simultaneous equations.
b. To convert a general eigen-value problem to a special eigen-value problem.

4.3.1 Solution of Simultaneous Equations by Cholesky Decomposition

If it is required to solve a set of simultaneous equations given by

$$KX = F \tag{i}$$

The steps to be followed are as follows.

Step 1: Factorise the stiffness matrix as

$$K = U^T U \tag{ii}$$

$$\therefore \; U^T U X = F \tag{iii}$$

Let an intermediate variable Y be defined as

$$UX = Y \tag{iv}$$

Substituting equation (iv) in equation (iii)

$$U^T Y = F \tag{v}$$

Step 2: Solve equation (v) for Y by forward substitution.

Step 3: Using the value of Y, solve equation (iv) by back substitution.

4.3.1.1 Cholesky Factorisation: Example

The procedure for solving a set of simultaneous equations using Cholesky factorisation is illustrated by an example.

Solve the equation $KX = F$, where

$$K = \begin{bmatrix} 14 & -6 & 0 & 0 \\ -6 & 10 & -4 & 0 \\ 0 & -4 & 8 & -4 \\ 0 & 0 & -4 & 4 \end{bmatrix}, \quad F = \begin{bmatrix} 8 \\ 0 \\ 0 \\ 0 \end{bmatrix}$$

Step 1: Using the equations derived in Section 4.1.1, determine the matrix U.

$$u_{11} = \sqrt{k_{11}} = \sqrt{14} = 3.741657$$

$$u_{12} = \frac{k_{12}}{u_{11}} = \frac{-6}{3.741657} = -1.603567$$

$$u_{22} = \sqrt{(k_{22} - u_{12}^2)} = \sqrt{(10 - (-1.603567)^2)} = 2.725541$$

$$u_{13} = \frac{k_{13}}{u_{11}} = 0, \; u_{23} = \frac{(k_{23} - u_{12}\,u_{13})}{u_{22}} = \frac{(-4-0)}{2.725541} = -1.467599$$

$$u_{33} = \sqrt{(k_{33} - u_{113}^2 - u_{23}^2)} = \sqrt{(8 - 0 - (-1.467599)^2)} = 2.417882$$

$$u_{14} = \frac{k_{14}}{u_{11}} = 0 \; u_{24} = \frac{(k_{24} - u_{12}\,u_{14})}{u_{22}} = 0$$

$$u_{34} = \frac{(k_{34} - u_{13}\,u_{14} - u_{23}\,u_{24})}{u_{33}} = \frac{(-4-0-0)}{2.417882} = -1.654340$$

$$u_{44} = \sqrt{(k_{44} - u_{14}^2 - u_{24}^2 - u_{34}^2)} = \sqrt{(4 - 0 - 0 - (-1.654340)^2)} = 1.123903$$

$$U = \begin{bmatrix} 3.741657 & -1.603567 & 0 & 0 \\ 0 & 2.725541 & -1.467599 & 0 \\ 0 & 0 & 2.417882 & -1.654341 \\ 0 & 0 & 0 & 1.123903 \end{bmatrix}$$

$$U^T = \begin{bmatrix} 3.741657 & 0 & 0 & 0 \\ -1.603567 & 2.725541 & 0 & 0 \\ 0 & -1.467599 & 2.417882 & 0 \\ 0 & 0 & -1.654341 & 1.123903 \end{bmatrix}$$

Step 2: Solve for Y from $U^T Y = F$.

$$U^T Y = \begin{bmatrix} 3.741657 & 0 & 0 & 0 \\ -1.603567 & 2.725541 & 0 & 0 \\ 0 & -1.467599 & 2.417882 & 0 \\ 0 & 0 & -1.654341 & 1.123903 \end{bmatrix} \begin{bmatrix} y_1 \\ y_2 \\ y_3 \\ y_4 \end{bmatrix} = \begin{bmatrix} 8 \\ 0 \\ 0 \\ 0 \end{bmatrix}$$

Clearly, one can determine Y as follows.
From the first equation, $3.741657\, y_1 = 8,\; y_1 = 2.138090$
From the second equation, $-1.603567\, y_1 + 2.725541\, y_2 = 0$

Knowing y_1, $y_2 = 1.257941$
From the third equation, $-1.467599\, y_2 + 2.417882\, y_3 = 0$
Knowing y_1 and y_2, $y_3 = 0.763542$
From the last equation, $-1.654341\, y_3 + 1.123903\, y_4 = 0$
Knowing y_1, y_2 and y_3, $y_4 = 1.123903$
This process is known as ***forward substitution***.

Clearly it is useful to replace the vector F by vector Y. Each element of F can be replaced by the corresponding element of Y as soon as it is calculated. This saves the need to store both F and Y.

Step 3: Knowing Y, solve for X from $U X = Y$

$$UX = \begin{bmatrix} 3.741657 & -1.603567 & 0 & 0 \\ 0 & 2.725541 & -1.467599 & 0 \\ 0 & 0 & 2.417882 & -1.654341 \\ 0 & 0 & 0 & 1.123903 \end{bmatrix} \begin{bmatrix} x_1 \\ x_2 \\ x_3 \\ x_4 \end{bmatrix} = \begin{bmatrix} 2.138090 \\ 1.257941 \\ 0.763542 \\ 1.123903 \end{bmatrix}$$

From the last equation, $1.123903\, x_4 = 1.123903$, $x_4 = 1.0$
From the third equation, $-1.654341 x_4 + 2.417882\, x_3 = 0.763542$, $x_3 = 1.0$
From the second equation, $-1.467599\, x_3 + 2.725541\, x_2 = 1.257941$, $x_2 = 1.0$
From the first equation, $-1.603567\, x_2 + 3.741657\, x_1 = 2.138090$, $x_1 = 1.0$
This process is known as ***back substitution***.

Clearly it is useful to replace the vector Y by vector X. Each element of Y can be replaced by the corresponding element of X as soon as it is calculated. This saves the need to store both Y and X.

This shows that once the triangular decomposition is done, then using forward and then backward substitution, simultaneous equation can be solved. Clearly at the end of the computations, K is replaced by U and F by X. This procedure will be used in later stages of eigen-value computations.

4.3.2 Converting a General to a Special Eigen-value Problem

In the determination of natural frequencies and corresponding mode shapes of structures, one needs to solve the eigen-value problem given by equation (3.2). However many eigen-value solution procedures are tailored to solve the special eigen-value problem given by equation (3.5). As explained in Section 3.1, the conversion can be done so as to preserve both the bandedness and symmetry by using equation (3.9). The general eigen-value problem is given by

$$K X = \omega^2 M X \tag{4.1}$$

where K is the structural stiffness matrix, M = mass matrix, ω = natural frequency in radians per second, X = corresponding mode shape.

It is assumed in the following that the stiffness matrix K is symmetric and positive definite.

Equation (4.1) is converted to a special eigen-value problem

$$A Y = \lambda Y \tag{4.2}$$

where

$$A = U^{T} M U^{1},\ Y = UX,\ \lambda = 1/\omega^{2},\ K = U^{T} U \tag{4.3}$$

The general eigen-value problem (4.1) has been transformed in to a special eigen-value problem (4.3). The main advantage of equation (4.3) is it is symmetric and banded.

In the solution procedure commonly used to calculate the eigen-vectors, it is required to evaluate, for a given value of Y, the product A Y. The main advantage of equation (4.3), is that this can be achieved without having to invert any matrix as follows.

It is required to evaluate, for a given vector Y, the matrix product A Y. A Y is given by

$$A\,Y = U^{T} M U^{1} Y \tag{4.4}$$

The product is evaluated in a series of steps as follows.
Step 1: Using back substitution, calculate the value of Z which is an intermediate variable.

$$U^{1} Y = Z \text{ or } UZ = Y$$

Step 2: Calculate R, which is an intermediate variable, from the matrix product

$$M Z = R$$

Step 3: Using forward substitution, calculate the value of S, which is equal to AY.

$$U^{T} R = S \text{ or } U^{T} S = R$$

Thus the product $U^{T} M U^{1} Y$ is determined without resorting to the evaluation of any inverse. This procedure will be used extensively in a later Section.

4.3.2.1 General to a Special Eigen-value Problem: Example

Evaluate the product $A\,Y = U^{T} M U^{1} Y$, given that

$$K = \begin{bmatrix} 14 & -6 & 0 & 0 \\ -6 & 10 & -4 & 0 \\ 0 & -4 & 8 & -4 \\ 0 & 0 & -4 & 4 \end{bmatrix},\ M = \begin{bmatrix} 4 & 0 & 0 & 0 \\ 0 & 4 & 0 & 0 \\ 0 & 0 & 2 & 0 \\ 0 & 0 & 0 & 2 \end{bmatrix},\ Y = \begin{bmatrix} 1 \\ 1 \\ 1 \\ 1 \end{bmatrix}$$

From Section 4.1.1.1, $K = U^{T} U$, where

$$U = \begin{bmatrix} 3.741657 & -1.603567 & 0 & 0 \\ 0 & 2.725541 & -1.467599 & 0 \\ 0 & 0 & 2.417882 & -1.654341 \\ 0 & 0 & 0 & 1.123903 \end{bmatrix},\ Y = \begin{bmatrix} 1 \\ 1 \\ 1 \\ 1 \end{bmatrix}$$

Step 1: Use back substitution to solve for Z from the equation, $U Z = Y$

$$UZ = \begin{bmatrix} 3.741657 & -1.603567 & 0 & 0 \\ 0 & 2.725541 & -1.467599 & 0 \\ 0 & 0 & 2.417882 & -1.654341 \\ 0 & 0 & 0 & 1.123903 \end{bmatrix} \begin{bmatrix} z_1 \\ z_2 \\ z_3 \\ z_4 \end{bmatrix} = Y = \begin{bmatrix} 1 \\ 1 \\ 1 \\ 1 \end{bmatrix}$$

$$1.123903\, z_4 = 1,\ z_4 = 0.889757$$
$$2.417882\, z_3 - 1.654341\, z_4 = 1,\ z_3 = 1.022366$$
$$2.725541\, z_2 - 1.467599\, z_3 = 1,\ z_2 = 0.917405$$
$$3.741657\, z_1 - 1.603567\, z_2 = 1,\ z_1 = 0.660435$$

Step 2: Calculate the product, $M\,Z = R$

$$MZ = \begin{bmatrix} 4 & 0 & 0 & 0 \\ 0 & 4 & 0 & 0 \\ 0 & 0 & 2 & 0 \\ 0 & 0 & 0 & 2 \end{bmatrix} \begin{bmatrix} 0.660435 \\ 0.917405 \\ 1.022366 \\ 0.889757 \end{bmatrix} = \begin{bmatrix} 2.641740 \\ 3.669620 \\ 2.044732 \\ 1.779514 \end{bmatrix} = R$$

Step 3 : Use forward substitution to solve for S from the equation, $U^T S = R$,

$$U^T S = \begin{bmatrix} 3.741657 & 0 & 0 & 0 \\ -1.603567 & 2.725541 & 0 & 0 \\ 0 & -1.467599 & 2.417882 & 0 \\ 0 & 0 & -1.654341 & 1.123903 \end{bmatrix} \begin{bmatrix} s_1 \\ s_2 \\ s_3 \\ s_4 \end{bmatrix} = R = \begin{bmatrix} 2.641740 \\ 3.669620 \\ 2.044732 \\ 1.779514 \end{bmatrix}$$

$$3.741657\, s_1 = 2.641740,\ s_1 = 0.706035$$
$$-1.603567\, s_1 + 2.725541\, s_2 = 3.669620,\ s_2 = 1.761777$$
$$-1.467599 s_2 + 2.417882\, s_3 = 2.044732,\ s_3 = 1.915029$$
$$-1.654341\, s_3 + 1.123903\, s_4 = 1.779514 \quad s_4 = 4.402181$$

$$S = U^{-T} M\, U^{-1} Y = \begin{bmatrix} 0.706035 \\ 1.761777 \\ 1.915029 \\ 4.402181 \end{bmatrix}$$

4.4 CROUT FACTORISATION

While calculating the U matrix by Cholesky factorisation of the stiffness matrix K, it was noticed in Section 4.3 that the calculation of diagonal terms involves the extraction of the square root. If the stiffness matrix is positive definite, then the extraction of the square root does not cause any problems. If this is not so, then Cholesky factorisation cannot be carried out. Fortunately, instead of factorising the stiffness matrix K as $K = U^T U$, if the matrix is factorised as $K = U^T D\, U$, where D

is a diagonal matrix and U is an upper triangular matrix with unity on the diagonals, there will be no need to calculate square roots. This is known as Crout factorisation. In the analysis of stable elastic structures, Cholesky factorisation presents no problems.

The general formulae for Crout factorisation are developed in a straightforward manner.

$$A = \begin{bmatrix} a_{11} & a_{12} & a_{13} & a_{14} \\ & a_{22} & a_{23} & a_{24} \\ \textit{symmetric} & & a_{33} & a_{34} \\ & & & a_{44} \end{bmatrix}$$

$$A = U^T \; D \; U = \begin{bmatrix} 1 & 0 & 0 & 0 \\ u_{12} & 1 & 0 & 0 \\ u_{13} & u_{23} & 1 & 0 \\ u_{14} & u_{24} & u_{34} & 1 \end{bmatrix} \begin{bmatrix} d_{11} & 0 & 0 & 0 \\ 0 & d_{22} & 0 & 0 \\ 0 & 0 & d_{33} & 0 \\ 0 & 0 & 0 & d_{44} \end{bmatrix} \begin{bmatrix} 1 & u_{12} & u_{13} & u_{14} \\ 0 & 1 & u_{23} & u_{24} \\ 0 & 0 & 1 & u_{34} \\ 0 & 0 & 0 & 1 \end{bmatrix}$$

Multiplying through

$$a_{11} = d_{11} \;, u_{12}\, d_{11} = a_{12} \;, \; u_{13}\, d_{11} = a_{13} \;, \; u_{14}\, d_{11} = a_{14}$$

$$\therefore d_{11} = a_{11} \quad , u_{1j} = \frac{a_{1j}}{d_{11}}, j = 2,4$$

$$a_{22} = u_{12}^2\, d_{11} + d_{22} \;, \; a_{23} = u_{12}\, u_{13}\, d_{11} + d_{22}\, u_{23} \quad , a_{24} = u_{12}\, u_{14}\, d_{11} + u_{24} d_{22}$$

$$\therefore d_{22} = a_{22} - u_{12}^2\, d_{11}, \; u_{2j} = \frac{a_{2j} - u_{12}\, u_{1j}\, d_{11}}{d_{22}} \;, \; j = 3,4$$

$$a_{33} = u_{13}^2\, d_{11} + u_{23}^2\, d_{22} + d_{33} \;, \quad a_{34} = u_{13}\, u_{14}\, d_{11} + u_{23}\, u_{24}\, d_{22} + d_{33}\, u_{34}$$

$$\therefore d_{33} = a_{33} - u_{13}^2\, d_{11} - u_{23}^2\, d_{22} \;, \; u_{3j} = \frac{a_{34} - \sum_{k=1}^{2} u_{k3}\, u_{kj} d_{kk}}{d_{33}}, j = 4$$

$$a_{44} = u_{14}^2\, d_{11} + u_{24}^2\, d_{22} + u_{34}^2\, d_{33} + d_{44}$$

$$d_{44} = a_{44} - u_{14}^2\, d_{11} - u_{24}^2\, d_{22} - u_{34}^2\, d_{33}$$

Clearly, the general formula will be

$$d_{11} = a_{11}, \; u_{ii} = 1.0, \;, \; d_{ii} = a_{ii} - \sum_{k=1}^{i-1} u_{ki}^2\, d_{kk} \;, \; i \geq 2$$

$$u_{ij} = \frac{a_{ij} - \sum_{k=1}^{i-1} u_{ki}\, u_{kj}\, d_{kk}}{d_{ii}}, j > i \textit{ and } u_{ij} = 0 \;, \; j < i$$

4.4.1 Solution of Simultaneous Equations by Crout Decomposition

If it is required to solve a set of simultaneous equations given by

$$K X = F \tag{i}$$

The steps to be followed are as follows.
Step 1: Factorise the stiffness matrix as

$$K = U^T D U \tag{ii}$$

$$\therefore \; U^T D U X = F \tag{iii}$$

Let an intermediate variable Y be defined as

$$D U X = Y \tag{iv}$$

$$U X = D^{-1} Y \tag{v}$$

Substituting equation (iv) in equation (iii)

$$U^T Y = F \tag{vi}$$

Step 2: Solve equation (vi) for Y by forward substitution.
Step 3: Calculate $D^{-1} Y$. Since D is a diagonal matrix, this step involves dividing y_i by d_i.
Step 4: Using the value of $D^{-1} Y$, solve for X by back substitution, using equation (v).

4.4.1.1 Crout Factorisation: Example

The procedure for solving a set of simultaneous equations using Crout factorisation is illustrated by an example.
Solve the equation $K X = F$, where

$$K = \begin{bmatrix} 14 & -6 & 0 & 0 \\ -6 & 10 & -4 & 0 \\ 0 & -4 & 8 & -4 \\ 0 & 0 & -4 & 4 \end{bmatrix}, \quad F = \begin{bmatrix} 8 \\ 0 \\ 0 \\ 0 \end{bmatrix}$$

Step 1: Using the equations derived in Section 4.4, determine the matrices D and U.

$$d_{11} = a_{11} = 14, \quad u_{12} = \frac{a_{12}}{d_{11}} = \frac{-6}{14} = -0.4286, \quad u_{13} = u_{14} = 0$$

$$d_{22} = a_{22} - u_{12}^2 d_{11} = 7.4286, \quad u_{23} = \frac{a_{23} - u_{12} u_{13} d_{11}}{d_{22}} = -0.5385, u_{24} = 0$$

$$d_{33} = a_{33} - u_{13}^2 d_{11} - u_{23}^2 d_{22} = 5.8458, \; u_{34} = \frac{a_{34} - u_{13} u_{14} d_{11} - u_{23} u_{24} d_{22}}{d_{33}} = -0.6843$$

$$d_{44} = a_{44} - u_{14}^2 d_{11} - u_{24}^2 d_{22} - u_{34}^2 d_{33} = 1.2630$$

$$U = \begin{bmatrix} 1.0 & -0.4286 & 0 & 0 \\ 0 & 1.0 & -0.5385 & 0 \\ 0 & 0 & 1.0 & -0.6843 \\ 0 & 0 & 0 & 1.0 \end{bmatrix}$$

$d_{11} = 14.0$, $d_{22} = 7.4286$, $d_{33} = 5.8458$, $d_{44} = 1.2630$

Step 2: Solve for Y from $U^T Y = F$.

$$U^T Y = \begin{bmatrix} 1.0 & 0 & 0 & 0 \\ -0.4286 & 1.0 & 0 & 0 \\ 0 & -0.5385 & 1.0 & 0 \\ 0 & 0 & -0.6843 & 1.0 \end{bmatrix} \begin{bmatrix} y_1 \\ y_2 \\ y_3 \\ y_4 \end{bmatrix} = \begin{bmatrix} 8 \\ 0 \\ 0 \\ 0 \end{bmatrix}$$

Determine *Y* by forward substitution.
From the first equation, $y_1 = 8$
From the second equation, $-0.4286\, y_1 + y_2 = 0$
Knowing y_1, $y_2 = 3.4288$
From the third equation, $-0.5385\, y_2 + y_3 = 0$
Knowing y_1 and y_2, $y_3 = 1.8464$
From the last equation, $-0.6843\, y_3 + y_4 = 0$
Knowing y_1, y_2 and y_3, $y_4 = 1.2635$
Clearly it is useful to replace the vector *F* by vector *Y*. Each element of *F* can be replaced by the corresponding element of Y as soon as it is calculated. This saves the need to store both *F* and *Y*.
Step 3: Calculate $D^{-1}Y = [8/14, 3.4288/7.4286, 1.8464/5.8458, 1.2635/1.2630]^T$
$= [0.5714, 0.4616, 0.3159, 1.0]^T$
Step 4: Knowing $D^{-1} Y$, using back substitution, solve for *X* from $U X = D^{-1} Y$

$$UX = \begin{bmatrix} 1.0 & -0.4286 & 0 & 0 \\ 0 & 1.0 & -0.5385 & 0 \\ 0 & 0 & 1.0 & -0.6843 \\ 0 & 0 & 0 & 1.0 \end{bmatrix} \begin{bmatrix} x_1 \\ x_2 \\ x_3 \\ x_4 \end{bmatrix} = \begin{bmatrix} 0.5714 \\ 0.4616 \\ 0.3159 \\ 1.0 \end{bmatrix}$$

From the last equation, $y_4 = 1.0$
From the third equation, $-0.6843y_4 + y_3 = 0.3159$, $y_3 = 1.0$
From the second equation, $-0.5385\, y_3 + y_2 = 0.4616$, $y_2 = 1.0$
From the first equation, $-0.4286\, y_2 + y_1 = 0.5714$, $y_1 = 1.0$
Clearly it is useful to replace the vector $D^{-1}Y$ by vector *X*. This saves the need to store both *Y* and *X*.

4.5 GAUSSIAN ELIMINATION METHOD

Gaussian elimination method is another method normally used for the solution of simultaneous equations. The method basically reduces a system of N simultaneous equations in N unknowns into a set of N simultaneous equations with the difference that the first equation, say, has N unknowns, the second has only (N - 1) unknowns, the third has (N - 2) unknowns and so on until the last equation has only one unknown. Using the last modified equation, the value of the Nth unknown can be determined. Having determined the Nth unknown, using the modified $(N - 1)^{th}$ equation, the $(N - 1)^{th}$ unknown can be determined. Continuing the process with the modified equations, the first unknown can be determined from the first equation. Consider the set of four equations.

$$\begin{bmatrix} a_{11} & a_{12} & a_{13} & a_{14} \\ a_{21} & a_{22} & a_{23} & a_{24} \\ a_{31} & a_{32} & a_{33} & a_{34} \\ a_{41} & a_{42} & a_{43} & a_{44} \end{bmatrix} \begin{bmatrix} x_1 \\ x_2 \\ x_3 \\ x_4 \end{bmatrix} = \begin{bmatrix} f_1 \\ f_2 \\ f_3 \\ f_4 \end{bmatrix}$$

Step 1: Eliminate x_1 from equations 2 to 4 using equation 1.
To eliminate x_1 from equation 2, subtract from equation 2, (a_{21}/a_{11}) times the first equation. The new second equation now becomes

$$(a_{22} - a_{12} \times a_{21}/a_{11})\, x_2 + (a_{23} - a_{13} \times a_{21}/a_{11})\, x_3 + (a_{24} - a_{14} \times a_{21}/a_{11})\, x_4 = (f_2 - f_1 \times a_{21}/a_{11})$$

Repeat the process with equation 3. Subtract from equation 3, (a_{31}/a_{11}) times the first equation. The new third equation becomes

$$(a_{32} - a_{12} \times a_{31}/a_{11})\, x_2 + (a_{33} - a_{13} \times a_{31}/a_{11})\, x_3 + (a_{34} - a_{14} \times a_{31}/a_{11})\, x_4 = (f_3 - f_1 \times a_{31}/a_{11})$$

Finally, subtract from equation 4, (a_{41}/a_{11}) times the first equation. The new fourth equation becomes

$$(a_{42} - a_{12} \times a_{41}/a_{11})\, x_2 + (a_{43} - a_{13} \times a_{41}/a_{11})\, x_3 + (a_{44} - a_{14} \times a_{41}/a_{11})\, x_4 = (f_4 - f_1 \times a_{41}/a_{11})$$

Thus at the end of the first cycle of elimination, the set of equations becomes

$$\begin{bmatrix} a_{11} & a_{12} & a_{13} & a_{14} \\ & b_{22} & b_{23} & b_{24} \\ & b_{32} & b_{33} & b_{34} \\ & b_{42} & b_{43} & b_{44} \end{bmatrix} \begin{bmatrix} x_1 \\ x_2 \\ x_3 \\ x_4 \end{bmatrix} = \begin{bmatrix} f_1 \\ g_2 \\ g_3 \\ g_4 \end{bmatrix}$$

where b_{ij} and g_i are the modified values at the end of the first cycle.
Step 2: In step 2, x_2 is eliminated from equations 3 and 4, using equation 2.
Subtract from equation 3, (b_{32}/b_{22}) times the second equation. The new equation 3 becomes

$$(b_{33} - b_{23} \times b_{32}/b_{22})\, x_3 + (b_{34} - b_{24} \times b_{32}/b_{22}) x_4 = (g_3 - g_2 \times b_{32}/b_{22})$$

Similarly subtract from equation 4, (b_{42}/b_{22}) times the second equation. The new equation 4 becomes

$$(b_{43} - b_{23} \times b_{42}/b_{22})\, x_3 + (b_{44} - b_{24} \times b_{42}/b_{22}) x_4 = (g_4 - g_2 \times b_{42}/b_{22})$$

Thus at the end of the second cycle of elimination, the set of equations becomes

$$\begin{bmatrix} a_{11} & a_{12} & a_{13} & a_{14} \\ & b_{22} & b_{23} & b_{24} \\ & & c_{33} & c_{34} \\ & & c_{43} & c_{44} \end{bmatrix} \begin{bmatrix} x_1 \\ x_2 \\ x_3 \\ x_4 \end{bmatrix} = \begin{bmatrix} f_1 \\ g_2 \\ h_3 \\ h_4 \end{bmatrix}$$

where c_{ij} and h_i are the further modified values at the end of the second cycle.

Step 3: In step 4, using equation 3, x_3 is eliminated from equation 4.
Subtract from equation 4, (c_{43}/c_{33}) times the third equation. The new equation 4 becomes
$(c_{44} - c_{34} \times c_{43}/c_{33})\, x_4 = (h_4 - h_3 \times c_{43}/c_{33})$
Thus at the end of the third cycle of elimination, the set of equations becomes

$$\begin{bmatrix} a_{11} & a_{12} & a_{13} & a_{14} \\ & b_{22} & b_{23} & b_{24} \\ & & c_{33} & c_{34} \\ & & & d_{44} \end{bmatrix} \begin{bmatrix} x_1 \\ x_2 \\ x_3 \\ x_4 \end{bmatrix} = \begin{bmatrix} f_1 \\ g_2 \\ h_3 \\ j_4 \end{bmatrix}$$

where d_{ij} and j_i are the modified values at the end of the third cycle.

Clearly, x_4 can be calculated from the modified fourth equation. Knowing x_4, x_3 can be calculated from the third equation, and so on until all the unknowns are determined.

The general formulae for carrying out Gaussian reduction are

$$a_{ij} = a_{ij} - a_{kj} \times (a_{ik}/a_{kk})$$
$$f_i = f - f_k \times (a_{ik}/a_{kk})$$
$$k = 1 \text{ to } (N - 1)$$
$$i = (k + 1) \text{ to } N$$
$$j = (k + 1) \text{ to } N$$

where N = number of unknowns, k = k^{th} unknown to be eliminated from the ith equations, where i goes from (k+1) to N equations.

It is worth noting that if the original matrix is symmetric, then the reduced matrix also remains symmetric. Clearly this fact can be taken advantage of in numerical computations.

The back substitution phase is given by

$$x_N = f_N/a_{NN}$$

$$x_i = \frac{[f_i - \sum_{j=i+1}^{N} a_{ij}\, x_j\,]}{a_{ii}}, \quad i = (N-1) \text{ to } 1$$

4.5.1 Gaussian Elimination Method: Example

Gaussian elimination is illustrated by a simple example. It is required to solve the set of equations defined by $KX = F$, where

$$K = \begin{bmatrix} 14 & -6 & 0 & 0 \\ -6 & 10 & -4 & 0 \\ 0 & -4 & 8 & -4 \\ 0 & 0 & -4 & 4 \end{bmatrix}, \quad F = \begin{bmatrix} 8 \\ 0 \\ 0 \\ 0 \end{bmatrix}$$

Step 1: As x_1 is absent from equations 3 and 4, it needs to be eliminated from equation 2 only. Subtract from equation 2, (-6/14) times equation 1. Therefore equation 2 becomes

$$[10 - (-6) \times (-6/14)]\, x_2 - 4x_3 = [0 - 8 \times (-6/14)]$$

Simplifying

$$7.4286\, x_2 - 4\, x_3 = 3.4286$$

The modified set of equations become

$$\begin{bmatrix} 14 & -6 & 0 & 0 \\ 0 & 7.4286 & -4 & 0 \\ 0 & -4 & 8 & -4 \\ 0 & 0 & -4 & 4 \end{bmatrix} \begin{bmatrix} x_1 \\ x_2 \\ x_3 \\ x_4 \end{bmatrix} = \begin{bmatrix} 8.0 \\ 3.4286 \\ 0 \\ 0 \end{bmatrix}$$

Note that ignoring the first row and column, the rest of the matrix is symmetric. This fact can be taken advantage of in working on only the upper triangular matrix.
Step 2: As x_2 is absent from equations 4, it needs to be eliminated from equation 3 only. Subtract from equation 3, (-4/7.4286) times equation 2. Therefore equation 3 becomes

$$[8 - (-4) \times (-4/7.4286)]\, x_3 - 4\, x_4 = [0 - 3.4286 \times (-4/7.4286)]$$

Simplifying

$$5.8462\, x_3 - 4\, x_4 = 1.8462$$

The modified set of equations become,

$$\begin{bmatrix} 14 & -6 & 0 & 0 \\ 0 & 7.4286 & -4 & 0 \\ 0 & 0 & 5.8462 & -4 \\ 0 & 0 & -4 & 4 \end{bmatrix} \begin{bmatrix} x_1 \\ x_2 \\ x_3 \\ x_4 \end{bmatrix} = \begin{bmatrix} 8.0 \\ 3.4286 \\ 1.8462 \\ 0 \end{bmatrix}$$

Note that ignoring the first two rows and columns, the rest of the matrix is symmetric.
Step 3: Subtract from equation 4, (-4/5.8462) times equation 3. Therefore equation 4 becomes

$$[4 - (-4) \times (-4/5.8462)]\, x_4 = [0 - 1.8462 \times (-4/5.8462)]$$

Simplifying

$$1.2632\, x_4 = 1.2632$$

The modified set of equations become

$$\begin{bmatrix} 14 & -6 & 0 & 0 \\ 0 & 7.4286 & -4 & 0 \\ 0 & 0 & 5.8462 & -4 \\ 0 & 0 & 0 & 1.2632 \end{bmatrix} \begin{bmatrix} x_1 \\ x_2 \\ x_3 \\ x_4 \end{bmatrix} = \begin{bmatrix} 8.0 \\ 3.4286 \\ 1.8462 \\ 1.2632 \end{bmatrix}$$

Step 4: Using back substitution, calculate the value of the unknowns.
From equation 4, 1.2632 x_4 = 1.2632, x_4 = 1.0
From equation 3, 5.8462 x_3 - 4.0 (x_4 = 1) = 1.8462, x_3 = 1.0
From equation 2, 7.4286 x_2 – 4.0 (x_3 = 1) = 3.4286, x_2 = 1.0
From the first equation, 14.0 x_1 – 6.0 (x_2 = 1) = 8.0, x_1 = 1.0

4.6 DETERMINANT OF A MATRIX

In the solution of eigen-value problems, it is frequently required to calculate the determinant of a matrix. In Section 3.2.1, general formulae were given for the calculation of the determinant of a square matrix. It was also pointed out that the general formula such as equation (3.3) were unsuitable for practical computation, except in the case of upper or lower triangular matrices. As shown in Section 3.2.1.1, for upper and lower triangular matrices, the determinant is given by the continued product of the diagonal terms of the matrix.

4.6.1 Determinant from Gaussian Elimination

The Gaussian elimination method is not only useful for the solution of simultaneous equations, it also gives a simple method for finding the determinant of a square matrix. One of the important properties of determinants is stated as follows.

If a matrix B is obtained from a square matrix A by adding one row (column) vector of A number α times a different row (column) vector, then the determinant of A and B are equal.

Gaussian reduction process basically involves modifying a given square matrix into an upper triangular matrix by simply adding a constant times of one equation to another equation. Therefore the determinant of the original square matrix is equal to the determinant of the modified upper triangular matrix. The determinant of the latter is given by the product of all diagonal terms of the reduced matrix. For the example in Section 4.2.1, the determinant is given by

$$|A| = 14 \times 7.4286 \times 5.8462 \times 1.2632 = 768.0$$

4.6.2 Determinant from Cholesky Factorisation

Like the Gaussian elimination method, Cholesky factorisation also enables the determinant of square matrices to be determined conveniently. One of the important properties of determinants is that if A and B are two square matrices of same size, then the determinant of (AB) = determinant of A x Determinant of B.

Cholesky factorisation of the stiffness matrix K results in the relationship

$$K = U^T U$$

$$\therefore \; |K| = |U^T \; U| = |U^T| \, |U| = |U|^2$$

For the example in Section 4.3.1.1, using the diagonal elements of U, the determinant of U is given by

$$|U| = 3.741657 \times 2.725541 \times 2.417882 \times 1.123903 = 27.712814$$

Hence the determinant of K is given by

$$|K| = |U|^2 = (27.712814)^2 = 768.0$$

4.7 EFFICIENT STORAGE SCHEMES

Figure 4.1 shows a symmetric structural stiffness matrix with 15 unknowns. In Fig. 4.1, D represents a diagonal element, x a non-zero element. N represents an element, that is zero to start with but will not remain zero during the elimination process. All other elements are zero to start with and will remain so throughout the elimination process. From an efficient storage point of view, all one needs to do is simply have storage space for only those elements, which can be non-zero.

4.7.1 Rectangular Format

One possible storage scheme is to employ a rectangular storage scheme, as shown in Fig. 4.2. The diagonal elements occupy the first column and only the upper triangular part of the matrix is stored. The number of columns is simply equal to half the maximum width of the band in which non-zero elements lie. For example, in the above example the maximum bandwidth of six occurs only in equations three and six. Thus the total storage requirement = 15 x 6 = 90. An element a_{ij} in the original scheme is in position $a_{i,(j-i+1)}$ in the new scheme. No element exists if (j – I + 1) is greater than the half bandwidth of 6. As can be seen, there are many locations that are not used in the computations. Thus although this is a reasonably efficient scheme, it can certainly be improved upon.

4.7.2 Vector or Skyline Format

A storage scheme, which is even more efficient than the rectangular scheme, is simply to store only those elements that need to be stored. In the above example the number of locations required can be verified to be equal to 60. In other words, even for a small problem, compared to the rectangular format, savings can be quite significant.

$$
\begin{bmatrix}
D_1 & x & x & x & & & & & & & & & & & \\
x & D_2 & x & x & & & & & & & & & & & \\
x & x & D_3 & x & x & x & x & x & & & & & & & \\
x & x & x & D_4 & x & x & x & N & & & & & & & \\
 & & x & x & D_5 & x & x & x & x & x & & & & & \\
 & & x & x & x & D_6 & x & x & x & N & & & & & \\
 & & x & x & x & x & D_7 & x & x & x & x & & & & \\
 & & x & N & x & x & x & D_8 & x & x & x & & & & \\
 & & & & x & x & x & x & D_9 & x & x & x & x & & \\
 & & & & x & N & x & x & x & D_{10} & x & x & x & & \\
 & & & & & & x & x & x & x & D_{11} & x & x & & \\
 & & & & & & & & x & x & x & D_{12} & x & x & x \\
 & & & & & & & & x & x & x & x & D_{13} & x & x \\
 & & & & & & & & & & & x & x & D_{14} & x \\
 & & & & & & & & & & & x & x & x & D_{15}
\end{bmatrix}
$$

Fig. 4.1 Structural stiffness matrix

$$
K = \begin{bmatrix}
D_1 & a_{12} & a_{13} & a_{14} & & \\
D_2 & a_{23} & a_{24} & & & \\
D_3 & a_{34} & a_{35} & a_{36} & a_{37} & a_{38} \\
D_4 & a_{45} & a_{46} & a_{47} & a_{48} & \\
D_5 & a_{56} & a_{57} & a_{58} & a_{59} & a_{5,10} \\
D_6 & a_{67} & a_{68} & a_{69} & a_{6,10} & \\
D_7 & a_{78} & a_{79} & a_{7,10} & a_{7,11} & \\
D_8 & a_{89} & a_{8,10} & a_{8,11} & & \\
D_9 & a_{9,10} & a_{9,11} & a_{9,12} & a_{9,13} & \\
D_{10} & a_{10,11} & a_{10,12} & a_{10,13} & & \\
D_{11} & a_{11,12} & a_{11,13} & & & \\
D_{12} & a_{12,13} & a_{12,14} & a_{12,15} & & \\
D_{13} & a_{13,14} & a_{13,15} & & & \\
D_{14} & a_{14,15} & & & & \\
D_{15} & & & & &
\end{bmatrix}
$$

Fig. 4.2 Rectangular storage scheme

The required elements are stored as a column vector, as shown in Fig. 4.3. In each column of the original stiffness matrix the elements stored are in the order: diagonal element, the element one row above the diagonal, the element two rows above the diagonal and so on until the last element in the column which needs to be stored. Fig. 4.3 shows the elements stored in the new storage scheme, generally known as the skyline or vector format.

In order to map the elements from the old scheme to the new scheme, it is necessary to know the location of each diagonal element and the number of non-zero elements above each diagonal. For this example, Table 4.1 gives the position of the diagonal elements and also the height of the non-zero column of numbers above each diagonal element. The position of any element a_{ij} in the upper triangle of the original scheme is in the skyline format at location Diagonal (j) + (j - i), j > i. In addition, (j - i + 1) $\leq$ Column(j).

4.7.2.1 Calculation of Position of Diagonal Elements and Column Height

In finite element analysis, it is fairly simple to calculate the values of DIAGONAL(i) and COLUMN(i). All one has to do is to start with DIAGONAL(1) = 1, DIAGOANL(2) = 2 and COLUMN(1) =1. Then for the ith equation, one examines all the elements to check which element(s) contribute the smallest variable to that equation. The difference between the equation in question and the smallest variable is the column length COLUMN(i) for that equation. This added to the value of DIAGONAL(i) becomes the value of DIAGONAL(i + 1).

$$
\begin{aligned}
K^T = [&D_1\ D_2\ a_{12}\ D_3\ a_{23}\ a_{13}\ D_4\ a_{34}\ a_{24}\ a_{14}\ D_5\ a_{45}\ a_{35}\ D_6\ a_{56}\ a_{46}\ a_{36} \\
&D_7\ a_{67}\ a_{57}\ a_{47}\ a_{37}\ D_8\ a_{78}\ a_{68}\ a_{58}\ a_{48}\ a_{38}\ D_9\ a_{89}\ a_{79}\ a_{69}\ a_{59} \\
&D_{10}\ a_{9,10}\ a_{8,10}\ a_{7,10}\ a_{6,10}\ a_{5,10}\ D_{11}\ a_{10,11}\ a_{9,11}\ a_{8,11}\ a_{7,11} \\
&D_{12}\ a_{11,12}\ a_{10,12}\ a_{9,12}\ D_{13}\ a_{12,13}\ a_{11,13}\ a_{10,13}\ a_{9,13}\ D_{14}\ a_{13,14}\ a_{12,14} \\
&D_{15}\ a_{14,15}\ a_{13,15}\ a_{12,15}\ ]
\end{aligned}
$$

Fig 4.3 Vector format

Table 4.1
Values of DIAGONAL(i) and CLUMN(i)

Column number, I	1	2	3	4	5	6	7	8	9	10	11	12	13	14	15
Location of the diagonal element DIAGONAL(I)	1	2	4	7	11	14	18	23	29	34	40	45	49	54	57
Height above the diagonal COLUMN(I)	1	2	3	4	3	4	5	6	5	6	5	4	5	3	4

4.8 FORTRAN-90 PROGRAMS

In this Section the print outs of fairly elementary computer programs written in FORTRAN 90 code are given. The programs have been intentionally kept simple so that the reader can see how the 'theory' is translated in to programs. A study of these programs will greatly enhance the understanding of the more complex programs to be discussed in the later Sections of the book.

4.8.1 Program CHOLFACT

```
!**********************************************************
!SOLUTION OF SIMULTANEOUS EQUATIONS USING CHOLESKY FACTORISATION
! MATRIX IS ASSUMED TO BE SYMMETRIC AND POSITIVE DEFINITE
!*******************************************************************
      PROGRAM CHOLFACT
!*******VARIABLE DECLARATION*************************************
      IMPLICIT NONE
      REAL:: SUM,FACTOR
      INTEGER:: I,J,IM1,IP1,K,N,NM1
      REAL,ALLOCATABLE:: A(:,:),U(:,:), F(:),Y(:),X(:)
!*******FILE MANAGEMENT******************************************
      CHARACTER(LEN=40)::FILEIN, FILEOUT
      PRINT*,'NAME OF INPUT FILE'
      READ*,FILEIN
      PRINT*,'NAME OF OUTPUT FILE'
      READ*,FILEOUT
      OPEN(UNIT=1, FILE =FILEIN,STATUS='OLD',ACTION='READ')
      OPEN(UNIT=2, FILE =FILEOUT,STATUS='REPLACE',ACTION='WRITE')
!*******INPUT DATA***********************************************
      READ(1,*) N
      ALLOCATE(A(N,N),U(N,N),F(N),Y(N),X(N))
!*******INPUT MATRIX TO BE FACTORISED. INPUT THE DATA BY ROWS
      READ(1,*) ((A(I,J),J=I,N),I=1,N)
!*******INPUT THE LOAD VECTOR************************************
      READ(1,*)(F(I),I=1,N)
!*******PRINT THE INPUT MATRIX***********************************
      WRITE(2,*)' MATRIX TO BE FACTORISED'
      DO I = 1,N
      WRITE(2,'(6(4X,F10.4))')(A(I,J), J = I,N)
      END DO
!*******PRINT THE LOAD VECTOR************************************
      WRITE(2,*)' LOAD VECTOR'
      WRITE(2,'(6(4X,F10.4))')(F(I), I = 1,N)
!*******CALCULATE THE UPPER TRIANGULAR MATRIX U*************
!*******CALCULATE THE FIRST ROW OF U*****************************
      U(1,1) = SQRT(A(1,1))
      FACTOR=1/U(1,1)
      DO J=2,N
      U(1,J)=A(1,J)*FACTOR
      END DO
!*******CALCULATE THE ELEMENTS OF ROWS 2 TO N***************
      DO I = 2,N
      IP1=I+1
```

```
!*******CALCULATE THE DIAGONAL ELEMENT**********************
      IM1=I-1
      SUM=0
      DO K= 1,IM1
      SUM = SUM + U(K,I)*U(K,I)
      END DO
      U(I,I)=SQRT(A(I,I)-SUM)
      IF(I==N)EXIT
!*******CALCULATE THE OFF DIAGONAL ELEMENTS******************
      FACTOR=1.0/U(I,I)
      DO J= IP1,N
      SUM = 0
      DO K = 1,IM1
      SUM = SUM + U(K,I)*U(K,J)
      END DO
      U(I,J)=(A(I,J)-SUM)*FACTOR
      END DO
      END DO
!*******PRINT U MATRIX*******************************************
      WRITE(2,*) 'U MATRIX'
      DO I = 1,N
      WRITE(2,*)(U(I,J), J=I,N)
      END DO
!*******USE FORWARD SUBSTITUTION TO SOLVE FOR Y*************
      Y(1)=F(1)/U(1,1)
      DO I = 2,N
      SUM = 0
      IM1=I-1
      DO J= 1,IM1
      SUM = SUM + U(J,I)*Y(J)
      END DO
      Y(I)= (F(I)-SUM)/U(I,I)
      END DO
!*******USE BACK SUBSTITUTION TO SOLVE FOR X******************
      X(N)=Y(N)/U(N,N)
      NM1=N-1
      DO I= NM1,1,-1
      SUM = 0
      IP1=I+1
      DO J= N,IP1,-1
      SUM=SUM + U(I,J)*X(J)
      END DO
      X(I)=(Y(I)-SUM)/U(I,I)
      END DO
!*******PRINT DISPALCEMENTS**************************************
      WRITE(2,*)'DISPLACEMENTS'
      WRITE(2,'(6(4X,F10.4))')(X(I), I = 1,N)
          END PROGRAM CHOLFACT
```

4.8.2 Program CROUT

```
!*****************************************************************
! SOLUTION OF SIMULTANEOUS EQUATIONS USING CROUT
! FACTORISATION
! A =U TRANSFORM *D*U
! A IS ASSUMED TO BE SYMMETRIC BUT NEED NOT BE POSITIVE DEFINITE
```

```
!*******************************************************************
      PROGRAM CROUT
!******VARIABLE DECLARATION*************************************
      IMPLICIT NONE
      REAL:: SUM,FACTOR
      INTEGER:: I,J,JM1,IM1,IP1,K,N,NM1
      REAL,ALLOCATABLE:: A(:,:),U(:,:), F(:),X(:),Y(:)
!*******FILE MANAGEMENT*****************************************
      CHARACTER(LEN=40)::FILEIN, FILEOUT
      PRINT*,'NAME OF INPUT FILE'
      READ*,FILEIN
      PRINT*,'NAME OF OUTPUT FILE'
      READ*,FILEOUT
      OPEN(UNIT=1, FILE =FILEIN,STATUS='OLD',ACTION='READ')
      OPEN(UNIT=2, FILE =FILEOUT,STATUS='REPLACE',ACTION='WRITE')
!******INPUT DATA************************************************
      READ(1,*) N
      NM1 = N-1
      ALLOCATE(A(N,N),U(N,N),F(N),X(N),Y(N))
!*******INPUT MATRIX TO BE FACTORISED*************************
      READ(1,*) ((A(I,J), J=I,N),I=1,N)
!*******INPUT LOAD VECTOR*************************************
      READ(1,*)(F(I),I=1,N)
!*******PRINT THE INPUT MATRIX*********************************
      WRITE(2,*)' MATRIX TO BE FACTORISED'
      DO I = 1,N
      WRITE(2,'(6(4X,F10.4))')(A(I,J), J = I,N)
      END DO
!*******PRINT LOAD VECTOR*************************************
      WRITE(2,'(6(4X,F10.4))')(F(I), I = 1,N)
!*******SINCE DIAGONAL ELEMENTS OF U ARE ALL EQUAL TO UNITY,
!*******ELEMENT D(I,I) IS STORED IN U(I,I)*************************
!*******CALCULATE THE UPPER TRIANGULAR MATRIX U*************
!*******CALCULATE THE FIRST ROW OF U***************************
      U(1,1) = A(1,1)
      FACTOR=1/U(1,1)
      DO J=2,N
      U(1,J)=A(1,J)*FACTOR
      END DO
!*******CALCULATE THE ELEMENTS OF ROWS 2 TO N**********
      DO I = 2,N
      IP1=I+1
!*******CALCULATE THE DIAGONAL ELEMENT D(I,I) *************
!*******AND STORE IT IN U(I,I)**********************************
      IM1=I-1
      SUM=0
      DO K= 1,IM1
!*******REMEMBER THAT ELEMENT D(K,K) IS STORED IN U(K,K)******
      SUM = SUM + U(K,I)*U(K,I)*U(K,K)
      END DO
      U(I,I)= A(I,I)-SUM
      IF(I==N)EXIT
!*******CALCULATE THE OFF DIAGONAL ELEMENTS**************
!*******REMEMBER THAT ELEMENT D(I,I) IS STORED IN U(I,I)*********
      FACTOR=1.0/U(I,I)
      DO J= IP1,N
      SUM = 0
```

```
      DO K = 1,IM1
      SUM = SUM + U(K,I)*U(K,J)*U(K,K)
      END DO
      U(I,J)=(A(I,J)-SUM)*FACTOR
      END DO
      END DO
!*******PRINT U MATRIX*******************************************
      WRITE(2,*) 'U MATRIX'
      WRITE(2,*) 'NOTE DIAGONAL ELEMENTS OF U ARE UNITY BUT AS
         PRINTED'
       WRITE(2,*)'THEY ARE THE ELEMETS OF D'
      DO I = 1,N
      WRITE(2,*)(U(I,J), J=I,N)
      END DO
!*******USE FORWARD SUBSTITUTION TO SOLVE U TRANSFORM Y = F**
      Y(1)=F(1)
      DO I= 2,N
      SUM = 0
      IM1=I-1
      DO J= 1,IM1
      SUM = SUM + U(J,I)*Y(J)
      END DO
      Y(I)=F(I)-SUM
      END DO
!*******CALCULATE D INVERSE Y. ********************************
!*******DIAGONALS OF D IN DIAGONALS OF U MATRIX ************
      DO I = 1,N
      Y(I)=Y(I)/U(I,I)
      END DO
!*******USE BACK SUBSTITUTION TO SOLVE U X = D INVERSE Y******
      X(N)=Y(N)
      NM1=N-1
      DO I = NM1,1,-1
      SUM = 0
      IP1=I+1
      DO J = N,IP1,-1
      SUM = SUM +U(I,J)*X(J)
      END DO
      X(I)=Y(I)-SUM
      END DO
!*******PRINT DISPLACEMENTS***********************************
      WRITE(2,*)'DISPLACEMENTS'
      WRITE(2,'(6(4X,F10.4))')(X(I),I=1,N)
      END PROGRAM CROUT
```

4.8.3 Program GAUSS

```
!******************************************************************
!*****THIS PROGRAM SOLVES SIMULTANEOUS EQUATIONS BY GAUSSIAN
!      ELIMINATION  METHOD
!******************************************************************
!*****VARIABLE DECLARATION
      IMPLICIT NONE
      INTEGER::I,J,K,N,NM1,KP1,IP1
      REAL::FACTOR, SUM,DETERMINANT
      REAL,ALLOCATABLE::A(:,:),F(:),X(:)
```

```
!*******FILE MANAGEMENT*******************************************
      CHARACTER(LEN=40)::FILEIN, FILEOUT
      PRINT*,'NAME OF INPUT FILE'
      READ*,FILEIN
      PRINT*,'NAME OF OUTPUT FILE'
      READ*,FILEOUT
      OPEN(UNIT=1, FILE =FILEIN,STATUS='OLD',ACTION='READ')
      OPEN(UNIT=2, FILE =FILEOUT,STATUS='REPLACE',ACTION='WRITE')
!*******INPUT THE NUMBER OF UNKNOWNS*********************
      READ(1,*)N
      ALLOCATE(A(N,N),F(N),X(N))
!******INPUT THE MATRIX ROW BY ROW**************************
      READ(1,*)((A(I,J),J=1,N),I=1,N)
!******INPUT THE LOAD VECTOR********************************
      READ(1,*)(F(I),I=1,N)
!*******ECHO THE INPUT******************************************
      WRITE(2,'(/,A)')'INPUT MATRIX'
      DO I = 1,N
      WRITE(2,'(6(F10.4,4X))')(A(I,J),J=1,N)
      END DO
      WRITE(2,'(/,A)')'INPUT LOAD VECTOR'
      DO I = 1,N
      WRITE(2,'(F10.4)')F(I)
      END DO
!*******GAUSSIAN REDUCTION PHASE****************************
      NM1=N-1
!*******CHOOSE THE EQUATION WITH WHICH TO ELIMINATE********
!******* THE VARIABLES******************************************
      DO K=1,NM1
      KP1 = K+1
      DO I =KP1,N
      FACTOR = A(I,K)/A(K,K)
      F(I)=F(I)-F(K)*FACTOR
      DO J = KP1,N
      A(I,J) = A(I,J) - A(K,J)*FACTOR
      END DO
      END DO
      END DO
!*******BACK SUBSTITUTION PHASE********************************
      X(N) = F(N)/A(N,N)
      DO I = NM1,1,-1
      IP1=I+1
      SUM = 0
      DO J = IP1,N
      SUM = SUM + A(I,J)*X(J)
      END DO
      X(I)=(F(I)-SUM)/A(I,I)
      END DO
!******CALCULATE THE DETERMINANT************************
      DETERMINANT = 1
      DO I = 1,N
      DETERMINANT = DETERMINANT*A(I,I)
      END DO
!*******PRINT OUT THE RESULTS********************************
      WRITE(2,'(/,A,E14.6)')'DETERMINANT = ',DETERMINANT
      WRITE(2,'(/,A)')' VARIABLE NO.    VALUE'
      DO I = 1,N
```

```
WRITE(2,'(I5,4X,F10.4)')I,X(I)
END DO
END PROGRAM GAUSS
```

CHAPTER 5

SOLUTION METHODS FOR THE GENERAL EIGEN-VALUE PROBLEM

5.1 INTRODUCTION

Discussion in Chapter 3 has highlighted the importance of the eigen-values and the corresponding eigen-vectors in the solution of linear problems in structural dynamics. In this chapter, many different procedures for the determination of the eigen-pairs of large-scale problems defined by equation (3.2) for a general eigen-value problem (GEP) will be discussed. The conversion, if required, from general to special eigen-value problem is effected through equation (3.9).

There are basically four approaches to the solution of the eigen-value problem. They are

- Vector Iterative Methods: These are best suited for determining the eigen-pairs at the lower end of the spectrum. An example of this is the inverse iteration method discussed in Section 5.3.
- Transformation Methods: These are used for determining all the eigen-pairs of a problem. In practice, these are mainly used for the complete solution of a reduced version of the original problem. Examples of this are the general Jacobi method and Lanczos method, discussed in Sections 5.6 and 5.8 respectively.
- Sturm Sequence Method: This method is basically a search for the roots of the characteristic equation using Sturm sequence property. This method is discussed in Section 5.9.
- Factorisation Method: The most well known of this class is the QR factorisation. This is discussed in Section 5.10.

5.2 FORWARD ITERATION

One of the most important properties of the eigen-value problem is that successive iterations on an arbitrary vector will converge to the eigen-vector corresponding to the highest eigen-value. This important property is the basis of many methods used in the solution of large-scale eigen-value problems.

Consider the special eigen-value problem

$$A\,Y = \lambda\,Y \tag{5.1}$$

Let Y_l be an arbitrary vector. Since Y_l is an arbitrary vector, as shown in Section 3.4.3, it can be expressed in terms of the eigen-vectors as

$$Y_1 = \sum_{1}^{N} C_i \ \phi_i \tag{5.2}$$

where, C_i are constants and φ_i are the eigen-vectors. The subscript 1 on Y indicates that it is the value of a vector Y at the start of a set of iterations.
Pre-multiplying both sides of equation (5.2) by A

$$AY_1 = \sum_{i=1}^{N} C_i \ A\Phi_i \tag{5.3}$$

Since φ_i is an eigen-vector

$$A\Phi_i = \lambda_i \ \Phi_i \tag{5.4}$$

Substituting equation (5.4) in equation (5.3)

$$AY_1 = \sum_{i=1}^{N} C_i \ \lambda_i \ \Phi_i \tag{5.5}$$

Taking λ_1, the largest eigen-value as a common factor, equation (5.5) becomes

$$AY_1 = \lambda_1 \sum_{i=1}^{N} C_i \frac{\lambda_i}{\lambda_1} \ \Phi_i \tag{5.6}$$

This completes the first iteration. For the second iteration, the arbitrary vector Y_2 is chosen as equal to AY_1, after normalising it to unit length. Following the same steps as before, at the end of second iteration

$$AY_2 = \lambda_1 \sum_{i=1}^{N} C_i \left[\frac{\lambda_i}{\lambda_1}\right]^2 \Phi_i \tag{5.7}$$

If this process is continued, then after m iterations

$$AY_m = \lambda_1 \sum_{i=1}^{N} C_i \left[\frac{\lambda_i}{\lambda_1}\right]^m \Phi_i \tag{5.8}$$

Since λ_1 is the highest eigen-value

$$m \to \infty, \quad \left[\frac{\lambda_i}{\lambda_1}\right]^m \to 0 \tag{5.9}$$

and

$$Y_m \to \phi_1 \tag{5.10}$$

This method of obtaining the highest eigen-vector is also called power or Stodola method.

Because of the fact that in vibration problems, the interest lies in the lower end of eigen-frequencies, all operations are done on the matrix

$$A = U^T M \, U, \; K = U^T U, \; \lambda = 1/\omega^2 \tag{5.11}$$

When the forward iteration is operated on equation (5.11), it is known as inverse iteration. The highest eigen-value $\lambda_1 = 1/\omega^2$ will be equal to the lowest frequency.

5.2.1 Forward Iteration: Example

Determine the smallest value of ω^2 for the three-storey shear frame shown in Fig. 3.1. It is given that

$$K = \begin{bmatrix} 14 & -6 & 0 \\ -6 & 10 & -4 \\ 0 & -4 & 4 \end{bmatrix}, \quad M = \begin{bmatrix} 4 & 0 & 0 \\ 0 & 4 & 0 \\ 0 & 0 & 2 \end{bmatrix}$$

Using the Cholesky factorisation discussed in Section 4.3, $K = U^T U$.

$$u_{11} = \sqrt{k_{11}} = \sqrt{14} = 3.7417, \; u_{12} = \frac{k_{12}}{u_{11}} = -1.6036, \; u_{13} = \frac{k_{13}}{u_{11}} = 0$$

$$u_{22} = \sqrt{(k_{22} - u_{12}^2)} = 2.7255, \; u_{23} = \frac{(k_{23} - u_{12}\, u_{13})}{u_{22}} = -1.4676$$

$$u_{33} = \sqrt{(k_{33} - u_{13}^2 - u_{23}^2)} = 1.3587$$

$$U = \begin{bmatrix} 3.741657 & -1.603568 & 0 \\ 0 & 2.725541 & -1.467600 \\ 0 & 0 & 1.358732 \end{bmatrix}$$

$$U^T = \begin{bmatrix} 3.741657 & 0 & 0 \\ -1.603568 & 2.725541 & 0 \\ 0 & -1.467600 & 1.358732 \end{bmatrix}$$

<u>Iteration 1:</u>

Step 1: Let the arbitrary vector be

$$Y = \begin{bmatrix} 1 \\ 1 \\ 1 \end{bmatrix} = \sqrt{3} \begin{bmatrix} 0.5774 \\ 0.5774 \\ 0.5774 \end{bmatrix}$$

Step 2: Using the normalised value of Y^T =[0.5774, 0.5774, 0.5774], solve for Z from $U\,Z = Y$ by using back substitution as explained in Section 4.3.2.1.

$$\begin{bmatrix} 3.741657 & -1.603568 & 0 \\ 0 & 2.725541 & -1.467600 \\ 0 & 0 & 1.358732 \end{bmatrix} \begin{bmatrix} z_1 \\ z_2 \\ z_3 \end{bmatrix} = \begin{bmatrix} 0.5774 \\ 0.5774 \\ 0.5774 \end{bmatrix}, \; Z = \begin{bmatrix} 0.34318 \\ 0.44067 \\ 0.42496 \end{bmatrix}$$

Step 3: Calculate $R = M\,Z$

$$R = \begin{bmatrix} 4 & 0 & 0 \\ 0 & 4 & 0 \\ 0 & 0 & 2 \end{bmatrix} \begin{bmatrix} 0.34318 \\ 0.44067 \\ 0.42496 \end{bmatrix} = \begin{bmatrix} 1.37270 \\ 1.76268 \\ 0.84992 \end{bmatrix}$$

Step 4: Using forward substitution, solve for S from $U^T S = R$ as explained in Section 4.3.2.1.

$$\begin{bmatrix} 3.741657 & 0 & 0 \\ -1.603568 & 2.725541 & 0 \\ 0 & -1.467600 & 1.358732 \end{bmatrix} \begin{bmatrix} s_1 \\ s_2 \\ s_3 \end{bmatrix} = \begin{bmatrix} 1.37270 \\ 1.76268 \\ 0.84992 \end{bmatrix}, S = \begin{bmatrix} 0.36687 \\ 0.86257 \\ 1.55721 \end{bmatrix}$$

Step 5: Normalise S to unit length. This gives the first estimate of eigen-vector and eigen-value.

$$S = \begin{bmatrix} 0.36687 \\ 0.86257 \\ 1.55721 \end{bmatrix} = 1.8176 \begin{bmatrix} 0.2019 \\ 0.4748 \\ 0.8568 \end{bmatrix}$$

The first estimate of $\lambda = 1/\omega^2$ is 1.8176 or $\omega^2 = 0.5502$. Compared with the exact value of $\omega^2 = 0.4606$ determined in Section 3.2.4, the value at first iteration is about 20% too high.

Iteration 2:

Step 1: Start with the estimated vector $Y^T = [0.2019 \quad 0.4748 \quad 0.8568]$ as the arbitrary vector.

Step 2: Calculate Z from, $U Z = Y$

$$\begin{bmatrix} 3.741657 & -1.603568 & 0 \\ 0 & 2.725541 & -1.467600 \\ 0 & 0 & 1.358732 \end{bmatrix} \begin{bmatrix} z_1 \\ z_2 \\ z_3 \end{bmatrix} = \begin{bmatrix} 0.2019 \\ 0.4748 \\ 0.8568 \end{bmatrix}, Z = \begin{bmatrix} 0.27414 \\ 0.51375 \\ 0.63059 \end{bmatrix}$$

Step 3: Calculate $R = M Z$

$$R = \begin{bmatrix} 4 & 0 & 0 \\ 0 & 4 & 0 \\ 0 & 0 & 2 \end{bmatrix} \begin{bmatrix} 0.27414 \\ 0.51375 \\ 0.63059 \end{bmatrix} = \begin{bmatrix} 1.09656 \\ 2.05500 \\ 1.26118 \end{bmatrix}$$

Step 4: Solve $U^T S = R$

$$U^T P = \begin{bmatrix} 3.741657 & 0 & 0 \\ -1.603568 & 2.725541 & 0 \\ 0 & -1.467600 & 1.358732 \end{bmatrix} \begin{bmatrix} s_1 \\ s_2 \\ s_3 \end{bmatrix} = \begin{bmatrix} 1.09656 \\ 2.05500 \\ 1.26118 \end{bmatrix}, S = \begin{bmatrix} 0.29307 \\ 0.92641 \\ 1.92884 \end{bmatrix}$$

Step 5: Normalise S to unit length. This gives the second estimate of eigen-vector and eigen-value.

$$S = \begin{bmatrix} 0.29307 \\ 0.92641 \\ 1.92884 \end{bmatrix} = 2.1598 \begin{bmatrix} 0.1357 \\ 0.4289 \\ 0.8931 \end{bmatrix}$$

The second estimate of $\lambda = 1/\omega^2$ is 2.1598 or $\omega^2 = 0.4630$. Compared with the exact value of $\omega^2 = 0.4606$ determined in Section 3.2.4, the value at second iteration is about 0.50 % too high.

Iteration 3:

Step 1: Start with $Y^T = [0.1357 \quad 0.4289 \quad 0.8931]$.

Step 2: Solve $U Z = Y$

$$\begin{bmatrix} 3.741657 & -1.603568 & 0 \\ 0 & 2.725541 & -1.467600 \\ 0 & 0 & 1.358732 \end{bmatrix} \begin{bmatrix} z_1 \\ z_2 \\ z_3 \end{bmatrix} = \begin{bmatrix} 0.1357 \\ 0.4289 \\ 0.8931 \end{bmatrix}, Z = \begin{bmatrix} 0.25540 \\ 0.51130 \\ 0.65730 \end{bmatrix}$$

Step 3: Calculate $R = M Z$

$$M Z = \begin{bmatrix} 4 & 0 & 0 \\ 0 & 4 & 0 \\ 0 & 0 & 2 \end{bmatrix} \begin{bmatrix} 0.25540 \\ 0.51130 \\ 0.65730 \end{bmatrix} = \begin{bmatrix} 1.02158 \\ 2.04519 \\ 1.31461 \end{bmatrix}$$

Step 4: Solve $U^T S = R$

$$\begin{bmatrix} 3.741657 & 0 & 0 \\ -1.603568 & 2.725541 & 0 \\ 0 & -1.467600 & 1.358732 \end{bmatrix} \begin{bmatrix} s_1 \\ s_2 \\ s_3 \end{bmatrix} = \begin{bmatrix} 1.02158 \\ 2.04519 \\ 1.31461 \end{bmatrix}, S = \begin{bmatrix} 0.27303 \\ 0.91102 \\ 1.95154 \end{bmatrix}$$

Step 5: Normalise S to unit length and calculate the third estimate of the eigen-value.

$$S = \begin{bmatrix} 0.27303 \\ 0.91102 \\ 1.95154 \end{bmatrix} = 2.1710 \begin{bmatrix} 0.1258 \\ 0.4196 \\ 0.8989 \end{bmatrix}$$

The third estimate of $\lambda = 1/\omega^2$ is 2.1710 or $\omega^2 = 0.4606$. This compares with the exact value of 0.46056.

Calculation of eigen-vector φ: Y corresponds to the eigen-vector of A. The eigen-vectors of $A\ Y = \lambda\ Y$ are related to eigen-vectors of $K\ X = \omega^2\ M\ X$, as explained in Section 4.3.2, through $U\ X = Y$. Therefore using the value of Y^T = [0.1258, 0.4196, 0.8989], solve for X using back substitution on $U X = Y$.

$$\begin{bmatrix} 3.741657 & -1.603568 & 0 \\ 0 & 2.725541 & -1.467600 \\ 0 & 0 & 1.358732 \end{bmatrix} \begin{bmatrix} x_1 \\ x_2 \\ x_3 \end{bmatrix} = \begin{bmatrix} 0.1258 \\ 0.4196 \\ 0.8989 \end{bmatrix}, X = \begin{bmatrix} 0.25227 \\ 0.51018 \\ 0.66157 \end{bmatrix}$$

Finally, M normalise X to obtain the eigen-vector φ.

$$\begin{bmatrix} 0.25227 & 0.51018 & 0.66157 \end{bmatrix} \begin{bmatrix} 4 & 0 & 0 \\ 0 & 4 & 0 \\ 0 & 0 & 2 \end{bmatrix} \begin{bmatrix} 0.25227 \\ 0.51018 \\ 0.66157 \end{bmatrix} = 2.17105$$

$$\phi = \frac{1}{\sqrt{2.17105}} \begin{bmatrix} 0.25227 \\ 0.51018 \\ 0.66157 \end{bmatrix} = \begin{bmatrix} 0.1712 \\ 0.3463 \\ 0.4490 \end{bmatrix}$$

After only three iterations, the value of the vector φ^T = [0.1712, 0.3463, 0.4490]. This compares very well with the exact value of φ^T = [0.1708, 0.3461, 0.4496] calculated in Section 3.2.5.1.

5.3 INVERSE ITERATION AND HIGHER FREQUENCIES

If one is interested in evaluating higher frequencies, then it is necessary to operate on a vector that is orthogonal to the vector corresponding to the highest eigen-value. If this is not done, then in inverse iteration, the iteration simply converges to the highest eigen-value. Fortunately, given an arbitrary vector, it is possible to generate another vector, which is normal to the given vector. The procedure is known as Gram-Schmidt method of orthogonalisation.

5.3.1 Gram-Schmidt Method of Orthogonalisation

One of the most commonly employed methods of generating a vector, which is normal to a given set of vector, is the Gram-Schmidt method. The necessary equations can be derived as follows.

Let Y_1 be a given unit vector. It is required to generate a vector Y_2, orthogonal to Y_1. Let W be an arbitrary vector. Let the three vectors be linearly related by

$$Y_2 = W + \alpha Y \tag{5.12}$$

Where α is a constant to be determined.
Since, Y_1 and Y_2 are orthogonal, $Y_1^T Y_2 = 0$. Therefore

$$Y_1^T Y_2 = Y_1^T W + \alpha Y_1^T Y_1 = 0 \tag{5.13}$$

Since Y_1 is a unit vector, $Y_1^T Y_1 = 1$. Therefore, from equation (5.13)

$$\alpha = - Y_1^T W \tag{5.14}$$

If we have to generate a vector that is orthogonal to a set of orthogonal unit vectors Y_i, i = 1 to m, let

$$Y_{m+1} = W + \Sigma \alpha_i Y_i \tag{5.15}$$

If $Y_i^T Y_i = 1$ and $Y_i^T Y_i = 0$, $i \neq j$ then

$$\alpha_i = - Y_i^T W \tag{5.16}$$

5.3.1.1 *Orthogonalisation: Example*

Use of the Gram-Schmidt procedure is illustrated by two examples.
Example 1: Given a unit vector $Y_1^T = [0.1258, 0.4196, 0.8989]$, generate a vector Y_2 orthogonal to Y_1.
The steps to be followed are as follows.
Step 1: Choose an arbitrary vector , $W^T = [1, 1, 1]$.
Step 2: Calculate α.

$$\alpha = -Y_1^T W = -\begin{bmatrix}0.1258 & 0.4196 & 0.8989\end{bmatrix}\begin{bmatrix}1\\1\\1\end{bmatrix} = -1.4443$$

Step 3: Calculate Y_2.

$$Y_2 = W + \alpha\ Y_1 = \begin{bmatrix}1\\1\\1\end{bmatrix} - 1.4443\begin{bmatrix}0.1258\\0.4196\\0.8989\end{bmatrix} = \begin{bmatrix}0.8183\\0.3940\\-0.2983\end{bmatrix}$$

Step 4: Normalise Y_2 to unit length.

$$Y_2^T\ Y_2 = \begin{bmatrix}0.8183 & 0.3940 & -0.2983\end{bmatrix}\begin{bmatrix}0.8183\\0.3940\\-0.2983\end{bmatrix} = 0.9138$$

$$Y_2 = \frac{1}{\sqrt{0.9138}}\begin{bmatrix}0.8183\\0.3940\\-0.2983\end{bmatrix} = \begin{bmatrix}0.8560\\0.4122\\-0.3121\end{bmatrix}$$

Step 5: Check the orthogonality of Y_1 and Y_2.

$$\begin{bmatrix}0.1258 & 0.4196 & 0.8989\end{bmatrix}\begin{bmatrix}0.8560\\0.4122\\-0.3121\end{bmatrix} = 0$$

The vector Y_1 and Y_2 are orthogonal to within the limits of accuracy desired.

<u>Example 2</u>: Generate a vector Y_3, orthogonal to the unit orthogonal vectors Y_1.and Y_2 given in Example 1 above.

Step 1: Start with an arbitrary vector W, different from the one used before for the calculation of Y_2. Let $W^T = [1, -1, 1]$.

Step 2: Calculate α_1 and α_2.

$$\alpha_1 = -Y_1^T\ W = -\begin{bmatrix}0.1258 & 0.4196 & 0.8989\end{bmatrix}\begin{bmatrix}1\\-1\\1\end{bmatrix} = -0.6051$$

$$\alpha_2 = -Y_2^T\ W = -\begin{bmatrix}0.8545 & 0.4141 & -0.3135\end{bmatrix}\begin{bmatrix}1\\-1\\1\end{bmatrix} = -0.1269$$

Step 3: Calculate $Y_3 = W + \alpha_1\ Y_1 + \alpha_2\ Y_2$

$$Y_3 = \begin{bmatrix}1\\-1\\1\end{bmatrix} - 0.6051\begin{bmatrix}0.1258\\0.4196\\0.8989\end{bmatrix} - 0.1269\begin{bmatrix}0.8545\\0.4141\\-0.3135\end{bmatrix} = \begin{bmatrix}0.81544\\-1.30645\\0.49586\end{bmatrix}$$

Step 4: Normalise Y_3 to unit length.

$$Y_3^T\ Y_3 = \begin{bmatrix}0.81544 & -1.30645 & 0.49586\end{bmatrix}\begin{bmatrix}0.81544\\-1.30645\\0.49586\end{bmatrix} = 2.61763$$

$$Y_3 = \frac{1}{\sqrt{2.61763}}\begin{bmatrix} 0.81544 \\ -1.30645 \\ 0.49586 \end{bmatrix} = \begin{bmatrix} 0.5040 \\ -0.8075 \\ 0.3065 \end{bmatrix}$$

Therefore the three unit orthogonal vectors are given by

$$Y_3 = \begin{bmatrix} 0.5040 \\ -0.8075 \\ 0.3065 \end{bmatrix},\ Y_2 = \begin{bmatrix} 0.8545 \\ 0.4141 \\ -0.3135 \end{bmatrix},\ Y_1 = \begin{bmatrix} 0.1258 \\ 0.4196 \\ 0.8989 \end{bmatrix}$$

Clearly the procedure can be continued deriving as many orthogonal vectors as required.

5.4 CALCULATION OF HIGHER FREQUENCIES

The procedure for calculating the second lowest frequency is identical to the procedure used for calculating the lowest frequency described in Section (5.2.1). The only difference is that after choosing an arbitrary vector, another vector orthogonal to the eigenvector corresponding to the lowest frequency is generated and this is used as the vector for iteration purposes.

5.4.1 Calculation of Higher Frequencies: Example

The procedure for calculating the higher natural frequencies is illustrated by an example.

Example : It is required to calculate the second frequency for the three-storey shear frame considered in Section 5.2.1. It is given that the eigen-vector for the first frequency is equal to $Y_1^T = [0.1258,\ 0.4196, 0.8989]$.

Iteration 1:

Step 1: Following the Gram-Schmidt orthogonalisation procedure in Section 5.3.1, using an arbitrary vector $W^T = [1,\ 1,\ 1]$, generate Y_2 orthogonal to Y_1. Using the calculation in Section 5.3.1.1,

$$\alpha = -Y_1^T W = -[0.1258 \quad 0.4196 \quad 0.8989]\begin{bmatrix} 1 \\ 1 \\ 1 \end{bmatrix} = -1.4443$$

$$Y_2 = W + \alpha\, Y_1 = \begin{bmatrix} 1 \\ 1 \\ 1 \end{bmatrix} - 1.4443\begin{bmatrix} 0.1258 \\ 0.4196 \\ 0.8989 \end{bmatrix} = \begin{bmatrix} 0.8183 \\ 0.3940 \\ -0.2983 \end{bmatrix},\ Y_2^T\, Y_2 = 0.9138$$

$$Y_2 = \frac{1}{\sqrt{0.9138}}\begin{bmatrix} 0.8183 \\ 0.3940 \\ -0.2983 \end{bmatrix} = \begin{bmatrix} 0.8560 \\ 0.4122 \\ -0.3121 \end{bmatrix}$$

Step 2: Using the normalised value of Y_2, calculate Z from $UZ = Y$

$$\begin{bmatrix} 3.741657 & -1.603568 & 0 \\ 0 & 2.725541 & -1.467600 \\ 0 & 0 & 1.358732 \end{bmatrix} \begin{bmatrix} z_1 \\ z_2 \\ z_3 \end{bmatrix} = \begin{bmatrix} 0.8560 \\ 0.4122 \\ -0.3121 \end{bmatrix}, Z = \begin{bmatrix} 0.2406 \\ 0.0276 \\ -0.2297 \end{bmatrix}$$

Step 3: Calculate $R = MZ$

$$R = \begin{bmatrix} 4 & 0 & 0 \\ 0 & 4 & 0 \\ 0 & 0 & 2 \end{bmatrix} \begin{bmatrix} 0.2406 \\ 0.0276 \\ -0.2297 \end{bmatrix} = \begin{bmatrix} 0.9623 \\ 0.1104 \\ -0.4594 \end{bmatrix}$$

Step 4: Calculate S from $U^T S = R$

$$\begin{bmatrix} 3.741657 & 0 & 0 \\ -1.603568 & 2.725541 & 0 \\ 0 & -1.467600 & 1.358732 \end{bmatrix} \begin{bmatrix} s_1 \\ s_2 \\ s_3 \end{bmatrix} = \begin{bmatrix} 0.9623 \\ 0.1104 \\ -0.4594 \end{bmatrix}, \ S = \begin{bmatrix} 0.2572 \\ 0.1918 \\ -0.1309 \end{bmatrix}$$

Step 5: Normalise S to unit length which gives the first estimate of eigen-vector and eigen-value.

$$S^T S = 0.1201, \ S = \frac{1}{\sqrt{0.1201}} \begin{bmatrix} 0.2572 \\ 0.1918 \\ -0.1309 \end{bmatrix} = \begin{bmatrix} 0.7422 \\ 0.5535 \\ -0.3777 \end{bmatrix}, \ \sqrt{0.1201} = 0.34655$$

The first estimate of $\lambda = 1/\omega^2$ is 0.34655 or $\omega^2 = 2.8856$. Compared with the exact value of $\omega^2 = 2.6820$ determined in Section 3.2.4, the value at first iteration is about 8% too high.

<u>Iteration 2:</u>

Step 1: Use the normalised estimated vector from the first iteration as the arbitrary vector W and generate another vector orthogonal to Y_1.

$$Y_1 = \begin{bmatrix} 0.1258 \\ 0.4196 \\ 0.8989 \end{bmatrix}, \ W = \begin{bmatrix} 0.7422 \\ 0.5535 \\ -0.3777 \end{bmatrix}, \ \alpha = -Y_1^T W = 0.0139$$

$$Y_2 = W + \alpha Y_1 = \begin{bmatrix} 0.7440 \\ 0.5594 \\ -0.3652 \end{bmatrix}$$

Step 2: Solve $U Z = Y_2$

$$\begin{bmatrix} 3.741657 & -1.603568 & 0 \\ 0 & 2.725541 & -1.467600 \\ 0 & 0 & 1.358732 \end{bmatrix} \begin{bmatrix} z_1 \\ z_2 \\ z_3 \end{bmatrix} = \begin{bmatrix} 0.7440 \\ 0.5594 \\ -0.3652 \end{bmatrix}, \ Z = \begin{bmatrix} 0.2248 \\ 0.0605 \\ -0.2688 \end{bmatrix}$$

Step 3: Calculate $R = M Z$

$$R = \begin{bmatrix} 4 & 0 & 0 \\ 0 & 4 & 0 \\ 0 & 0 & 2 \end{bmatrix} \begin{bmatrix} 0.2248 \\ 0.0605 \\ -0.2688 \end{bmatrix} = \begin{bmatrix} 0.8992 \\ 0.2420 \\ -0.5376 \end{bmatrix}$$

Step 4: Solve $U^T S = R$

$$\begin{bmatrix} 3.741657 & 0 & 0 \\ -1.603568 & 2.725541 & 0 \\ 0 & -1.467600 & 1.358732 \end{bmatrix} \begin{bmatrix} s_1 \\ s_2 \\ s_3 \end{bmatrix} = \begin{bmatrix} 0.8992 \\ 0.2420 \\ -0.5376 \end{bmatrix}, \; S = \begin{bmatrix} 0.2404 \\ 0.2302 \\ -0.1470 \end{bmatrix}$$

Step 5: Normalise S to unit length.

$$S^T S = 0.1324, \sqrt{0.1324} = 0.36386, S = \frac{1}{0.36386} \begin{bmatrix} 0.2404 \\ 0.2302 \\ -0.1470 \end{bmatrix} = \begin{bmatrix} 0.6606 \\ 0.6327 \\ -0.4040 \end{bmatrix}$$

The second estimate of $\lambda = 1/\omega^2$ is 0.36386 or $\omega^2 = 2.7483$. Compared with the exact value of $\omega^2 = 2.6820$ determined in Section 3.2.4, the value at second iteration is about 3 % too high.

Iteration 3:

Step 1: Use the normalised estimated vector at the end of second iteration as the arbitrary vector W and generate another vector orthogonal to Y_1.

$$Y_1 = \begin{bmatrix} 0.1258 \\ 0.4196 \\ 0.8989 \end{bmatrix}, \; W = \begin{bmatrix} 0.6606 \\ 0.6327 \\ -0.4040 \end{bmatrix}, \; \alpha = -Y_1^T W = 0.0146$$

$$Y_2 = W + \alpha Y_1 = \begin{bmatrix} 0.6624 \\ 0.6388 \\ -0.3909 \end{bmatrix}$$

Step 2: Solve $U Z = Y_2$

$$\begin{bmatrix} 3.741657 & -1.603568 & 0 \\ 0 & 2.725541 & -1.467600 \\ 0 & 0 & 1.358732 \end{bmatrix} \begin{bmatrix} z_1 \\ z_2 \\ z_3 \end{bmatrix} = \begin{bmatrix} 0.6624 \\ 0.6388 \\ -0.3909 \end{bmatrix}, \; Z = \begin{bmatrix} 0.2111 \\ 0.0795 \\ -0.2877 \end{bmatrix}$$

Step 3: Calculate $R = M Z$

$$R = \begin{bmatrix} 4 & 0 & 0 \\ 0 & 4 & 0 \\ 0 & 0 & 2 \end{bmatrix} \begin{bmatrix} 0.2111 \\ 0.0795 \\ -0.2877 \end{bmatrix} = \begin{bmatrix} 0.8444 \\ 0.3180 \\ -0.5754 \end{bmatrix}$$

Step 4: Solve $U^T S = R$

$$\begin{bmatrix} 3.741657 & 0 & 0 \\ -1.603568 & 2.725541 & 0 \\ 0 & -1.467600 & 1.358732 \end{bmatrix} \begin{bmatrix} s_1 \\ s_2 \\ s_3 \end{bmatrix} = \begin{bmatrix} 0.8444 \\ 0.3180 \\ -0.5754 \end{bmatrix}, \; S = \begin{bmatrix} 0.2257 \\ 0.2495 \\ -0.1541 \end{bmatrix}$$

Step 5: Normalise S to unit length.

$$S^T S = 0.1369, \sqrt{0.1369} = 0.3700, S = \frac{1}{0.3700}\begin{bmatrix} 0.2257 \\ 0.2495 \\ -0.1541 \end{bmatrix} = \begin{bmatrix} 0.6099 \\ 0.6743 \\ -0.4165 \end{bmatrix}$$

The third estimate of $\lambda = 1/\omega^2$ is 0.3700 or $\omega^2 = 2.7025$. Compared with the exact value of $\omega^2 = 2.6820$ determined in Section 3.2.4, the value at the end of third iteration is less than 1% in error.

The corresponding value of eigen-vector can be calculated from the relationship U X = Y.

$$\begin{bmatrix} 3.741657 & -1.603568 & 0 \\ 0 & 2.725541 & -1.467600 \\ 0 & 0 & 1.358732 \end{bmatrix}\begin{bmatrix} x_1 \\ x_2 \\ x_3 \end{bmatrix} = \begin{bmatrix} 0.6099 \\ 0.6743 \\ -0.4165 \end{bmatrix}, X = \begin{bmatrix} 0.1983 \\ 0.0823 \\ -0.3065 \end{bmatrix}$$

After M normalisation, the vector X is given by

$$X^T MX = 0.3723, \sqrt{0.3723} = 0.6101, X = \frac{1}{\sqrt{0.3723}}\begin{bmatrix} 0.1983 \\ 0.0823 \\ -0.3065 \end{bmatrix} = \begin{bmatrix} 0.3250 \\ 0.1349 \\ -0.5024 \end{bmatrix}$$

The exact eigen-vectors is $X_2^T = [0.3115 \quad 0.1699 \quad -0.4982]$ which is perhaps not as accurate as the eigen-value. This is quite a common observation. Greater number of iterations are required to obtain the correct eigen-vector than the corresponding eigen-value.

5.4.2 Simultaneous Iteration

The procedure discussed in the previous Section clearly leads to the possibility of iterating at any one time not on just one vector but simultaneously on a number of orthogonal vectors. This procedure is known for obvious reasons as simultaneous iteration method. This will be discussed at greater length in Chapter 6.

5.5 SHIFT TECHNIQUE

In the previous sections, inverse iteration was used to solve the general eigen-value problem. If an estimate of an eigen-value is available, using a technique called the shift technique, convergence can be speeded up substantially. In practice, very often such estimates might be available either from experience with similar structures or from an analysis based on a fairly crude modelling of the problem. Several rules of thumb are often used in practice at the initial design stage. For examples for a building of height H in metres, the lowest natural frequency f is approximately equal to 46/H. If H = 50 m, then f = 0.9 Hertz. Similarly for bridges, the first natural frequency is approximately given by 100/L, where L is the span in metres. Many such formulae are available for estimating the lowest frequency of many types of bridges, towers, floors, etc.

The idea of shift is quite simple. Consider the special eigen-value problem

$$A\,y = \lambda\,y \qquad (5.17)$$

Subtracting $\mu\,Y$, where μ is a scalar, from both sides of equation (5.17),

$$(A - \mu\,I)\,y = (\lambda - \mu)\,y \qquad (5.18)$$

Equation (5.18) shows that if λ is the eigen-values of matrix A, then the eigen-value of matrix $(A - \mu\,I)$ is equal to $(\lambda - \mu)$.

In a similar manner, for the general eigen-value problem, if

$$K\,x = \omega^2\,M\,x \qquad (5.19)$$

Subtracting $\mu\,M$, where μ is a scalar, from both sides of the equation (5.19),

$$(K - \mu\,M)\,x = (\omega^2 - \mu)\,M\,x \qquad (5.20)$$

There are three reasons for using a shift.

a. If K is singular, as can happen when dealing with structures not fixed to any foundation, then using the shift technique one needs to use $(K - \mu\,M)$ rather than K in carrying out the inverse iterations. Matrix $(K - \mu\,M)$ in general will not be singular.
b. If an approximation μ to ω^2 is known, and this is used as the shift, inverse iteration will converge to $\lambda_1 = 1/(\omega^2 - \mu)$, which can be very large. Since the larger the value of λ_1, the smaller the ratio of λ_i/λ_1, faster will be the convergence.
c. There are methods of solution of eigen-value problems that yield only the eigen-values but not eigen-vectors. In such cases, inverse iteration with shift can be used to obtain the corresponding eigen-vectors quite rapidly.

5.5.1 Shift Technique: Example

Using a shift of μ = 0.46, obtain the exact lowest frequency by inverse iteration. Use the data for the three-storey shear frame from Section 5.2.1.

$$K = \begin{bmatrix} 14 & -6 & 0 \\ -6 & 10 & -4 \\ 0 & -4 & 4 \end{bmatrix}, \quad M = \begin{bmatrix} 4 & 0 & 0 \\ 0 & 4 & 0 \\ 0 & 0 & 2 \end{bmatrix}$$

Step 1: Using a shift value of μ = 0.46, form the matrix [K - 0.46 M].

$$K - \mu\,M = \begin{bmatrix} 14-0.46\times4 & -6 & 0 \\ -6 & 10-0.46\times4 & -4 \\ 0 & -4 & 4-0.46\times2 \end{bmatrix} = \begin{bmatrix} 12.16 & -6 & 0 \\ -6 & 8.16 & -4 \\ 0 & -4 & 3.08 \end{bmatrix}$$

Step 2: Using the procedure explained in Section 4.3, calculate the Cholesky factor U, where $[K - \mu\,M] = U^T\,U$.

$$u_{11} = \sqrt{k_{11}} = \sqrt{12.16} = 3.4871,\; u_{12} = \frac{k_{12}}{u_{11}} = -1.7206,\; u_{13} = \frac{k_{13}}{u_{11}} = 0$$

$$u_{22} = \sqrt{(k_{22} - u_{12}^2)} = 2.2802,\; u_{23} = \frac{(k_{23} - u_{12}\,u_{13})}{u_{22}} = -1.7542$$

$$u_{33} = \sqrt{(k_{33} - u_{13}^2 - u_{23}^2)} = 0.0528$$

$$U = \begin{bmatrix} 3.487119 & -1.720618 & 0 \\ 0 & 2.280236 & -1.754205 \\ 0 & 0 & 0.052588 \end{bmatrix}$$

$$U^T = \begin{bmatrix} 3.487119 & 0 & 0 \\ -1.720618 & 2.280236 & 0 \\ 0 & -1.754205 & 0.052588 \end{bmatrix}$$

Iteration 1:

Step 1: Let the arbitrary vector be

$$Y = \begin{bmatrix} 1 \\ 1 \\ 1 \end{bmatrix} = \sqrt{3} \begin{bmatrix} 0.5774 \\ 0.5774 \\ 0.5774 \end{bmatrix}$$

Step 2: Using the normalised value of Y, solve $U\,Z = Y$

$$\begin{bmatrix} 3.487119 & -1.720618 & 0 \\ 0 & 2.280236 & -1.754205 \\ 0 & 0 & 0.052588 \end{bmatrix} \begin{bmatrix} z_1 \\ z_2 \\ z_3 \end{bmatrix} = \begin{bmatrix} 0.5774 \\ 0.5774 \\ 0.5774 \end{bmatrix}, \; Z = \begin{bmatrix} 4.45833 \\ 8.69999 \\ 10.97969 \end{bmatrix}$$

Step 3: Calculate $R = M\,Z$

$$R = \begin{bmatrix} 4 & 0 & 0 \\ 0 & 4 & 0 \\ 0 & 0 & 2 \end{bmatrix} \begin{bmatrix} 4.45833 \\ 8.69999 \\ 10.97969 \end{bmatrix} = \begin{bmatrix} 17.83332 \\ 34.79996 \\ 21.95938 \end{bmatrix}$$

Step 4: Solve $U^T S = R$

$$\begin{bmatrix} 3.487119 & 0 & 0 \\ -1.720618 & 2.280236 & 0 \\ 0 & -1.754205 & 0.052588 \end{bmatrix} \begin{bmatrix} s_1 \\ s_2 \\ s_3 \end{bmatrix} = \begin{bmatrix} 17.83332 \\ 34.79996 \\ 21.95938 \end{bmatrix}, S = \begin{bmatrix} 5.1141 \\ 19.1205 \\ 1055.39 \end{bmatrix}$$

Step 5: Normalising S to unit length gives the first estimate of eigen-vector and eigen-value.

$$S^T S = 1.1142 \times 10^6, \sqrt{S^T S} = 1055.57, S = \frac{1}{1055.5726} \begin{bmatrix} 5.114056 \\ 19.120519 \\ 1055.386973 \end{bmatrix} = \begin{bmatrix} 0.0049 \\ 0.0181 \\ 0.9998 \end{bmatrix}$$

The first estimate of $\lambda = 1055.5726$. Therefore $\omega^2 = \mu + 1/\lambda = 0.46 + 0.000547 = 0.460547$. The 'exact' value calculated in Section 3.2.4 is $\omega^2 = 0.46056$. The eigen-vector can now be determined by solving $U\,X = Y$

$$\begin{bmatrix} 3.487119 & -1.720618 & 0 \\ 0 & 2.280236 & -1.754205 \\ 0 & 0 & 0.052588 \end{bmatrix} \begin{bmatrix} x_1 \\ x_2 \\ x_3 \end{bmatrix} = \begin{bmatrix} 4.844817x10^{-3} \\ 0.018114 \\ 0.999824 \end{bmatrix},$$

$$X = \begin{bmatrix} 7.181346 \\ 14.551378 \\ 18.904553 \end{bmatrix}$$

After M normalisation, the vector X is given by

$$X^T MX = 1768.02156, \quad X = \frac{1}{\sqrt{1768.02156}} \begin{bmatrix} 7.181346 \\ 0.08189 \\ -0.30625 \end{bmatrix} = \begin{bmatrix} 0.17079 \\ 0.34607 \\ 0.44960 \end{bmatrix}$$

The correct vector is X^T = [0.1712 0.3463 0.4490]. As can be seen, in only one iteration, the correct value of both the eigen-value as well as eigen-vector can be obtained.

5.5.2 Higher Frequencies and Shift Technique

When higher frequencies are required, then a similar procedure can be used. The only difference is that if Cholesky factorisation fails due to square root problem, then one has to use Crout factorisation as explained in Section 4.4.

If Crout factorisation is used, the conversion from general to special eigen-value problem proceeds in a way similar to that in Section 4.3.2. The following equations are similar to equations (4.2) to (4.8).

The general eigen-value problem is given by

$$K\,X = \omega^2 M\,X \tag{5.21}$$

Using Crout factorisation

$$K = U^T D\,U \tag{5.22}$$

Substituting equation (5.22) in equation (5.21),

$$U^T D\,U\,X = \omega^2 M\,[U^{1}\,U]\,X \tag{5.23}$$

Setting $U\,X = Y, \lambda = 1/\omega^2$, equation (5.23) becomes

$$\lambda\,Y = D^{-1}\,U^T M\,U^{1}\,Y \tag{5.24}$$

5.5.2.1 *Higher Frequencies and Shift Technique: Example*

Using a shift of μ = 2.6, obtain the exact second lowest frequency by inverse iteration. Use the data for the three-storey shear frame from Section 5.2.1.

$$K - \mu\,M = \begin{bmatrix} 14-2.6\times4 & -6 & 0 \\ -6 & 10-2.6\times4 & -4 \\ 0 & -4 & 4-2.6\times2 \end{bmatrix} = \begin{bmatrix} 3.6 & -6 & 0 \\ -6 & -0.40 & -4 \\ 0 & -4 & -1.2 \end{bmatrix}$$

The determinants of 1 x 1, 2 x 2 and 3 x 3 sub-matrices of the above matrix are equal to 3.6, -37.44 and –12.672. This indicates that the matrix is not positive definite. If Cholesky factorisation is attempted, using the equations from Section 4.3, we have

$$u_{11} = \sqrt{3.6},\ u_{12} = a_{12}/u_{11} = -3.1623,\ u_{13} = a_{13}/u_{11} = 0.$$

$u_{22} = \sqrt{\{a_{22} - u_{12}^2\}} = \sqrt{(-0.40 - 10)}$, which cannot be evaluated. Therefore Cholesky factorisation fails.

Using Crout factorisation, and using the equations from Section 4.4,

$$d_{11} = k_{11} = 3.6, u_{12} = \frac{k_{12}}{d_{11}} = \frac{-6}{3.6} = -1.6667, u_{13} = \frac{k_{13}}{d_{11}} = 0$$

$$d_{22} = k_{22} - u_{12}^2\, d_{11} = -0.40 - (-1.6667)^2\, 3.6 = -10.40,$$

$$u_{23} = \frac{k_{23} - u_{12}\, u_{13}\, d_{11}}{d_{22}} = \frac{-4-0}{-10.4} = 0.3846$$

$$d_{33} = k_{33} - u_{13}^2\, d_{11} - u_{23}^2\, d_{22} = -1.2 - 0 - (0.3846)^2 (-1.4) = 0.3385$$

$$U = \begin{bmatrix} 1.0 & -1.666667 & 0 \\ 0 & 1.0 & 0.384615 \\ 0 & 0 & 1.0 \end{bmatrix},\ D = \begin{bmatrix} 3.6 & 0 & 0 \\ 0 & -10.4 & 0 \\ 0 & 0 & 0.338462 \end{bmatrix}$$

$$U^T = \begin{bmatrix} 1.0 & 0 & 0 \\ -1.666667 & 1.0 & 0 \\ 0 & 0.384615 & 1.0 \end{bmatrix},\ D^{-1} = \begin{bmatrix} 0.277778 & 0 & 0 \\ 0 & -0.096154 & 0 \\ 0 & 0 & 2.954541 \end{bmatrix}$$

<u>Iteration 1:</u>

Step 1: Let the arbitrary vector Y be

$$Y = \begin{bmatrix} 1 \\ 1 \\ 1 \end{bmatrix} = \sqrt{3} \begin{bmatrix} 0.577350 \\ 0.577350 \\ 0.577350 \end{bmatrix}$$

Step 2: Using the normalised value of Y, solve $U\,Z = Y$

$$\begin{bmatrix} 1.0 & -1.666667 & 0 \\ 0 & 1.0 & 0.384615 \\ 0 & 0 & 1.0 \end{bmatrix} \begin{bmatrix} z_1 \\ z_2 \\ z_3 \end{bmatrix} = \begin{bmatrix} 0.577350 \\ 0.577350 \\ 0.577350 \end{bmatrix},\ Z = \begin{bmatrix} 1.169505 \\ 0.355293 \\ 0.577350 \end{bmatrix}$$

Step 3: Calculate $R = M\,Z$

$$R = \begin{bmatrix} 4 & 0 & 0 \\ 0 & 4 & 0 \\ 0 & 0 & 2 \end{bmatrix} \begin{bmatrix} 1.169505 \\ 0.355293 \\ 0.577350 \end{bmatrix} = \begin{bmatrix} 4.678020 \\ 1.421172 \\ 1.154701 \end{bmatrix}$$

Step 4: Solve $U^T S = R$

$$\begin{bmatrix} 1.0 & 0 & 0 \\ -1.666667 & 1.0 & 0 \\ 0 & 0.384615 & 1.0 \end{bmatrix} \begin{bmatrix} s_1 \\ s_2 \\ s_3 \end{bmatrix} = \begin{bmatrix} 4.678020 \\ 1.421172 \\ 1.154701 \end{bmatrix}, S = \begin{bmatrix} 4.678020 \\ 9.217874 \\ -2.390631 \end{bmatrix}$$

Step 5: Calculate $P = D^{-1} S$. Since D is a diagonal matrix, this step involves dividing the elements of S by the corresponding diagonal elements of D.

$$P = D^{-1}\, S = \begin{bmatrix} 1.299451 \\ -0.886336 \\ -7.063219 \end{bmatrix}, \sqrt{P^T P} = 7.236244, \quad P = \frac{1}{7.236244} \begin{bmatrix} 0.179575 \\ -0.122486 \\ -0.976089 \end{bmatrix}$$

The first estimate of λ = 7.236244. Therefore $\omega^2 = \mu + 1/\lambda = 2.6 + 0.1382 = 2.7382$. Compared with the exact value of 2.6820 from Section 3.2.4, the approximate value is 2% too high.

The eigen-vector can now be determined by solving U X = Y

$$\begin{bmatrix} 1.0 & -1.666667 & 0 \\ 0 & 1.0 & 0.384615 \\ 0 & 0 & 1.0 \end{bmatrix} \begin{bmatrix} x_1 \\ x_2 \\ x_3 \end{bmatrix} = \begin{bmatrix} 0.179575 \\ -0.122486 \\ -0.976089 \end{bmatrix}, \; X = \begin{bmatrix} 0.601129 \\ 0.252933 \\ -0.976089 \end{bmatrix}$$

After M normalisation, the vector X is given by

$$X^T M\, X = 3.606824, \quad X = \frac{1}{\sqrt{3.606824}} \begin{bmatrix} 0.601129 \\ 0.252933 \\ -0.976089 \end{bmatrix} = \begin{bmatrix} 0.316523 \\ 0.133181 \\ -0.513957 \end{bmatrix}$$

The correct eigen-vector is X^T = [0.3115 0.1699 -0.4982] and the eigen-value is 2.68199 As can be seen, the value of both the eigen-value as well as eigen-vector can be obtained to reasonable accuracy in only one iteration and without any need for orthogonalisation of the arbitrary vector with respect to eigen-vectors corresponding to lower values of eigen-values. The reason for this is that the appropriate value of shift makes the shifted eigen-value the largest.

5.6 TRANSFORMATION METHODS

Discussions so far have been confined to the inverse iteration method mainly aimed at obtaining the eigen-value and the corresponding eigen-vectors at the lower end of the spectrum, which are of interest to the designers. In general, with very large-scale problems, the eigen-pairs of interest tend to be a very small percentage, in most cases much less than 10%, of the total pairs of the system. However there are many methods specifically designed to determine all the eigen-pairs. The importance of these methods lies in the fact that these methods are also used in practice to determine all the eigen-pairs of a reduced system, which is much smaller than the original system. They all use an important concept known as similarity transformation.

5.7 JACOBI DIAGONALISATION OF THE GENERAL EIGEN-VALUE PROBLEM

The basic idea of similarity transformation is to transform iteratively the given stiffness and mass matrices into another pair of diagonal matrices, such that the general eigen-value problem defined by the modified matrices has exactly the same eigen-values as the original pair.

Consider the general eigen-value problem

$$K\,Y = M\,Y\,\Lambda \tag{5.25}$$

where K and M are stiffness and mass matrices respectively, Λ is a diagonal matrix whose elements are the eigen-values and Y is a matrix whose columns are the eigen-vectors. The object is to transform equation (5.25) in a series of steps to

$$\bar{K}\,X = \bar{M}\,X\,\Lambda \tag{5.26}$$

where both $\bar{K}$ and $\bar{M}$ are diagonal matrices. The eigen-values are obtained from

$$\bar{M}^{-1}\,\bar{K} = \Lambda \tag{5.27}$$

This transformation is accomplished using an as yet undefined matrix P.
Multiplying both sides of equation (5.25) by P^T

$$P^T\,K\,Y = P^T\,M\,Y\,\Lambda \tag{5.28}$$

Substituting Y = P X in equation (5.28)

$$P^T\,K\,P\,X = P^T\,M\,P\,X\,\Lambda \tag{5.29}$$

This shows that the eigen-values of equation (5.25) and (5.28) are identical. The eigen-vectors are related by $Y = P\,X$.

The matrix P is chosen such that the modified matrices $P^T\,K\,P$ and $P^T\,M\,P$ become increasingly diagonal.

Consider a 7 x 7 symmetric matrix A. Let it be required to make the off diagonal term $b_{63} = b_{36} = 0$, where $B = P^T\,A\,P$. Let the matrix P be a modified identity matrix with two undetermined terms α and β.

$$P = \begin{bmatrix} 1 & & & & & & \\ & 1 & & & & & \\ & & 1 & & & \alpha & \\ & & & 1 & & & \\ & & & & 1 & & \\ & & \beta & & & 1 & \\ & & & & & & 1 \end{bmatrix}$$

Tables 5.1 and 5.2 show the details of matrix multiplications involved in calculating the matrix B. From Table 5.2

$$b_{63} = b_{36} = a_{66}\,\beta + a_{33}\alpha + a_{36}(1+\alpha\,\beta) \tag{5.30a}$$

Table 5.1 Matrix $P^T A$

$$P^T A = \begin{bmatrix} a_{11} & a_{12} & a_{13} & a_{14} & a_{15} & a_{16} & a_{17} \\ a_{21} & a_{22} & a_{23} & a_{24} & a_{25} & a_{26} & a_{27} \\ a_{31}+a_{61}\beta & a_{32}+a_{62}\beta & a_{33}+a_{63}\beta & a_{34}+a_{64}\beta & a_{35}+a_{65}\beta & a_{36}+a_{66}\beta & a_{37}+a_{67}\beta \\ a_{41} & a_{42} & a_{43} & a_{44} & a_{45} & a_{46} & a_{47} \\ a_{51} & a_{52} & a_{53} & a_{54} & a_{55} & a_{56} & a_{57} \\ a_{31}\alpha+a_{61} & a_{32}\alpha+a_{62} & a_{33}\alpha+a_{63} & a_{34}\alpha+a_{64} & a_{35}\alpha+a_{65} & a_{36}\alpha+a_{66} & a_{37}\alpha+a_{67} \\ a_{71} & a_{72} & a_{73} & a_{74} & a_{75} & a_{76} & a_{77} \end{bmatrix}$$

Table 5.2 Matrix $P^T AP$

$$B = P^T A P = \begin{bmatrix} a_{11} & a_{12} & a_{13}+a_{16}\beta & a_{14} & a_{15} & a_{13}\alpha+a_{16} & a_{17} \\ a_{21} & a_{22} & a_{23}+a_{26}\beta & a_{24} & a_{25} & a_{23}\alpha+a_{26} & a_{27} \\ a_{31}+a_{61}\beta & a_{32}+a_{62}\beta & b_{33} & a_{34}+a_{64}\beta & a_{35}+a_{65}\beta & b_{36} & a_{37}+a_{67}\beta \\ a_{41} & a_{42} & a_{43}+a_{46}\beta & a_{44} & a_{45} & a_{43}\alpha+a_{46} & a_{47} \\ a_{51} & a_{52} & a_{53}+a_{56}\beta & a_{54} & a_{55} & a_{53}\alpha+a_{56} & a_{57} \\ a_{31}\alpha+a_{61} & a_{32}\alpha+a_{62} & b_{63} & a_{34}\alpha+a_{64} & a_{35}\alpha+a_{65} & b_{66} & a_{37}\alpha+a_{67} \\ a_{71} & a_{72} & a_{73}+a_{76}\beta & a_{74} & a_{75} & a_{73}\alpha+a_{76} & a_{77} \end{bmatrix}$$

If we have to zero the term b_{36} in both $P^T\ K\ P$ and $P^T\ M\ P$, then there are two equations to determine the parameters α and β. In the product $B = P^T A\ P$, the only elements different from that of A are in the rows 3 and 6, and in the columns 3 and 6. The elements of symmetric matrix B can be calculated explicitly from the elements of A and the parameters α and β. Symmetry can be exploited to reduce storage and there is no need to store P.

In general, if the element b_{ij} needs to be zeroed, then $p_{ij} = \alpha$ and $p_{ji} = \beta$. The element $b_{ij} = b_{ji}$ is given by

$$b_{ij} = b_{ji} = a_{jj}\,\beta + a_{ii}\alpha + a_{ij}(1+\alpha\,\beta) \tag{5.30b}$$

In order to make the stiffness and mass matrices increasingly diagonal, the b_{ij} element of both P^T KP and P^T M P must be made equal to zero. Therefore

$$k_{ii}\,\alpha + k_{jj}\,\beta + k_{ij}\,(1 + \alpha\,\beta) = 0 \tag{5.31}$$

$$m_{ii}\,\alpha + m_{jj}\,\beta + m_{ij}\,(1 + \alpha\,\beta) = 0 \tag{5.32}$$

Eliminating α β from the two equations (5.31) and (5.32)

$$\alpha\,(k_{ii}\,m_{ij} - m_{ii}\,k_{ij}) + \beta(k_{jj}\,m_{ij} - m_{jj}\,k_{ij}) = 0 \tag{5.33}$$

Equation (5.33) can be written as

$$\alpha\,p + \beta\,q = 0 \tag{5.34}$$

where

$$p = (k_{ii}\,m_{ij} - m_{ii}\,k_{ij}),\quad q = (k_{jj}\,m_{ij} - m_{jj}\,k_{ij}) \tag{5.35}$$

If $p \neq 0$, then, from equation (5.34), $\alpha = -(q/p)\,\beta$. Substituting $\alpha = -(q/p)\,\beta$ in equation (5.31),

$$q\,k_{ij}\,\beta^2 + [q\,k_{ii} - p\,k_{jj}]\,\beta - p\,k_{ij} = 0 \tag{5.36}$$

Using equation (5.35), the term inside the square bracket in equation (5.36) can be written as

$$[q\,k_{ii} - p\,k_{jj}] = -k_{ij}\, r \text{, where } r = k_{ii}\, m_{jj} - m_{ii}\, k_{jj} \tag{5.37}$$

Substituting equation (5.37) in equation (5.36) and simplifying

$$q\beta^2 - r\beta - p = 0 \tag{5.38}$$

Solving the quadratic equation (5.38)

$$\beta = \frac{1}{q}\left(\frac{r}{2} \pm \sqrt{[\{\frac{r}{2}\}^2 + p\,q]}\right) \tag{5.39}$$

Substituting for β in equation (5.34)

$$\alpha = -\beta\frac{q}{p} = -\frac{1}{p}\left(\frac{r}{2} \pm \sqrt{[\{\frac{r}{2}\}^2 + p\,q]}\right) \tag{5.40}$$

Noting that

$$\left(\frac{r}{2} \pm \sqrt{[\{\frac{r}{2}\}^2 + p\,q]}\right)\left(\frac{r}{2} \mp \sqrt{[\{\frac{r}{2}\}^2 + p\,q]}\right) = -pq$$

we can write another expression for α and β as

$$\beta = -\frac{p}{\left(\frac{r}{2} \mp \sqrt{[\{\frac{r}{2}\}^2 + p\,q]}\right)}$$

$$\alpha = \frac{q}{\left(\frac{r}{2} \mp \sqrt{[\{\frac{r}{2}\}^2 + p\,q]}\right)}$$

In order to prevent the denominator from becoming zero, the denominator is taken as

$$\left(\frac{r}{2} + (sign\ of\ r)\sqrt{[\{\frac{r}{2}\}^2 + p\,q]}\right)$$

Therefore the values α and β are given by the expressions,

$$\beta = -\frac{p}{\left(\frac{r}{2} + (sign\,of\ r)\sqrt{[\{\frac{r}{2}\}^2 + p\,q]}\right)} \tag{5.41}$$

$$\alpha = \frac{q}{\left(\frac{r}{2} + (sign\,of\ r)\sqrt{[\{\frac{r}{2}\}^2 + p\,q]}\right)} \tag{5.42}$$

If the element being zeroed is a_{ij}, then the modified elements are

$$b_{ii} = a_{ii} + 2a_{ij}\,\beta + a_{jj}\,\beta^2 \tag{5.43a}$$

$$b_{jj} = a_{ii}\ \alpha^2 + 2a_{ij}\ \alpha + \ a_{jj} \tag{5.43b}$$

$$b_{ik} = a_{ik} + a_{jk}\ \beta,\ \ b_{jk} = a_{ik}\ \alpha + a_{jk},\ k = 1, N \tag{5.43c}$$

5.7.1 Calculation Procedure for the Jacobi Method

The process of converting the matrices into diagonal form is done iteratively. The procedure is systematic. The operations detailed below are done for both mass and stiffness matrices simultaneously.

i. Choose the term next to the diagonal in the first row and if necessary reduce it to zero. Calculate the modified matrix B = P^T A P. Using the modified matrix B, choose the next term and repeat the operation until the operation on the last term of the first row is completed.

ii. Choose the second row and repeat the operations as in (i) above.

iii. Continue the operations until the last but one row is completed.

This is called one sweep. It will be noticed that terms zeroed previously become non-zero again. Therefore repeat further sweeps until all the off diagonal terms are acceptably small. Since some of the terms to be zeroed can be very small, there can be savings in effort if only those terms that are greater than a certain limiting value are zeroed. Similarly for accuracy, in the initial stages of iteration, the limiting value of a term needed to be zeroed can be much higher than at later stages in iteration.

From a computational point of view, as shown by Table 5.2, only four temporary vectors are needed to store the two rows and two columns that are affected by the transformation.

5.7.2 General Jacobi Method: Example

The calculation procedure detailed in Section 5.7.1, is illustrated by an example.

<u>Example</u>: Determine the eigen-values of the general eigen-value problem, considered in Section 5.2.1. The stiffness and mass matrices are given by

$$K = \begin{bmatrix} 14 & -6 & 0 \\ -6 & 10 & -4 \\ 0 & -4 & 4 \end{bmatrix},\quad M = \begin{bmatrix} 4 & 0 & 0 \\ 0 & 4 & 0 \\ 0 & 0 & 2 \end{bmatrix}$$

<u>Sweep 1:</u>

Step 1: Set the modified element (1,2) in the stiffness and mass matrix to be zero.

$k_{11} = 14$, $k_{22} = 10$, $k_{12} = k_{21} = -6$

$m_{11} = 4$, $m_{22} = 4$, $m_{12} = m_{21} = 0$

$p = k_{11}\, m_{12} - m_{11}\, k_{12} = 24$

$q = k_{22}\, m_{12} - m_{22}\, k_{12} = 24$

$r = k_{11}\, m_{22} - m_{11}\, k_{22} = 16$, $r/2 = 8$, r is positive

$x = 0.5\, r + \sqrt{(0.25\, r^2 + pq)} = 8 + \sqrt{(8^2 + 24 \times 24)} = 33.2982$

$\alpha = q/x = 0.7208$, $\beta = p/x = -0.7208$

Only the elements in rows 1 and 2 and columns 1 and 2 will be affected.
For K (for M), the elements become
$b_{12} = b_{21} = 0$
$b_{11} = a_{11} + a_{22}\,\beta^2 + 2\,a_{12}\,\beta = 27.8441(6.0780)$
$b_{22} = a_{11}\,\alpha^2 + a_{22} + 2\,a_{12}\,\alpha = 8.6238(6.0780)$
Off diagonal terms for M but for K are
$b_{k1} = b_{1k} = a_{1k} + a_{k2}\,\beta$
$k = 3,\ b_{13} = 2.8830$
$b_{k2} = b_{2k} = a_{1k}\,\alpha + a_{2k}$
$k = 3,\ b_{23} = -4.0$
At the end of iteration 1, the matrices K and M have been modified to

$$K = \begin{bmatrix} 27.8441 & 0 & 2.8830 \\ symm & 8.6238 & -4.0 \\ & -4.0 & 4.0 \end{bmatrix},\ M = \begin{bmatrix} 6.0780 & 0 & 0 \\ symm & 6.0780 & 0 \\ & 0 & 2.0 \end{bmatrix}$$

Step 2: Set element (1,3) in the stiffness and mass matrix to be zero.
$k_{11} = 27.8441,\ k_{33} = 4.0,\ k_{31} = k_{13} = 2.8830$
$m_{11} = 6.0780,\ m_{22} = 2.0,\ m_{13} = m_{31} = 0$
$p = k_{11}\,m_{13} - m_{11}\,k_{13} = -17.5247$
$q = k_{33}\,m_{13} - m_{33}\,k_{13} = -5.7664$
$r = k_{11}\,m_{33} - m_{11}\,k_{33} = 31.3774$, $r/2 = 15.6887$, r is positive
$x = 0.5\,r + \sqrt{(0.25\,r^2 + pq)} = 15.6887 + \sqrt{(15.6887^2 + 17.5247 \times 5.7664)} = 34.3217$
$\alpha = q/x = -0.1680,\ \beta = p/x = 0.5106$
Only the elements in rows 1 and 3 and columns 1 and 3 will be affected. For *K* (for *M*), the elements become
$b_{13} = b_{31} = 0$
$b_{11} = a_{11} + a_{33}\,\beta^2 + 2\,a_{13}\,\beta = 31.8308(6.5995)$
$b_{33} = a_{11}\,\alpha^2 + a_{33} + 2\,a_{13}\,\alpha = 3.8172(2.1716)$
Off diagonal terms for M but for K are
$b_{k1} = b_{1k} = a_{1k} + a_{3k}\,\beta$
$k = 2,\ b_{12} = -2.0423$
$b_{k3} = b_{3k} = a_{1k}\,\alpha + a_{3k}$
$k = 2,\ b_{23} = 3.8172$

$$K = \begin{bmatrix} 31.8308 & -2.0423 & 0 \\ symm & 8.6238 & -4.0 \\ & & 3.8172 \end{bmatrix},\ M = \begin{bmatrix} 6.5995 & 0 & 0 \\ symm & 6.0780 & 0 \\ & 0 & 2.1716 \end{bmatrix}$$

Step 3: Set element (2,3) in the stiffness and mass matrix to be zero.
$k_{22} = 8.6238,\ k_{33} = 3.8172,\ k_{23} = k_{32} = -4.0$
$m_{22} = 6.0780,\ m_{33} = 2.1716,\ m_{13} = m_{23} = 0$
$p = k_{22}\,m_{23} - m_{22}\,k_{23} = 24.3128$
$q = k_{33}\,m_{23} - m_{33}\,k_{23} = 8.6864$
$r = k_{22}\,m_{33} - m_{22}\,k_{33} = -4.4736$, $r/2 = -2.2368$, r is negative
$x = 0.5\,r + \sqrt{(0.25\,r^2 + pq)} = -2.2368 - \sqrt{(2.2378^2 + 24.3128 \times 8.6864)} = -16.9403$
$\alpha = q/x = -0.5128,\ \beta = p/x = 1.4352$

Only the elements in rows 2 and 3 and columns 2 and 3 will be affected.
$b_{32} = b_{23} = 0$
$b_{22} = a_{22} + a_{33}\,\beta^2 + 2\,a_{23}\,\beta = 5.0048(10.5509)$
$b_{33} = a_{11}\,\alpha^2 + a_{33} + 2\,a_{13}\,\alpha = 10.1868(3.7696)$
Off diagonal terms for K are
$b_{k2} = b_{2k} = a_{2k} + a_{k3}\,\beta$
$k = 1,\ b_{12} = -2.0423$
$b_{k3} = b_{3k} = a_{2k}\,\alpha + a_{3k}$
$k = 1,\ b_{13} = 1.0472$

At the end of iteration 1, the matrices K and M have been modified to

$$K = \begin{bmatrix} 31.8308 & -2.0423 & 1.0472 \\ symm & 5.0548 & 0 \\ symm & 0 & 10.1868 \end{bmatrix},\ M = \begin{bmatrix} 6.5994 & 0 & 0 \\ 0 & 10.5508 & 0 \\ 0 & 0 & 3.7696 \end{bmatrix}$$

Although the matrices K and M are not fully diagonalized, simply using the ratios of the diagonal elements as estimates of eigen-values, we have the three eigen-values as
ω^2 =(31.8308/6.5994= 4.8233, 5.0048/10.5508 = 0.4744, 10.1868/3.7696 = 2.7024)
The exact eigen-values calculated in Section 3.2.4 are 4.8575, 0.4606 and 2.6820. From a comparison between the exact and approximate eigen-values, one notices that even after only one 'sweep' in this simple problem, eigen-values are reasonable estimates of the 'exact' values.
After completing the first sweep, calculations are repeated for the second sweep. For the second sweep, we start with the matrix B at the end of first sweep.

Sweep 2:
Step 1: Set element (1,2) in the stiffness and mass matrix to be zero.
$k_{11} = 31.8308,\ k_{22} = 5.0048,\ k_{12} = k_{21} = -2.0423$
$m_{11} = 6.5994,\ m_{22} = 10.5508,\ m_{12} = m_{21} = 0$
$p = k_{11}\,m_{12} - m_{11}\,k_{12} = 13.4790$
$q = k_{22}\,m_{12} - m_{22}\,k_{12} = 21.5500$
$r = k_{11}\,m_{22} - m_{11}\,k_{22} = 302.8398$, $r/2 = 151.4199$, r is positive
$x = 0.5\,r + \sqrt{(0.25\,r^2 + pq)} = 151.4199 + \sqrt{(151.4199^2 + 13.4790 \times 21.5500)} =$ 303.7960
$\alpha = q/x = 0.0709,\ \beta = p/x = -0.0444$
Only the elements in rows 1 and 2 and columns 1 and 2 will be affected.
$b_{12} = b_{21} = 0$
$b_{11} = a_{11} + a_{22}\,\beta^2 + 2\,a_{12}\,\beta = 32.0219(6.6201)$
$b_{22} = a_{11}\,\alpha^2 + a_{22} + 2\,a_{12}\,\alpha = 4.8752(10.5840)$
$b_{k1} = b_{1k} = a_{1k} + a_{k2}\,\beta$
$k = 3,\ b_{13} = 1.0472$
$b_{k2} = b_{2k} = a_{1k}\,\alpha + a_{2k}$
$k = 3,\ b_{23} = 0.0743$

$$K = \begin{bmatrix} 32.0219 & 0 & 1.0472 \\ symm & 4.8752 & 0.0743 \\ & & 10.1868 \end{bmatrix}, \quad M = \begin{bmatrix} 6.6201 & 0 & 0 \\ symm & 10.5840 & 0 \\ & & 3.7696 \end{bmatrix}$$

Step 2: Set element (1,3) in the stiffness and mass matrix to be zero.
$k_{11} = 32.0219$, $k_{33} = 10.1868$, $k_{31} = k_{13} = 1.0472$
$m_{11} = 6.0201$, $m_{22} = 3.7696$, $m_{13} = m_{31} = 0$
$p = k_{11} m_{12} - m_{11} k_{12} = -6.9336$
$q = k_{22} m_{12} - m_{22} k_{12} = -3.9483$
$r = k_{11} m_{22} - m_{11} k_{22} = 53.2839$, $r/2 = 53.2839$, r is positive
$x = 0.5\, r + \sqrt{(0.25\, r^2 + pq)} = 53.2839 + \sqrt{(53.2839^2 + 6.9336 \times 3.9483)} = 53.7928$
$\alpha = q/x = -0.0734$, $\beta = p/x = 0.1289$
Only the elements in rows 1 and 3 and columns 1 and 3 will be affected.
$b_{12} = b_{21} = 0$
$b_{11} = a_{11} + a_{22} \beta^2 + 2\, a_{13} \beta = 32.4611\ (6.6828)$
$b_{33} = a_{11} \alpha^2 + a_{33} + 2\, a_{13} \alpha = 10.2056\ (3.8053)$
$b_{k1} = b_{1k} = a_{1k} + a_{k3} \beta$
$k = 2$, $b_{12} = 0.0096$
$b_{k3} = b_{3k} = a_{1k} \alpha + a_{3k}$
$k = 2$, $b_{23} = 0.0743$

$$K = \begin{bmatrix} 32.4611 & 0.0096 & 0 \\ symm & 8.8752 & 0.0743 \\ & & 10.2056 \end{bmatrix}, \quad M = \begin{bmatrix} 6.6828 & 0 & 0 \\ 0 & 10.5840 & 0 \\ 0 & 0 & 3.8053 \end{bmatrix}$$

Step 3: Set element (2,3) in the stiffness and mass matrix to be zero.
$k_{22} = 8.8752$, $k_{33} = 10.2056$, $k_{23} = k_{32} = 0.0743$
$m_{22} = 10.5840$, $m_{33} = 3.8053$, $m_{32} = m_{23} = 0$
$p = k_{22} m_{23} - m_{22} k_{23} = -0.7864$
$q = k_{33} m_{23} - m_{33} k_{23} = -0.2828$
$r = k_{22} m_{33} - m_{22} k_{33} = -87.4726$, $r/2 = -44.7363$, r is negative
$x = 0.5\, r - \sqrt{(0.25\, r^2 + pq)} = -44.7363 - \sqrt{(-44.7363^2 + 0.7864 \times 0.2828)} = -89.4749$
$\alpha = q/x = 0.0032$, $\beta = p/x = -0.0088$
Only the elements in rows 2 and 3 and columns 2 and 3 will be affected.
$b_{32} = b_{23} = 0$
$b_{22} = a_{22} + a_{33} \beta^2 + 2\, a_{23} \beta = 4.8747\ (10.5843)$
$b_{33} = a_{11} \alpha^2 + a_{33} + 2\, a_{13} \alpha = 10.2061\ (3.8054)$
$b_{k2} = b_{2k} = a_{2k} + a_{k3} \beta$
$k = 1$, $b_{12} = 0.0096$
$b_{k3} = b_{3k} = a_{2k} \alpha + a_{3k}$
$k = 1$, $b_{13} = 0.0003$

$$K = \begin{bmatrix} 32.4611 & 0.0096 & 0.0003 \\ symm & 4.8747 & 0 \\ & & 10.2061 \end{bmatrix}, \quad M = \begin{bmatrix} 6.6828 & 0 & 0 \\ symm & 10.5843 & 0 \\ & & 3.8054 \end{bmatrix}$$

Although the matrices K and M are not fully diagonalised, simply using the ratios of the diagonal elements as estimates of eigen-values, the three eigen-values are ω^2 = 32.4611/6.6828 = 4.8574, 4.8747/10.5843 = 0.4606 and 10.2061/3.8054 = 2.6820. The exact eigen-values from Section 3.2.4 are (4.8575, 0.4606 and 2.6820). From a comparison between the exact and approximate eigen-values, one notices that even after only two 'sweeps' in this simple problem, eigen-values are very close estimates of the 'exact' values.

5.7.3 Determination of Eigen-Vectors

The eigen-vectors are determined at the same time as the eigen-values. Starting with a unit matrix (a matrix with unit value on the diagonals and zero else where), the matrix is repeatedly post-multiplied by the matrix P. At the end of the iterations, the matrix contains the eigen-vectors stored by columns. If the matrix P is designed to zero the $(i,j)^{th}$ element, then post-multiplication by P only affects the i and j columns of the eigen-vector matrix.

5.7.3.1 Determination of Eigen-Vectors: Example

Taking the values from the Example in Section 5.7.2, start with the eigen-vector matrix as

$$EIGENVEC = \begin{bmatrix} 1 & 0 & 0 \\ 0 & 1 & 0 \\ 0 & 0 & 1 \end{bmatrix}$$

Iteration 1:

$i = 1, j = 2, \alpha = 0.7208, \beta = -0.7208$

$$P = \begin{bmatrix} 1 & 0.7208 & 0 \\ -0.7208 & 1 & 0 \\ 0 & 0 & 1 \end{bmatrix}$$

Post-multiplying EIGENVEC by P

$$EIGENVEC = \begin{bmatrix} 1 & 0 & 0 \\ 0 & 1 & 0 \\ 0 & 0 & 1 \end{bmatrix} \begin{bmatrix} 1 & 0.7208 & 0 \\ -0.7208 & 1 & 0 \\ 0 & 0 & 1 \end{bmatrix} = \begin{bmatrix} 1 & 0.7208 & 0 \\ -0.7208 & 1 & 0 \\ 0 & 0 & 1 \end{bmatrix}$$

$i = 1, j = 3, \alpha = -0.1680, \beta = 0.5106$

$$P = \begin{bmatrix} 1 & 0 & -0.1680 \\ 0 & 1 & 0 \\ 0.5106 & 0 & 1 \end{bmatrix}$$

Post-multiplying EIGENVEC by P

$$EIGENVEC = \begin{bmatrix} 1 & 0.7208 & 0 \\ -0.7208 & 1 & 0 \\ 0 & 0 & 1 \end{bmatrix} \begin{bmatrix} 1 & 0 & -0.1680 \\ 0 & 1 & 0 \\ 0.5106 & 0 & 1 \end{bmatrix}$$

$$= \begin{bmatrix} 1 & 0.7208 & -0.1680 \\ -0.7208 & 1 & 0.1211 \\ 0.5106 & 0 & 1 \end{bmatrix}$$

i = 2, j = 3, α = -0.5128, β = 1.4352

$$P = \begin{bmatrix} 1 & 0 & 0 \\ 0 & 1 & -0.5128 \\ 0 & 1.4352 & 1 \end{bmatrix}$$

Post-multiplying EIGENVEC by P

$$EIGENVEC = \begin{bmatrix} 1 & 0.7208 & -0.1680 \\ -0.7208 & 1 & 0.1211 \\ 0.5106 & 0 & 1 \end{bmatrix} \begin{bmatrix} 1 & 0 & 0 \\ 0 & 1 & -0.5128 \\ 0 & 1.4352 & 1 \end{bmatrix}$$

$$= \begin{bmatrix} 1 & 0.4796 & -0.5376 \\ -0.7208 & 1.1738 & -0.3917 \\ 0.5106 & 1.4352 & 1 \end{bmatrix}$$

<u>Iteration 2:</u>

i = 1, j = 2, α = 0.0709, β = -0.0444

$$P = \begin{bmatrix} 1 & 0.0709 & 0 \\ -0.0444 & 1 & 0 \\ 0 & 0 & 1 \end{bmatrix}$$

Post-multiplying EIGENVEC by P

$$EIGENVEC = \begin{bmatrix} 1 & 0.4796 & -0.5376 \\ -0.7208 & 1.1738 & -0.3917 \\ 0.5106 & 1.4352 & 1 \end{bmatrix} \begin{bmatrix} 1 & 0.0709 & 0 \\ -0.0444 & 1 & 0 \\ 0 & 0 & 1 \end{bmatrix}$$

$$= \begin{bmatrix} 0.9787 & 0.5506 & -0.5376 \\ -0.7728 & 1.1227 & -0.3917 \\ 0.4469 & 1.4714 & 1 \end{bmatrix}$$

i = 1, j = 3, α =- 0.0734, β = 0.1289

$$P = \begin{bmatrix} 1 & 0 & -0.0734 \\ 0 & 1 & 0 \\ 0.1289 & 0 & 1 \end{bmatrix}$$

Post-multiplying EIGENVEC by P

$$EIGENVEC = \begin{bmatrix} 0.9787 & 0.5506 & -0.5376 \\ -0.7728 & 1.1227 & -0.3917 \\ 0.4469 & 1.4714 & 1 \end{bmatrix} \begin{bmatrix} 1 & 0 & -0.0734 \\ 0 & 1 & 0 \\ 0.1289 & 0 & 1 \end{bmatrix}$$

$$= \begin{bmatrix} 0.9094 & 0.5506 & -0.6094 \\ -0.8233 & 1.1227 & -0.3350 \\ 0.5758 & 1.4714 & 0.9672 \end{bmatrix}$$

$i = 2, j = 3, \alpha = 0.0032, \beta = -0.0088$

$$P = \begin{bmatrix} 1 & 0 & 0 \\ 0 & 1 & 0.0032 \\ 0 & -0.0088 & 1 \end{bmatrix}$$

Post-multiplying EIGENVEC by P

$$EIGENVEC = \begin{bmatrix} 0.9094 & 0.5506 & -0.6094 \\ -0.8233 & 1.1227 & -0.3350 \\ 0.5758 & 1.4714 & 0.9672 \end{bmatrix} \begin{bmatrix} 1 & 0 & 0 \\ 0 & 1 & 0.0032 \\ 0 & -0.0088 & 1 \end{bmatrix}$$

$$= \begin{bmatrix} 0.9094 & 0.5559 & -0.6077 \\ -0.8233 & 1.1256 & -0.3314 \\ 0.5758 & 1.4629 & 0.9719 \end{bmatrix}$$

At the end of the second iteration, M normalising the approximate eigen-vectors

$$EIGENVEC = \begin{bmatrix} 0.3520 & 0.1692 & -0.3124 \\ -0.3187 & 0.3451 & -0.1717 \\ 0.2229 & 0.4523 & 0.4958 \end{bmatrix}$$

The 'exact' values are

$$EIGENVEC = \begin{bmatrix} 0.3518 & 0.1708 & -0.3116 \\ -0.3184 & 0.3461 & -0.1699 \\ 0.2229 & 0.4496 & 0.4982 \end{bmatrix}$$

As can be seen the two sets of results are fairly close.

5.8 LANCZOS METHOD OF TRI-DIAGONALISATION

The Lanczos method falls into a class of solution methods, that do not attempt to solve the eigen-value problem directly, but like the Jacobi method of Section 5.7, transform the given matrix into a form whose eigen-values are the same as that of the original system, except that the solution of the modified system is much simpler to achieve. Lanczos transforms the given matrix into a tri-diagonal form. In a tri-diagonal matrix, there are only three elements in each row, the diagonal element and one element on either side of the diagonal. The theory of the method is as follows.

Let the eigen-value problem to be solved be defined by

$$A\,Y = Y\Lambda \tag{5.44}$$

Let X be a matrix whose columns are vectors $x_1, x_2, \ldots x_N$, which are mutually orthogonal unit vectors. Therefore, $X^T X = I$, where I is an identity matrix.

Multiplying equation (5.44) by X^T and replacing Y by X V, equation (5.44) transforms to

$$X^T A\,X\,V = X^T X\,V\Lambda \tag{5.45}$$

However because $X^T X = I$, equation (5.45) can be simplified as

$$X^T A\,X\,V = V\Lambda \tag{5.46}$$

Let

$$X^T A\,X = T \tag{5.47}$$

where T is a symmetric tri-diagonal matrix.

$$T = \begin{bmatrix} \alpha_1 & \beta_1 & & & & & & \\ \beta_1 & \alpha_2 & \beta_2 & & & & & \\ & \beta_2 & \alpha_3 & \beta_3 & & & & \\ & & \beta_3 & \alpha_4 & \beta_4 & & & \\ \cdot & \cdot & \cdot & \cdot & \cdot & \cdot & \cdot & \cdot \\ \cdot & \cdot & \cdot & \cdot & \cdot & \cdot & \cdot & \cdot \\ & & & & & \beta_{N-2} & \alpha_{N-1} & \beta_{N-1} \\ & & & & & & \beta_{N-1} & \alpha_N \end{bmatrix} \tag{5.48}$$

Equation (5.46) can be written as

$$T\,V = V\Lambda \tag{5.49}$$

This shows that the eigen-values of problems defined by equations (5.44) and (5.49) are identical and that T has the same eigen-values as A.

Pre-multiplying equation (5.47) by X

$$X\,X^T A\,X = X\,T \tag{5.50}$$

Noting that $X\,X^T = I$, equation (5.50) simplifies to

$$A\,X = X\,T \tag{5.51}$$

Carrying out the multiplication, $AX = XT$, leads to a series of linear simultaneous equations in α_i, β_i.

$$Ax_1 = \alpha_1 x_1 + \beta_1\, x_2 \tag{5.52}$$

$$Ax_2 = \beta_1\, x_1 + \alpha_2\, x_2 + \beta_2\, x_3 \tag{5.53}$$

$$A\,x_i = \beta_{i-1}\, x_{i-1} + \alpha_i\, x_i + \beta_i\, x_{i+1} \tag{5.54}$$

$$A\,x_N = \beta_{N-1}\, x_{N-1} + \alpha_N\, x_N \tag{5.55}$$

Pre-multiplying equation (5.52) by x_1^T

$$x_1^T\, A\,x_1 = \alpha_1\, x_1^T\, x_1 + \beta_1\, x_1^T\, x_2 \tag{5.56}$$

However since x_1 and x_2 are orthogonal, $x_1^T\, x_2 = 0$, equation (5.56) simplifies to

$$A\begin{bmatrix}x_1 & x_2 & x_3 & x_4 & \cdot & \cdot & x_{N-1} & x_N\end{bmatrix} =$$

$$\begin{bmatrix}x_1 & x_2 & x_3 & x_4 & \cdot & \cdot & x_{N-1} & x_N\end{bmatrix}\begin{bmatrix}\alpha_1 & \beta_1 & & & & & & \\ \beta_1 & \alpha_2 & \beta_2 & & & & & \\ & \beta_2 & \alpha_3 & \beta_3 & & & & \\ & & \beta_3 & \alpha_4 & \beta_4 & & & \\ \cdot & \cdot & \cdot & \cdot & \cdot & \cdot & \cdot & \cdot \\ \cdot & \cdot & \cdot & \cdot & \cdot & \cdot & \cdot & \cdot \\ & & & & & \beta_{N-2} & \alpha_{N-1} & \beta_{N-1} \\ & & & & & & \beta_{N-1} & \alpha_N\end{bmatrix}$$

$$x_1^T A x_1 = \alpha_1 \tag{5.57}$$

Therefore starting the computations with an arbitrary vector x_1 of unit length, α_1 can be determined. Knowing the value of α_1, from equation (5.52)

$$\bar{x}_2 = \beta_1 x_2 = A x_1 - \alpha_1 x_1 \tag{5.58}$$

$$\bar{x}_2^T \bar{x}_2 = \beta_1^2 x_2^T x_2 \tag{5.59}$$

Since x_2 is a vector of unit length

$$\therefore \quad \bar{x}_2^T \bar{x}_2 = \beta_1^2 \tag{5.60}$$

Knowing the value of β_1

$$x_2 = \frac{\bar{x}_2}{\beta_1} \tag{5.61}$$

Pre-multiplying equation (5.48) by x_2^T

$$x_2^T A x_2 = \beta_1 x_2^T x_1 + \alpha_2 x_2^T x_2 + \beta_2 x_2^T x_3 \tag{5.62}$$

Having known α_1 and β_1, and making use of the orthogonality conditions and the fact that x_i's are all vectors of unit length, α_2 can be determined from equation (5.62).

$$x_2^T A x_2 = \alpha_2 x_2^T x_2 = \alpha_2 \tag{5.63}$$

Proceeding in a similar manner, the general equations can be determined .

$$\alpha_i = x_i^T A x_i \tag{5.64}$$

$$A x_i = \beta_{i-1} x_{i-1} + \alpha_i x_i + \beta_i x_{i+1} \tag{5.65}$$

$$\bar{x}_{i+1} = \beta_i x_{i+1} = A x_i - \beta_{i-1} x_{i-1} - \alpha_i x_i \tag{5.66}$$

$$\beta_i^2 = \bar{x}_{i+1}^T \bar{x}_{i+1} \tag{5.67}$$

$$x_{i+1} = \frac{\bar{x}_{i+1}}{\beta_i} \tag{5.68}$$

In the case of the general eigen-value problem, $A = U^T M U^1$

5.8.1 Lanczos Method of Tri-diagonalisation: Example

Determine the tri-diagonal matrix for the three-storey shear frame used in Section 5.3.

$$K = \begin{bmatrix} 14 & -6 & 0 \\ -6 & 10 & -4 \\ 0 & -4 & 4 \end{bmatrix}, \quad M = \begin{bmatrix} 4 & 0 & 0 \\ 0 & 4 & 0 \\ 0 & 0 & 2 \end{bmatrix}$$

$$K = U^T\ U\ , U = \begin{bmatrix} 3.741657 & -1.603568 & 0 \\ 0 & 2.725541 & -1.467600 \\ 0 & 0 & 1.358732 \end{bmatrix}$$

$$U^T = \begin{bmatrix} 3.741657 & 0 & 0 \\ -1.603568 & 2.725541 & 0 \\ 0 & -1.467600 & 1.358732 \end{bmatrix}$$

Step 1: Let the arbitrary starting vector be $x_1^T = \{1\ , 0, 0\}$

Step 2: Solve $U\,Z = x_1$

$$\begin{bmatrix} 3.741657 & -1.603568 & 0 \\ 0 & 2.725541 & -1.467600 \\ 0 & 0 & 1.358732 \end{bmatrix} \begin{bmatrix} z_1 \\ z_2 \\ z_3 \end{bmatrix} = \begin{bmatrix} 1.0 \\ 0 \\ 0 \end{bmatrix}, \ Z = \begin{bmatrix} 0.267261 \\ 0 \\ 0 \end{bmatrix}$$

Step 3: Calculate $R = M\,Z$

$$R = \begin{bmatrix} 4 & 0 & 0 \\ 0 & 4 & 0 \\ 0 & 0 & 2 \end{bmatrix} \begin{bmatrix} 0.267261 \\ 0 \\ 0 \end{bmatrix} = \begin{bmatrix} 1.069045 \\ 0 \\ 0 \end{bmatrix}$$

Step 4: Solve $U^T\,S = R$

$$\begin{bmatrix} 3.741657 & 0 & 0 \\ -1.603568 & 2.725541 & 0 \\ 0 & -1.467600 & 1.358732 \end{bmatrix} \begin{bmatrix} s_1 \\ s_2 \\ s_3 \end{bmatrix} = \begin{bmatrix} 1.069045 \\ 0 \\ 0 \end{bmatrix}, \ S = \begin{bmatrix} 0.285714 \\ 0.168100 \\ 0.181569 \end{bmatrix}$$

Step 5: Since $S = A\ x_1$, α_1 can be determined

$$\alpha_1 = x_1^T\ A\ x_1 = x_1^T S = [1.0\ \ 0\ \ 0] \begin{bmatrix} 0285714 \\ 0.168100 \\ 0.181569 \end{bmatrix} = 0.285714$$

Step 6: Determine the product $\beta_1\ x_2$

$$\bar{x}_2 = \beta_1\, x_2 = Ax_1 - \alpha_1\, x_1 = \begin{bmatrix} 0.285714 \\ 0.168100 \\ 0.181569 \end{bmatrix} - 0.285714 \begin{bmatrix} 1.0 \\ 0 \\ 0 \end{bmatrix} = \begin{bmatrix} 0 \\ 0.168100 \\ 0.181569 \end{bmatrix}$$

Step 7: Determine β_1 and x_2

$$\beta_1^2 = \bar{x}_2^T \bar{x}_2 = 0.061225\ ,\ \beta_1 = 0.247436,\quad x_2 = \frac{\bar{x}_2}{\beta_1} = \begin{bmatrix} 0 \\ 0.679366 \\ 0.733800 \end{bmatrix}$$

Step 8: Repeat steps similar to steps 2-7 and calculate α_2 and β_2. Solve for Z from $U\,Z = x_2$

$$\begin{bmatrix} 3.741657 & -1.603568 & 0 \\ 0 & 2.725541 & -1.467600 \\ 0 & 0 & 1.358732 \end{bmatrix} \begin{bmatrix} z_1 \\ z_2 \\ z_3 \end{bmatrix} = \begin{bmatrix} 0 \\ 0.679366 \\ 0.733800 \end{bmatrix},$$

$$Z = \begin{bmatrix} 0.231455 \\ 0.540062 \\ 0.540062 \end{bmatrix}$$

Calculate $R = M\,Z$

$$R = \begin{bmatrix} 4 & 0 & 0 \\ 0 & 4 & 0 \\ 0 & 0 & 2 \end{bmatrix} \begin{bmatrix} 0.231455 \\ 0.540062 \\ 0.540062 \end{bmatrix} = \begin{bmatrix} 0.925821 \\ 2.160248 \\ 1.080124 \end{bmatrix}$$

Solve for R from $U^T\,S = R$

$$\begin{bmatrix} 3.741657 & 0 & 0 \\ -1.603568 & 2.725541 & 0 \\ 0 & -1.467600 & 1.358732 \end{bmatrix} \begin{bmatrix} s_1 \\ s_2 \\ s_3 \end{bmatrix} = \begin{bmatrix} 0.925821 \\ 2.160248 \\ 1.080124 \end{bmatrix},$$

$$Ax_2 = S = \begin{bmatrix} 0.247436 \\ 0.938173 \\ 1.808294 \end{bmatrix}$$

$$\alpha_2 = x_2^T\,A\,x_2 = x_2^T S = \begin{bmatrix} 0 & 0.679366 & 0.733800 \end{bmatrix} \begin{bmatrix} 0.247436 \\ 0.938173 \\ 1.808294 \end{bmatrix} = 1.964289$$

Calculate the product $\beta_2\,x_3$

$$\bar{x}_3 = \beta_2\,x_3 = A\,x_2 - \beta_1\,x_1 - \alpha_2\,x_2$$

$$= \begin{bmatrix} 0.247436 \\ 0.938173 \\ 1.808294 \end{bmatrix} - 0.247436 \begin{bmatrix} 1.0 \\ 0 \\ 0 \end{bmatrix} - 1.964289 \begin{bmatrix} 0 \\ 0.679366 \\ 0.733800 \end{bmatrix} = \begin{bmatrix} 0 \\ -0.396298 \\ 0.366898 \end{bmatrix}$$

Calculate β_2 and x_3

$$\beta_2^2 = \bar{x}_3^T \bar{x}_3 = 0.291667, \quad \beta_2 = 0.540062$$

$$x_3 = \frac{\bar{x}_3}{\beta_2} = \begin{bmatrix} 0 \\ -0.733801 \\ 0.679362 \end{bmatrix}$$

Step 9: Repeat steps similar to step 8 and calculate α_3 and β_3. Solve for Z from $U Z = x_3$

$$\begin{bmatrix} 3.741657 & -1.603568 & 0 \\ 0 & 2.725541 & -1.467600 \\ 0 & 0 & 1.358732 \end{bmatrix} \begin{bmatrix} z_1 \\ z_2 \\ z_3 \end{bmatrix} = \begin{bmatrix} 0 \\ -0.733801 \\ 0.679362 \end{bmatrix},$$

$$Z = \begin{bmatrix} -0.000001 \\ -0.000002 \\ 0.499997 \end{bmatrix}$$

Calculate $R = M Z$

$$R = \begin{bmatrix} 4 & 0 & 0 \\ 0 & 4 & 0 \\ 0 & 0 & 2 \end{bmatrix} \begin{bmatrix} -0.000001 \\ -0.000002 \\ 0.499997 \end{bmatrix} = \begin{bmatrix} -0.000004 \\ -0.000008 \\ 0.999994 \end{bmatrix}$$

Solve for S from $U^T S = R$

$$\begin{bmatrix} 3.741657 & 0 & 0 \\ -1.603568 & 2.725541 & 0 \\ 0 & -1.467600 & 1.358732 \end{bmatrix} \begin{bmatrix} s_1 \\ s_2 \\ s_3 \end{bmatrix} = \begin{bmatrix} -0.000004 \\ -0.000008 \\ 0.999994 \end{bmatrix}$$

$$Ax_3 = S = \begin{bmatrix} -0.000001 \\ -0.000004 \\ 0.735972 \end{bmatrix}$$

$$\alpha_3 = x_3^T \, A \, x_3 = x_3^T \, S = \begin{bmatrix} 0 & -0.733801 & 0.679362 \end{bmatrix} \begin{bmatrix} -0.000001 \\ -0.000004 \\ 0.735972 \end{bmatrix} = 0.499991$$

This completes the calculations to determine the tri-diagonal matrix T

$$T = \begin{bmatrix} 0.285714 & 0.247436 & \\ 0.247436 & 1.964289 & 0.540062 \\ & 0.540062 & 0.499991 \end{bmatrix}$$

The details of how the eigen-values of a tri-diagonal matrix T can be determined by QR factorisation or the method of bisection will be explained in Sections 5.9.5 and 5.10. The eigen-values can be shown to be $\lambda = 1/\omega^2 = 0.205869$, 0.372857 and 2.1712698. The eigen-vectors are best calculated using inverse iteration with shift as explained in Section 5.5.1.

5.8.2 Comments on Convergence of Lanczos Method

Figure 5.1 shows an eleven-storey shear frame. The stiffness and mass matrices are shown in Tables 5.3 and 5.4 respectively.

Table 5.3 Stiffness matrix

4	-4									
-4	10	-6								
	-6	14	-8							
		-8	18	-10						
			-10	22	-12					
				-12	24	-12				
					-12	26	-14			
						-14	28	-14		
							-14	30	-16	
								-16	32	-16
									-16	32

Table 5.4 Mass matrix

1										
	2									
		2								
			2							
				2						
					3					
						3				
							3			
								4		
									4	
										4

The tri-diagonal matrix is shown in Table 5.5. Table 5.6 shows the eigen-values of all the leading sub-matrices of the tri-diagonal matrix. What is clear from the data in Table 5.6 is that the eigen-values of sub-matrices converge towards the eigen-values of the whole structure and the convergence takes place such that the values at the higher end converge from the left and the values at the lower end converge from the right. Since in any practical situation, the interest generally lies at the lower end of the eigen-values or some times in the values lying in a certain band, it is not necessary for the entire T matrix to be evaluated and then its eigen-values

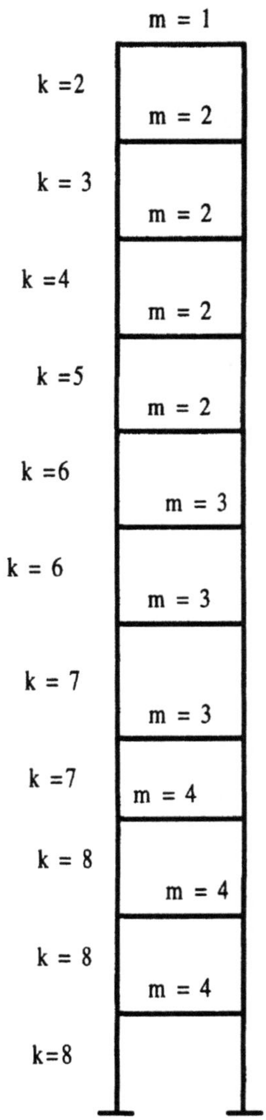

Fig. 5.1 Eleven-storey shear frame

Table 5.5 The Tri-diagonal Matrix

0.2500	0.4714									
0.4714	6.6160	2.8512								
	2.8512	2.5260	0.5085							
		0.5085	0.6661	0.1855						
			0.1855	0.3486	0.0885					
				0.0885	0.1895	0.0617				
					0.0617	3.6079	3.9508			
						3.9508	4.6414	0.0428		
							0.0428	0.1033	0.0683	
								0.0683	1.2238	0.4390
									0.4390	0.2570

Table 5.6 Eigen-values of sub-matrices.

0.2500										
6.6507	0.2153									
8.1022	1.1144	0.1755								
8.1094	1.3865	0.4157	0.1468							
8.1094	1.3957	0.5331	0.2427	0.1259						
8.1094	1.3957	0.5419	0.2816	0.1618	0.1058					
8.1094	3.6090	1.3957	0.5419	0.2813	0.1613	0.1055				
8.1094	8.1093	1.3957	0.5420	0.2860	0.1833	0.1306	0.0891			
8.1094	8.1094	1.3957	0.5420	0.2661	0.1868	0.1382	0.1046	0.0766		
8.1094	8.1094	1.3957	1.2280	0.5420	0.2661	0.1886	0.1375	0.1029	0.0749	
8.1094	8.1094	1.3965	1.3957	0.5420	0.2661	0.1869	0.1398	0.1134	0.0886	0.0617

calculated. The Lanczos method can be iterative in the sense that after a certain number of vectors are calculated and the corresponding tri-diagonal matrix is calculated, the eigen-values of the sub-matrices are calculated. If the indications are that sufficient eigen-values to sufficient accuracy have been calculated, then it is not necessary to compute any further. As a general rule of thumb, if m eigen-values are required, then calculations should be continued until the T matrix is at least of size 2 m x 2m. A certain amount of doubt persists in exactly identifying the equivalence of the eigen-values of matrices T and the partial eigen-values of A. However, Sturm sequence checks to be discussed in Section 5.9 can be used to ensure that all the necessary eigen-values of interest are calculated.

In the calculation of the T-matrix, it has been assumed that all the vectors are orthogonal. Unfortunately, one of the difficulties, which has been noticed in the application of the Lanczos method, is that due to the accumulation of round off errors in large-scale computation, loss of orthogonality occurs. The simplest solution is to make sure using the Gram-Schmidt orthogonalisation procedure (see Section 5.3.1 that the vectors are orthogonal. Unfortunately, in large-scale problems, this can be expensive. The alternative is to use what is called 'selective orthogonalisation'.

5.8.3 Selective Orthogonalisation

Selective orthogonalisation means orthogonalising the vectors only when the error is above an acceptable limit. This requires an inexpensive procedure for detecting the shift away from orthogonality. This can be as follows.

Consider equation (5.54)

$$A x_i = \beta_i \, x_{i-1} + \alpha_i \, x_i + \beta_{i+1} \, x_{i+1} \tag{5.54}$$

Pre-multiplying equation (5.54) by x_j^T

$$x_j^T A \, x_i = \beta_i \, h_{j,i-1} + \alpha_i \, h_{j,i} + \beta_{i+1} \, h_{j,i+1} \tag{5.69}$$

where $h_{j,i} = h_{i,j} = x_i^T x_j = x_j^T x_i$

Similarly, using j instead of i as a subscript in equation (5.54)

$$A x_j = \beta_j \, x_{j-1} + \alpha_j \, x_j + \beta_{j+1} \, x_{j+1} \tag{5.70}$$

Pre-multiplying equation (5.70) by x_i^T

$$x_i^T A \, x_j = \beta_j \, h_{i,j-1} + \alpha_j \, h_{i,j} + \beta_{j+1} \, h_{i,j+1} \tag{5.71}$$

Equating equations (5.69) and (5.70) and rearranging

$$\beta_{j+1} \, h_{i,j+1} = \beta_i \, h_{j,i-1} + \alpha_i \, h_{i,j} + \beta_{i+1} \, h_{j,i+1} - \alpha_j \, h_{i,j} - \beta_j \, h_{i,j-1} \tag{5.72}$$

In matrix form, equation (5.72) can be expressed as

$$\beta_{j+1}\begin{bmatrix}\eta_{1,j+1}\\ \eta_{2,j+1}\\ \eta_{3,j+1}\\ \cdot\\ \eta_{i,j+1}\\ \cdot\\ \eta_{j-1,j+1}\end{bmatrix}=\begin{bmatrix}\alpha_1 & \beta_2 & & & & & \\ \beta_2 & \alpha_2 & \beta_3 & & & & \\ & \beta_3 & \alpha_3 & \beta_4 & & & \\ & & \cdot & \cdot & \cdot & & \\ & & & \beta_i & \alpha_i & \beta_{i+1} & \\ & & & & & \cdot & \cdot \\ & & & & & \beta_{j-1} & \alpha_{j-1}\end{bmatrix}\begin{bmatrix}\eta_{1,j}\\ \eta_{2,j}\\ \eta_{3,j}\\ \cdot\\ \eta_{i,j}\\ \cdot\\ \eta_{j-1,j}\end{bmatrix}-\alpha_j\begin{bmatrix}\eta_{1,j}\\ \eta_{2,j}\\ \eta_{3,j}\\ \cdot\\ \eta_{i,j}\\ \cdot\\ \eta_{j-1,j}\end{bmatrix}-\beta_j\begin{bmatrix}\eta_{1,j-1}\\ \eta_{2,j-1}\\ \eta_{3,j-1}\\ \cdot\\ \eta_{i,j-1}\\ \cdot\\ \eta_{j-1,j-1}\end{bmatrix} \tag{5.73}$$

$$j\ge 2\,,\quad 1\le i\le (j-1)$$

$$\eta_{ij}=\eta_{ji}=q_i^T\;q_j=q_j^T\;q_i\,,i\ne j$$

$$\eta_{ii}=\eta_{jj}=0$$

This recursion formula in equation (5.73) is used as follows.
Assume that $q_i^T q_i = 1.0$ and $\eta_{j,j-1} = \varepsilon$ for $j \geq 2$, where the maximum error tolerable in orthogonalisation is ε.
Step 1: Assume $\eta_{12} = \varepsilon$.
Step 2: From the recursion formula

$$j=2,\ i=1:\ \beta_3\eta_{13}=\alpha_1\eta_{12}-\alpha_2\,\eta_{12}-\beta_2\eta_{11}$$

Therefore η_{13} is known and η_{23} is assumed to be ε.
Step 3: From the recursion formula

$$j=3,\ 1\le i\le 2:\ \beta_4\begin{bmatrix}\eta_{14}\\ \eta_{24}\end{bmatrix}=\begin{bmatrix}\alpha_1 & \beta_2\\ \beta_2 & \alpha_2\end{bmatrix}\begin{bmatrix}\eta_{14}\\ \eta_{24}\end{bmatrix}-\alpha_3\begin{bmatrix}\eta_{13}\\ \eta_{23}\end{bmatrix}-\beta_3\begin{bmatrix}\eta_{12}\\ \eta_{22}\end{bmatrix}$$

Therefore η_{14} and η_{24} can be determined and assume $\eta_{34} = \varepsilon$.
Step 4: From the recursion formula

$$j=4\,,1\le i\le 3:\ \beta_5\begin{bmatrix}\eta_{15}\\ \eta_{25}\\ \eta_{35}\end{bmatrix}=\begin{bmatrix}\alpha_1 & \beta_2 & 0\\ \beta_2 & \alpha_2 & \beta_3\\ 0 & \beta_3 & \alpha_3\end{bmatrix}\begin{bmatrix}\eta_{14}\\ \eta_{24}\\ \eta_{34}\end{bmatrix}-\alpha_4\begin{bmatrix}\eta_{14}\\ \eta_{24}\\ \eta_{34}\end{bmatrix}-\beta_4\begin{bmatrix}\eta_{13}\\ \eta_{23}\\ \eta_{33}\end{bmatrix}$$

Therefore η_{15}, η_{25} and η_{35} can be determined and assume $\eta_{45} = \varepsilon$.

Clearly this process can be continued and whenever the error exceeds acceptable limits, full orthogonalisation with respect to all the previous vectors can be undertaken.

5.9 STURM SEQUENCE

Many iterative methods such as the simultaneous iteration are available for the determination of the first few eigen-values and eigen-vectors of a large matrix. However, because of the difficulty of always ensuring that only the first few eigen-pairs have been determined, it is good practice to use a test that ensures that this has been done. Although many methods of checking exist, the Sturm sequence check is the most suitable one for this purpose. In order to establish the basic ideas of the method, consider a symmetric 4 x 4 matrix A_4

$$A_4 = \begin{bmatrix} a_{11} & a_{12} & a_{13} & a_{14} \\ a_{12} & a_{22} & a_{23} & a_{24} \\ a_{13} & a_{23} & a_{33} & a_{34} \\ a_{14} & a_{24} & a_{34} & a_{44} \end{bmatrix} \tag{5.74}$$

Let the eigen-values of A_4 be

$$^4\lambda_1 > {}^4\lambda_2 > {}^4\lambda_3 > {}^4\lambda_4$$

Note that the superscript 4 to the left of λ indicates that the eigen-values belong to the matrix A_4. The characteristic equation is given by equating to zero, the determinant of $(A_4 - \mu I)$.

$$f_4(\lambda) = ({}^4\lambda_1 - \lambda)\ ({}^4\lambda_2 - \lambda)({}^4\lambda_3 - \lambda)({}^4\lambda_4 - \lambda)$$

Let A_3 be a sub-matrix derived by deleting the fourth row and column from A_4

$$A_3 = \begin{bmatrix} a_{11} & a_{12} & a_{13} \\ a_{12} & a_{22} & a_{23} \\ a_{13} & a_{23} & a_{33} \end{bmatrix} \tag{5.75}$$

Let the eigen-values of A_3 be

$$^3\lambda_1 > {}^3\lambda_2 > {}^3\lambda_3$$

Let the eigen-vectors of A_3 corresponding to $^3\lambda_1$, $^3\lambda_2$ and $^3\lambda_3$ be unit length vectors $^3q_1$, $^3q_2$ and $^3q_3$, respectively. Let the matrices ^3Q and $^3\Lambda$ contain the complete set of eigen-vectors and eigen-values, respectively.

$$^3Q = \begin{bmatrix} ^3q_{11} & ^3q_{12} & ^3q_{13} \\ ^3q_{21} & ^3q_{22} & ^3q_{23} \\ ^3q_{31} & ^3q_{32} & ^3q_{33} \end{bmatrix},\quad ^3\Lambda = \begin{bmatrix} ^3\lambda_1 & & \\ & ^3\lambda_2 & \\ & & ^3\lambda_3 \end{bmatrix}$$

The three matrices A_3, ^3Q and $^3\Lambda$ are related by the special eigen-value problem as follows

$$A_3\ {}^3Q = {}^3Q\ {}^3\Lambda \tag{5.76}$$

Pre-multiplying equation (5.76), by $^3Q^T$

$$^3Q^T\ A_3\ {}^3Q = {}^3Q^T\ {}^3Q\ {}^3\Lambda \tag{5.77}$$

Remembering that the eigen-vectors are mutually orthogonal and they are of unit length

$$\because\ {}^3Q^T\ {}^3Q = I, \therefore\ {}^3Q^T\ A_3\ {}^3Q = {}^3\Lambda \tag{5.78}$$

In order to ensure that the correct eigen-values in a given interval have been calculated, one needs to establish some limits within which the eigen-values lie. One way of doing this is to investigate how the eigen-values of A_3 and A_4 are related. Let a matrix P be defined as

$$P = \begin{bmatrix} & {}^3Q & & 0 \\ & & & 0 \\ & & & 0 \\ 0 & 0 & 0 & 1 \end{bmatrix}, P^T = \begin{bmatrix} & {}^3Q^T & & 0 \\ & & & 0 \\ & & & 0 \\ 0 & 0 & 0 & 1 \end{bmatrix}, P^T P = \begin{bmatrix} & {}^3Q\ {}^3Q^T & & 0 \\ & & & 0 \\ & & & 0 \\ 0 & 0 & 0 & 1 \end{bmatrix} = I$$

Considering the eigen-value problem

$$A_4\ {}^4Q = {}^4Q\ {}^4\Lambda \tag{5.79}$$

Pre-multiplying equation (5.79) by P^T

$$P^T\ A_4\ {}^4Q = P^T\ {}^4Q\ {}^4\Lambda \tag{5.80}$$

Setting ${}^4Q = P\,{}^4Y$

$$P^T\ A_4 P\ {}^4Y = P^T\ P\ {}^4Y\ {}^4\Lambda \tag{5.81}$$

Since $P^T P = I$, and letting $G_4 = P^T A_4 P$, equation (5.81) simplifies to

$$G_4\ {}^4Y = {}^4Y\ {}^4\Lambda \tag{5.82}$$

Comparison of equations (5.79) and (5.81) shows that the eigen-values of G_4 are the same as the eigen-values of A_4. Substituting for matrices P and A_4, G_4 is given by

$$G_4 = \begin{bmatrix} & {}^3Q^T & & 0 \\ & & & 0 \\ & & & 0 \\ 0 & 0 & 0 & 1 \end{bmatrix} \begin{bmatrix} & A_3 & & a_{14} \\ & & & a_{24} \\ & & & a_{34} \\ a_{14} & a_{24} & a_{34} & a_{44} \end{bmatrix} \begin{bmatrix} & {}^3Q & & 0 \\ & & & 0 \\ & & & 0 \\ 0 & 0 & 0 & 1 \end{bmatrix} \tag{5.83}$$

Multiplying through, equation (5.83) becomes

$$G_4 = \begin{bmatrix} {}^3Q^T\ A_3\ {}^3Q & Q^3 V \\ V^T\ {}^3Q & a_{44} \end{bmatrix} \tag{5.84}$$

where $V^T = [a_{14},\ a_{24},\ a_{34}]$

Substituting equation (5.78) in equation (5.84)

$$G_4 = \begin{bmatrix} {}^3\Lambda & C \\ C^T & a_{44} \end{bmatrix},\quad C = Q^3 V = \begin{bmatrix} c_1 \\ c_2 \\ c_3 \end{bmatrix}, c_i = \sum_{j=1}^{3} a_{j4}\ {}^3q_{ji}$$

In order to find the eigen-values of G_4 (or A_4) form

$$G - \mu I = \begin{bmatrix} {}^3\lambda_1 - \mu & 0 & 0 & c_1 \\ 0 & {}^3\lambda_2 - \mu & 0 & c_2 \\ 0 & 0 & {}^3\lambda_3 - \mu & c_3 \\ c_1 & c_2 & c_3 & a_{44} - \mu \end{bmatrix}$$

Successively substituting for $\mu = {}^3\lambda_i$, i = 1 to 3, the determinants are

$$|G_4 - {}^3\lambda_1 I| = (-1)^{4+1} c_1 \begin{bmatrix} 0 & {}^3\lambda_2 - {}^3\lambda_1 & 0 \\ 0 & 0 & {}^3\lambda_3 - {}^3\lambda_1 \\ c_1 & c_2 & c_3 \end{bmatrix} = ({}^3\lambda_2 - {}^3\lambda_1) c_1 \begin{bmatrix} 0 & {}^3\lambda_3 - {}^3\lambda_1 \\ c_1 & c_3 \end{bmatrix}$$

$$= -({}^3\lambda_2 - {}^3\lambda_1)({}^3\lambda_3 - {}^3\lambda_1) c_1^2$$

$$|G_4 - {}^3\lambda_2 I| = ({}^3\lambda_1 - {}^3\lambda_2) \begin{bmatrix} 0 & 0 & c_2 \\ 0 & {}^3\lambda_3 - {}^3\lambda_2 & c_3 \\ c_2 & c_3 & a_{44} - {}^3\lambda_2 \end{bmatrix} + (-1)^{4+1} c_1 \begin{bmatrix} 0 & 0 & 0 \\ 0 & 0 & {}^3\lambda_3 - {}^3\lambda_1 \\ c_1 & c_2 & c_3 \end{bmatrix}$$

$$= ({}^3\lambda_1 - {}^3\lambda_2) c_2 \begin{bmatrix} 0 & {}^3\lambda_3 - {}^3\lambda_2 \\ c_2 & c_3 \end{bmatrix} = -({}^3\lambda_1 - {}^3\lambda_2)({}^3\lambda_3 - {}^3\lambda_2) c_2^2 = ({}^3\lambda_1 - {}^3\lambda_2)({}^3\lambda_2 - {}^3\lambda_3) c_2^2$$

$$|G_4 - {}^3\lambda_3 I| = ({}^3\lambda_1 - {}^3\lambda_3) \begin{bmatrix} {}^3\lambda_2 - {}^3\lambda_3 & 0 & c_2 \\ 0 & 0 & c_3 \\ c_2 & c_3 & a_{44} - {}^3\lambda_3 \end{bmatrix} + (-1)^{4+1} c_1 \begin{bmatrix} 0 & {}^3\lambda_2 - {}^3\lambda_3 & 0 \\ 0 & 0 & 0 \\ c_1 & c_2 & c_3 \end{bmatrix}$$

$$= ({}^3\lambda_1 - {}^3\lambda_3)({}^3\lambda_2 - {}^3\lambda_3) \begin{bmatrix} 0 & c_3 \\ c_3 & a_{44} - {}^3\lambda_3 \end{bmatrix} = -({}^3\lambda_1 - {}^3\lambda_3)({}^3\lambda_2 - {}^3\lambda_3) c_3^2$$

Since ${}^3\lambda_1 > {}^3\lambda_2 > {}^3\lambda_3$, and c_i^2 are all positive, the three determinants can be expressed as

$$|G_4 - {}^3\lambda_1 I| = -({}^3\lambda_1 - {}^3\lambda_2)({}^3\lambda_1 - {}^3\lambda_3) c_1^2 \text{ is negative}$$

$$|G_4 - {}^3\lambda_2 I| = +({}^3\lambda_1 - {}^3\lambda_2)({}^3\lambda_2 - {}^3\lambda_3) c_2^2 \text{ is positive}$$

Therefore, one root of A_4 must lie between ${}^3\lambda_1$ and ${}^3\lambda_2$. Similarly

$$|G_4 - {}^3\lambda_2 I| = +({}^3\lambda_1 - {}^3\lambda_2)({}^3\lambda_2 - {}^3\lambda_3) c_2^2 \text{ is positive}$$

$$|G_4 - {}^3\lambda_3 I| = -({}^3\lambda_1 - {}^3\lambda_3)({}^3\lambda_2 - {}^3\lambda_3) c_3^2 \text{ is negative}$$

Therefore, one root of A_4 must lie between ${}^3\lambda_2$ and ${}^3\lambda_3$.

Since for any stable structure, the determinant of G_4 will be positive, we have

$$|G_4 - 0I| \text{ is positive}$$

$$|G_4 - {}^3\lambda_1 I| = -({}^3\lambda_1 - {}^3\lambda_2)({}^3\lambda_1 - {}^3\lambda_3) c_1^2 \text{ is negative}$$

Therefore, one root of A_4 must lie between 0 and ${}^3\lambda_1$.

This shows that the eigen-values of A_3 interleave the eigen-values of A_4. In a similar way, considering the eigen-values of sub-matrices A_3 and A_2, A_2 and A_1, it can be shown that the eigen-values of A_2 interleave the eigen-values of A_3 and that of A_1 interleave A_2.

5.9.1 Interleaving of Eigen-values: Example

The following stiffness and mass matrices refer to the four-storey shear frame shown in Fig. 3.4. Calculate the eigen-values of all its sub-matrices and show that the eigen-values interleave as discussed in Section 5.9.

$$K = \begin{bmatrix} 14 & -6 & 0 & 0 \\ -6 & 10 & -4 & 0 \\ 0 & -4 & 8 & -4 \\ 0 & 0 & -4 & 4 \end{bmatrix}, \quad M = \begin{bmatrix} 4 & 0 & 0 & 0 \\ 0 & 4 & 0 & 0 \\ 0 & 0 & 2 & 0 \\ 0 & 0 & 0 & 2 \end{bmatrix}$$

Form the matrix $[K - \omega^2 M]$ of all the leading sub-matrices and determine their eigen-values.

$$A_4 = K - \omega^2 M = \begin{bmatrix} 14-4\omega^2 & -6 & 0 & 0 \\ -6 & 10-4\omega^2 & -4 & 0 \\ 0 & -4 & 8-2\omega^2 & -4 \\ 0 & 0 & -4 & 4-2\omega^2 \end{bmatrix}$$

$$|A_4| = f_4(\omega^2) = 64\omega^8 - 768\omega^6 + 2848\,\omega^4 - 3328\omega^2 + 768$$

$$f_4(\omega^2) = 0, \textit{ when } \omega^2 = 0.303146,\ 1.591674,\ 4.241005 \textit{ and } 5.864176$$

$$A_3 = \begin{bmatrix} 14-4\omega^2 & -6 & 0 \\ -6 & 10-4\omega^2 & -4 \\ 0 & -4 & 8-2\omega^2 \end{bmatrix}$$

$$|A_3| = f_3(\omega^2) = -32\omega^6 + 320\,\omega^4 - 912\omega^2 + 608$$

$$f_3(\omega^2) = 0, \textit{ when } \omega^2 = 0.957622,\ 3.746381 \textit{ and } 5.295998$$

$$A_2 = \begin{bmatrix} 14-4\omega^2 & -6 \\ -6 & 14-4\omega^2 \end{bmatrix}$$

$$|A_2| = f_2(\omega^2) = 16\,\omega^4 - 96\,\omega^2 + 104$$

$$f_2(\omega^2) = 0, \textit{ when } \omega^2 = 1.418861 \textit{ and } 4.581139$$

$$A_1 = \left[14 - 4\omega^2\right]$$

$$|A_1| = f_1(\omega^2) = 14 - 4\,\omega^2$$

$$f_1(\omega^2) = 0, \textit{ when } \omega^2 = 3.50$$

Table 5.7 shows the eigen-values of all the leading sub-matrices and it is clear that they do interleave as shown in Section 5.9. Fig. 5.2 shows a plot of $f_i(\omega^2)$, i = 1 to 4.

Table 5.7 Eigen-values of sub-matrices

			3.50			
		1.4189		4.5811		
	0.9576		3.7464		5.2960	
0.3032		1.5917		4.2410		5.8642

5.9.2 Sturm Sequence Property and Sign Counts

The property that the roots of $f_{i-1}(\omega^2)$ interleave the roots of $f_i(\omega^2)$ can be made use of to determine for any value of ω^2, how many eigen-values are greater than the value chosen. For the example considered in Section 5.9.1, let the value of $f_i(\omega^2)$, i= 0 to 4 be calculated for $\omega^2 = 1.2$. Assuming that $f_o(1.2)$ is arbitrarily equal to 1, the values of the determinants of leading sub-matrices are
$f_0(1.2) = +1$, $f_1(1.2) = +9.2$, $f_2(1.2) = +11.84$, $f_3(1.2) = -80.90$, $f_4(1.2) = -318.87$
If the attention is focussed on the signs of these, they are

$$f_0(1.2) = +,\ f_1(1.2) = +,\ f_2(1.2) = +,\ f_3(1.2) = -\ \text{and}\ f_4(1.2) = -$$

This is known as Sturm sequence. If we compare the agreements in the signs of the consecutive members of the above sequence, we have one agreement between the signs of $f_0(1.2)$ and $f_1(1.2)$, as they are both positive. Similarly there is another agreement between the signs of $f_1(1.2)$ and $f_2(1.2)$. However, between the signs of $f_2(1.2)$ and $f_3(1.2)$ there is no agreement, as one is positive and the other one is negative. Finally, between the signs of $f_3(1.2)$ and $f_4(1.2)$, there is one more agreement, because both of them are negative. The total number of agreements in the sign is therefore equal to three. Thus there are three eigen-values of A_4 which are greater than $\omega^2 = 1.2$. These are, from Table 5.7, $\omega^2 = 1.5917$, 4.2410 and 5.8642.

As another check, consider ω^2 = say 3.6. Assuming that $f_o(3.6)$ is arbitrarily equal to 1, the values of the determinants of leading sub-matrices are
$f_0(3.6) = +1$, $f_1(3.6) = -0.40$, $f_2(3.6) = -34.24$, $f_3(3.6) = -20.99$ and $f_4(3.6) = 615.01$
The signs of these are

$$f_0(3.6) = +,\ f_1(3.6) = -,\ f_2(3.6) = -,\ f_3(3.6) = -\ \text{and}\ f_4(3.6) = +$$

If we compare the agreements of the signs of the consecutive members of the above sequence, we have between the signs of $f_0(3.6)$ and $f_1(3.6)$ no agreement, between the signs of $f_1(3.6)$ and $f_2(3.6)$ one agreement, between the signs of $f_2(3.6)$ and $f_3(3.6)$ another agreement and, finally, between the signs of $f_3(3.6)$ and $f_4(3.6)$ no agreement. The total number of agreements in sign is therefore equal to two. Thus there are two eigen-values of A_4 which are greater than $\omega^2 = 3.6$. These are, from Table 5.7, $\omega^2 = 4.2410$ and 5.8642.

Finally, as a last check, consider ω^2 = say 6.0. Assuming that $f_o(6.0)$ is arbitrarily equal to 1, the values of the determinants of leading sub-matrices are
$f_0(6) = +1$, $f_1(6) = -10$, $f_2(6) = 104$, $f_3(6) = -256$ and $f_4(6) = 384$
The signs of these are

$$f_0(6) = +,\ f_1(6) = -,\ f_2(6) = +,\ f_3(6) = -\ \text{and}\ f_4(6) = +$$

If we compare the agreements in the signs of the consecutive members of the above sequence, we find that there are no agreements. Thus the total number of agreements in the sign is zero, indicating that there are no eigen-values of A_4 which are greater than $\omega^2 = 6.0$. From the previous calculations, the highest eigen-value is equal to $\omega^2 = 5.8642$, which accords with the conclusion from the count of sign agreement.

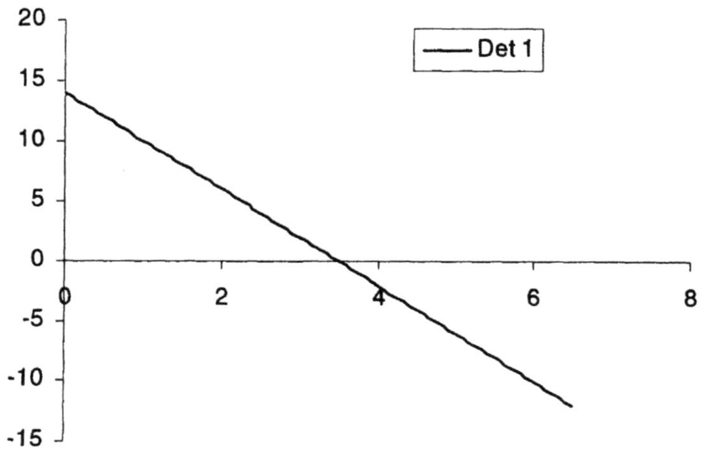

Fig. 5.2a Plot of $f_1(\omega^2)$

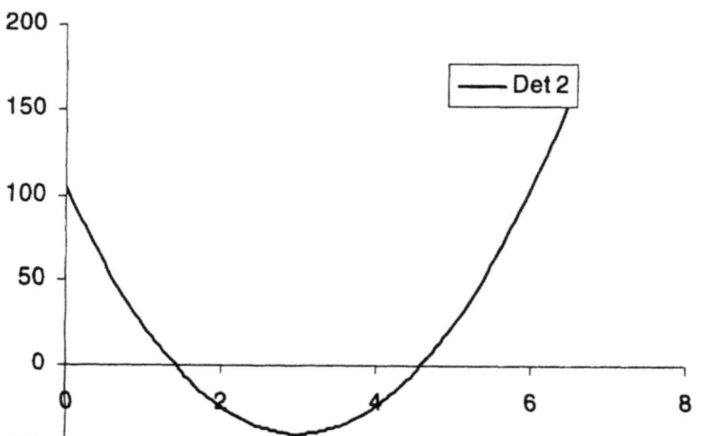

Fig. 5.2b Plot of $f_2(\omega^2)$

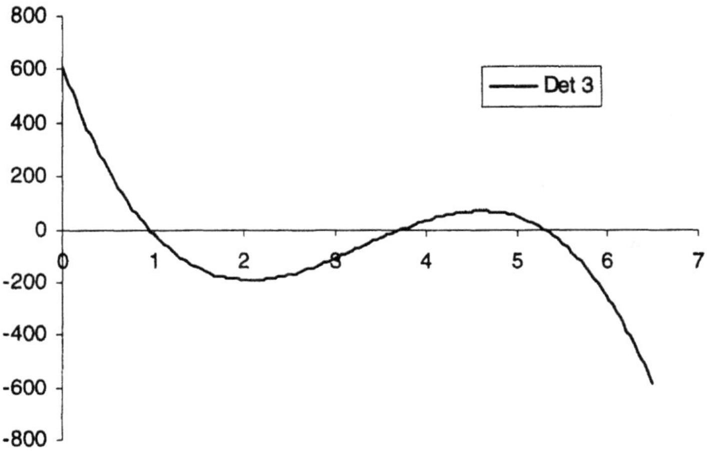

Fig. 5.2c Plot of $f_3(\omega^2)$

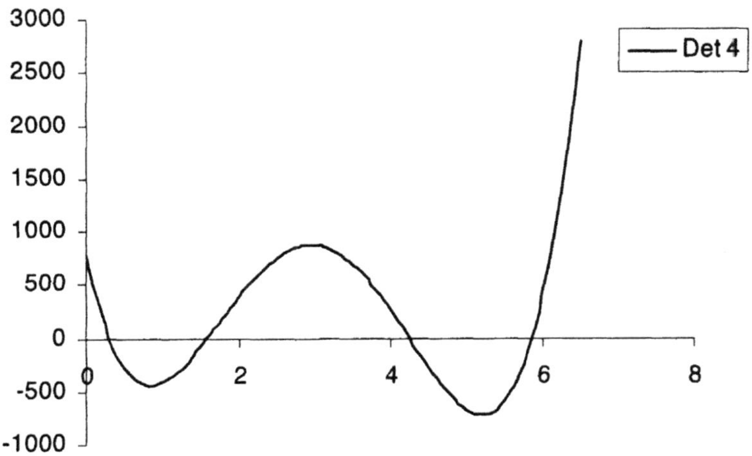

Fig. 5.2d Plot of $f_4(\omega^2)$

The Sturm sequence property is extremely useful for two reasons.

- It is useful in determining the number of eigen-values in a given band. For example, if the lower limit of the band is λ_{lower} and the upper limit is λ_{upper}, then the difference in the number of sign agreements calculated at $\lambda = \lambda_{lower}$ and $\lambda = \lambda_{upper}$ will be the number of eigen-values lying in the given band. In many practical applications, if the frequency band within which the forces acting on a structure is known, then all that the designer would like to know is if there are any significant natural frequencies of the structure in that range, which might cause problems of resonance.

- It is also useful in isolating a specific eigen-value. All that one has to do is to ensure that the difference in the number of sign agreements calculated at $\lambda = \lambda_{lower}$ and $\lambda = \lambda_{upper}$ is equal to one. Iteratively, the difference between the upper and lower limits can be made as small as required.

5.9.3 Sign Counts: Example

The ideas of Section 5.9.3 are illustrated by an example.

Example: For the four-storey shear frame discussed in Section 5.9.1, calculate the number of eigen-values lying in the range of $1.5 < \omega^2 < 5.0$.

The solution procedure consists of forming $[K - \omega^2 M]$ at $\omega^2 = 1.5$ and $\omega^2 = 5.0$ and using the Gaussian elimination procedure, $[K - \omega^2 M]$ is reduced to the upper triangular form. Then the determinants of the sub-matrices are determined by simply considering the continued product of the diagonal terms. From Section 5.9.1

$$K - \omega^2 M = \begin{bmatrix} 14-4\omega^2 & -6 & 0 & 0 \\ -6 & 10-4\omega^2 & -4 & 0 \\ 0 & -4 & 8-2\omega^2 & -4 \\ 0 & 0 & -4 & 4-2\omega^2 \end{bmatrix}$$

At $\omega^2 = 5$

$$K - 5.0M = \begin{bmatrix} -6 & -6 & 0 & 0 \\ -6 & -10 & -4 & 0 \\ 0 & -4 & -2 & -4 \\ 0 & 0 & -4 & -6 \end{bmatrix}$$

After Gaussian elimination

$$\begin{bmatrix} -6 & -6 & 0 & 0 \\ 0 & -4 & -4 & 0 \\ 0 & 0 & 2 & -4 \\ 0 & 0 & 0 & -14 \end{bmatrix}$$

The determinants of the submatrices are

$$f_1(5) = -6,\ f_2(5) = -6(-4),\ f_3(5) = -6(-4)(2),\ f_4(5) = -6(-4)(2)(-14)$$

The signs of $f_0(5) = +$, $f_1(5) = -$, $f_2(5) = +$, $f_3(5) = +$ and $f_4(5) = -$. Therefore the number of sign agreements is equal to 1.

In fact if we simply count the number of positive diagonal terms, then this will be equal to the number of sign agreements. In this case only 2 is positive.

Similarly

$$K - 1.5M = \begin{bmatrix} 8 & -6 & 0 & 0 \\ -6 & 4 & -4 & 0 \\ 0 & -4 & 5 & -4 \\ 0 & 0 & -4 & 1 \end{bmatrix}$$

After Gaussian elimination

$$\begin{bmatrix} 8 & -6 & 0 & 0 \\ 0 & -0.5 & -4 & 0 \\ 0 & 0 & 37 & -4 \\ 0 & 0 & 0 & 0.568 \end{bmatrix}$$

The determinants of the sub-matrices are

$f_1(1.5) = 8,\ f_2(1.5) = 8(-0.5),\ f_3(1.5) = 8(-0.5)(37),\ f_4(1.5) = 8(-0.5)(37)(0.568)$

The signs of $f_0(1.5) = +$, $f_1(1.5) = +$, $f_2(1.5) = -$, $f_3(1.5) = -$ and $f_4(1.5) = -$.

Therefore the number of sign agreements is equal to 3. This is also equal to the number of positive diagonal terms.

The difference in the number of sign agreements is equal to 3 - 1 = 2. There must be two eigen–values in the range of $1.5 < \omega^2 < 5.0$.

5.9.4 Isolating a Specific Eigen-value by Root Bi-section: Example

Sturm sequence property is very useful for isolating a specific eigen-value. This is demonstrated by an example.

Example: Using the Sturm sequence procedure, isolate the value of the eigen-value at the lower end of the range of $1.5 < \omega^2 < 5.0$.

From the Example in Section 5.9.3, there are three eigen-values above $\omega^2 = 1.5$ and one below $\omega^2 = 5.0$. Therefore we need to find another value of ω^2, above which there are two eigen-values.

Start the calculations at ω^2 equal to the average of upper and lower limits. Therefore form $[K - \omega^2 M]$ at $\omega^2 = (1.5 + 5.0)/2.0 = 3.25$.

$$K - 3.25M = \begin{bmatrix} 1 & -6 & 0 & 0 \\ -6 & -3 & -4 & 0 \\ 0 & -4 & 1.5 & -4 \\ 0 & 0 & -4 & -2.5 \end{bmatrix}$$

After Gaussian elimination

$$\begin{bmatrix} 1 & -6 & 0 & 0 \\ 0 & -39 & -4 & 0 \\ 0 & 0 & 1.91 & -4 \\ 0 & 0 & 0 & -4.41 \end{bmatrix}$$

The number of positive diagonal terms are only two, indicating that there are therefore two eigen-values greater than 3.25. The upper limit has been reduced from 5.0 to 3.25.

Thus only one eigen-value is in the range $1.5 < \omega^2 < 3.25$.

The range can be narrowed by repeatedly bisecting the range until the desired isolation of the eigen-value is obtained. Forming $[K - \omega^2 M]$ at $\omega^2 = (1.5 + 3.25)/2 = 2.375$.

$$K - 2.375M = \begin{bmatrix} 4.5 & -6 & 0 & 0 \\ -6 & 0.5 & -4 & 0 \\ 0 & -4 & 3.25 & -4 \\ 0 & 0 & -4 & -0.75 \end{bmatrix}$$

After Gaussian elimination

$$\begin{bmatrix} 4.5 & -6 & 0 & 0 \\ 0 & -7.5 & -4 & 0 \\ 0 & 0 & 5.38 & -4 \\ 0 & 0 & 0 & -3.72 \end{bmatrix}$$

There are only two positive diagonal terms and hence two eigen-values greater than 2.375. The upper limit has been reduced until further from 3.15 to 2.375. Thus only one eigen-value is in the range $1.5 < \omega^2 < 2.375$.

This is a very powerful procedure. It is not necessary to work to very great accuracy. After the eigen-values have been isolated to sufficient accuracy, inverse iteration with shift can be used to calculate accurately both the eigen-value and the eigen-vector.

5.10 QR FACTORISATION

QR factorisation is a procedure for obtaining all the eigen-values of a symmetric matrix, defining the special eigen-value problem. It is particularly effective when applied to tri-diagonal matrices. The method is used mainly to determine the eigen-values. Although eigen-vectors can also be determined, computationally it is more efficient to determine them using inverse iteration using shifts almost equal to the predetermined eigen-values. The theory behind the method is as follows.

The special eigen-value problem is defined by

$$A\,Y = Y\Lambda \tag{5.85}$$

Let the given matrix A be factorised into

$$A = Q\,R \tag{5.86}$$

where Q = an orthogonal matrix i.e. $Q^T = Q^{-1}$, R is an upper triangular matrix.

Pre-multiplying equation (5.86), by Q^{-1}

$$Q^{-1} A = R \tag{5.87}$$

Post-multiplying equation (5.87) by Q

$$Q^{-1} A\,Q = R\,Q \tag{5.88}$$

Substituting Q^T for Q^{-1} in equation (5.88)

$$Q^T A\, Q = R\, Q \tag{5.89}$$

As shown in equations (5.78) to (5.82), since Q is an orthogonal matrix, R Q has the same eigen-values as A.

In QR factorisation, Q is built as a product of a series of orthogonal matrices N_i. For an n × n matrix

$$Q = N_1\ N_2\ N_3\ ...N_{n-2}\ N_{n-1} \tag{5.90a}$$

$$Q^T = N_{n-1}^T\ N_{n-2}^T\ ...N_3^T\ N_2^T\ N_1^T \tag{5.90b}$$

$$R = Q^T\ A = N_{n-1}^T\ N_{n-2}^T\ ...N_3^T\ N_2^T\ N_1^T\ A \tag{5.90c}$$

$$Q^T\ AQ = R\ Q\ = R N_1\ N_2\ N_3\ ...N_{n-2}\ N_{n-1} \tag{5.90d}$$

The (n - 1) orthogonal matrices N_i are so chosen that each is designed to eliminate one sub-diagonal element $a_{(i+1),\,i}$. To illustrate the idea, consider a 7 x 7 symmetric tri-diagonal matrix shown in equation (5.91). The object is to eliminate the element in the (2,1) position. Let N_1 be as shown in equation (5.92). It is a modified unit diagonal matrix.

$$A = \begin{bmatrix} a_{11} & a_{12} & & & & & \\ a_{21} & a_{22} & a_{23} & & & & \\ & a_{32} & a_{33} & a_{34} & & & \\ & & a_{43} & a_{44} & a_{45} & & \\ & & & a_{54} & a_{55} & a_{56} & \\ & & & & a_{65} & a_{66} & a_{67} \\ & & & & & a_{76} & a_{77} \end{bmatrix} \tag{5.91}$$

$$N_1^T = \begin{bmatrix} c & s & & & & & \\ -s & c & & & & & \\ & & 1 & & & & \\ & & & 1 & & & \\ & & & & 1 & & \\ & & & & & 1 & \\ & & & & & & 1 \end{bmatrix} \tag{5.92}$$

where c =cos θ, s = sinθ

$$N_1^T A = \begin{bmatrix} a_{11}c + a_{21}s & a_{12}c + a_{22}c & a_{23}s & & & & \\ -a_{11}s + a_{21}c & -a_{12}\ s + a_{22}c & a_{23}c & & & & \\ & a_{32} & a_{33} & a_{34} & & & \\ & & a_{43} & a_{44} & a_{45} & & \\ & & & a_{54} & a_{55} & a_{56} & \\ & & & & a_{65} & a_{66} & a_{67} \\ & & & & & a_{76} & a_{77} \end{bmatrix} \tag{5.93}$$

The element in position (2,1) in equation (5.88) can be made zero, if we set

$$-a_{11}\,s + a_{21}\,c = 0,\ \tan\theta = s/c = a_{21}/a_{11}$$

Note that only elements in rows 1 and 2 are affected and the bandwidth does not increase.

In the next stage letting the matrix $N_1^T A = B$, the object is to choose N_2 such that the element in position (3,2) is eliminated.

$$B = \begin{bmatrix} b_{11} & b_{12} & b_{23} & & & & \\ & b_{22} & b_{23} & & & & \\ & a_{32} & a_{33} & a_{34} & & & \\ & & a_{43} & a_{44} & a_{45} & & \\ & & & a_{54} & a_{55} & a_{56} & \\ & & & & a_{65} & a_{66} & a_{67} \\ & & & & & a_{76} & a_{77} \end{bmatrix}$$

$$N_2^T = \begin{bmatrix} 1 & & & & & & \\ & c & s & & & & \\ & -s & c & & & & \\ & & & 1 & & & \\ & & & & 1 & & \\ & & & & & 1 & \\ & & & & & & 1 \end{bmatrix}$$

$$N_2^T N_1^T A = N_2^T B = \begin{bmatrix} b_{11} & b_{12} & b_{23} & & & & \\ & b_{22}c + a_{32}s & b_{23}c + a_{33}s & a_{34}s & & & \\ & -b_{22}s + a_{32}c & -b_{23}s + a_{33}c & a_{34}c & & & \\ & & a_{43} & a_{44} & a_{45} & & \\ & & & a_{54} & a_{55} & a_{56} & \\ & & & & a_{65} & a_{66} & a_{67} \\ & & & & & a_{76} & a_{77} \end{bmatrix}$$

Clearly for $-b_{22}\,s + a_{32}c = 0$, $-a_{11}\,s + a_{21}\,c = 0$, $\tan\theta = a_{21}/b_{22}$. Once again, only the elements in rows 2 and 3 are affected. The process is continued, until the matrix is reduced to an upper diagonal form.

5.10.1 QR Factorisation: Example

The QR factorisation method is illustrated by an Example. Using the tri-diagonal matrix from the Section 5.8.1

$$B = T = \begin{bmatrix} 0.285714 & 0.247436 & \\ 0.247436 & 1.964289 & 0.540062 \\ & 0.540062 & 0.499991 \end{bmatrix}$$

the QR factorisation is applied to the matrix T.
Step 1: Using an orthogonal matrix N_1, reduce the element (2,1) in the product $N_1^T B$ to zero.

$$N_1^T = \begin{bmatrix} \cos\theta & \sin\theta & 0 \\ -\sin\theta & \cos\theta & 0 \\ 0 & 0 & 1 \end{bmatrix}$$

$$N_1^T B = \begin{bmatrix} b_{11}c + b_{21}s & b_{12}c + b_{22}s & b_{13}c + b_{23}s \\ -b_{11}s + b_{21}c & -b_{12}s + b_{22}c & -b_{13}s + b_{23}c \\ 0 & b_{32} & b_{33} \end{bmatrix},$$

$$c = \cos\theta, s = \sin\theta$$

To make $(-b_{11}\, s + b_{21}\, c) = 0$, then, $\tan\theta_1 = s/c = b_{21}/b_{11} = 0.866027$, $\theta_1 = 40.89344$, $c = \cos\theta_1 = 0.755928$, $s = \sin\theta_1 = 0.654654$

$$N_1^T = \begin{bmatrix} 0.755928 & 0.654654 & 0 \\ -0.654654 & 0.755928 & 0 \\ 0 & 0 & 1.0 \end{bmatrix}$$

$$N_1^T B = \begin{bmatrix} 0.377964 & 1.472973 & 0.353553 \\ 0 & 1.322875 & 0.408247 \\ 0 & 0.540061 & 0.5 \end{bmatrix}$$

Step 2: Using an orthogonal matrix N_2, reduce the element (3,1) in the product $N_2^T N_1^T B$ to zero.

$$N_2^T = \begin{bmatrix} 1 & 0 & 0 \\ 0 & \cos\theta & \sin\theta \\ 0 & -\sin\theta & \cos\theta \end{bmatrix}$$

$$N_2^T N_1^T B = \begin{bmatrix} 0.377964 & 1.472973 & 0.353553 \\ 0 & 1.322875c + 0.540061s & 0.408247c + 0.5s \\ 0 & -1.322875s + 0.540061c & -0.408247s + 0.5c \end{bmatrix}$$

To make element (3,2) to be zero, $\tan\theta_2 = s/c = 0.540061/1.322875 = 0.408248$, $\theta_2 = 22.207640$, $c = \cos\theta_2 = 0.925820$, $s = \sin\theta_2 = 0.377964$

$$N_2^T = \begin{bmatrix} 1 & 0 & 0 \\ 0 & 0.925820 & 0.377964 \\ 0 & -0.377964 & 0.925820 \end{bmatrix}$$

$$N_2^T N_1^T B = \begin{bmatrix} 0.377964 & 1.472973 & 0.353553 \\ 0 & 1.428868 & 0.566945 \\ 0 & 0 & 0.308607 \end{bmatrix}$$

$R = N_2^T N_1^T B$, which is an upper triangular matrix.
Step 3: The next stage is to obtain $R\,Q$ in a series of steps. Obtain $R\,N_1$

$$N_1 = \begin{bmatrix} 0.755928 & -0.654654 & 0 \\ 0.654654 & 0.755928 & 0 \\ 0 & 0 & 1.0 \end{bmatrix}$$

$$RN_1 = \begin{bmatrix} 1.250001 & 0.866026 & 0.353553 \\ 0.935414 & 1.080121 & 0.566945 \\ 0 & 0 & 0.308607 \end{bmatrix}$$

Step 4: Calculate $R\,N_1\,N_2$

$$N_2 = \begin{bmatrix} 1 & 0 & 0 \\ 0 & 0.925820 & -0.377964 \\ 0 & 0.377964 & 0.925820 \end{bmatrix}$$

$$R\,N_1\,N_2 = \begin{bmatrix} 1.250001 & 0.935414 & 0 \\ 0.935414 & 1.214282 & 0.116642 \\ 0 & 0.116642 & 0.285715 \end{bmatrix}$$

This completes the first cycle of iteration. For the second cycle of iteration, we use the matrix $R\,N_1\,N_2$ at the end of the first iteration as the starting matrix, B. We continue to iterate until the off diagonal elements are sufficiently small. The calculations are summarised below.

Iteration 2:

$$B = \begin{bmatrix} 1.250005 & 0.935416 & 0 \\ symmetric & 1.214282 & 0.116642 \\ & & 0.285715 \end{bmatrix}$$

$$\theta_1 = 36.8086, \ \theta_2 = 15.8163$$

$$R = \begin{bmatrix} 1.561255 & 1.476462 & 0.069885 \\ 0 & 0.427959 & 0.167726 \\ 0 & 0 & 0.249445 \end{bmatrix}$$

$$RN_1\,N_2 = \begin{bmatrix} 2.134618 & 0.256409 & 0 \\ symmetric & 0.375384 & 0.067987 \\ & & 0.240001 \end{bmatrix}$$

Iteration 3:

$$B = \begin{bmatrix} 2.134618 & 0.256409 & 0 \\ symmetric & 0.375384 & 0.067987 \\ & & 0.240001 \end{bmatrix}$$

$$\theta_1 = 6.84951, \ \theta_2 = 11.2394$$

$$R' = \begin{bmatrix} 2.149963 & 0.299348 & 0.008108 \\ 0 & 0.348815 & 0.112986 \\ 0 & 0 & 0.222241 \end{bmatrix}$$

$$RN_1\ N_2 = \begin{bmatrix} 2.170319 & 0.041600 & 0 \\ symmetric & 0.361705 & 0.043317 \\ & & 0.217979 \end{bmatrix}$$

If at the end of the third iteration, the eigen-values are assumed to be approximately given by the diagonal values, then $\lambda = 1/\omega^2 = 2.170319$, 0.361705 and 0.217979, or $\omega^2 = 0.460762$, 2.764684 and 4.587598.

The exact eigen-pairs calculated in Section 3.2.4, using the determinant of the matrix $[K - \omega^2 M]$ are given by $\omega^2 = 0.4606$, 2.6820 and 4.8575. Even the largest eigen-value is in error by less than 6%. Iterations can clearly be continued to any desired accuracy.

From a computational point of view, it is worth noting that it is not necessary to neither form nor store the matrices, N_1, N_2, etc. If the element in position (i + 1, i) is being eliminated, then two temporary vectors are needed to store the modified elements of the row i and (i + 1) and it is necessary to store only the values of θ_1, θ_2, etc. for later use.

5.11 GERSCHGORIN LIMITS ON EIGEN-VALUES

When using the method of root bisection, it is useful to know the limits within which all the eigen-values of a symmetric matrix lie. Gerschgorin's limits provide a method for establishing the limits. If the symmetric eigen-value problem is defined by A X = λX, it can be shown that the smallest eigen-value λ_1 and the largest eigen-value λ_n must lie within the following limits.

$$(a_{ii})_{\max} \le \lambda_n \le \{a_{kk} + \sum_{j\ne k} |a_{jk}|\}_{\max} \tag{5.94}$$

$$\{a_{kk} - \sum_{j\ne k} |a_{jk}|\}_{\min} \le \lambda_1 \le (a_{ii})_{\min} \tag{5.95}$$

k = row number, a_{jk} = an element of row k.

5.11.1 Gerschgorin Limits: Example

As a simple example on the calculation of Gerschgorin limits, consider the tri-diagonal matrix from Section 5.8.1.

$$A = T = \begin{bmatrix} 0.285714 & 0.247436 & 0 \\ 0.247436 & 1.964289 & 0.540062 \\ 0 & 0.540062 & 0.499991 \end{bmatrix}$$

Calculation of the limits on the largest eigen-value:
$(a_{ii})_{max}$ = Largest diagonal term = 1.964289
Similarly,

$$\{a_{kk} + \sum_{j\neq k}|a_{jk}|\} = 0.285714 + |0.247436| = 0.53315 \textit{ from Row1}$$

$$= 1.964289 + |0.247436| + |0.540062| = 2.751787 \textit{ from Row2}$$

$$= 0.499991 + |0.540062| = 1.040053 \textit{ from Row3}$$

$$\therefore \{a_{kk} + \sum_{j\neq k}|a_{jk}|\}_{max} = 2.751787$$

Gerschgorin's limits indicates that the largest eigen-value must lie between the limits

$$[(a_{ii})_{max} = 1.964289] \leq \lambda_n \leq [\{a_{kk} + \sum_{j\neq k}|a_{jk}|\}_{max} = 2.751787]$$

Actual value of λ_n = 2.170318
Similarly
Calculation of the limits on the smallest eigen-value:
$(a_{ii})_{min}$ = *Smallest diagonal term* = 0.285714

$$\{a_{kk} - \sum_{j\neq k}|a_{jk}|\} = 0.285714 - |0.247436| = 0.038278 \textit{ from Row1}$$

$$= 1.964289 - |0.247436| - |0.540062| = 1.176791 \textit{ from Row2}$$

$$= 0.499991 - |0.540062| = -0.040071 \textit{ from Row3}$$

$$\therefore \{a_{kk} - \sum_{j\neq k}|a_{jk}|\}_{min} = -0.040071$$

Gerschgorin's limits indicates that the smallest eigen-value must lie between the limits

$$[\{a_{kk} - \sum_{j\neq k}|a_{jk}|\}_{min} = -0.040071] \leq \lambda_1 \leq [\ (a_{ii})_{min} = 0.285714]$$

However, because of the fact that for a positive definite or semi-definite matrix, all eigen-values are positive, the lower limit must be zero. Therefore we can write

$$0 \leq \lambda_1 \leq 0.285714$$

Actual value of λ_1 = 0.2059.

5.12 FORTRAN-90 PROGRAMS

A series of fairly intentionally elementary programs are listed. The programs follow quite closely the theory developed in the previous Sections. A large number of print statements to output intermediate steps has been included. This is to assist the reader in checking their hand calculations, if so required, when learning a particular method. No effort has been made to economise on the storage requirements or to introduce subroutines. This is with the intention of making the programs as 'transparent' as possible.

5.12.1 Program LANCZGEN

```
!*******************************************************************
!*        LANCZOS METHOD FOR TRI-DIAGONALISATION                   *
!* EIGEN-VALUES DETERMINED BY BISECTION AND STURM SEQUENCE *
!*SIGN COUNT. EIGEN-VECTORS DETERMINED BY INVERSE ITERATION*
!* WITH SHIFT*******************************************************
          PROGRAM LANCZGEN
!*******VARIABLE DECLARATION***********************************
          IMPLICIT NONE
          REAL:: MAX1, MAX2, MIN1, MIN2,TEMP1,TEMP2,SUM1, SUM2,&
          &UPPER, LOWER, ROOT, FACTOR, BETA1,BETAJ, BETAO,&
          &BETAJM1, ALPHAJ,C1,SUM,NROOT,SHIFT,RAYLEIGH1,&
          &RAYLEIGH2,ROOT1,BETA,NREAL
          INTEGER::I,J,K,L,N,NP1,NM1,IP1,IM1,KM1,KM2,KP1,KROOT,&
          &COUNT,ITERATION,NJ,NJP1,JM1
          REAL,ALLOCATABLE::STIF(:,:),MASS(:,:),A(:,:),V(:),Z(:),X(:,:),MZ(:),&
          &B(:,:),LAMDA(:,:),DETVEC(:),TEMP(:),U(:,:),VBAR(:)
!*******FILE MANAGEMENT****************************************
      CHARACTER(LEN=40)::FILEIN, FILEOUT
      PRINT*,'NAME OF INPUT FILE'
      READ*,FILEIN
      PRINT*,'NAME OF OUTPUT FILE'
      READ*,FILEOUT
      OPEN(UNIT=1, FILE =FILEIN,STATUS='OLD',ACTION='READ')
      OPEN(UNIT=2, FILE =FILEOUT,STATUS='REPLACE',ACTION='WRITE')
!*******INPUT DATA*********************************************
          READ(1,*) N
          NP1 = N+1
          NM1= N-1
          ALLOCATE(STIF(N,N),MASS(N,N),A(N,N),V(N),Z(N),X(N,N),MZ(N),&
          &B(N,2),LAMDA(N,2),DETVEC(NP1),TEMP(N),U(N,N),VBAR(N))
!*******INPUT SYMMETRIC UPPER HALF OF STIFFNESS AND MASS ***
!         MATRICES*********
          READ(1,*) ((STIF(I,J),J=I,N),I=1,N)
          READ(1,*) ((MASS(I,J),J=I,N),I=1,N)
!*******COMPLETE THE SYMMETRIC UPPER HALF*******************
          DO I = 2,N
          IM1 = I-1
          DO J = 1,IM1
          STIF(I,J) = STIF(J,I)
          MASS(I,J)=MASS(J,I)
          END DO
          END DO
!*******PRINT THE INPUT MATRICES****************************
          WRITE(2,*)'STIFFNESS  MATRIX'
          DO I = 1,N
           WRITE(2,'(6(4X,F10.4))')(STIF(I,J), J = 1,N)
           END DO
          WRITE(2,*)'MASS  MATRIX'
   DO I = 1,N
          WRITE(2,'(6(4X,F10.4))')(MASS(I,J), J = 1,N)
          END DO
!******************************************************************
!******CALCULATE CHOLESKY FACTORISATION OF STIFFNESS MATRIX
!         U TRANSFORM × U = STIF
!*******************************************************************
```

```
        U(1,1)=SQRT(STIF(1,1))
!*******CALCULATE THE FIRST ROW****************************
        DO J=2,N
        U(1,J)=STIF(1,J)/U(1,1)
        END DO
!******CALCULATE THE REST OF MATRIX*************************
        DO I = 2,N
!*******CALCULATE THE DIAGONAL ELEMENT*******************
        SUM=0
        IM1=I-1
        IP1=I+1
        DO K= 1,IM1
        SUM=SUM+U(K,I)*U(K,I)
        END DO
        U(I,I)=SQRT(STIF(I,I)-SUM)
        DO J = IP1,N
        IF(J>N)EXIT
        SUM=0
        DO K=1,IM1
        SUM=SUM+U(K,I)*U(K,J)
        END DO
        U(I,J)=(STIF(I,J)-SUM)/U(I,I)
        END DO
        END DO
!******PRINT U MATRIX********************************
        WRITE(2,*)'U MATRIX'
        DO I = 1,N
        WRITE(2,'(6(4X,F10.4))')(U(I,J), J = I,N)
        END DO
!**************************************************
!       LANCZOS ITERATION
!**************************************************
!*******SET THE INITIAL VECTOR X1*********************
        DO K = 1,N
        X(K,1) = 0
        END DO
        X(1,1)=1
!*******START LANCZOS ITERATIONS*********************
        DO I=1,N
        IP1=I+1
        IM1=I-1
        WRITE(2,*)'X VECTOR'
        WRITE(2,'(6(4X,F10.4))')(X(J,I), J = 1,N)
!*******CALCULATE THE PRODUCT A × XI,**********************
!*******USE BACK  SUBSTITUTION TO SOLVE U×Z = XI***********
        Z(N)=X(N,I)/U(N,N)
        DO J=1,NM1
        SUM=0.0
        NJ=N-J
        NJP1=NJ+1
        DO K=NJP1,N
        SUM=SUM+U(NJ,K)*Z(K)
        END DO
        Z(NJ)=(X(NJ,I)-SUM)/U(NJ,NJ)
        END DO
!*******CALCULATE THE PRODUCT M × Z ****************************
        DO J = 1,N
```

```
        MZ(J)=0
        DO K=1,N
        MZ(J)=MZ(J)+MASS(J,K)*Z(K)
        END DO
        END DO
!******USE FORWARD SUBSTITUTION TO SOLVE UT ×V = MZ*******
        V(1)=MZ(1)/U(1,1)
        DO J=2,N
        SUM=0.0
        JM1=J-1
        DO K=1,JM1
        SUM=SUM+U(K,J)*V(K)
        END DO
        V(J)=(MZ(J)-SUM)/U(J,J)
        END DO
        WRITE(2,*)'  A* XI = V  VECTOR'
        WRITE(2,'(6(4X,F10.4))')(V(J), J = 1,N)
!*******CALCULATE ALPHA AND BETA AND STORE THEM IN MATRIX B
!*******MULTIPLY XI ×TRANSPOSE A × XI*****************************
        SUM=0.0
        DO K=1,N
        SUM=SUM+X(K,I)*V(K)
        END DO
        B(I,1) = SUM
        IF(I==N)EXIT
!*******CALCULATE THE NOT YET NORMALISED XI+1  VECTOR*****
!*******CALCULATE BETA*******************************************
        DO J=1,N
        X(J,IP1)=V(J)-B(I,1)*X(J,I)
        IF(I > 1)THEN
        X(J,IP1)=X(J,IP1)-B(IM1,2)*X(J,IM1)
        END IF
        END DO
        SUM=0
        DO K=1,N
        SUM=SUM+X(K,IP1)*X(K,IP1)
        END DO
        BETA=SQRT(SUM)
        DO K=1,N
        X(K,IP1)=X(K,IP1)/BETA
        END DO
        B(I,2)=BETA
        END DO
!*******PRINT X VECTORS******************************************
        WRITE(2,*)'X VECTORS'
        DO I = 1,N
        WRITE(2,'(6(4X,F10.4))')(X(I,J), J = 1,N)
        END DO
!*******PRINTING OUT THE COEFFICIENTS OF TRIDIAGONAL MATRIX**
        WRITE(2,*)'COEFFICIENTS OF TRIDIAGONAL MATRIX'
        WRITE(2,*)'ALPHA       BETA'
        DO I = 1,N
        IF(I == N) THEN
        WRITE(2,'(F10.4)')B(I,1)
        ELSE
        WRITE(2,'(F10.4,4X,F10.4)')B(I,1), B(I,2)
        END IF
```

```
        END DO
!*******GERSHGORIN LIMITS ON EIGEN-VALUES**********************
        MAX1 = 0
        MAX2 = 0
        MIN1 = 0
        MIN2 = 0
        DO I = 1,N
        IM1=I-1
        IF(MAX1 < B(I,1)) MAX1 = B(I,1)
        IF(MIN1 > B(I,1))  MIN1 = B(I,1)
        SUM1 = B(I,1)
        SUM2 = B(I,1)
        IF(I ==1)THEN
        TEMP1=ABS(B(1,2))
        TEMP2= 0.0
        ELSE
        TEMP1= ABS(B(IM1,2))
        TEMP2=ABS(B(I,2))
        END IF
        SUM1= SUM1+TEMP1+TEMP2
        SUM2= SUM2-TEMP1-TEMP2
        IF(MAX2 < SUM1) MAX2 = SUM1
        IF(MIN2 >  SUM2) MIN2 = SUM2
        END DO
        UPPER = MAX(MAX1,MAX2)
        LOWER = MIN(MIN1,MIN2)
!*******PUT THE UPPER AND LOWER LIMITS FOR ALL EIGEN-VALUES**
        DO I = 1,N
        LAMDA(I,1) = LOWER
        LAMDA(I,2) = UPPER
        END DO
!*******BISECTION AND STURM SEQUENCE COUNT************
        DO KROOT = 1, N
        DO ITERATION = 1, 500
        ROOT = 0.5*(LAMDA(KROOT,1) + LAMDA(KROOT,2))
        DETVEC(1) = 1
        DETVEC(2) =B(1,1)-ROOT
        COUNT = 0
        IF (DETVEC(2) < 0 )THEN
        COUNT = 1
        END IF
        DO K = 3,NP1
        KM1 = K-1
        KM2= K-2
        DETVEC(K) = (B(KM1,1)- ROOT)*DETVEC(KM1)-&
        &DETVEC(KM2)*B(KM2,2)**2
        FACTOR = DETVEC(KM1)*DETVEC(K)
        IF (FACTOR <0 )THEN
        COUNT = COUNT +1
        END IF
        END DO
        IF (COUNT ==0) THEN
        DO I = 1, N
        IF(LAMDA(I,1) < ROOT) LAMDA(I,1) = ROOT
        END DO
        END IF
        IF (COUNT /= 0) THEN
```

```
        DO I = 1, COUNT
        IF(LAMDA(I,2) > ROOT) LAMDA(I,2) = ROOT
        END DO
        DO I = COUNT+1 , N
        IF(LAMDA(I,1) < ROOT) LAMDA(I,1) = ROOT
        END DO
        END IF
        IF (ABS(LAMDA(KROOT,1)-LAMDA(KROOT,2)) < 1.0E-6) EXIT
        END DO
        END DO
        DO I = 1, N
        LAMDA(I,1)=1.0/LAMDA(I,1)
        LAMDA(I,2)=1.0/LAMDA(I,2)
        END DO
!*******************************************************************
!*CALCULATE THE EIGEN-VECTORS BY INVERSE ITERATION AND SHIFT
        NREAL=N
        DO L = 1,N
!*******SET THE ARBITRARY VECTOR V WITH VALUES EQUAL TO ***
!*******SQUARE ROOT OF N, WHICH GIVES A UNIT LENGTH*********
        DO I = 1,N
        V(I)=1/SQRT(NREAL)
        END DO
!*******FORM A=STIF- SHIFT × MASS******************************
        A= STIF-0.99*LAMDA(L,1)*MASS
!*******FACTORISE A USING CROUT FACTORISATION****************
********CALCULATE THE FIRST ROW OF U*************************
        U(1,1)=A(1,1)
        FACTOR=1.0/U(1,1)
        DO J = 2,N
        U(1,J)=A(1,J)*FACTOR
        END DO
!*******CALCULATE THE ELEMENTS OF ROWS 2 TO N***************
        DO I = 2,N
        IP1=I+1
!**CALCULATE THE DIAGONAL ELEMENT D(I,I) AND STORE IT IN U(I,I)*
        IM1=I-1
        SUM=0
        DO K=1,IM1
        SUM = SUM+ U(K,K)*U(K,I)**2
        END DO
        U(I,I)=A(I,I)-SUM
        IF(I==N) EXIT
!*******CALCULATE THE OFF DIAGONAL ELEMENTS*************
        FACTOR=1.0/U(I,I)
        DO J = IP1,N
        SUM = 0
        DO K = 1,IM1
        SUM = SUM +U(K,I)*U(K,J)*U(K,K)
        END DO
        U(I,J)=(A(I,J)-SUM)*FACTOR
        END DO
        END DO
!*******PRINT A = STIF - LAMDA ×MASS***************************
        WRITE(2,*)'A-MATRIX'
        DO I = 1,N
        WRITE(2,*)(A(I,J),J=1,N)
```

```
        END DO
!DIAGONAL ELEMENTS OF U REPRESENTING D(I,I)*******************
        WRITE(2,*)'U MATRIX'
        DO I = 1,N
        WRITE(2,*)(U(I,J),J=1,N)
        END DO
!******DO ONLY TWO ITERATIONS OF INVERSE ITERATION************
        DO ITERATION = 1,2
!*******CALCULATE THE PRODUCT A× V, ****************************
!******USE BACK  SUBSTITUTION TO SOLVE U×Z = V******************
        Z(N)=V(N)
        DO J=1,NM1
        SUM=0.0
        NJ=N-J
        NJP1=NJ+1
        DO K=NJP1,N
        SUM=SUM+U(NJ,K)*Z(K)
        END DO
        Z(NJ)=V(NJ)-SUM
        END DO
!*****CALCULATE, M×Z *******************************************
        DO J = 1,N
        MZ(J)=0
        DO K=1,N
        MZ(J)=MZ(J)+MASS(J,K)*Z(K)
        END DO
        END DO
!*******USE FORWARD SUBSTITUTION TO SOLVE UT ×V=MZ*******
        VBAR(1)=MZ(1)
        DO J=2,N
        SUM=0.0
        JM1=J-1
        DO K=1,JM1
        SUM=SUM+U(K,J)*VBAR(K)
        END DO
        VBAR(J)=(MZ(J)-SUM)
        END DO
!*****DIVIDE ELEMENTS OF VBAR BY THE DIAGONAL ELEMENTS OF D*
        DO I = 1,N
        VBAR(I)=VBAR(I)/U(I,I)
        END DO
!******CONVERT VBAR TO UNIT LENGTH******************
        SUM = 0
        DO J = 1,N
        SUM = SUM+VBAR(J)**2
        END DO
        SUM = 1.0/SQRT(SUM)
        DDO I = 1,
        V(I)=VBAR(I)*SUM
        END DO
        END DO
!*******CALCULATE THE EIGEN-VECTORS OF GENERAL EIGEN-VALUE
!PROBLEM.  USE BACK  SUBSTITUTION TO SOLVE U×X = V********
        X(N,L)=V(N)
        DO J=1,NM1
        SUM=0.0
        NJ=N-J
```

```
          NJP1=NJ+1
          DO K=NJP1,N
          SUM=SUM+U(NJ,K)*X(K,L)
          END DO
          X(NJ,L)=V(NJ)-SUM
          END DO
          END DO
!******M NORMALISE THE  EIGEN-VECTOR**********************
          DO L = 1,N
          SUM1=0
          DO J = 1,N
          SUM2=0
          DO K = 1,N
          SUM2=SUM2+MASS(J,K)*X(K,L)
          END DO
          SUM1=SUM1+SUM2*X(J,L)
          END DO
          SUM1=1.0/SQRT(SUM1)
          DO I = 1,N
          X(I,L)=X(I,L)*SUM1
          END DO
          END DO
          WRITE(2,*)'UPPER AND LOWER LIMITS OF OMEGA SQUARED'
          DO I = 1, N
          WRITE(2,'(2(4X,F10.4))') LAMDA(I,1), LAMDA(I,2)
          END DO
!*******PRINT EIGEN-VECTORS*********************************
          WRITE(2,*)'EIGEN-VECTORS'
          DO I = 1,N
          WRITE(2,'(6(4X,F10.4))')(X(I,J),J=1,N)
          END DO
          END PROGRAM LANCZGEN
```

5.12.2 Program GENJACOBI

```
!************************************************************
!*****PROGRAM FOR THE SOLUTION OF GENERAL EIGEN-VALUE       *
!*      PROBLEM BY JACOBI DIAGONALISATION                   *
!************************************************************
          PROGRAM GENJACOBI
!*******VARIABLE DECLARATIONS*******************************
          IMPLICIT NONE
          REAL:: ALPHA,BETA,DIAGONALKI,DIAGONALKJ,DIAGONALMI,&
          &DIAGONALMJ,C1,C2,K11,K22,K12,M11,M22,M12,P,Q,R&
          &DENOM,SUM,SUM1
          INTEGER:: I,J,K,L,N,NM1,IP1,IM1,ITERATION,FLAG,ISIGN
          REAL, ALLOCATABLE::
          STIF(:,:),MASS(:,:),ROWKI(:),ROWKJ(:),COLUMNKI(:),&
          &COLUMNKJ(:),ROWMI(:),ROWMJ(:),COLUMNMI(:),COLUMNMJ(:),&
          &EIGENVEC(:,:),EIGENI(:),EIGENJ(:),OMEGA2(:)
!*******FILE MANAGEMENT****************************************
       CHARACTER(LEN=40)::FILEIN, FILEOUT
       PRINT*,'NAME OF INPUT FILE'
```

```
    READ*,FILEIN
    PRINT*,'NAME OF OUTPUT FILE'
    READ*,FILEOUT
    OPEN(UNIT=1, FILE =FILEIN,STATUS='OLD',ACTION='READ')
    OPEN(UNIT=2, FILE =FILEOUT,STATUS='REPLACE',ACTION='WRITE')
!*******INPUT DATA'**********************************************
        READ(1,*) N
        ALLOCATE(STIF(N,N),MASS(N,N),ROWKI(N),ROWKJ(N),&
        &COLUMNKI(N),COLUMNKJ(N),ROWMI(N),ROWMJ(N),&
        &COLUMNMI(N),COLUMNMJ(N),EIGENVEC(N,N),EIGENI(N),&
        &EIGENJ(N),OMEGA2(N))
!*******READ SYMMETRIC UPPER HALF OF STIFFNESS AND MASS
!******* MATRICES**********************************************
        READ(1,*)(( STIF(I,J), J=I,N),I=1,N)
        READ(1,*) ((MASS(I,J), J=I,N),I=1,N)
!*******COMPLETE THE SYMMETRIC UPPER HALF********************
        NM1= N-1
        DO I = 2, N
        IM1 = I-1
        DO J = 1,IM1
        STIF(I,J) = STIF(J,I)
        MASS(I,J) = MASS(J,I)
        END DO
        END DO
!******PRINT THE INPUT MATRICES********************************
        WRITE(2,*)'STIFFNESS MATRIX'
        DO I = 1,N
        WRITE(2,'(6(4X,F10.6))')(STIF(I,J), J = 1,N)
        END DO
        WRITE(2,*)'MASS MATRIX'
        DO I = 1,N
        WRITE(2,'(6(4X,F10.6))')(MASS(I,J), J = 1,N)
        END DO
!******START WITH A UNIT DIAGONAL MATRIX FOR EIGEN-VECTOR***
        EIGENVEC=0
        DO I = 1,N
        EIGENVEC(I,I)=1.0
        END DO
!*******JACOBI REDUCTION ITERATION CYCLE*********************
        ITERATIONS: DO ITERATION = 1, 500
        FLAG=0
        ROWLOOP:  DO I = 1,N
        IP1=I+1
        COLUMNLOOP: DO J = IP1,N
        K11=STIF(I,I)
        K22=STIF(J,J)
        K12=STIF(I,J)
        M11=MASS(I,I)
        M22=MASS(J,J)
        M12=MASS(I,J)
        P=K11*M12-M11*K12
        Q=K22*M12-M22*K12
        R=K11*M22-K22*M11
        ISIGN=R/ABS(R)
        DENOM=1.0/(0.5*(R+ISIGN*SQRT(R**2+4*P*Q)))
        BETA=-P*DENOM
        ALPHA=Q*DENOM
```

```
          C1=ABS(ALPHA)
          C2=ABS(BETA)
          IF(C1 > 1.0E-6.OR. C2 > 1.0E-6) THEN
          FLAG=1
!*****TEMPORARY STORAGE OF MODIFIED ELEMENTS OF ITH AND JTH
!***** ROWS AND COLUMNS*************************************
          DO K=1,N
!*******MODIFIED ITH ROW*************************************
          ROWKI(K)=STIF(I,K)+STIF(J,K)*BETA
          ROWMI(K)=MASS(I,K)+MASS(J,K)*BETA
!*******MODIFIED JTH ROW**************************************
          ROWKJ(K)=STIF(I,K)*ALPHA+STIF(J,K)
          ROWMJ(K)=MASS(I,K)*ALPHA+MASS(J,K)
!******MODIFIED ITH COLUMN**********************************
          COLUMNKI(K)=STIF(K,I)+STIF(K,J)*BETA
          COLUMNMI(K)=MASS(K,I)+MASS(K,J)*BETA
!******MODIFIED JTH COLUMN**********************************
          COLUMNKJ(K)=STIF(K,I)*ALPHA+STIF(K,J)
          COLUMNMJ(K)=MASS(K,I)*ALPHA+MASS(K,J)
!*******MODIFIED COLUMNS OF EIGEN-VECTOR*******************
          EIGENI(K)=EIGENVEC(K,I)+EIGENVEC(K,J)*BETA
          EIGENJ(K)=EIGENVEC(K,I)*ALPHA+EIGENVEC(K,J)
          END DO
!*******MODIFIED DIAGONAL ELEMENTS CORRESPONDING TO I AND J*
          DIAGONALKI= STIF(I,I)+STIF(J,J)*BETA**2+&
          &2*STIF(I,J)*BETA
          DIAGONALKJ= STIF(I,I)*ALPHA**2+STIF(J,J)+&
          &2*STIF(I,J)*ALPHA
          DIAGONALMI= MASS(I,I)+MASS(J,J)*BETA**2+&
          &2*MASS(I,J)*BETA
          DIAGONALMJ= MASS(I,I)*ALPHA**2+MASS(J,J)+&
          &2*MASS(I,J)*ALPHA
!*******MODIFY ITH AND JTH ROWS AND COLUMNS OF STIFFNESS AND
!******* MASS MATRICES*****************************************
          DO K = 1,N
          STIF(I,K)=ROWKI(K)
          STIF(J,K)=ROWKJ(K)
          STIF(K,I)=COLUMNKI(K)
          STIF(K,J)=COLUMNKJ(K)
          STIF(I,I)=DIAGONALKI
          STIF(J,J)=DIAGONALKJ
          STIF(I,J)=0
          STIF(J,I)=0
          MASS(I,K)=ROWMI(K)
          MASS(J,K)=ROWMJ(K)
          MASS(K,I)=COLUMNMI(K)
          MASS(K,J)=COLUMNMJ(K)
          MASS(I,I)=DIAGONALMI
          MASS(J,J)=DIAGONALMJ
          MASS(I,J)=0
          MASS(J,I)=0
!*******MODIFY ITH AND JTH COLUMNS OF EIGEN-VECTOR MATRIX**
          EIGENVEC(K,I)=EIGENI(K)
          EIGENVEC(K,J)=EIGENJ(K)
          END DO
          WRITE(2,'("ITERATION   = ",I4)')ITERATION
          WRITE(2,*)'MODIFIED STIFFNESS AND MASS MATRICES'
```

```
          WRITE(2,'(I4,4X,I4)')I,J
          WRITE(2,'("ALPHA = ",E14.6,"   BETA = ",E14.6)')ALPHA,BETA
          WRITE(2,*)'MODIFIED STIFFNESS MATRIX'
          DO K = 1,N
          WRITE(2,'(6(4X,6F10.6))')(STIF(K,L), L = 1,N)
          END DO
          WRITE(2,*)'MODIFIED MASS MATRIX'
          DO K = 1,N
          WRITE(2,'(6(4X,6F10.6))')(MASS(K,L), L = 1,N)
          END DO
          END IF
          END DO  COLUMNLOOP
          END DO  ROWLOOP
          IF(FLAG==0)EXIT
          END DO ITERATIONS
          DO I = 1, N
          OMEGA2(I)=STIF(I,I)/MASS(I,I)
          END DO
!*******OBTAIN THE ORIGINAL MASS MATRIX***********************
!*******READ SYMMETRIC UPPER HALF OF MASS MATRIX**********
          READ(1,*) ((MASS(I,J), J=I,N),I=1,N)
!*******COMPLETE THE SYMMETRIC UPPER HALF******************
          DO I = 2, N
          IM1 = I-1
          DO J = 1,IM1
          MASS(I,J) = MASS(J,I)
          END DO
          END DO
!*******M NORMALISE THE EIGEN-VECTORS**********************
          DO I = 1,N
          SUM=0
          DO J= 1,N
          SUM1=0
          DO K = 1,N
          SUM1 = SUM1+MASS(J,K)*EIGENVEC(K,I)
          END DO
          SUM=SUM+SUM1*EIGENVEC(J,I)
          END DO
          DENOM=1.0/SQRT(SUM)
          DO K = 1,N
          EIGENVEC(K,I)=EIGENVEC(K,I)*DENOM
          END DO
          END DO
          WRITE(2,*)'EIGEN VALUES'
          DO I = 1, N
          WRITE(2,'(I4,4X,F10.6)')I,OMEGA2(I)
          END DO
          WRITE(2,*)'EIGEN-VECTOR MATRIX'
          DO K = 1,N
          WRITE(2,'(6(4X,6F10.6))')(EIGENVEC(K,L), L = 1,N)
          END DO
          END PROGRAM GENJACOBI
```

5.12.3 Program QRLANCZ

```
!******************************************************************
!       LANCZOS METHOD FOR TRI-DIAGONALISATION                   *
!       EIGEN-VALUES DETERMINED BY QR FACTORISATION              *
!   EIGEN-VECTORS DETERMINED BY INVERSE ITERATION WITH SHIFT*
!******************************************************************
        PROGRAM QRLANCZ
!******VARIABLE DECLARATION**********************************
        IMPLICIT NONE
        REAL:: TEMP1,TEMP2,SUM1, SUM2,FACTOR,SUM,&
        &BETA,NREAL,THETA,COSTHETA,SINTHETA
        INTEGER:: I,J,K,L,N,NP1,NM1,IP1,IM1,COUNT,ITERATION,NMJ,&
        &NJP1,JM1,JP1,FLAG
        REAL,ALLOCATABLE:: STIF(:,:),MASS(:,:),A(:,:),&V(:),Z(:),X(:,:),&
        &MZ(:),B(:,:),LAMDA(:),U(:,:),VBAR(:),UCOS(:),USIN(:)
!*******FILE MANAGEMENT*****************************************
     CHARACTER(LEN=40)::FILEIN, FILEOUT
     PRINT*,'NAME OF INPUT FILE'
     READ*,FILEIN
     PRINT*,'NAME OF OUTPUT FILE'
     READ*,FILEOUT
     OPEN(UNIT=1, FILE =FILEIN,STATUS='OLD',ACTION='READ')
     OPEN(UNIT=2, FILE =FILEOUT,STATUS='REPLACE',ACTION='WRITE')
!******INPUT DATA***********************************************
        READ(1,*) N
        NP1 = N+1
        NM1= N-1
        ALLOCATE(STIF(N,N),MASS(N,N),A(N,N),V(N),Z(N),X(N,N),MZ(N),&
        &B(N,2),LAMDA(N),U(N,N),VBAR(N),UCOS(N),USIN(N))
!*INPUT SYMMETRIC UPPER HALF OF STIFFNESS AND MASS MATRICES*
        READ(1,*) ((STIF(I,J),J=I,N),I=1,N)
        READ(1,*) ((MASS(I,J),J=I,N),I=1,N)
!********COMPLETE THE SYMMETRIC LOWER HALF******************
        DO I = 2,N
        IM1 = I-1
        DO J = 1,IM1
        STIF(I,J) = STIF(J,I)
        MASS(I,J)=MASS(J,I)
        END DO
        END DO
!*******PRINT THE INPUT MATRICES********************************
        WRITE(2,*)'STIFFNESS  MATRIX'
        DO I = 1,N
        WRITE(2,'(6(4X,F10.4))')(STIF(I,J), J = 1,N)
        END DO
        WRITE(2,*)'MASS  MATRIX'
        DO I = 1,N
        WRITE(2,'(6(4X,F10.4))')(MASS(I,J), J = 1,N)
        END DO
!******************************************************************
!******CHOLESKY FACTORISATION OF STIFFNESS MATRIX
!       LANCZOS ITERATION                                   *
```

THESE TWO SECTIONS SAME AS IN PROGRAM LANCZGEN

```
!******************************************************************
!       QR FACTORISATION                                    *
!******************************************************************
```

```
!*********TRANSFER ALPHA, BETA VALUES TO A MATRIX*************
        A = 0
        DO I = 1,N
        IP1=I+1
        A(I,I)=B(I,1)
        IF(IP1 <= N) THEN
        A(I,IP1)=B(I,2)
        A(IP1,I)=B(I,2)
        END IF
        END DO
        WRITE(2,*)'A  MATRIX TRANSFERRED FROM B - MATRIX'
        DO I = 1,N
        WRITE(2,'(6(4X,F10.4))')(A(I,J), J = 1,N)
        END DO
        DO ITERATION = 1,500
!*******FLAG = 1 MEANS REQUIRED ACCURACY NOT ATTAINED*******
        FLAG =0
        DO I = 1, NM1
        IP1=I+1
        IF(A(I,I) ==0) THEN
        THETA = ATAN(1.0)
        ELSE
        THETA = ATAN(A(IP1,I)/A(I,I))
        END IF
        COSTHETA= COS(THETA)
        SINTHETA = SIN(THETA)
        UCOS(I) = COSTHETA
        USIN(I) = SINTHETA
        DO J = I,N
        TEMP1 = A(I,J)
        TEMP2 = A(IP1,J)
        A(I,J) = COSTHETA*TEMP1 + SINTHETA*TEMP2
        A(IP1,J) = -SINTHETA*TEMP1  + COSTHETA*TEMP2
        END DO
        IF(ABS(THETA)>1.0E-6) FLAG = 1
        END DO
!*******CALCULATE THE  PRODUCT R×Q*****************************
        DO J = 1, NM1
        JP1=J+1
        COSTHETA= UCOS(J)
        SINTHETA = USIN(J)
        DO I = 1, JP1
        TEMP1 = A(I,J)
        TEMP2 = A(I,JP1)
        A(I,J) = COSTHETA*TEMP1 + SINTHETA*TEMP2
        A(I,JP1) = -SINTHETA*TEMP1  + COSTHETA*TEMP2
        END DO
        END DO
        IF (FLAG ==0) EXIT
        END DO
        WRITE(2,'("NO. OF EIGEN-VALUE        EIGEN-VALUE")')
        DO I = 1, N
        LAMDA(I)=1.0/A(I,I)
        WRITE(2,'(I4,4X,F10.6)')I,LAMDA(I)
        END DO
!******************************************************************
!*CALCULATE THE EIGEN-VECTORS BY INVERSE ITERATION AND SHIFT
```

```
!*******FORM A=STIF- SHIFT × MASS**********************************
        NREAL=N
        DO L = 1,N
!******SET THE ARBITRARY VECTOR V WITH VALUES EQUAL TO ******
!   SQUARE ROOT OF N, WHICH GIVES A UNIT LENGTH****************
        DO I = 1,N
        V(I)=1/SQRT(NREAL)
        END DO
        A= STIF-0.99*LAMDA(L)*MASS
!*******FACTORISE A USING CROUT FACTORIZATION, A = UT× D× U***
!********CALCULATE THE FIRST ROW OF U**************************
        U(1,1)=A(1,1)
        FACTOR=1.0/U(1,1)
        DO J = 2,N
        U(1,J)=A(1,J)*FACTOR
        END DO
!*******CALCULATE THE ELEMENTS OF ROWS 2 TO N**************
        DO I = 2,N
        IP1=I+1
!******* STORE IN U(I,I) THE DIAGONAL ELEMENT D(I,I) ************
        IM1=I-1
        SUM=0
        DO K=1,IM1
        SUM = SUM+ U(K,K)*U(K,I)**2
        END DO
        U(I,I)=A(I,I)-SUM
        IF(I==N) EXIT
!******CALCULATE THE OFF DIAGONAL ELEMENTS*******************
        FACTOR=1.0/U(I,I)
        DO J = IP1,N
        SUM = 0
        DO K = 1,IM1
        SUM = SUM +U(K,I)*U(K,J)*U(K,K)
        END DO
        U(I,J)=(A(I,J)-SUM)*FACTOR
        END DO
        END DO
!*******PRINT A = STIF – LAMDA × MASS******************************
        WRITE(2,*)'A-MATRIX'
        DO I = 1,N
        WRITE(2,*)(A(I,J),J=1,N)
        END DO
!******PRINT THE U MATRIX.  THE DIAGONAL ELEMENTS ARE D(I,I)***
        WRITE(2,*)'U MATRIX'
        DO I = 1,N
        WRITE(2,*)(U(I,J),J=1,N)
        END DO
!******DO ONLY TWO ITERATIONS OF INVERSE ITERATION***********
        DO ITERATION = 1,2
!*******CALCULATE THE PRODUCT AXI,  A=D-1× U-T ×M× U-1*********
!*******USE BACK  SUBSTITUTION TO SOLVE UZ = XI
        Z(N)=V(N)
        DO J=1,NM1
        SUM=0.0
        NMJ=N-J
        NJP1=NMJ+1
        DO K=NJP1,N
```

```
        SUM=SUM+U(NMJ,K)*Z(K)
        END DO
        Z(NMJ)=V(NMJ)-SUM
        END DO
!******CALCULATE M×Z*****************************************
     DO J = 1,N
        MZ(J)=0
        DO K=1,N
        MZ(J)=MZ(J)+MASS(J,K)*Z(K)
        END DO
        END DO
!*******USE FORWARD SUBSTITUTION TO SOLVE UT ×V = M×Z*********
        VBAR(1)=MZ(1)
        DO J=2,N
        SUM=0.0
        JM1=J-1
        DO K=1,JM1
        SUM=SUM+U(K,J)*VBAR(K)
        END DO
        VBAR(J)=(MZ(J)-SUM)
        END DO
!******DIVIDE ELEMENTS OF VBAR BY THE DIAGONAL ELEMENTS OF D*
        DO I = 1,N
        VBAR(I)=VBAR(I)/U(I,I)
        END DO
!*******CONVERT VBAR TO UNIT LENGTH***************************
        SUM = 0
        DO J = 1,N
        SUM = SUM+VBAR(J)**2
        END DO
        SUM = 1.0/SQRT(SUM)
        DO I = 1,N
        V(I)=VBAR(I)*SUM
        END DO
        END DO
!******CALCULATE THE EIGEN-VECTORS OF GENERAL EIGEN-VALUE
!PROBLEM , U * X = V
!******USE BACK  SUBSTITUTION TO SOLVE U ×X = V*****************
        X(N,L)=V(N)
        DO J=1,NM1
        SUM=0.0
        NMJ=N-J
        NJP1=NMJ+1
        DO K=NJP1,N
        SUM=SUM+U(NMJ,K)*X(K,L)
        END DO
        X(NMJ,L)=V(NMJ)-SUM
        END DO
        END DO*
!*******M NORMALISE THE EIGEN-VECTOR*************************
        DO L = 1,N
        SUM1=0
        DO J = 1,N
        SUM2=0
        DO K = 1,N
        SUM2=SUM2+MASS(J,K)*X(K,L)
        END DO
```

```
        SUM1=SUM1+SUM2*X(J,L)
        END DO
        SUM1=1.0/SQRT(SUM1)
        DO I = 1,N
        X(I,L)=X(I,L)*SUM1
        END DO
        END DO
         WRITE(2,*)'OMEGA SQUARED'
        DO I = 1, N
        WRITE(2,'(2(4X,F10.4))') LAMDA(I)
        END DO
!*******PRINT EIGEN-VECTORS************************************
        WRITE(2,*)'EIGEN-VECTORS'
        DO I = 1,N
        WRITE(2,'(6(4X,F10.4))')(X(I,J),J=1,N)
        END DO
        END PROGRAM QRLANCZ
!*****************************************************************
```

CHAPTER 6

SOLUTION OF LARGE-SCALE EIGEN-VALUE PROBLEMS

6.1 INTRODUCTION

In Chapter 5, different methods for the solution of eigen-value problems were described. In this chapter, two popular methods used for the solution of large-scale problems will be described in detail. The two are

- the simultaneous iteration method
- the sub-space iteration method

It has to be remembered, that in most practical cases, only a very small percentage of the full eigen-pairs are needed. These methods are therefore geared to address this specific requirement.

6.2 SIMULTANEOUS ITERATION METHOD

It was mentioned in Section 5.4.2, that inverse iteration on a set of mutually orthogonal vectors can be used to determine the partial solution to the eigen-value problem. This is the basis of the simultaneous iteration method, except that additional steps are taken to speed up convergence.

The method will be presented for the solution of the special eigen-value problem

$$A\,x = \lambda\,x \tag{6.1}$$

The general eigen-value problem $Kx = \omega^2 Mx$ is handled by operating on the matrix

$$A = U^T MU^1, \text{ where } K = U^T U \tag{6.2}$$

6.2.1 Theoretical Basis

The special eigen-value problem is

$$A\,X = X\,\Lambda \tag{6.3}$$

where A is an n×n symmetric positive definite matrix, X contains a set of n mutually orthogonal unit length vectors and Λ is a diagonal matrix containing the n eigen-values.

Because the vectors are of unit length and mutually orthogonal

$$X^T X = I \tag{6.4}$$

If one is interested in the first m eigen-pairs, where m << n, the main step is to chose arbitrarily a set of m mutually orthogonal unit length trial vectors u_i, i = 1, m.

$$U = \{u_1\ u_2\ u_3 \ldots u_m\},\ U^T U = I \tag{6.5}$$

As explained in Section 3.4.3, any arbitrary vector can be expressed in terms of eigen-vectors. Therefore

$$U = X\,C, \quad U^T = C^T X^T \tag{6.6}$$

where C is a matrix of constants of size n x m

$$C = \begin{bmatrix} c_{11} & c_{21} & c_{31} & . & . & c_{m1} \\ c_{12} & c_{22} & c_{32} & . & . & c_{m2} \\ c_{13} & c_{23} & c_{33} & . & . & c_{m3} \\ . & . & . & . & . & . \\ . & . & . & . & . & . \\ c_{1n} & c_{2n} & c_{3n} & . & . & c_{mn} \end{bmatrix} \tag{6.7}$$

Since U contains mutually orthogonal unit length vectors

$$U^T\,U = I = C^T(X^T X)\,C = C^T\,C \tag{6.8}$$

Let X be partitioned as X_a and X_b, where X_a corresponds to the desired m vectors and X_b corresponds to the remaining (n-m) vectors

$$X = \{\,X_a \quad X_b\} \tag{6.9}$$

Similarly let C also be portioned as C_a and C_b

$$C = \begin{bmatrix} C_a \\ C_b \end{bmatrix} \tag{6.10}$$

If U is a close estimate of the first m eigen-vectors contained in X_a, then C_b will be approximately zero. Hence

$$U \approx X_a C_a, \; C_a^T\,C_a \approx I \tag{6.11}$$

$$C_a = \begin{bmatrix} c_{11} & c_{12} & . & c_{1m} \\ c_{21} & c_{22} & . & c_{2m} \\ . & . & . & . \\ c_{m1} & c_{m2} & . & c_{mm} \end{bmatrix}, \; C_a^T\;C_a = \begin{bmatrix} \sum_{i=1}^{m} c_{i1}^2 & \sum_{i=1}^{m} c_{i1}\,c_{i2} & . & \sum_{i=1}^{m} c_{i1}\,c_{im} \\ & \sum_{i=1}^{m} c_{i2}^2 & . & \sum_{i=1}^{m} c_{i2}\,c_{im} \\ & & . & . \\ \text{symmetric} & & & \sum_{i=1}^{m} c_{im}^2 \end{bmatrix} \tag{6.12}$$

If U is a close estimate of the first m eigen-vectors, then from equation (6.11), the diagonal terms of C_a will be much larger than the off diagonal terms. Therefore if terms containing the products of off diagonal terms are ignored, then

$$C_a^T\;C_a = \begin{bmatrix} c_{11}^2 & c_{11}\,c_{12} + c_{21}\,c_{22} & . & c_{11}\,c_{1m} + c_{m1}\,c_{mm} \\ & c_{22}^2 & . & c_{22}\,c_{2m} + c_{m2}\,c_{mm} \\ & & . & . \\ & & & c_{mm}^2 \end{bmatrix} \tag{6.13}$$

Since $C_a^T\;C_a$ is approximately an identity matrix, the diagonal terms have to be equal to unity. Therefore $c_{ii} = 1$, i = 1 to m. Substituting this in equation (6.13)

$$C_a^T\ C_a = \begin{bmatrix} 1 & c_{12}+c_{21} & . & c_{1m}+c_{m1} \\ & 1 & . & c_{2m}+c_{m2} \\ & & . & . \\ & & & 1 \end{bmatrix} \tag{6.14}$$

In order that the off diagonal terms become zero

$$c_{ij} + c_{ji} = 0 \tag{6.15}$$

Therefore the C_a matrix becomes

$$C_a = \begin{bmatrix} 1 & -c_{12} & . & -c_{1m} \\ c_{12} & 1 & . & -c_{2m} \\ . & . & . & . \\ c_{1m} & c_{2m} & . & 1 \end{bmatrix} \tag{6.16}$$

In other words, as the set of vectors defined by U converge towards the first m eigen-vectors, the matrix C_a becomes skew symmetric with unit values on the diagonal.

In inverse iteration, the main step is the computation of the product of A and U.

$$V = A\ U \approx A\ X_a\ C_a \tag{6.17}$$

From equation (6.3)

$$A\ X_a = X_a\ \Lambda_a \tag{6.18}$$

$$\therefore\ V = X_a\ \Lambda_a\ C_a \tag{6.19}$$

$$\Lambda_a = \begin{bmatrix} \lambda_1 & & & \\ & \lambda_2 & & \\ & & . & \\ & & & \lambda_m \end{bmatrix} \tag{6.20}$$

$$\Lambda_a\ C_a = \begin{bmatrix} \lambda_1 & -c_{12}\lambda_1 & . & -c_{1m}\lambda_1 \\ c_{12}\lambda_2 & \lambda_2 & . & -c_{2m}\lambda_2 \\ . & . & . & . \\ c_{1m}\lambda_m & c_{2m}\lambda_m & . & \lambda_m \end{bmatrix}. \tag{6.21}$$

In the standard inverse iteration, the normalised value of vector V will be used for the next iteration cycle. However in the simultaneous iteration method, a so called interaction matrix B defined as follows is constructed

$$B = U^T\ V \tag{6.22}$$

Substituting for $U = X_a\ C_a$ from equation (6.11) and $V = X_a\ \Lambda_a\ C_a$ from equation (6.17) in equation (6.22)

$$B = C_a^{\ T}\ X_a^{\ T}\ X_a\ \Lambda_a\ C_a \tag{6.23}$$

Substituting for $X_a^{\ T}\ X = I$ from equation (6.4) in equation (6.23)

$$B = C_a^{\ T}\ \Lambda_a\ C_a \tag{6.24}$$

Substituting equation (6.16) for C_a and ignoring terms containing products of off diagonal terms of C_a

$$B = \begin{bmatrix} \lambda_1 & -c_{12}(\lambda_1 - \lambda_2) & . & -c_{1m}(\lambda_1 - \lambda_m) \\ & \lambda_2 & . & -c_{2m}(\lambda_2 - \lambda_m) \\ & & . & \\ symmetric & & & \lambda_m \end{bmatrix} \tag{6.25}$$

Clearly, the diagonal elements of B converge towards the required eigen-values. From equation (6.22), the coefficients of C_a can be calculated

$$b_{11} = \lambda_1 \text{ , } b_{22} = \lambda_2, \ldots b_{mm} = \lambda_m \tag{6.26}$$

$$b_{1m} = -c_{1m}(\lambda_1 - \lambda_m)$$

$$= -c_{1m}(b_{11} - b_{mm}) \tag{6.27}$$

$$\therefore c_{1m} = -\frac{b_{1m}}{(b_{11} - b_{mm})}$$

In general

$$c_{ij} = -\frac{b_{ij}}{(b_{ii} - b_{jj})} \text{ , } i \neq j$$

$$c_{ii} = 1 \tag{6.28}$$

Using equations (6.11) and (6.19)

$$V C_a^T = X_a \Lambda_a C_a C_a^T = X_a \Lambda_a \tag{6.29}$$

Clearly, instead of using $V = A\ U$ as the matrix for the next stage of iteration as in standard inverse iteration, it is better to use $V\ C_a^T$ as the matrix for the next iteration, because the product converges towards X_a and the eigen-values Λ_a.

It has been observed from empirical evidence that for faster convergence it is better to use instead c_{ij}^*, rather than c_{ij} from equation (6.28), where

$$c_{ij}^* = -\frac{2b_{ij}}{[(b_{ii} - b_{jj}) + s\sqrt{(b_{ii} - b_{jj})^2 + 16b_{ij}^2}]} \text{ , } i \neq j \text{ , } s = sign\ of\ (b_{ii} - b_{jj})$$

$$c_{ii}^* = 1 \tag{6.30}$$

This also avoids the problem of zero denominators in the case of repeated eigen-values.

6.2.2 Implementation of the Simultaneous Iteration Method

In the case of undamped vibration problems, defined by

$$K X = M X \Omega^2 \tag{6.31}$$

the following steps need to be followed. For inverse iteration, equation (6.2) is used.

i. Use Cholesky factorisation to calculate the matrix U where $U^T U = K$, where K is a symmetric positive definite matrix of size n x n.

ii. Start with m set of arbitrary orthogonal unit length vectors. Let them be contained in the matrix VECTOR of size n x m.

iii. Calculate the matrix Z, by solving the equation $U Z = VECTOR$. This requires only back substitution.

iv. Calculate the matrix R, by multiplying $R = M Z$.

v. Calculate the matrix V, by solving the equation $U^T V = R$. This requires only forward substitution.

vi. Form the interaction matrix $B = VECTOR^T \times V$. From the matrix B matrix, form the C_a matrix using equation (6.30).

vii. Form matrix $W = V\times C_a$. Sort out the columns of W into the correct order of dominance as dictated by the corresponding diagonal value of B.

viii. Convert the columns of W into a set of mutually orthogonal vectors. This is done using the Gram-Schmidt orthogonalisation procedure as explained in Section 5.3.1.

ix. Use the orthogonalised vectors as the new set of vectors and go to step (iii).

x. Repeat steps (iii) to (ix) until satisfactory convergence is obtained.

xi. Columns of matrix VECTOR are the eigen-vectors of the matrix A. The eigen-vectors X of the general eigen-value problem are obtained by $U\times X = VECTOR$. X is obtained by using back substitution. The vectors need to be finally M normalised.

It is worth remembering that the necessary coefficients b_{ii}, b_{jj} and b_{ij}, and hence c_{ij}^*, can be calculated without explicitly forming the B and the C_a matrices as will be explained in the next Section.

6.2.2.1 *Calculation of W matrix*

As explained in Section 6.2.2, the calculation of W matrix involves three steps as follows.

- Form the interaction matrix $B = VECTOR^T\times V$.
- From the matrix B form the C_a matrix using equation (6.30).
- Form matrix $W = V\times C_a^T$.

In order to understand how the matrix W is formed, consider the case of a matrix of size n = 6, m = 4

$$W = V\,C_a^{*T} = \begin{bmatrix} v_{11} & v_{12} & v_{13} & v_{14} \\ v_{21} & v_{22} & v_{23} & v_{24} \\ v_{31} & v_{32} & v_{33} & v_{34} \\ v_{41} & v_{42} & v_{43} & v_{44} \\ v_{51} & v_{52} & v_{53} & v_{54} \\ v_{61} & v_{62} & v_{63} & v_{64} \end{bmatrix} \begin{bmatrix} 1 & c_{12} & c_{13} & c_{14} \\ -c_{12} & 1 & c_{23} & c_{24} \\ -c_{13} & -c_{23} & 1 & c_{34} \\ -c_{14} & -c_{24} & -c_{34} & 1 \end{bmatrix} \tag{6.32}$$

A general term of w_{ij} can be written as

$$w_{ij} = \sum_{k=1}^{m} v_{ik}\, c_{jk},\; i=1 \textit{ to } n \textit{ and } j = 1 \textit{ to } m \tag{6.33}$$

Since the diagonal terms of C_a^* are unity, initially the matrix V can be copied into W. The terms of W affected by a term c_{jk} are

$$w_{lj} = -v_{lk}\, c_{jk}\;,\; w_{lk} = +v_{lj}\, c_{jk}\;,\; l = 1,n \tag{6.34}$$

In other words, as soon the term c_{jk} is formed, elements of the jth column of W are decreased as follows

$$w_{lj} = w_{lj} - v_{lk}\, c_{jk}\;,\quad l = 1,n$$

Similarly elements of the kth column of W are increased as follows

$$w_{lk} = w_{lk} + v_{lj}\, c_{jk}\;,\quad l = 1,n$$

The calculations for the formation of matrix W proceeds as follows

- Copy matrix V into W
- Start the loop: j = 1, m
- Start the loop: k = 1, m
- Skip if j = k, otherwise form b_{jj}, b_{kk}, b_{jk}, where

$$b_{jj} = \sum_{p=1}^{N} vector_{pj} \times v_{pj},\; b_{kk} = \sum_{p=1}^{N} vector_{pk} \times v_{pk},\; b_{jk} = \sum_{p=1}^{N} vector_{pj} \times v_{pk}$$

- Form c_{jk}^*

$$c_{jk}^* = -\frac{2b_{jk}}{[(b_{jj} - b_{kk}) + s\sqrt{(b_{jj} - b_{kk})^2 + 16b_{jk}^2}\,]},\; j \neq k,\, s = \textit{sign of } (b_{jj} - b_{kk})$$

- Modify all elements of the columns j and k of W as previously explained

$$w_{lj} = w_{lj} - v_{lk}\, c_{jk}\;,\quad l = 1,n$$

$$w_{lk} = w_{lk} + v_{lj}\, c_{jk}\;,\quad l = 1,n$$

6.2.3 Example of the Simultaneous Iteration Method

Calculate the eigen-pairs for the four-storey shear frame with the following data.

$$K = \begin{bmatrix} 14 & -6 & 0 & 0 \\ & 10 & -4 & 0 \\ symmetric & & 8 & -4 \\ & & & 4 \end{bmatrix}, \; M = \begin{bmatrix} 4 & 0 & 0 & 0 \\ & 4 & 0 & 0 \\ symmetric & & 2 & 0 \\ & & & 2 \end{bmatrix}$$

Step 1: Calculate the Cholesky factorisation of K using the procedure described in Section 4.3.

$$K = U^T \; U \; , \; U = \begin{bmatrix} 3.741657 & -1.603567 & 0 & 0 \\ 0 & 2.725541 & -1.467599 & 0 \\ 0 & 0 & 2.417882 & -1.654341 \\ 0 & 0 & 0 & 1.123903 \end{bmatrix}$$

Iteration cycle 1:

Step 2: Start with a set of orthogonal vectors of unit length. Here it is assumed that m, the number of vectors required, is equal to 3. The orthogonal unit vector is chosen as

$$VECTOR = \begin{bmatrix} 1 & 0 & 0 \\ 0 & 1 & 0 \\ 0 & 0 & 1 \\ 0 & 0 & 0 \end{bmatrix}$$

Step 3: Use back substitution and calculate the matrix Z, from $U\,Z = VECTOR$

$$Z = \begin{bmatrix} 0.2673 & 0.1572 & 0.0954 \\ 0 & 0.3699 & 0.2227 \\ 0 & 0 & 0.4136 \\ 0 & 0 & 0 \end{bmatrix}$$

Step 4: Calculate $R = M\,Z$

$$R = \begin{bmatrix} 1.0690 & 0.6290 & 0.3818 \\ 0 & 1.4676 & 0.8908 \\ 0 & 0 & 0.8272 \\ 0 & 0 & 0 \end{bmatrix}$$

Step 5: Use forward substitution to calculate the matrix V, where $U^T\,V = R$

$$V = \begin{bmatrix} 0.2857 & 0.1681 & 0.1020 \\ 0.1681 & 0.6374 & 0.3869 \\ 0.1020 & 0.3869 & 0.5769 \\ 0.1502 & 0.5694 & 0.8492 \end{bmatrix}$$

Step 6: Calculate the interaction matrix $B = VECTOR^T\,V$. Also calculate the skew-symmetric matrix C_a with unit values on the diagonal.

The explicit form of B and C_a matrices is given purely for reference. In practical calculation, it is not necessary to store the matrices B and C_a as explained in Section 6.2.2.

$$B = \begin{bmatrix} 0.2857 & 0.1681 & 0.1020 \\ 0.1681 & 0.6374 & 0.3869 \\ 0.1020 & 0.3869 & 0.5769 \end{bmatrix}$$

$$C_a = \begin{bmatrix} 1 & 0.3028 & 0.2575 \\ & 1 & -0.4809 \\ skew-symmetric & & 1 \end{bmatrix}$$

The estimated eigen-values are given by the diagonal values of B. Thus the eigen-values are 0.6374, 0.5769 and 0.2857. They are in positions 2, 3 and 1.

Step 7: Calculate $W = V \times C_a$

$$W = \begin{bmatrix} 0.2086 & 0.3037 & 0.0948 \\ -0.1245 & 0.8743 & 0.1237 \\ -0.1636 & 0.6952 & 0.4172 \\ -0.2409 & 1.0233 & 0.6141 \end{bmatrix}$$

Rearrange W such that the second column which corresponds to the eigen-value of 0.6374 is in the first column, the third column which corresponds to the eigen-value of 0.5769 is in the second column and finally the first column which corresponds to the eigen-value of 0.2857 is in the third column. The rearranged W matrix is therefore

$$W = \begin{bmatrix} 0.3037 & 0.0948 & 0.2086 \\ 0.8743 & 0.1237 & -0.1245 \\ 0.6952 & 0.4172 & -0.1636 \\ 1.0233 & 0.6141 & -0.2409 \end{bmatrix}$$

Step 8: Orthogonalise the columns of the rearranged W and normalise them to unit length. Use the Gram-Schmidt procedure explained in Section 5.3.1.

Orthogonalise column 2 with respect to column 1

$$w_1^T w_1 = 2.3871,\ w_1^T w_2 = 1.0554,\ \alpha = -1.0554/2.3871 = -0.4421$$

Orthogonalised $w_2 = w_2 + \alpha\, w_1 = [-0.0395\ \ -0.2628\ \ 0.1098\ \ 0.1617]^T$

Orthogonalise column 3 with respect to columns 1 and 2. For column 2, use the new orthogonalised value.

$$w_1^T w_1 = 2.3871,\ w_1^T w_3 = -0.4058,\ \alpha_1 = -\ (-0.4058)/2.3871 = 0.1700$$

$$w_2^T w_2 = 0.1088,\ w_2^T w_3 = -0.0324,\ \alpha_2 = -\ (-0.0324)/0.1088 = 0.2978$$

Orthogonalised w_3 is

$$w_3 = w_3 + \alpha_1 w_1 + \alpha_2 w_2 = [0.2484\ \ -0.0542\ \ -0.0127\ \ -0.0188]^T$$

Convert the orthogonalised vectors to unit length.

$$w_1^T w_1 = 2.3871,\ w_2^T w_2 = 0.1088,\ w_3^T w_3 = 0.0652$$

Divide the vectors in columns 1, 2 and 3 respectively by $\sqrt{2.3871}$,$\sqrt{0.1088}$ and $\sqrt{0.0652}$. The orthogonalised, normalised vectors are used as the starting vectors for the next iteration.

$$VECTOR = W = \begin{bmatrix} 0.1966 & -0.1197 & 0.9732 \\ 0.5659 & -0.7967 & -0.2123 \\ 0.4500 & 0.3329 & -0.0499 \\ 0.6623 & 0.4900 & -0.0735 \end{bmatrix}$$

Iteration 2:

Step 3: Use back substitution and calculate the matrix Z, from $U\,Z = VECTOR$

$$Z = \begin{bmatrix} 0.2755 & -0.0566 & 0.2116 \\ 0.5249 & -0.0575 & -0.1131 \\ 0.5893 & 0.4360 & -0.0654 \\ 0.5893 & 0.4360 & -0.0654 \end{bmatrix}$$

Step 4: Calculate $R = M\,Z$

$$R = \begin{bmatrix} 1.1100 & -0.2266 & 0.8465 \\ 2.0998 & -0.2301 & -0.4524 \\ 1.1786 & 0.8720 & -0.1308 \\ 1.1786 & 0.8720 & -0.1308 \end{bmatrix}$$

Step 5: Use forward substitution to calculate the matrix V, where $U^T\,V = R$

$$V = \begin{bmatrix} 0.2967 & -0.0606 & 0.2262 \\ 0.9450 & -0.1201 & -0.0329 \\ 1.0610 & 0.2878 & -0.0741 \\ 2.6105 & 1.1995 & -0.2254 \end{bmatrix}$$

Step 6: Calculate the interaction matrix $B = VECTOR^T\,V$. Also calculate the skew-symmetric matrix C_a with unit values on the diagonal.

$$B = \begin{bmatrix} 2.7994 & 0.8441 & -0.1568 \\ 0.8441 & 0.7865 & -0.1360 \\ -0.1568 & -0.1360 & 0.2474 \end{bmatrix}$$

$$C_a = \begin{bmatrix} 1 & -0.2840 & 0.0605 \\ & 1 & 0.2084 \\ skew-symmetric & & 1 \end{bmatrix}$$

The estimated eigen-values are given by the diagonal values of B. Thus the eigen-values are 2.7994, 0.7865 and 0.2474. They are in positions 1, 2 and 3. Therefore there is no need to rearrange W.

Step 7: Calculate $W = V \times C_a$

$$W = \begin{bmatrix} 0.2658 & -0.1920 & 0.2316 \\ 0.9128 & -0.3816 & -0.0007 \\ 1.1472 & 0.0019 & 0.0501 \\ 2.9648 & 0.5050 & 0.1826 \end{bmatrix}$$

Step 8: Using the Gram-Schmidt procedure, orthogonalise the columns of W and normalise them to unit length.

Orthogonalise column 2 with respect to column 1

$$w_1^T w_1 = 11.01,\ w_1^T w_2 = 1.10,\ \alpha = -1.10/11.01 = -0.10$$

Orthogonalised w_2

$$w_2 = w_2 + \alpha\, w_1 = [-0.2185 \ \ -0.4728 \ \ -0.1128 \ \ 0.2088]^T$$

Orthogonalise column 3 with respect to columns 1 and 2. For column 2, use the new orthogonalised value

$$w_1^T w_1 = 11.01,\ w_1^T w_3 = 0.6598,\ \alpha_1 = -0.6598/11.01 = -0.0599$$
$$w_2^T w_2 = 0.3276,\ w_2^T w_3 = -0.0178,\ \alpha_2 = -(-0.0178)/0.3276 = 0.0543$$

Orthogonalised w_3

$$w_3 = w_3 + \alpha_1 w_1 + \alpha_2 w_2 = [0.2038 \ \ -0.0811 \ \ -0.0247 \ \ 0.0163]^T$$

Convert the orthogonalised vectors to unit length

$$w_1^T w_1 = 11.01,\ w_2^T w_2 = 0.3276,\ w_3^T w_3 = 0.0490$$

Divide the vectors in columns 1, 2 and 3 respectively by $\sqrt{11.01}$, $\sqrt{0.3276}$ and $\sqrt{0.0490}$. The orthogonalised, normalised vectors are used as the starting vectors for the next iteration.

$$VECTOR = W = \begin{bmatrix} 0.0801 & -0.3818 & 0.9208 \\ 0.2751 & -0.8260 & -0.3664 \\ 0.3457 & -0.1970 & -0.1118 \\ 0.8935 & 0.3648 & 0.0735 \end{bmatrix}$$

At the end of the second iteration, the predicted eigen-values are $\omega^{-2} = 2.7994$, 0.7865 and 0.2474. The first three correct values computed in Section 3.2.4 are $\omega^{-2} = 3.287$, 0.6283 and 0.2358.

<u>Iteration 3:</u>

Step 3: Use back substitution and calculate the matrix Z, from $U\,Z = VECTOR$

$$Z = \begin{bmatrix} 0.2232 & -0.1995 & 0.1881 \\ 0.4708 & -0.2274 & -0.1352 \\ 0.6869 & 0.1406 & -0.0015 \\ 0.7950 & 0.3246 & -0.0654 \end{bmatrix}$$

Step 4: Calculate $R = M\,Z$

$$R = \begin{bmatrix} 0.8928 & -0.7979 & 0.7525 \\ 1.8833 & -0.9095 & -0.5409 \\ 1.3739 & 0.2812 & -0.0029 \\ 1.5900 & 0.6492 & 0.1308 \end{bmatrix}$$

Step 5: Use forward substitution to calculate the matrix V, where $U^T\,V = R$

$$V = \begin{bmatrix} 0.2386 & -0.2133 & 0.2011 \\ 0.8314 & -0.4592 & -0.0801 \\ 1.0728 & -0.1624 & -0.0499 \\ 2.9939 & 0.3386 & 0.0430 \end{bmatrix}$$

Step 6: Calculate the interaction matrix $B = VECTOR^T V$. Also calculate the skew-symmetric matrix C_a with unit values on the diagonal.

$$B = \begin{bmatrix} 3.2939 & 0.1030 & 0.0153 \\ 0.1030 & 0.6162 & 0.0149 \\ 0.0153 & 0.0149 & 0.2233 \end{bmatrix}$$

$$C_a = \begin{bmatrix} 1 & -0.0382 & -0.0050 \\ & 1 & -0.0378 \\ skew-symmetric & & 1 \end{bmatrix}$$

The estimated eigen-values are given by the diagonal values of B. Thus the eigen-values are 3.2939, 0.6162 and 0.2233. They are in positions 1, 2 and 3. Therefore there is no need to rearrange W.

Step 7: Calculate $W = V \times C_a$

$$W = \begin{bmatrix} 0.2315 & -0.2148 & 0.2080 \\ 0.8134 & -0.4940 & -0.0669 \\ 1.0664 & -0.2053 & -0.0491 \\ 3.0071 & 0.2257 & 0.0153 \end{bmatrix}$$

Step 8: Using the Gram-Schmidt procedure, orthogonalise the columns of W and normalise them to unit length.

Orthogonalise column 2 with respect to column 1

$$w_1^T w_1 = 10.8951, \; w_1^T w_2 = 0.0008, \; \alpha = -0.0008/10.8951 = -0.00008$$

Orthogonalised w_2

$$w_2 = w_2 + \alpha w_1 = [-0.2149 \;\; -0.4946 \;\; -0.2061 \;\; 0.2234]^T$$

Orthogonalise column 3 with respect to columns 1 and 2. For column 2, use the new orthogonalised value

$$w_1^T w_1 = 10.8951, \; w_1^T w_3 = 0.0001, \; \alpha_1 = -0.0001/10.8951 = 0$$

$$w_2^T w_2 = 0.3832, \; w_2^T w_3 = 0.0019, \; \alpha_2 = -(0.0019)/0.3832 = -0.0050$$

Orthogonalised w_3

$$w_3 = w_3 + \alpha_1 w_1 + \alpha_2 w_2 = [0.2093 \;\; -0.0635 \;\; -0.0468 \;\; 0.0177]^T$$

Convert the orthogonalised vectors to unit length.

$$w_1^T w_1 = 10.8951, \; w_2^T w_2 = 0.3832, \; w_3^T w_3 = 0.0499$$

Divide the vectors in columns 1, 2 and 3 respectively by $\sqrt{10.8951}$, $\sqrt{0.3832}$ and $\sqrt{0.0499}$. The orthogonalised, normalised vectors are used as the starting vectors for the next iteration.

$$VECTOR = W = \begin{bmatrix} 0.0701 & -0.3472 & 0.9329 \\ 0.2464 & -0.7989 & -0.2830 \\ 0.3231 & -0.3329 & -0.2085 \\ 0.9110 & 0.3609 & 0.0797 \end{bmatrix}$$

At the end of the third iteration, the predicted eigen-values are ω^{-2} = 3.2939, 0.6162 and 0.2233. The first three correct values computed in Section 3.2.4 are ω^{-2} = 3.287, 0.6283 and 0.2358.

Clearly the values are converging towards the exact values. Table 6.1 summarises the results of ten iterations.

Table 6.1 Convergence study

Iteration no.	$1/\omega_1^2$	$1/\omega_2^2$	$1/\omega_3^2$
1	0.6374	0.5759	0.2837
2	2.7994	0.7865	0.2474
3	3.2939	0.6162	0.2233
4	3.2987	0.6269	0.2283
5	3.2987	0.6282	0.2316
6	3.2987	0.6283	0.2335
7	3.2987	0.6283	0.2346
8	3.2987	0.6283	0.2352
9	3.2987	0.6283	0.2355
10	3.2987	0.6283	0.2356
Exact values	3.2987	0.6283	0.2358

Although at the end of iteration 3 convergence has not occurred, an approximate value of the eigen-vectors can be calculated using the VECTOR at the end of the third iteration.

$$VECTOR = \begin{bmatrix} 0.0701 & -0.3472 & 0.9329 \\ 0.2464 & -0.7989 & -0.2830 \\ 0.3231 & -0.3329 & -0.2085 \\ 0.9110 & 0.3609 & 0.0797 \end{bmatrix}$$

Using back substitution, solving X from $U \times X = VECTOR$

$$X = \begin{bmatrix} 0.2163 & -0.1995 & 0.1960 \\ 0.4610 & -0.2490 & -0.1245 \\ 0.6882 & 0.0820 & -0.0383 \\ 0.8106 & 0.3211 & 0.0700 \end{bmatrix}$$

M normalising X, the estimated eigen-vectors are

$$EIGENVECTORS = \begin{bmatrix} 0.1191 & -0.2520 & 0.4101 \\ 0.2538 & -0.3145 & -0.2605 \\ 0.3789 & 0.1036 & -0.0802 \\ 0.4463 & 0.4056 & 0.1465 \end{bmatrix}$$

The converged eigen-vectors are

$$CORRECT\ VALUES = \begin{bmatrix} 0.1191 & -0.2469 & 0.3947 \\ 0.2537 & -0.3141 & -0.1959 \\ 0.3788 & 0.0850 & -0.2487 \\ 0.4465 & 0.4165 & 0.2232 \end{bmatrix}$$

Clearly, even at the end of the third iteration, the first two sets of eigen-vectors have almost converged to their correct values.

6.3 SUBSPACE ITERATION METHOD

The subspace iteration method solves the first m eigen-pairs of the general eigen-value problem given by equation (6.31). The basic steps of the method are as follows.

- Start with a matrix *Y*, which contains a set of m arbitrary vectors, where m is the desired number of first m sets of eigen-pairs to be determined.
- Solve for *YBAR*, where $K \times YBAR = M \times Y$. This is nothing but inverse iteration applied to the general eigen-value problem using m sets of vectors simultaneously. If inverse iteration alone were employed, then for the next iteration, YBAR would be used instead of Y and further iterations are carried out, until convergence is reached. In subspace iteration, convergence is speeded up by the next step.
- Form the m×m subspace stiffness and mass matrices *KBAR* and *MBAR*, where

$KBAR = YBAR^T \times K \times YBAR$, $MBAR = YBAR^T \times M \times YBAR$

- Solve the general eigen-value problem

$KBAR \times Q = MBAR \times Q \times \Omega$, Q = eigen-vectors, Ω = eigen-values

- For the next iteration, **instead of using** *YBAR* as the starting vector, **use** $YBAR \times Q$

In order to understand the reason for using the modified starting vector, it is necessary to understand an important idea known as Rayleigh quotient.

6.3.1 Rayleigh Quotient

Consider the general eigen-value problem with the eigen-vectors M normalised

$$K \times X = M \times X \times \Lambda,\ X^T \times M \times X = I,\ X^T \times K \times X = \Lambda \tag{6.35}$$

If q is an arbitrary vector, Rayleigh quotient is defined as the ratio

$$\rho(q) = \frac{q^T K q}{q^T M q} \tag{6.36}$$

As explained in Section 3.2.3, q can be expressed in terms of the eigen-vectors X as $q = X\,C$, where the constants c_i are given by $c_i = x_i^T M\,q$. Therefore

$$q^T \times M \times q = C^T \times [X^T \times M \times X] \times C = C^T \times C = \sum_{I=1}^{n} c_i^2 \tag{6.37}$$

$$q^T \times K \times q = C^T \times [X^T \times K \times X] \times C = C^T \times \Lambda \times C = \sum_{I=1}^{n} \lambda_i\ c_i^2 \tag{6.38}$$

$$\rho(q) = \frac{\sum_{i=1}^{n} \lambda_i\, c_i^2}{\sum_{i=1}^{n} c_i^2} \tag{6.39}$$

$$If\ \lambda_1 < \lambda_2 \ldots < \lambda_n\ ,\ \frac{\rho(q)}{\lambda_1} = \frac{\sum_{i=1}^{n} \frac{\lambda_i}{\lambda_1}\ c_i^2}{\sum_{i=1}^{n} c_i^2} > 1,\ \frac{\rho(q)}{\lambda_n} = \frac{\sum_{i=1}^{n} \frac{\lambda_i}{\lambda_n}\ c_i^2}{\sum_{i=1}^{n} c_i^2} < 1 \tag{6.40}$$

$$\therefore\ \lambda_1 < \rho(q) < \lambda_n \tag{6.41}$$

If the vector q is M orthogonal to first j eigen-vectors, then $c_i = 0$, $i = 1, 2, \ldots j$.

$$\frac{\rho(q)}{\lambda_{j+1}} = \frac{\sum_{i=j+1}^{n} \frac{\lambda_i}{\lambda_{j+1}}\ c_i^2}{\sum_{i=j+1}^{n} c_i^2} > 1,\ \frac{\rho(q)}{\lambda_n} = \frac{\sum_{i=j+1}^{n} \frac{\lambda_i}{\lambda_n}\ c_i^2}{\sum_{i=j+1}^{n} c_i^2} < 1 \tag{6.42}$$

$$\therefore\ \lambda_{j+1} < \rho(q) < \lambda_n \tag{6.43}$$

Clearly for all q, satisfying the condition $x_i^T M\, q = 0$, $i = 1, 2, \ldots j$, the minimum of $\rho(q) = \lambda_{j+1}$.

6.3.1.1 *Rayleigh Quotient: Example*

Consider the four-storey shear frame shown in Fig. 3.4. The stiffness and mass matrices are

$$K = \begin{bmatrix} 14 & -6 & 0 & 0 \\ & 10 & -4 & 0 \\ symmetric & & 8 & -4 \\ & & & 4 \end{bmatrix},\ M = \begin{bmatrix} 4 & 0 & 0 & 0 \\ & 4 & 0 & 0 \\ symmetric & & 2 & 0 \\ & & & 2 \end{bmatrix}$$

The exact eigen-pairs were shown in Sections 3.2.4 and 3.2.5 to be $\omega^2 = 0.3032$, 1.5917, 4.2410, 5.8642 and

$$X = \begin{bmatrix} 0.1191 & 0.2469 & 0.39383 & 0.1406 \\ 0.2537 & 0.3141 & -0.1946 & -0.2216 \\ 0.3788 & -0.0850 & -0.2520 & 0.5345 \\ 0.4465 & -0.4165 & 0.2249 & -0.2767 \end{bmatrix}$$

Consider two cases. In case 1, q is a general vector while in case 2 it is a vector M orthogonal to x_1 and x_2.
Case 1: Let $q = [0.1 \;\; 0.2 \;\; 0.3 \;\; 0.4]^T$
$q^T K q = 0.22$, $q^T M q = 0.7$, $\rho = 0.3143$
$(\lambda_1 = 0.3032) < (\rho = 0.3143) < (\lambda_4 = 5.8642)$
Case 2: Let $q = [0.3194 \;\; -0.1869 \;\; -0.0846 \;\; 0.1141]^T$
$q^T K q = 2.7297$, $q^T M q = 0.5881$, $\rho = 4.6416$
$(\lambda_3 = 4.2410) < (\rho = 4.6416) < (\lambda_4 = 5.8642)$

6.3.1.2 *Differentiation of a matrix product $q^T K q$*

Consider for simplicity the following matrix product.

$$q^T K q = [q_1 \quad q_2 \quad q_3 \quad q_4] \begin{bmatrix} k_{11} & k_{12} & k_{13} & k_{14} \\ & k_{22} & k_{23} & k_{24} \\ & & k_{33} & k_{34} \\ symm. & & & k_{44} \end{bmatrix} \begin{bmatrix} q_1 \\ q_2 \\ q_3 \\ q_4 \end{bmatrix} \tag{6.44}$$

Carrying out the multiplication

$$f(q) = f(q_1, q_2, q_3) = q^T K q = \sum_{i=1}^{4} \sum_{j=1}^{4} k_{ij} \, q_i \, q_j \tag{6.45}$$

$f(q) = q^T K q$ is known as a pure quadratic form, because it contains only second order terms in q_i. Partially differentiating f(q), with respect to q_1, q_2, q_3 and q_4

$$\frac{\partial f}{\partial q} = \begin{bmatrix} \frac{\partial f}{\partial q_1} \\ \frac{\partial f}{\partial q_2} \\ \frac{\partial f}{\partial q_3} \\ \frac{\partial f}{\partial q_4} \end{bmatrix} = 2 \begin{bmatrix} k_{11} & k_{12} & k_{13} & k_{14} \\ & k_{22} & k_{23} & k_{24} \\ & & k_{33} & k_{34} \\ symm & & & k_{44} \end{bmatrix} \begin{bmatrix} q_1 \\ q_2 \\ q_3 \\ q_4 \end{bmatrix} = 2Kq \tag{6.46}$$

Consider the general eigen-value problem in equation (6.35). It is required to determine the first m eigen-pairs of the problem. The iteration starts with a matrix *Y*, containing m arbitrary vectors $y_1, y_2, \ldots y_m$. Solve for *YBAR* from

$$K \times YBAR = M \times Y \tag{6.47}$$

Form the Rayleigh quotient $\rho(z)$ using the matrix $Z = YBAR \times Q$, where the matrix *Q* of size m x m, is yet to be defined.

$$\rho(z_1, z_2, \ldots z_m) = \frac{Z^T \times K \times Z}{Z^T \times M \times Z} = \frac{Q^T \times YBAR^T \times K \times YBAR \times Q}{Q^T \times YBAR^T \times M \times YBAR \times Q} \tag{6.48}$$

Denoting $YBAR^T \times K \times YBAR = KBAR$ and $YBAR^T \times M \times YBAR = MBAR$,

$$\rho(z_1, z_2, \ldots z_m) = \frac{Q^T \times KBAR \times Q}{Q^T \times MBAR \times Q} \tag{6.49}$$

The stationary value of the Rayleigh coefficient (6.49) is when

$$\frac{\partial \rho(Z)}{\partial Q} = 0 \tag{6.50}$$

Using the usual rules for the differentiation of a quotient, and noting that

$$\frac{\partial}{\partial Q}[Q^T \times KBAR \times Q] = 2 \times KBAR \times Q \tag{6.51}$$

$$\frac{\partial}{\partial Q}[Q^T \times MBAR \times Q] = 2 \times MBAR \times Q \tag{6.52}$$

$$\because \frac{\partial \rho(Z)}{\partial Q} = 0$$

$$\therefore \{2 \times KBAR \times Q\} Q^T \times MBAR \times Q - \{2 \times MBAR \times Q\} \times Q^T \times KBAR \times Q = 0 \tag{6.53}$$

If the matrix Q is interpreted as the eigen-vectors of the general eigen-value problem

$$KBAR \times Q = MBAR \times Q \times \Omega \tag{6.54}$$

Where the diagonal matrix Ω contains the eigen-values, then

$$Q^T \times KBAR \times Q = Q^T \times MBAR \times Q \times \Omega \tag{6.55}$$

Substituting for $Q^T \times KBAR \times Q = Q^T \times MBAR \times Q \times \Omega$ in equation (6.53)

$$\{2 \times KBAR \times Q\} \times Q^T \times MBAR \times Q - \{2 \times MBAR \times Q\} \times Q^T \times MBAR \times Q \times \Omega = 0 \tag{6.56}$$

Simplifying, equation (6.56),

$$KBAR \times Q - MBAR \times Q \times \Omega = 0 \tag{6.57}$$

This shows that the stationary value of the Rayleigh quotient using the 'vector' $YBAR \times Q$ is reached when Q is the eigen-vector matrix of the general eigen-value problem connecting *KBAR* and *MBAR*.

6.3.2 Implementation of the Subspace Iteration Method

The solution procedure has some resemblance to the simultaneous iteration method discussed in Section 6.3). The steps in the method are as follows.

1. Use Cholesky factorisation to calculate the matrix U where $U^T \times U = K$, where K is a symmetric positive definite matrix of size n x n.
2. Start with a set of m arbitrary vectors. Let them be contained in the matrix *VECTOR* of size n x m.

3. Calculate the matrix product $MVEC = M \times VECTOR$.
4. Calculate the matrix $VECBAR$, by solving the equation $K \times VECBAR = MVEC$. The solution is achieved through Cholesky method. This is done in two stages.
5. Using forward substitution, solve for Y from $U^T \times Y = MVEC$.
6. Having calculated Y, using back substitution, calculate $VECBAR$ by solving $U \times VECBAR = Y$.
7. Calculate the subspace stiffness matrix
$$KBAR = VECBAR^T \times K \times VECBAR.$$
8. Since $K \times VECBAR = MVEC$, $KBAR$ is efficiently calculated as
$$KBAR = VECBAR^T \times MVEC$$
9. Calculate the subspace mass matrix $MBAR = VECBAR^T \times M \times VECBAR$. This is efficiently done in two stages as follows.
10. Calculate $MVECBAR = M \times VECBAR$.
11. Calculate $MBAR = VECBAR^T \times MVECBAR$
12. Solve the general eigen-value problem defined by
$$KBAR \times EIGENVEC = MBAR \times EIGENVEC \times \Lambda$$
13. Sort out the columns $EIGENVEC$ into the correct order of dominance as dictated by the corresponding eigen-values.
14. Use $VECTOR = VECBAR \times EIGENVEC$ as the new set of vectors and go to step 3. Since in order to start the iteration at step (4), it is not $VECTOR$ but the product $MVEC = M \times VECTOR = M \times VECBAR \times EIGENVEC$, that is required, and as the product $MVECBAR = M \times VECBAR$ has already been calculated, the new iteration is started with
$$MVEC = MVECBAR \times EIGENVEC$$
15. Repeat steps (4) to (14) until satisfactory convergence is obtained.

6.3.2.1 *Guard Vectors*

It is generally found that the smallest eigen-values converge first. A higher convergence rate can be obtained by using more than the number of required eigen-values. It is recommended that if m eigen-values are needed, then the number of vectors used for iteration must be at least the lesser of (2m, m + 8). The extra number of vectors are known as guard vectors.

6.3.2.2 *Starting vectors*

It is necessary that the starting vectors in X are not orthogonal to any of the required eigen-vectors. It is suggested that the first vectors be simply the diagonal elements of the mass matrix and the other vectors have unit values in the place with the smallest ratio k_{ii}/m_{ii}.

6.3.3 Subspace Iteration Method: Example

Calculate the eigen-pairs for the four-storey shear frame data for matrices *K*, *M* and *U* from Section 6.2.3.

Iteration 1:

Step 2: Start with a set of arbitrary vectors. Here it is assumed that m, the number of vectors required is equal to 3. Assume

$$VECTOR = \begin{bmatrix} 1 & 0 & 0 \\ 0 & 1 & 0 \\ 0 & 0 & 1 \\ 0 & 0 & 0 \end{bmatrix}$$

Step 3: Calculate $MVEC = M \times VECTOR$

$$MVEC = \begin{bmatrix} 4 & 0 & 0 \\ 0 & 4 & 0 \\ 0 & 0 & 2 \\ 0 & 0 & 0 \end{bmatrix}$$

Step 4: Calculate Y using forward substitution, $U^T \times Y = MVEC$

$$Y = \begin{bmatrix} 1.0690 & 0 & 0 \\ 0.6290 & 1.4676 & 0 \\ 0.3818 & 0.8908 & 0.8272 \\ 0.5620 & 1.3112 & 1.2176 \end{bmatrix}$$

Step 5: Calculate $VECBAR$ using back substitution, $U \times VECBAR = Y$

$$VECBAR = \begin{bmatrix} 0.5 & 0.5 & 0.25 \\ 0.5 & 1.1667 & 0.5833 \\ 0.5 & 1.1667 & 1.0833 \\ 0.5 & 1.1667 & 1.0833 \end{bmatrix}$$

Step 6: Calculate the subspace stiffness matrix $KBAR = VECBAR^T \times MVEC$

$$KBAR = \begin{bmatrix} 2.0 & 2.0 & 1.0 \\ & 4.6667 & 2.3333 \\ symm. & & 2.1667 \end{bmatrix}$$

Step 7: Calculate $MVECBAR = M \times VECBAR$

$$MVECBAR = \begin{bmatrix} 2.0 & 2.0 & 1.0 \\ 2.0 & 4.6667 & 2.3333 \\ 1.0 & 2.3333 & 2.1667 \\ 1.0 & 2.3333 & 2.1667 \end{bmatrix}$$

Step 8: Calculate the subspace mass matrix, $MBAR = VECBAR^T \times MVECBAR$

$$MBAR = \begin{bmatrix} 3.0 & 5.6667 & 3.8333 \\ & 11.8889 & 8.2778 \\ symm. & & 6.3057 \end{bmatrix}$$

Step 8: Diagonalise subspace stiffness and mass matrices using the generalised Jacobi diagonalisation explained in Section 5.7. Great accuracy is not required, as only an approximate prediction of eigen-pairs is needed.
The 'diagonalised' subspace stiffness matrix is

$$KBAR = \begin{bmatrix} 1.3194 & 0 & 0 \\ & 3.3963 & 0 \\ symm. & & 4.8808 \end{bmatrix}$$

The 'diagonalised' subspace mass matrix is

$$MBAR = \begin{bmatrix} 0.2820 & 0 & 0 \\ & 1.6982 & 0 \\ symm. & & 15.2264 \end{bmatrix}$$

Subspace eigen-vectors are

$$EIGENVEC = \begin{bmatrix} 0.9631 & 0.7524 & 0.1603 \\ -0.7573 & 0.7524 & 0.3397 \\ 0.4116 & -1.5047 & 1.0 \end{bmatrix}$$

Eigen-values obtained as the ratio of diagonal k_{ii}/m_{ii} of the 'diagonalised' subspace stiffness and mass matrices are 4.6794, 2.00 and 0.3206.
Step 8: Calculate $MVEC = MVECBAR \times EIGENVEC$

$$MVEC = \begin{bmatrix} 0.8233 & 1.5047 & 2.0001 \\ -0.6473 & 1.5047 & 4.2395 \\ 0.0880 & -0.7524 & 3.1198 \\ 0.0880 & -0.7524 & 3.1198 \end{bmatrix}$$

The matrix *MVEC* should be arranged such that the columns correspond to the ascending values of the estimated eigen-values. Therefore *MVEC* becomes

$$MVEC = \begin{bmatrix} 2.0001 & 1.5047 & 0.8233 \\ 4.2395 & 1.5047 & -0.6473 \\ 3.1198 & -0.7524 & 0.0880 \\ 3.1198 & -0.7524 & 0.0880 \end{bmatrix}$$

Iteration 2:
Using the MVEC calculated at the first iteration, start the calculations for the second iteration
Step 4: Calculate Y using forward substitution, $U^T \times Y = MVEC$

$$Y = \begin{bmatrix} 0.5345 & 0.4022 & 0.2200 \\ 1.8700 & 0.7887 & -0.1081 \\ 2.4253 & 0.1676 & -0.0292 \\ 6.3458 & -0.4228 & 60.0353 \end{bmatrix}$$

Step 5: Calculate *VECBAR* using back substitution, $U \times VECBAR = Y$

$$VECBAR = \begin{bmatrix} 1.5599 & 0.1881 & 0.0440 \\ 3.3064 & 0.1881 & -0.0346 \\ 4.8663 & -0.1881 & 0.0094 \\ 5.6462 & -0.3762 & 0.0314 \end{bmatrix}$$

Step 6: Calculate the subspace stiffness matrix $KBAR = VECBAR^T \times MVEC$

$$KBAR = \begin{bmatrix} 49.9342 & -0.5868 & 0.0686 \\ & 0.9906 & -0.5868 \\ symm. & & 0.0622 \end{bmatrix}$$

Step 7: Calculate $MVECBAR = M \times VECBAR$

$$MVECBAR = \begin{bmatrix} 0.6.2396 & 0.7524 & 0.1759 \\ 13.2256 & 0.7524 & -0.1383 \\ 9.7326 & -0.3762 & 0.0188 \\ 11.2925 & -0.7524 & 0.0628 \end{bmatrix}$$

Step 8: Calculate subspace mass matrix, $MBAR = VECBAR^T \times MVECBAR$

$$MBAR = \begin{bmatrix} 164.5840 & -2.4174 & 0.2630 \\ & 0.6368 & -2.4174 \\ symm. & & 0.0147 \end{bmatrix}$$

Step 8: Diagonalise subspace stiffness and mass matrices using the generalised Jacobi diagonalisation.

The 'diagonalised' subspace stiffness matrix is

$$KBAR = \begin{bmatrix} 50.2072 & 0 & 0 \\ & 0.9976 & 0 \\ symm. & & 0.0621 \end{bmatrix}$$

The 'diagonalised' subspace mass matrix is

$$MBAR = \begin{bmatrix} 165.5770 & 0 & 0 \\ & 0.6113 & 0 \\ symm. & & 0.0138 \end{bmatrix}$$

The subspace eigen-vectors are

$$EIGENVEC = \begin{bmatrix} 1.0000 & 0.0158 & -0.0011 \\ -0.1813 & 0.9999 & 0.0332 \\ 0.1601 & -0.2751 & 1.0 \end{bmatrix}$$

Eigen-values obtained as the ratio of diagonal terms k_{ii} and m_{ii} respectively of the 'diagonalised' subspace stiffness and mass matrices are 0.3032, 1.6319 and 4.4908.
Step 8: Calculate *MVEC* = *MVECBAR* × *EIGENVEC*

$$MVEC = \begin{bmatrix} 6.1316 & 0.8055 & 0.1940 \\ 13.0675 & 0.9995 & -0.1281 \\ 9.8041 & -0.2274 & -0.0046 \\ 11.4393 & -0.5910 & 0.0251 \end{bmatrix}$$

At the end of the second iteration, the predicted eigen-values (= ω^2) are equal to 0.3032, 1.6319 and 4.4908. The corresponding exact values computed in Section 3.2.4 are 0.3032, 1.5917 and 4.2410 respectively. Clearly all the frequencies are tending to converge towards the correct values. Table 6.2 summarises the results of ten iterations. It is clear that the smallest eigen-value converges first and the highest eigen-value converges last. This is similar to what was observed in the case of simultaneous iterations as well. While this is a very useful result if one is interested in all the eigen-values from the lowest to a certian value, unfortunately this is not very helpful if one is interested in the eigen-values lying in a given range, which is often the case in practical problems.

Table 6.2: Convergence study

Iteration no.	ω_1^2	ω_2^2	ω_3^2
1	0.3206	2.00	4.6794
2	0.3032	1.6319	4.4908
3	0.3032	1.5950	4.3813
4	0.3032	1.5919	4.3175
5	0.3032	1.5917	4.2819
6	0.3032	1.5917	4.2627
7	0.3032	1.5917	4.2524
8	0.3032	1.5917	4.2470
9	0.3032	1.5917	4.2441
10	0.3032	1.5917	4.2427
Exact values	0.3032	1.5917	4.2411

The eigen-vectors can be computed from *VECTOR* = *VECBAR* × *EIGENVEC*. Using the values at the end of second iteration

$$VECBAR = \begin{bmatrix} 1.5599 & 0.1881 & 0.0440 \\ 3.3064 & 0.1881 & -0.0346 \\ 4.8663 & -0.1881 & 0.0094 \\ 5.6462 & -0.3762 & 0.0314 \end{bmatrix}$$

$$EIGENVEC = \begin{bmatrix} 1.0000 & 0.0158 & -0.0011 \\ -0.1813 & 0.9999 & 0.0332 \\ 0.1601 & -0.2751 & 1.0 \end{bmatrix}$$

$$VECTOR = \begin{bmatrix} 1.5329 & 0.2006 & 0.0485 \\ 3.2669 & 0.2499 & -0.0320 \\ 4.9020 & -0.1137 & -0.0023 \\ 5.7196 & -0.2955 & 0.0126 \end{bmatrix}$$

M normalising *VECTOR* yields the eigen-values

$$EIGENVECTOR = \begin{bmatrix} 0.1191 & 0.2566 & 0.4123 \\ 0.2539 & 0.3196 & -0.2723 \\ 0.3810 & -0.1455 & -0.0195 \\ 0.4445 & -0.3780 & 0.1068 \end{bmatrix}$$

The exact values are

$$CORRECT\ VALUES = \begin{bmatrix} 0.1191 & 0.2469 & 0.3947 \\ 0.2537 & 0.3141 & -0.1959 \\ 0.3788 & -0.0850 & -0.2487 \\ 0.4465 & -0.4165 & 0.2232 \end{bmatrix}$$

The values for the first frequency are almost exact and for the other two frequencies the values are not yet converged.

6.4 RITZ VECTORS

Both the simultaneous iteration method and the subspace iteration method aim to solve the general eigen-value problem. The eigen-pairs are used for the solution of the problem by the mode superposition method. It has been argued that this might not necessarily be the most efficient method of solving problems in structural dynamics. The reason for this is that whether a particular eigen-vector will make a significant contribution to the solution depends very much on the properties of the dynamic loading. So there has been attempts to explore the use of other orthogonal vectors, which take account of the type of loading in the solution of the problem. Two such procedures are explained in detail in the next section.

6.4.1 Superposition of Ritz Vectors

In matrix format, the differential equation to be solved is given by

$$M\,\frac{d^2X}{dt^2} + C\,\frac{dX}{dt} + K\,X = F \tag{6.58}$$

Let it be assumed that the load F and displacement X can be expressed in matrix form as a product of a space variable and a time function

$$F = f \times g(t) \tag{6.59}$$

$$X = \Phi \times c(t) \tag{6.60}$$

The differential equation to be solved is given by

$$M\,\Phi \frac{d^2c(t)}{dt^2} + C\,\Phi \frac{dc(t)}{dt} + K\;\Phi\, c(t) = f\;g(t) \tag{6.61}$$

Pre-multiplying by Φ^T

$$\Phi^T M\,\Phi \frac{d^2c(t)}{dt^2} + \Phi^T C\,\Phi \frac{dc(t)}{dt} + \Phi^T K\;\Phi\, c(t) = \Phi^T\, f\;g(t) \tag{6.62}$$

Substituting

$$\Phi^T M\,\Phi\; = M^*\;,\;\; \Phi^T C\,\Phi = C^*\;,\; \Phi^T K\;\Phi\; = K^*\;,\; \Phi^T f\, = f^* \tag{6.63}$$

$$M^*\frac{d^2c(t)}{dt^2} + C^*\frac{dc(t)}{dt} + K^* c(t) = \; f^*\; g(t) \tag{6.64}$$

Φ is called Ritz vector.

6.4.2 Generation of Ritz Vectors

The Ritz vectors are generated as follows. The first vector $\bar{\phi}_1$ is obtained by solving the static problem

$$K\bar{\phi}_1 = f$$

The vector is M normalised such that $\phi_1 = \dfrac{\bar{\phi}_1}{\beta}$, $\beta = \sqrt{(\bar{\phi}_1^T\; M\,\bar{\phi}_1)}$

For generating the rest of the Ritz vectors, the following recurrence relationship is used.

$$K\bar{\phi}_i = M\,\phi_{i-1}\;,\; i = 2,3,...N$$

After calculating $\bar{\phi}_i$, it is made M-orthogonal to previously calculated vectors and is finally M normalised.

It is perhaps worth noticing that the vector $\bar{\phi}_1$ corresponds to the deflection shape under static loading. The loading $M\varphi_i$ clearly represents the inertial forces due displacement pattern φ_i. Thus the Ritz vectors can be visualised as trying to capture the real displacement under dynamic loading in a series of steps.

6.4.3 Generation of Ritz Vectors: Example

Example: Generate the Ritz vectors for the following data.

$$K = \begin{bmatrix} 14 & -6 & 0 \\ -6 & 10 & -4 \\ 0 & -4 & 4 \end{bmatrix}, M = \begin{bmatrix} 4 & 0 & 0 \\ 0 & 4 & 0 \\ 0 & 0 & 2 \end{bmatrix}, f = \begin{bmatrix} 1 \\ 0 \\ 0 \end{bmatrix}$$

Step 1: Cholesky factorisation of $K = U^T U$

Step 2: Solve $K\overline{\phi_1} = f$

$$\bar{\phi}_1 = [0.125 \quad 0.125 \quad 0.125]^T$$

M normalise $\overline{\phi_1}$

$$\beta = \sqrt{\bar{\phi}_1^T \; M \, \bar{\phi}_1} = 0.395285, \; \phi_1 = \frac{\bar{\phi}_1}{\beta} [0.316228 \quad 0.316228 \quad 0.316228]^T$$

Step 3: Solve $K\overline{\phi_2} = M\,\phi_1$

$$M\,\phi_1 = [1.264911 \quad 1.264911 \quad 0.632456]^T$$

$$\bar{\phi}_2 = [0.39529 \quad 0.71152 \quad 0.86964]^T$$

Using the Gram-Schmidt orthogonalisation procedure, using $\overline{\phi_2}$, generate a vector normal to ϕ_1.

$$\varphi_2 = \bar{\phi}_2 - \alpha\phi_1 \,, \; \alpha = \phi_1^T M \; \bar{\phi}_2 \; = 1.95004$$

$$\varphi_2 = [-0.22137 \quad 0.09486 \quad 0.25298]^T$$

M normalise φ_2

$$\beta = \sqrt{\varphi_2^T \; M \, \varphi_2} \; = 0.6, \; \phi_2 = \frac{\varphi_2}{\beta} = [-0.368932 \quad 0.158114 \quad 0.421637]^T$$

Step 4: Solve $K\overline{\phi_3} = M\,\phi_2$

$$M\,\phi_2 = [-1.47580 \quad 0.63240 \quad 0.84326]^T$$

$$\bar{\phi}_3 = [-0.00002 \quad 0.24592 \quad 0.45673]^T$$

Using the Gram-Schmidt orthogonalisation procedure, using $\overline{\phi_3}$, generate a vector normal to ϕ_1 and ϕ_2

$$\varphi_3 = \bar{\phi}_3 - \alpha_1\,\phi_1 - \alpha_2\,\phi_2 \,, \; \alpha_1 = \phi_1^T M \; \bar{\phi}_3 \; = 0.59991, \; \alpha_2 = \phi_2^T M \; \bar{\phi}_3 \; = 0.54068$$

$$\varphi_3 = [0.009754 \quad -0.02927 \quad 0.03906]^T$$

M normalise φ_3

$$\beta = \sqrt{\varphi_3^T \; M \, \varphi_3} \; = 0.8282, \; \phi_3 = \frac{\varphi_3}{\beta} = [0.11785 \quad -0.353554 \quad 0.471403]^T$$

Because of the fact that the vectors ϕ_i are M-orthogonal and M-normal, the matrix M^* will be an identity matrix. However K* and C* will not be.

$$K^* = \Phi^T K \Phi = \begin{bmatrix} 0.80 & -0.9333 & 0.2981 \\ -0.9333 & 3.0333 & -0.9690 \\ 0.2981 & -0.9690 & 4.1667 \end{bmatrix}$$

If Rayleigh damping is assumed, then $C = \alpha M + \beta K$. Therefore $C^* = \alpha I + \beta K^*$. The simultaneous equations (6.64) can be solved by numerical methods explained in Chapter 11 or by the mode superposition method. This will require calculating the eigen-pairs of the matrix K^*. It has to be remembered that in large-scale problems, the matrix K^* will be very much smaller than K and hence the effort involved in determining its eigen-pairs will therefore be considerably less.

6.5 LANCZOS VECTORS

The Lanczos method of tridiagonalisation described in Section 5.8 also generates a set of M-orthogonal vectors. Hence they can also be used instead of the Ritz vectors calculated in Section 6.4. Normally, as explained in Section 5.8, the calculations start with an arbitrary vector. However in generating Lanczos vectors, the first vector is calculated by solving the static problem.

As shown in Section 5.8, given an n×n symmetric matrix A, starting with a unit length arbitrary vector q_1, a set of orthogonal unit vectors $q_2, q_3, \ldots q_n$ can be generated from the recursion formula given by equation (5.64). In using this approach for generating the Lanczos vectors for the solution of equation (6.58), a similar approach is used, except that the vectors are made M orthogonal and M normal.

The steps can be stated as follows.

Step 1: Solve $K\, r_0 = f$ and M normalise r_0 to calculate q_1

$q_1 = r_0/\beta_1,\ \beta_1 = \sqrt{\{r_0^T M\, r_0\}}$

q_0 is assumed to be zero.

Step 2: As in the case of the Ritz vector generation, solve $K\, \bar{r}_j = M\, q_j$ and using $\bar{r}_j$, generate a vector r_j, M orthogonal to q_j. q_{j-1} is already M-orthogonal to q_j. Do this for j = 1, 2, 3…n.

$$\bar{r}_j = K^{-1} M\, q_j \tag{6.65}$$

$$r_j = \bar{r}_j - \alpha_j\, q_j - \beta_j\, q_{j-1} \tag{6.66}$$

$$\alpha_j = q_j^T\, M\, \bar{r}_j \tag{6.67}$$

$$\beta_{j+1} = \sqrt{\{r_j^T\, M\, r_j\}} \tag{6.68}$$

$$q_{j+1} = \frac{1}{\beta_{j+1}} r_j \tag{6.69}$$

This step will enable one to generate as many vectors as desired. Normally only a small number of vectors, much less than the size n of the matrix K are generated. Note that substituting equations (6.65) and (6.69) in equation (6.66)

$$\beta_{j+1}\, q_{j+1} = K^{-1}\, M\, q_j - \alpha_j\, q_j - \beta_j\, q_{j-1} \tag{6.70}$$

This equation is similar to equation (5.65), except that $A = K^{1} M$.

6.5.1 Generation of Lanczos Vectors: Example

Before going into the details of how these vectors can be used, the procedure is illustrated by an example.

$$K = \begin{bmatrix} 14 & -6 & 0 \\ -6 & 10 & -4 \\ 0 & -4 & 4 \end{bmatrix},\ M = \begin{bmatrix} 4 & 0 & 0 \\ 0 & 4 & 0 \\ 0 & 0 & 2 \end{bmatrix},\ f = \begin{bmatrix} 1 \\ 0 \\ 0 \end{bmatrix}$$

Step 1: Cholesky factorisation of $K = U^T\, U$

Step 2: Solve $Kr_0 = f$

$$r_0 = [0.125 \quad 0.125 \quad 0.125]^T$$

M normalise r_0

$$\beta_1 = \sqrt{\{r_0^T M\, r_0\}} = 0.39529,\ q_1 = \frac{1}{\beta_1} r_0 = [0.31623\ 0.31623\ 0.31623]^T$$

Step 3: Solve $K\, \bar{r}_1 = M\, q_1$, $\bar{r}_1 = [0.39529 \quad 0.71151 \quad 0.86963]^T$

$$\alpha_1 = q_1^T\, M\, \bar{r}_1 = 1.9500$$

$$r_1 = \bar{r}_1 - \alpha_1\, q_1 - \beta_1\, q_0 = [-0.22137\ 0.09486\ 0.25297]^T$$

$$\beta_2 = \sqrt{\{r_1^T\, M\, r_1\}} = 0.60$$

$$q_2 = \frac{1}{\beta_2} r_1 = [-0.36893\ 0.15811\ 0.42164]^T$$

Step 4: Solve $K\, \bar{r}_2 = M\, q_2$, $\bar{r}_2 = [0 \quad 0.24596 \quad 0.45677]^T$

$$\alpha_2 = q_2^T\, M\, \bar{r}_2 = 0.54074$$

$$r_2 = \bar{r}_2 - \alpha_2\, q_2 - \beta_2\, q_1 = [0.00976\ -0.02928\ 0.03904]^T$$

$$\beta_3 = \sqrt{\{r_2^T\, M\, r_2\}} = 0.0828$$

$$q_3 = \frac{1}{\beta_3} r_2 = [0.11785\ -0.35355\ 0.47141]^T$$

Step 5: Solve $K\, \bar{r}_3 = M\, q_3$, $\bar{r}_3 = [0 \quad -0.07857 \quad 0.15714]^T$

$$\alpha_3 = q_3^T\, M\, \bar{r}_3 = 0.25926$$

Step 6: Form the tridiagonal matrix

$$T_3 = \begin{bmatrix} \alpha_1 & \beta_2 & 0 \\ \beta_2 & \alpha_2 & \beta_3 \\ 0 & \beta_3 & \alpha_3 \end{bmatrix} = \begin{bmatrix} 1.95 & 0.60 & 0 \\ 0.60 & 0.5407 & 0.0828 \\ 0 & 0.0828 & 0.2593 \end{bmatrix}$$

6.5.2 Use of Lanczos Vectors

The Lanczos vectors are M-orthogonal. If Q_m is a matrix whose columns are the m vectors q_1, q_2,...q_m, then $Q_m^T M Q_m = I$, an identity matrix.
Combining all the Lanczos vectors obtained after n steps, the recursion formula in equation (6.70) can be used to show that

$$K^{-1} M Q_n - Q_n T_n = 0 \tag{6.71}$$

Where

$Q_n = [q_1 \ q_2 \ ... q_n]$

$$T_n = \begin{bmatrix} \alpha_1 & \beta_2 & & & & & \\ \beta_2 & \alpha_2 & \beta_3 & & & & \\ & \beta_3 & \alpha_3 & \beta_4 & & & \\ & & \cdot & \cdot & \cdot & & \\ & & & \cdot & \cdot & \cdot & \\ & & & & \beta_{n-1} & \alpha_{n-1} & \beta_n \\ & & & & & \beta_n & \alpha_n \end{bmatrix} \tag{6.72}$$

If however only m < n vectors are computed, then equation (6.71) is modified to

$$K^{-1} \times M \times Q_m - Q_m \times T_m = [0 \ 0 \ ... \beta_{m+1} q_{m+1}] \tag{6.73}$$

Pre-multiplying equation (6.73) by $Q_m^T M$

$$Q_m^T \times M \times K^{-1} \times M \times Q_m - Q_m^T \times M \times Q_m \times T_m = Q_m^T \times M \times [0 \ 0 \ ... \beta_{m+1} q_{m+1}] \tag{6.74}$$

However, since the vector q_{m+1} is M-orthogonal to the vectors $q_1 \ q_2 \ ...q_m$, the right-hand side becomes zero and because $Q_m^T M Q_m = I$, equation (6.74) simplifies to

$$Q_m^T \times M \times K^{-1} \times M \times Q_m - T_m = 0 \tag{6.75}$$

If in the differential equation (6.58), substituting

$$F = f \times g(t) \tag{6.76}$$

$$X = Q_m \times c(t) \tag{6.77}$$

The differential equation to be solved is given by

$$M Q_m \frac{d^2c(t)}{dt^2} + C Q_m \frac{dc(t)}{dt} + K Q_m c(t) = f \ g(t) \tag{6.78}$$

Pre-multiplying equation (6.78) by $Q_m^T \times M \times K^{-1}$

$$Q_m^T M K^{-1} M Q_m \frac{d^2c(t)}{dt^2} + Q_m^T M K^{-1} C Q_m \frac{dc(t)}{dt} + Q_m^T M K^{-1} K Q_m c(t) \\ = Q_m^T M K^{-1} f \; g(t) \tag{6.79}$$

But from equation (6.75) $T_m = Q_m^T M K^1 M Q_m$ and the M orthogonal property of q vectors ensures that $Q_m^T M Q_m = I$. Therefore equation (6.79) simplifies to

$$T_m \frac{d^2c(t)}{dt^2} + Q_m^T M K^{-1} C Q_m \frac{dc(t)}{dt} + c(t) = Q_m^T M K^{-1} f \; g(t) \tag{6.80}$$

If Rayleigh damping is used, then $C = a_0 M + a_1 K$ and

$$Q_m^T M K^{-1} C Q_m = a_0 T_m + a_1 I \tag{6.81}$$

Similarly the right-hand side of equation (6.80) can be simplified as follows. The vector q_1 is obtained by first solving $K r_0 = f$ and then M normalising r_0 to calculate q_1 as $q_1 = r_0/\beta_1$, $\beta_1 = \sqrt{\{r_0^T M r_0\}}$. Therefore

$$r_o = \beta_1 q_1 = K^{-1} f \tag{6.82}$$

$$Q_m^T M K^{-1} f = \beta_1 Q_m^T M q_1 = \beta_1 [1 \;\; 0 \;\; 0 \;\; 0 \; . \; . \; .]^T \tag{6.83}$$

Equation (6.80) simplifies to

$$T_m \frac{d^2c(t)}{dt^2} + (a_0 T_m + a_1 I) \frac{dc(t)}{dt} + c(t) = \beta_1 [1 \;\; 0 \;\; 0 \;\; 0 \; . \; . \; .]^T g(t) \tag{6.84}$$

Equation (6.84) can be solved by numerical integration methods described in Chapter 11 or by the mode superposition method by determining the eigen-pairs of the matrix T_m.

6.6 FORTRAN-90 PROGRAMS

As in Section 5.11, two elementary programs to illustrate the two solution methods for the general eigen-value problem are given.

6.6.1 Program SIMULT

```
!*******************************************************************
! GENERAL EIGEN-VALUE PROBLEM SOLVED BY SIMULTANEOUS
! ITERATION. STIFFNESS MATRIX IS ASSUMED TO BE SYMMETRIC
! AND POSITIVE DEFINITE.*****************************************
      PROGRAM SIMULT
!*******VARIABLE DECLARATION*********************************
      IMPLICIT NONE
      REAL:: SUM,FACTOR,DENOM,C,BIG,TEMP,SUM1,SUM2,DIF
      INTEGER:: N,M,I,J,IM1,IP1,K,NM1,ITERATION,S,COUNT,M1,FLAG
      REAL,ALLOCATABLE:: STIF(:,:),MASS(:,:),U(:,:), VECTOR(:,:),V(:,:),B(:,:),T(:,:)&
      &,W(:,:),Z(:,:),R(:,:),ALPHA(:),OLDEIGEN(:)
!*******FILE MANAGEMENT**********************************************
      CHARACTER(LEN=40)::FILEIN, FILEOUT
      PRINT*,'NAME OF INPUT FILE'
      READ*,FILEIN
```

```
      PRINT*,'NAME OF OUTPUT FILE'
      READ*,FILEOUT
      OPEN(UNIT=1, FILE =FILEIN,STATUS='OLD',ACTION='READ')
      OPEN(UNIT=2, FILE =FILEOUT,STATUS='REPLACE',ACTION='WRITE')
!******INPUT MATRIX SIZE'************************************
      READ(1,*) N
!******INPUT NUMBER OF EIGEN-VALUES REQUIRED'**************
      READ(1,*) M
      ALLOCATE(STIF(N,N),MASS(N,N),U(N,N),VECTOR(N,M),&
            &V(N,M),B(M,M),T(M,M),W(N,M),Z(N,M),R(N,M),ALPHA(M),&
            &OLDEIGEN(M))
!****INPUT THE UPPER HALF OF STIFFNESS AND MASS MATRICES**
!*******INPUT THE DATA ROW BY ROW**************************
      READ(1,*) ((STIF(I,J),J=I,N),I=1,N)
      READ(1,*) ((MASS(I,J),J=I,N),I=1,N)
!*******COMPLETE THE SYMMETRIC LOWER HALF***************
      DO I = 2,N
      IM1=I-1
      DO J =1,IM1
      STIF(I,J)=STIF(J,I)
      MASS(I,J)=MASS(J,I)
      END DO
      END DO
!*******PRINT THE INPUT MATRICES****************************
      WRITE(2,*)' STIFFNESS MATRIX'
      DO I = 1,N
      WRITE(2,'(6(4X,F10.4))')(STIF(I,J), J = 1,N)
      END DO
      WRITE(2,*)' MASS MATRIX'
      DO I = 1,N
      WRITE(2,'(6(4X,F10.4))')(MASS(I,J), J = 1,N)
      END DO
!****** CHOLESKY FACTORISATION, STIF = U TRANSFORM× U*****
!******CALCULATE THE FIRST ROW OF U*************************
      U(1,1) = SQRT(STIF(1,1))
      FACTOR=1/U(1,1)
      DO J=2,N
      U(1,J)=STIF(1,J)*FACTOR
      END DO
!******CALCULATE THE ELEMENTS OF ROWS 2 TO N*************
      DO I = 2,N
      IP1=I+1
!*******CALCULATE THE DIAGONAL ELEMENT******************
      IM1=I-1
      SUM=0
      DO K= 1,IM1
      SUM = SUM + U(K,I)*U(K,I)
      END DO
      U(I,I)=SQRT(STIF(I,I)-SUM)
      IF(I==N)EXIT
!*******CALCULATE THE OFF DIAGONAL ELEMENTS*************
      FACTOR=1.0/U(I,I)
      DO J= IP1,N
      SUM = 0
      DO K = 1,IM1
```

```
      SUM = SUM + U(K,I)*U(K,J)
      END DO
      U(I,J)=(STIF(I,J)-SUM)*FACTOR
      END DO
      END DO
!*******PRINT U MATRIX***************************************
      WRITE(2,*) 'U MATRIX'
      DO I = 1,N
      WRITE(2,*)(U(I,J), J=I,N)
      END DO
!******DEALLOCATE STIF AS IT IS NOT REQUIRED****************
      DEALLOCATE(STIF)
!*******START WITH UNIT ORTHOGONAL ARBITRARY VECTORS****
!*ZERO THE VECTOR AND INPUT UNITY ON THE DIAGONAL ONLY***
!*******ASSUME EIGEN-VALUES ARE ALL ZERO*******************
      VECTOR=0
      OLDEIGEN(I)=0
      DO I= 1,M
      VECTOR(I,I) = 1.0
      END DO
      ITERATION: DO ITERATION = 1,100
!*******USE BACK SUBSTITUTION TO SOLVE U ×Z = VECTOR*******
      DO K=1,M
      Z(N,K)=VECTOR(N,K)/U(N,N)
      NM1=N-1
      DO I= NM1,1,-1
      SUM = 0
      IP1=I+1
      DO J= N,IP1,-1
      SUM=SUM + U(I,J)*Z(J,K)
      END DO
      Z(I,K)=(VECTOR(I,K)-SUM)/U(I,I)
      END DO
      END DO
!*******MULTIPLY R = MASS×Z**********************************
      R=MATMUL(MASS,Z)
!*USE FORWARD SUBSTITUTION TO SOLVE, U TRANSFORM × V = R *
      DO K = 1,M
      V(1,K)=R(1,K)/U(1,1)
      DO I = 2,N
      SUM = 0
      IM1=I-1
      DO J= 1,IM1
      SUM = SUM + U(J,I)*V(J,K)
      END DO
      V(I,K)= (R(I,K)-SUM)/U(I,I)
      END DO
      END DO
!*******CALCULATE THE B MATRIX, B = V TRANSPOSE ×V********
      B=MATMUL(TRANSPOSE(VECTOR),V)
!*******CONSTRUCT MATRIX T WHICH IS SKEW SYMMETRIC*****
      DO I = 1,M
      DO J = I,M
      IF (I==J) THEN
      T(I,I) = 1.0
```

```
      ELSE
      C= B(I,I)-B(J,J)
      S= C/ABS(C)
      DENOM=C+S*SQRT(C*C+16*B(I,J)**2)
      T(I,J)=-2*B(I,J)/DENOM
      T(J,I)=-T(I,J)
      END IF
      END DO
      END DO
!*******CALCULATE W = V × T*************************************
      W=MATMUL(V,T)
!******PRINT INTERMEDIATE RESULTS**************************
      WRITE(2,*)'ITERATION NUMBER ',ITERATION
      WRITE(2,*)'  ARBITRARY VECTOR'
      DO I = 1,N
      WRITE(2,'(6(4X,F10.4))')(VECTOR(I,J), J = 1,M)
      END DO
      WRITE(2,*)'  Z OF U*Z = VECTOR'
      DO I = 1,N
      WRITE(2,'(6(4X,F10.4))')(Z(I,J), J = 1,M)
      END DO
      WRITE(2,*)'  R = M Z'
      DO I = 1,N
      WRITE(2,'(6(4X,F10.4))')(R(I,J), J = 1,M)
      END DO
      WRITE(2,*)'  V  OF U TRANSFORM * V = R'
      DO I = 1,N
      WRITE(2,'(6(4X,F10.4))')(V(I,J), J = 1,M)
      END DO
      WRITE(2,*)'  B MATRIX'
      DO I = 1,M
      WRITE(2,'(6(4X,F10.4))')(B(I,J), J = 1,M)
      END DO
      WRITE(2,*)'  T MATRIX'
      DO I = 1,M
      WRITE(2,'(6(4X,F10.4))')(T(I,J), J = 1,M)
      END DO
      WRITE(2,*)'  W MATRIX'
      DO I = 1,N
      WRITE(2,'(6(4X,F10.4))')(W(I,J), J = 1,M)
      END DO
!*******REARRANGE THE COLUMNS OF W ACCORDING TO THE ****
!*******ASCENDING VALUES OF THE DIAGONAL ELEMENTS OF B***
      M1=M-1
      DO I = 1,M1
      BIG=0
      COUNT = 0
      DO J = I,M
      IF(B(J,J) > BIG) THEN
      BIG=B(J,J)
      COUNT=J
      END IF
      END DO
      DO K = 1,N
      TEMP = W(K,I)
```

```
      W(K,I)=W(K,COUNT)
      W(K,COUNT)=TEMP
      END DO
      TEMP=B(I,I)
      B(I,I)= BIG
      B(COUNT,COUNT)=TEMP
      END DO
!*******PRINT THE ESTIMATED EIGEN-VALUES******************
      WRITE(2,*)' ESTIMATED EIGEN-VALUES'
      WRITE(2,'(6(4X,F10.4))')(B(I,I), I = 1,M)
!*******PRINT THE W MATRIX AFTER REARRANGING*************
      WRITE(2,*)'  REARRANGED W MATRIX'
      DO I = 1,N
      WRITE(2,'(6(4X,F10.4))')(W(I,J), J = 1,M)
      END DO
!*TEST CONVERGENCE OF EIGEN-VALUES.  IF SATISFACTORY, STOP*
      FLAG=0
      DO I = 1,M
      B(I,I)=1.0/B(I,I)
      DIF=ABS(OLDEIGEN(I)-B(I,I))
      IF(DIF > 1.0E-4)FLAG=1
      END DO
      IF (FLAG > 0) THEN
      DO J = 1,M
      OLDEIGEN(J)=B(J,J)
      END DO
      ELSE
      EXIT
      END IF
!*******USE GRAM-SCHMIDT ORTHOGONALISATION TO GENERATE*
!******* ORTHOGONAL VECTORS*******************************
      DO I = 2,M
      IM1=I-1
      DO J = 1,IM1
      SUM2=0
      SUM1=0
      DO K=1,N
      SUM1=SUM1+W(K,I)*W(K,J)
      SUM2=SUM2+W(K,J)*W(K,J)
      END DO
      ALPHA(J)=SUM1/SUM2
      END DO
      DO J = 1,IM1
      DO K = 1,N
      W(K,I)=W(K,I)-ALPHA(J)*W(K,J)
      END DO
      END DO
      END DO
!*******PRINT SOME MORE INTERMEDIATE RESULTS**************
!*******PRINT THE W MATRIX AFTER ORTHOGONALISING*********
      WRITE(2,*)'  ORTHOGONALIZED W MATRIX'
      DO I = 1,N
      WRITE(2,'(6(4X,F10.4))')(W(I,J), J = 1,M)
      END DO
!*******MAKE THE VECTORS OF UNIT LENGTH*******************
```

```
      DO I = 1,M
      SUM=0
      DO J = 1,N
      SUM=SUM+W(J,I)*W(J,I)
      END DO
      SUM=SQRT(SUM)
      DO J = 1,N
      VECTOR(J,I)=W(J,I)/SUM
      END DO
      END DO
!*******PRINT THE W MATRIX AFTER NORMALISING*************
      WRITE(2,*)' NORMALISED W MATRIX'
      DO I = 1,N
      WRITE(2,'(6(4X,F10.4))')(VECTOR(I,J), J = 1,M)
      END DO
      END DO   ITERATIONS
!*****EIGEN VECTORS OF GENERAL EIGEN-VALUE PROBLEM******
!*****USE BACK SUBSTITUTION TO SOLVE U × Z = VECTOR********
      DO K=1,M
      Z(N,K)=VECTOR(N,K)/U(N,N)
      NM1=N-1
      DO I= NM1,1,-1
      SUM = 0
      IP1=I+1
      DO J= N,IP1,-1
      SUM=SUM + U(I,J)*Z(J,K)
      END DO
      Z(I,K)=(VECTOR(I,K)-SUM)/U(I,I)
      END DO
      END DO
!*******M NORMALISE THE EIGEN-VECTORS********************
      DO K = 1,M
      SUM1=0
      DO I = 1,N
      SUM2=0
      DO J = 1,N
      SUM2=SUM2+MASS(I,J)*Z(J,K)
      END DO
      SUM1=SUM1+SUM2*Z(I,K)
      END DO
      SUM1=1.0/SQRT(SUM1)
      DO I = 1,N
      Z(I,K)=Z(I,K)*SUM1
      END DO
      END DO
!*******PRINT FINAL RESULTS*********************************
      WRITE(2,*)'EIGEN-VECTORS'
      DO I = 1,N
      WRITE(2,'(6(4X,F10.4))')(Z(I,J), J = 1,M)
      END DO
      WRITE(2,*)'FREQUENCY SQUARED'
      WRITE(2,'(6(4X,F10.4))')(B(I,I), I = 1,M)
      END PROGRAM SIMULT
```

6.6.2 Program SUBSPACE

```
!*****************************************************************
! GENERAL EIGEN-VALUE PROBLEM SOLVED BY SUBSPACE              *
! ITERATION ORTHOGONAL VECTORS GENERATED BY SOLVING A      *
!.SMALLER EIGEN-VALUE PROBLEM.   STIFFNESS MATRIX IS           *
! ASSUMED TO BE SYMMETRIC AND POSITIVE DEFINITE***********
      PROGRAM SUBSPACE
!*******VARIABLE DECLARATION*******************************
      IMPLICIT NONE
      REAL:: K11,K22,K12,M11,M22,M12,P,Q,R,SUM,FACTOR,&
      &DENOM,BETA,ALPHA,DIAGONALKI,DIAGONALKJ,&
      &DIAGONALMI,DIAGONALMJ,OMEGA2,CONST,SMALL,&
      &TEMP,SUM1,SUM2,DIF
      INTEGER:: N,M,I,J,IM1,IP1,K,NM1,ITERATION,M1,KP1,L,&
      &FLAG1,FLAG2,JACOBIROTS,LCASE,ISIGN,COUNT,IPRINT
      REAL,ALLOCATABLE:: STIF(:,:),MASS(:,:),VECTOR(:,:),&
      &MVEC(:,:),MVECBAR(:,:),VECBAR(:,:),STIFSUB(:,:)&
      &,MASSSUB(:,:),ROWKI(:),ROWKJ(:),ROWMI(:),ROWMJ(:),&
      &COLUMNKI(:),COLUMNKJ(:),COLUMNMI(:),COLUMNMJ(:),&
      &EIGENVEC(:,:),EIGENI(:),EIGENJ(:),U(:,:),Y(:,:),VALUE(:),&
      &OLDEIGEN(:)
!*******FILE MANAGEMENT***************************************
      CHARACTER(LEN=40)::FILEIN, FILEOUT
      PRINT*,'NAME OF INPUT FILE'
      READ*,FILEIN
      PRINT*,'NAME OF OUTPUT FILE'
      READ*,FILEOUT
      OPEN(UNIT=1, FILE =FILEIN,STATUS='OLD',ACTION='READ')
      OPEN(UNIT=2, FILE =FILEOUT,STATUS='REPLACE',ACTION='WRITE')
!*******INPUT MATRIX SIZE*********************************
      READ(1,*) N
!*******INPUT NUMBER OF EIGEN-VALUES REQUIRED*************
      READ(1,*) M
      M1=M-1
!******ALLOCATE ARRAYS*************************************
      ALLOCATE(STIF(N,N),MASS(N,N),VECTOR(N,M),MVEC(N,M),&
            &MVECBAR(N,M),VECBAR(N,M),STIFSUB(M,M),&
            &MASSSUB(M,M),ROWKI(M),ROWKJ(M),ROWMI(M),&
            &ROWMJ(M),COLUMNKI(M),COLUMNKJ(M),COLUMNMI(M),&
            &COLUMNMJ(M),EIGENVEC(M,M),EIGENI(M),EIGENJ(M),&
            &U(N,N),Y(N,M),VALUE(M),OLDEIGEN(M))
!*******INPUT STIFFNESS AND MASS  MATRICES.****************
!*******INPUT THE DATA ROW BY ROW OF UPPER HALF***********
      READ(1,*) ((STIF(I,J),J=I,N),I=1,N)
      READ(1,*) ((MASS(I,J),J=I,N),I=1,N)
!******COMPLETE THE SYMMETRIC LOWER HALF****************
      DO I = 2,N
      IM1=I-1
      DO J =1,IM1
      STIF(I,J)=STIF(J,I)
      MASS(I,J)=MASS(J,I)
      END DO
      END DO
!******PRINT THE INPUT MATRICES*****************************
```

```
      WRITE(2,*)' STIFFNESS MATRIX'
      DO I = 1,N
      WRITE(2,'(6(4X,F10.4))')(STIF(I,J), J = 1,N)
      END DO
      WRITE(2,*)' MASS MATRIX'
      DO I = 1,N
      WRITE(2,'(6(4X,F10.4))')(MASS(I,J), J = 1,N)
      END DO
!******USE CHOLESKY FACTORISATION.  STIF = U TRANSFORM U*
```

THIS SECTION IS SAME AS IN PROGRAM SIMULT

```
!******DEALLOCATE MATRIX STIF AS IT IS NO LONGER REQUIRED**
      DEALLOCATE(STIF)
!*******START WITH  ZERO EIGEN-VALUES **********************
      OLDEIGEN=0
!*******START WITH  ARBITRARY VECTORS**********************
!********ZERO THE VECTOR************************************
      VECTOR=0
!******* AND INPUT UNITY ON THE DIAGONAL ONLY*************
      DO I= 1,M
      VECTOR(I,I) = 1.0
      END DO
!*******PRINT STARTING VECTOR*******************************
      WRITE(2,*)' STARTING ARBITRARY VECTOR'
      DO I = 1,N
      WRITE(2,'(6(4X,F10.4))')(VECTOR(I,J), J = 1,M)
      END DO
!*******MULTIPLY MASS×ARBITRARY VECTOR*******************
      MVEC=MATMUL(MASS,VECTOR)
!*******START OF ITERATION CYCLES***************************
      ITERATIONS:  DO ITERATION = 1,100
!*******SOLVE FOR Y BY FORWARD SUBSTITUTION.**************
!********U TRANSFORM×Y= MVEC*******************************
      DO LCASE = 1,M
      Y(1,LCASE)=MVEC(1,LCASE)/U(1,1)
      DO I= 2,N
      SUM=0
      IM1=I-1
      DO J = 1,IM1
      SUM=SUM+U(J,I)*Y(J,LCASE)
      END DO
      Y(I,LCASE)=(MVEC(I,LCASE)-SUM)/U(I,I)
      END DO
      END DO
!******USE BACK SUBSTITUTION TO SOLVE FOR U×VECBAR=Y****
      DO LCASE = 1,M
      VECBAR(N,LCASE)=Y(N,LCASE)/U(N,N)
      NM1=N-1
      DO I = NM1,1,-1
      SUM = 0
      IP1=I+1
      DO J= N,IP1,-1
      SUM=SUM+U(I,J)*VECBAR(J,LCASE)
      END DO
      VECBAR(I,LCASE)=(Y(I,LCASE)-SUM)/U(I,I)
      END DO
```

```
      END DO
!*******CALCULATE SUBSPACE STIFFNESS MATRIX*************
      STIFSUB=MATMUL(TRANSPOSE(VECBAR),MVEC)
!******CALCULATE SUBSPACE MASS MATRIX IN TWO STAGES*****
!******CALCULATE MVECBAR=MASS×VECBAR*******************
      MVECBAR=MATMUL(MASS,VECBAR)
!******CALCULATE MASSSUB=VECBARTRANSPOSE ×MVECBAR****
      MASSSUB=MATMUL(TRANSPOSE(VECBAR),MVECBAR)
!****************************************************************
!         PRINT SOME INTERMEDIATE RESULTS
!****************************************************************
      WRITE(2,*)'ITERATION NUMBER ',ITERATION
      WRITE(2,*)' MVEC=MASS*VECTOR'
      DO IPRINT = 1,N
      WRITE(2,'(6(4X,F10.4))')(MVEC(IPRINT,J), J = 1,M)
      END DO
      WRITE(2,*)' Y FROM U TRANSFORM * Y= MVEC'
      DO IPRINT = 1,N
      WRITE(2,'(6(4X,F10.4))')(Y(IPRINT,J), J = 1,M)
      END DO
      WRITE(2,*)' VECBAR, U * VECBAR= Y'
      DO IPRINT = 1,N
      WRITE(2,'(6(4X,F10.4))')(VECBAR(IPRINT,J), J = 1,M)
      END DO
      WRITE(2,*)' STIFSUB BEFORE DIAGONALIZATION'
      DO IPRINT = 1,M
      WRITE(2,'(6(4X,E14.6))')(STIFSUB(IPRINT,J), J = 1,M)
      END DO
      WRITE(2,*)' MVECBAR = MASS * VECBAR'
      DO IPRINT = 1,N
      WRITE(2,'(6(4X,F10.4))')(MVECBAR(IPRINT,J), J = 1,M)
      END DO
      WRITE(2,*)' MASSSUB BEFORE DIAGONALIZATION'
      DO IPRINT = 1,M
      WRITE(2,'(6(4X,E14.6))')(MASSSUB(IPRINT,J), J = 1,M)
      END DO
!**************************************************************
!*******CALCULATE THE EIGEN-VALUES AND EIGENVECTORS OF **
!*******GENERAL EIGEN-VALUE PROBLEM INVOLVING STIFSUB ***
!*******AND MASSSUB*****************************************
!*******START WITH A UNIT MATRIX FOR THE EIGEN-VECTORS****
      EIGENVEC=0.0
      DO I = 1,M
      EIGENVEC(I,I)=1.0
      END DO
!*******JACOBI DIAGONALISATION CYCLE. GENERAL EIGEN-VALUE
!****** PROBLEM*********************************************
      JACOBI: DO JACOBIROTS=1,50
      DO I=1,M1
      IP1=I+1
      DO J=IP1,M
      FLAG1=0
!*******FLAG1 = 0 MEANS THE (I,J) TERM IS VERY SMALL.        *
!**THEREFORE SKIP THE CORRESPONDING CALCULATIONS FOR **
!******THAT VALUE OF J********** ****************************
```

```
      K11=STIFSUB(I,I)
      K22=STIFSUB(J,J)
      K12=STIFSUB(I,J)
      M11=MASSSUB(I,I)
      M22=MASSSUB(J,J)
      M12=MASSSUB(I,J)
      P=K11*M12-M11*K12
      Q=K22*M12-M22*K12
      R=K11*M22-K22*M11
      ISIGN=R/ABS(R)
      DENOM=1.0/(0.5*(R+ISIGN*SQRT(R**2+4*P*Q)))
      BETA=-P*DENOM
      ALPHA=Q*DENOM
      IF(ABS(ALPHA)>1.0E-6 .OR. ABS(BETA)>1.0E-6)FLAG1=1
      IF(FLAG1==0)THN
!*******TEMPORARY STORAGE FOR MODIFIED ELEMENTS OF ITH **
!********AND JITH ROWS AND COLUMNS***************************
      DO K = 1,M
!*******MODIFIED ITH ROW*****************************************
      ROWKI(K)=STIFSUB(I,K)+STIFSUB(J,K)*BETA
      ROWMI(K)=MASSSUB(I,K)+MASSSUB(J,K)*BETA
!*******MODIFIED JTH ROW*****************************************
      ROWKJ(K)=STIFSUB(I,K)*ALPHA+STIFSUB(J,K)
      ROWMJ(K)=MASSSUB(I,K)*ALPHA+MASSSUB(J,K)
!*******MODIFIED ITH COLUMN************************************
      COLUMNKI(K)=STIFSUB(K,I)+STIFSUB(K,J)*BETA
      COLUMNMI(K)=MASSSUB(K,I)+MASSSUB(K,J)*BETA
!*******MODIFIED JTH COLUMN************************************
      COLUMNKJ(K)=STIFSUB(K,I)*ALPHA+STIFSUB(K,J)
      COLUMNMJ(K)=MASSSUB(K,I)*ALPHA+MASSSUB(K,J)
!******MODIFIED COULMNS OF EIGEN-VECTORS******************
      EIGENI(K)=EIGENVEC(K,I)+EIGENVEC(K,J)*BETA
      EIGENJ(K)= EIGENVEC(K,I)*ALPHA+EIGENVEC(K,J)
      END DO
      DIAGONALKI=STIFSUB(I,I)+STIFSUB(J,J)*BETA**2+&
      &2*STIFSUB(I,J)*BETA
      DIAGONALKJ=STIFSUB(I,I)*ALPHA**2+STIFSUB(J,J)+&
      &2*STIFSUB(I,J)*ALPHA
      DIAGONALMI=MASSSUB(I,I)+MASSSUB(J,J)*BETA**2+&
      &2*MASSSUB(I,J)*BETA
      DIAGONALMJ=MASSSUB(I,I)*ALPHA**2+MASSSUB(J,J)+&
      &2*MASSSUB(I,J)*ALPHA
!*******MODIFY ITH AND JTH ROWS AND COLUMNS OF STIFSUB, ***
!*******MASSSUB AND EIGENVEC MATRICES**********************
      DO K = 1,M
      STIFSUB(I,K)=ROWKI(K)
      STIFSUB(J,K)=ROWKJ(K)
      STIFSUB(K,I)=COLUMNKI(K)
      STIFSUB(K,J)=COLUMNKJ(K)
      STIFSUB(I,I)=DIAGONALKI
      STIFSUB(J,J)=DIAGONALKJ
      STIFSUB(I,J)=0
      STIFSUB(J,I)=0
      MASSSUB(I,K)=ROWMI(K)
      MASSSUB(J,K)=ROWMJ(K)
```

```
      MASSSUB(K,I)=COLUMNMI(K)
      MASSSUB(K,J)=COLUMNMJ(K)
      MASSSUB(I,I)=DIAGONALMI
      MASSSUB(J,J)=DIAGONALMJ
      MASSSUB(I,J)=0
      MASSSUB(J,I)=0
      EIGENVEC(K,I)=EIGENI(K)
      EIGENVEC(K,J)=EIGENJ(K)
      END DO
      END IF
      END DO
      END DO
      END DO JACOBI
!******END JACOBI ITERATIONS*******************************
!*********************************************************
!*******PRINT SOME MORE INTERMEDIATE RESULTS             *
!*********************************************************
!*******PRINT STIFSUB AND MASSSUB AFTER REDUCTION**********
      WRITE(2,*)' STIFSUB AFTER JACOBI ITERATIONS'
      DO I = 1,M
      WRITE(2,'(6(4X,E14.6))')(STIFSUB(I,J), J = 1,M)
      END DO
      WRITE(2,*)' MASSSUB AFTER JACOBI ITERATIONS'
      DO I = 1,M
      WRITE(2,'(6(4X,F14.6))')(MASSSUB(I,J), J = 1,M)
      END DO
!******PRINT EIGEN-VECTORS AND EIGEN-VALUES OF SUBSPACE**
!********EIGEN-VALUE PROBLEM*******************************
      WRITE(2,*)' SUBSPACE EIGEN-VECTORS'
      DO I = 1,M
      WRITE(2,'(6(4X,F10.4))')(EIGENVEC(I,J), J = 1,M)
      END DO
      WRITE(2,*)'SUBSPACE EIGEN-VALUES'
      DO I = 1,M
      OMEGA2=STIFSUB(I,I)/MASSSUB(I,I)
      VALUE(I)=OMEGA2
      WRITE(2,'(I4,4X,E14.6)')I,OMEGA2
      END DO
!******CHECK FOR CONVERGENCE*******************************
      FLAG2=0
      DO I = 1,M
      DIF=ABS(OLDEIGEN(I)-VALUE(I))
      IF(DIF > 1.0E-4)FLAG2=1
      END DO
      IF(FLAG2==0)EXIT
!*******CALCULATE NEW STARTING MVEC VECTORS************
      MVEC=MATMUL(MVECBAR,EIGENVEC)
!*******REARRANGE THE VECTORS IN THE ASCENDING ORDER OF **
!*******EIGEN-VALUES***************************************
      M1=M-1
      DO I = 1,M1
      SMALL=1E6
      COUNT=0
      DO J = I,M
      IF(VALUE(J)<SMALL)THEN
```

```
      SMALL=VALUE(J)
      COUNT = J
      END IF
      END DO
      DO K = 1,N
      TEMP=MVEC(K,I)
      MVEC(K,I)=MVEC(K,COUNT)
      MVEC(K,COUNT)=TEMP
      END DO
      TEMP=VALUE(I)
      VALUE(I)=SMALL
      VALUE(COUNT)=TEMP
      END DO
      OLDEIGEN=VALUE
      END DO ITERATIONS
!*******ITERATIONS END******************************************
!*****CALCULATE VECTOR = VECBAR×EIGENVEC AND NORMALISE
      VECTOR=MATMUL(VECBAR,EIGENVEC)
      DO K = 1,M
      SUM1=0
      DO I = 1,N
      SUM2=0
      DO J = 1,N
      SUM2=SUM2+MASS(I,J)* VECTOR (J,K)
      END DO
      SUM1=SUM1+SUM2* VECTOR (I,K)
      END DO
      SUM1=1.0/SQRT(SUM1)
      DO I = 1,N
      VECTOR (I,K)= VECTOR (I,K)*SUM1
      END DO
      END DO
!*******PRINT FINAL EIGEN-VALUES AND EIGENVECTORS**********
      WRITE(2,*)' EIGEN-VECTORS'
      DO I = 1,M
      WRITE(2,'(6(4X,F10.4))')( VECTOR (I,J), J = 1,M)
      END DO
      WRITE(2,*)'EIGEN-VALUES'
      WRITE(2,'(6(4X,F10.4))')(VALUE(J), J = 1,M)
      END PROGRAM SUBSPACE
```

CHAPTER 7

LINE ELEMENTS

7.1 INTRODUCTION

Many types of 'skeletal' structures, such as two and three dimensional pin-jointed trusses, rigid-jointed plane and space frames and plane grid structures, are assembled from members which are; in general, much longer than their cross-sectional dimensions. The members are jointed only at joints (nodes). Depending on the type of structure, the nature of the forces they resist vary. For example, a member of a pin-jointed truss can resist axial forces only. On the other hand, members of a rigid-jointed frame resist not only axial forces but also bending moment and shear forces, and in a space frame torsional forces as well. A plane grid structure is a two dimensional structure subjected to loads normal to the plane, and the members resist bending moments, shear forces and torsional moments.

This chapter is concerned with the derivation of relevant force-deformation relationships under dynamic conditions for the members of the skeletal structures.

7.2 BAR ELEMENT: DIFFERENTIAL EQUATION

Figure 7.1 shows a bar element subjected to end forces and a distributed inertial force. Let u(x,t) be the axial displacement, E the Young's modulus and A the area of cross section. Making the usual assumptions of elementary strength of materials that the axial deformation and stress are constant in the cross section, the only force that exists at a cross section is an axial force given by stress multiplied by the cross-sectional area. Considering the equilibrium of an element of infinitesimal length, Δx, the forces on the two cut sections are $F(x,t)$ and $F(x,t) + \frac{\partial F(x,t)}{\partial x}\Delta x$.

The distributed inertial force f per unit length is equal to

$$f = -m\frac{\partial^2 u(x,t)}{\partial t^2} \tag{7.1}$$

Equilibrium in the axial direction requires that

$$-F(x,t) + \{F(x,t) + \frac{\partial F(x,t)}{\partial x}\Delta x\} - m\frac{\partial^2 u(x,t)}{\partial t^2}\Delta x = 0 \tag{7.2}$$

Simplifying

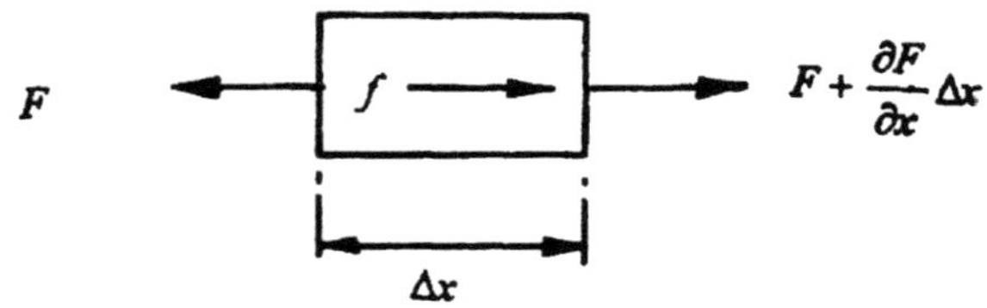

Fig. 7.1 Equilibrium of an element

$$\frac{\partial F(x,t)}{\partial x} - m\frac{\partial^2 u(x,t)}{\partial t^2} = 0 \tag{7.3}$$

The strain ε_x, the stress σ and the force F in the axial direction are given by

$$\varepsilon_x = \frac{\partial u}{\partial x},\ \sigma(x,t) = E\,\varepsilon_x = E\frac{\partial u}{\partial x},\ F = A\sigma \tag{7.4}$$

Substituting equation (7.4) in equation (7.3)

$$\frac{\partial}{\partial x}\{AE\frac{\partial u(x,t)}{\partial x}\} - m\frac{\partial^2 u(x,t)}{\partial t^2} = 0 \tag{7.5}$$

If AE is constant, then

$$AE\ \frac{\partial^2 u(x,t)}{\partial x^2} - m\frac{\partial^2 u(x,t)}{\partial t^2} = 0 \tag{7.6}$$

This is the basic differential equation governing the undamped free longitudinal vibration of a bar element. Since damping has been ignored and, in addition, no external forces act, the displacement u(x,t) must be simple harmonic and can be written as a product of the amplitude U (x) and cos ωt.

$$u(x,t) = U(x)\cos\omega t \tag{7.7}$$

Substituting equation (7.7) in equation (7.6) and simplifying

$$AE\ \frac{d^2 U(x)}{d x^2} + m\omega^2 U(x) = 0 \tag{7.8}$$

This is an ordinary differential equation with constant coefficients. The solution is obtained by assuming U (x) as $e^{\beta x}$. Substituting this expression in the differential equation (7.8)

$$(AE\beta^2 + m\omega^2)e^{\beta x} = 0 \tag{7.9}$$

Solving equation (7.9), for β,

$$\beta = \pm j\alpha,\ \ \alpha = \sqrt{\frac{m\omega^2}{AE}},\ j = \sqrt{-1} \tag{7.10}$$

The solution to equation (7.8), is therefore given by

$$U(x) = A_1\, e^{j\alpha x} + A_2\, e^{-j\alpha x} \tag{7.11}$$

where A_1 and A_2 are complex conjugate constants.

The complex constants can be avoided by writing U (x) as

$$U(x) = C_1 \cos\alpha x + C_2 \sin\alpha x \tag{7.12}$$

where C_1 and C_2 are real constants to be determined from the boundary conditions.

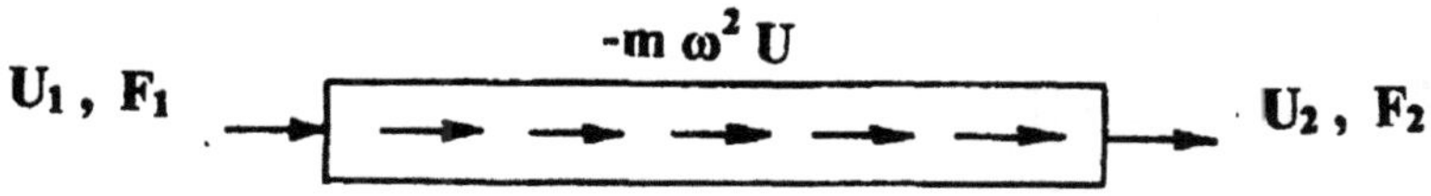

Fig. 7.2 Bar element with end forces and displacements

As shown in Fig. 7.2, the displacement boundary conditions are

$$U(x) = U_1 \text{ at } x = 0 \text{ and } U(x) = U_2 \text{ at } x = L \tag{7.13a}$$

Using equation (7.13a), equation (7.12) can be solved to determine the constants, C_1 and C_2.

$$C_1 = U_1\ , C_2 = \frac{\{U_2 - U_1 \cos\alpha L\}}{\sin\alpha L} \tag{7.13b}$$

Substituting for C_1 and C_2 in equation (7.12)

$$U(x) = \{\cos\alpha x - \frac{\cos\alpha L \sin\alpha x}{\sin\alpha L}\}\, U_1 + \{\frac{\sin\alpha x}{\sin\alpha L}\}\, U_2 \tag{7.14a}$$

Equation (7.14a) can be expressed as

$$U(x) = N_1(x) U_1 + N_2(x)\, U_2 \tag{7.14b}$$

where

$$N_1(x) = \{\cos\alpha x - \frac{\cos\alpha L \sin\alpha x}{\sin\alpha L}\}\ ,\ N_2(x) = \frac{\sin\alpha x}{\sin\alpha L} \tag{7.15}$$

The functions N_1 (x) and N_2 (x) are known as shape functions.

The axial force F (x) is given by

$$F(x) = AE\frac{dU(x)}{dx} = AE\,\alpha\ [\{-\sin\alpha x - \frac{\cos\alpha L \cos\alpha x}{\sin\alpha L}\}U_1 + \{\frac{\cos\alpha x}{\sin\alpha L}\}U_2] \tag{7.16}$$

The boundary conditions for the forces are

$$F(x) = -F_1, \text{ at } x = 0 \text{ and } F(x) = F_2, \text{ at } x = L \tag{7.17}$$

Substituting equation (7.17) in equation (7.16)

$$F_1 = AE[\{\frac{\cos\alpha L}{\sin\alpha L}\}\, U_1 - \{\frac{1}{\sin\alpha L}\}\, U_2] \tag{7.18a}$$

$$F_2 = AE\,[-\{\sin\alpha L + \frac{\cos^2\alpha L}{\sin\alpha L}\}\,U_1 + \{\frac{\cos\alpha L}{\sin\alpha L}\}\,U_2]$$
$$= AE[-\{\frac{1}{\sin\alpha L}\}\,U_1 + \{\frac{\cos\alpha L}{\sin\alpha L}\}\,U_2] \tag{7.18b}$$

Equations (7.18a and b) can be expressed in matrix form as

$$\begin{bmatrix} F_1 \\ F_2 \end{bmatrix} = \begin{bmatrix} k_{11} & k_{12} \\ k_{12} & k_{22} \end{bmatrix} \begin{bmatrix} U_1 \\ U_2 \end{bmatrix} \tag{7.19}$$

$$k_{11} = k_{22} = AE\,\cot\alpha L,\; k_{12} = -AE\frac{1}{\sin\alpha L} \tag{7.20}$$

k_{ij} are the coefficients of a dynamic stiffness matrix for a bar element.

7.3 NATURAL FREQUENCY AND MODE SHAPES OF BAR ELEMENT

The element dynamic stiffness matrix of the bar element in equations (7.19) and (7.20) can be used to determine the natural frequencies and corresponding mode shapes of bars with various boundary conditions. Fig. 7.3 shows a bar, which is fixed at end 1 and free at end 2. The boundary conditions are $U_1 = 0$ and $F_2 = 0$. But from equation (7.19), $F_2 = k_{22}\,U_2$. Since U_2 is not equal to zero, therefore $k_{22} = \cot\alpha L = 0$.

The value of $\alpha L = n\pi/2$, $n = 1,3,5....$ Substituting for αL from equation (7.10)

$$\alpha L = L\sqrt{\frac{m\omega^2}{AE}} = \frac{n\pi}{2} \tag{7.21a}$$

Equation (7.21a) can be solved to give

$$\omega = \frac{n\pi}{2}\sqrt{\frac{AE}{mL^2}}\;, n = 1,3,5... \tag{7.21b}$$

Note that mL = total mass on the bar and (AE/L) = axial stiffness of the bar. Thus

$$\omega = \frac{n\pi}{2}\sqrt{\frac{\textit{Axial stiffness}}{\textit{Total mass}}} \tag{7.22}$$

As can be seen, apart from the multiplying factor before the square root sign, the expression for the natural frequency is similar to equation (1.2), which corresponds to the natural frequency of a single spring-mass system.

Using equations (7.14) and (7.15), the displacement of the bar is given by

$$U(x) = U_2(x)N_2(x) \quad , N_2(x) = \frac{\sin \alpha x}{\sin \alpha L} = \frac{\sin (\frac{n\pi}{2}\frac{x}{L})}{\sin (\frac{n\pi}{2})} \tag{7.23}$$

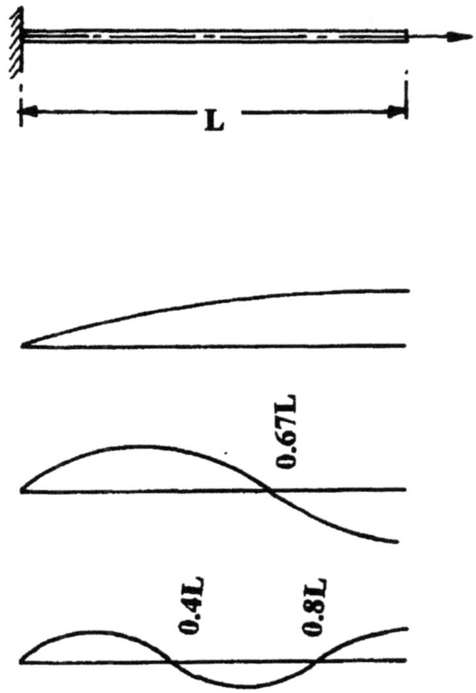

Fig. 7.3 Mode shapes for a fixed-free bar element

Figure 7.3 shows the first three natural frequencies and their corresponding mode shapes. In the lowest mode (n = 1), at any instant, all points are moving axially in the same direction. However for the next frequency (n = 3), the point at a distance of 0.67L from the fixed end does not move at all. It is a node and sections on either side of the node move in opposite directions. Similarly, for the next frequency (n = 5), there are nodes at a distance of 0.4L and 0.8L from the fixed end and sections on either side of a node move in opposite directions.

7.4 BEAM ELEMENT: FLEXURAL DEFORMATIONS ONLY

Figure 7.4 shows a beam element.

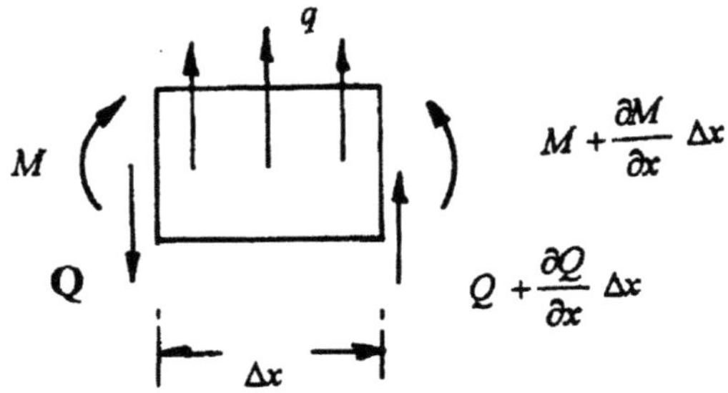

Fig. 7.4 A beam element

The internal 'forces' at a section are the bending moment M and shear force Q. Assuming that the external load q(x,t) is positive in the y-direction, bending moment M is positive sagging and shear force Q is positive in the positive y-direction on the face with the external normal along the positive x-axis, the basic equations of equilibrium are

$$-Q + \{Q + \frac{\partial Q}{\partial x}\Delta x\} + q\,\Delta x = 0 \qquad (7.24)$$

$$-M + \{M + \frac{\partial M}{\partial x}\Delta x\} + \{Q + \frac{\partial Q}{\partial x}\Delta x\}\Delta x + q\,\Delta x\frac{\Delta x}{2} = 0 \qquad (7.25)$$

Letting Δx tend to zero and simplifying, equations (7.24) and (7.25) become

$$\frac{\partial Q}{\partial x} + q = 0 \qquad (7.26a)$$

$$\frac{\partial M}{\partial x} + Q = 0 \qquad (7.26b)$$

Eliminating Q from equations (7.26a) and (7.26b)

$$\frac{\partial^2 M}{\partial x^2} - q = 0 \qquad (7.26c)$$

Assuming that the deformations are due to flexure only and that plane sections remain plane before and after bending, for an elastic beam, the relationship between the sagging moment M and the curvature is given by

$$EI\frac{\partial^2 v}{\partial x^2} = M \qquad (7.27)$$

where EI = flexural rigidity.
Substituting equation (7.27) in (7.26b)

$$Q = -\frac{\partial M}{\partial x} = -EI\frac{\partial^3 v}{\partial x^3} \tag{7.28}$$

Substituting equation (7.27) in equation (7.26c)

$$EI\frac{\partial^4 v(x,t)}{\partial x^4} - q(x,t) = 0 \tag{7.29}$$

When a beam is undergoing undamped, free vibration, the distributed load q corresponds to inertial force per unit length equal to $-m(\partial^2 v/\partial t^2)$, where m is the mass per unit length. Substituting the value of inertial force for q in equation (7.29), the differential equation for a beam undergoing undamped vibration is given by

$$EI\frac{\partial^4 v(x,t)}{\partial x^4} + m\frac{\partial^2 v(x,t)}{\partial t^2} = 0 \tag{7.30a}$$

In the case of undamped free vibration, the displacement v(x,t) can be expressed as

$$v(x,t) = V(x)\cos\omega t \tag{7.30b}$$

where ω = frequency of vibration.

Substituting equation (7.30b) in equation (7.30a) and simplifying

$$EI\frac{d^4V}{dx^4} - m\omega^2 V = 0 \tag{7.31}$$

This is an ordinary differential equation with constant coefficients. The solution is obtained by assuming V(x) as $e^{\beta x}$. Substituting this expression in equation (7.31)

$$(EI\beta^4 - m\omega^2)e^{\beta x} = 0 \tag{7.32}$$

Solving equation (7.32) for β

$$\beta^2 = \pm\sqrt{\frac{m\omega^2}{EI}}, \quad \beta = \pm\alpha, \pm j\alpha, \quad j = \sqrt{-1}, \quad \alpha = \sqrt[4]{\frac{m\omega^2}{EI}} \tag{7.33}$$

The solution to equation (7.32) is therefore given by

$$V(x) = A_1 e^{\alpha x} + A_2 e^{-\alpha x} + A_3 e^{j\alpha x} + A_4 e^{-j\alpha x} \tag{7.34}$$

where A_1 and A_2 are real constants but A_3 and A_4 are complex conjugate constants. As in the case of the solution for the bar element, the complex quantities can be avoided by writing V(x) as

$$V(x) = C_1\cos\alpha x + C_2\sin\alpha x + C_3\,Cosh\,\alpha x + C_4\,Sinh\,\alpha x \tag{7.35}$$

The constants C_1 to C_4 are determined from the boundary conditions. As shown in Fig. 7.5, the boundary conditions for displacements are

$$x = 0,\ V = V_1 \text{ and } \frac{dV}{dx} = -\theta_1, \quad x = L,\ V = V_2 \text{ and } \frac{dV}{dx} = -\theta_2 \tag{7.36}$$

Using equations (7.34) and (7.35), the constants are given by

$$C_1 = -\frac{(Ss - Cc + 1)}{2(1 - Cc)}V_1 - \frac{(Cs - Sc)}{2(1 - Cc)\beta L}\theta_1 L - \frac{(C - c)}{2(1 - Cc)}V_2 - \frac{(S - s)}{2(1 - Cc)\beta L}\theta_2 L$$

$$C_2 = -\frac{(Sc+Cs)}{2(1-Cc)}V_1 + \frac{(Ss + Cc -1)}{2(1-Cc)\,\beta L}\theta_1 L + \frac{(S+s)}{2(1-Cc)}V_2 + \frac{(C-c)}{2(1-Cc)\,\beta L}\theta_2 L$$

$$C_3 = \frac{(1 - Ss-Cc)}{2(1-Cc)}V_1 + \frac{(Cs - Sc)}{2(1-Cc)\,\beta L}\theta_1 L + \frac{(C-c)}{2(1-Cc)}V_2 + \frac{(S-s)}{2(1-Cc)\,\beta L}\theta_2 L$$

$$C_4 = \frac{(Sc+Cs)}{2(1-Cc)}V_1 - \frac{(Ss - Cc +1)}{2(1-Cc)\,\beta L}\theta_1 L - \frac{(S+s)}{2(1-Cc)}V_2 - \frac{(C-c)}{2(1-Cc)\,\beta L}\theta_2 L$$

$$s = \sin\alpha L,\ S = Sinh\,\alpha L,\ c = \cos\alpha L,\ C = Cosh\,\alpha L \qquad (7.37)$$

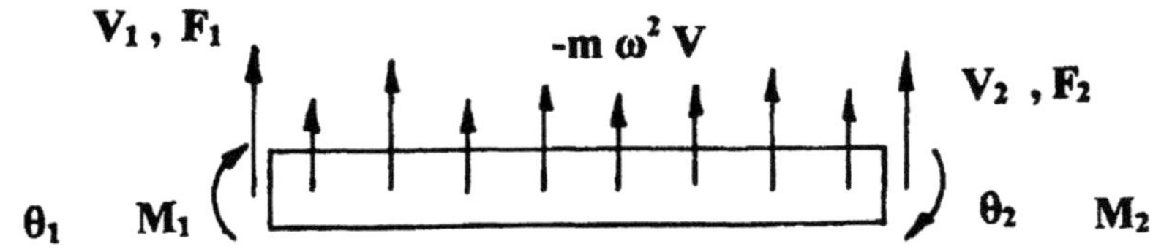

Fig. 7.5 Beam element

Substituting for the constants from equation (7.37), equation (7.35) can be written as

$$V = N_1(x)\,V_1 + N_2(x)\,\theta_1 L + N_3(x)\,V_2 + N_4(x)\,\theta_2 L \qquad (7.38)$$

The shape functions $N_1(x)$ to $N_4(x)$ are given by

$$N_1(x) = \frac{(Sc+Cs)(Sinh\,\alpha x - \sin\alpha x) - Ss(Cosh\,\alpha x - \cos\alpha x) + (1-Cc)(Cosh\,\alpha x + \cos\alpha x)}{2(1 - Cc)}$$

$$N_2(x) = \frac{-(Ss)(Sinh\alpha x - \sin\alpha x) - (1-Cc)(Sinh\alpha x + \sin\alpha x) + (Cs - Sc)(Cosh\alpha x - \cos\alpha x)}{2(1 - Cc)\beta L}$$

$$N_3(x) = \frac{-(S + s)(Sinh\alpha x - \sin\alpha x) + (C-c)(Cosh\alpha x - \cos\alpha x)}{2(1 - Cc)}$$

$$N_4(x) = \frac{-(C - c)(Sinh\alpha x - \sin\alpha x) + (S - s)(Cosh\alpha x - \cos\alpha x)}{2(1 - Cc)\beta L} \qquad (7.39)$$

The force boundary conditions are

$$x = 0:\ EI\frac{d^2v}{dx^2} = M_1,\ F_1 = EI\frac{d^3v}{dx^3}$$

$$x = L:\ -EI\frac{d^2v}{dx^2} = M_2,\ \ F_2 = -EI\frac{d^3v}{dx^3} \qquad (7.40)$$

Substituting equations (7.28) and (7.29) in equation (7.40), the dynamic stiffness matrix relationship between the amplitudes of end forces and end displacements is given by

$$\begin{bmatrix} F_1 \\ M_1 \\ F_2 \\ M_2 \end{bmatrix} = \frac{EI}{L} \begin{bmatrix} \frac{T_1}{L^2} & \frac{-Q_1}{L} & \frac{T_2}{L^2} & \frac{-Q_2}{L} \\ \frac{-Q_1}{L} & S_1 & \frac{Q_2}{L} & S_2 \\ \frac{T_2}{L^2} & \frac{Q_2}{L} & \frac{T_1}{L^2} & \frac{Q_1}{L} \\ \frac{-Q_2}{L} & S_2 & \frac{Q_1}{L} & S_1 \end{bmatrix} \begin{bmatrix} V_1 \\ \theta_1 \\ V_2 \\ \theta_2 \end{bmatrix} \tag{7.41}$$

$$S_1 = \frac{(Cs - Sc)\alpha L}{(1 - Cc)}, \; S_2 = \frac{(S - s)\alpha L}{(1 - Cc)} \tag{7.42a}$$

$$Q_1 = \frac{(Ss)(\alpha L)^2}{(1 - Cc)}, \; Q_2 = \frac{(C - c)(\alpha L)^2}{(1 - Cc)} \tag{7.42b}$$

$$T_1 = \frac{(Cs + Sc)(\alpha L)^3}{(1 - Cc)}, \; T_2 = \frac{(S + s)(\alpha L)^3}{(1 - Cc)} \tag{7.42c}$$

where, $s = \sin \alpha L$, $S = Sinh\,\alpha L$, $c = \cos \alpha L$, $C = Cosh\,\alpha L$.

7.5 NATURAL FREQUENCY AND MODE SHAPES OF BEAM ELEMENT

The element dynamic stiffness matrix can be used to determine the natural frequencies and corresponding mode shapes. This is illustrated by two examples.

7.5.1 Simply Supported Beam: Mode Shapes

Figure 7.6 shows a simply supported beam supporting a uniformly distributed mass and undergoing undamped free vibrations. The boundary conditions are $V_1 = V_2 = 0$ and $M_1 = M_2 = 0$. However θ_1 and θ_2 are not zero.

Substituting the boundary conditions in equation (7.41), the stiffness matrix is given by

$$\begin{bmatrix} 0 \\ 0 \end{bmatrix} = \frac{EI}{L} \begin{bmatrix} S_1 & S_2 \\ S_2 & S_1 \end{bmatrix} \begin{bmatrix} \theta_1 \\ \theta_2 \end{bmatrix} \tag{7.43}$$

A non-zero (or non-trivial) solution to θ_1 and θ_2 is possible only if the determinant is zero. Therefore $S_1^2 - S_2^2 = 0$. Substituting from equation (7.42a), for S_1 and S_2,

$(Cs - Sc)^2 - (S - s)^2 = 0$

Expanding the squares and noting that $(s^2 + c^2 = 1)$ and $(C^2 - S^2 = 1)$, the above equation can be simplified in a series of stages as follows

$$C^2 s^2 + S^2 c^2 - 2\, Cs\, Sc - S^2 - s^2 + 2Ss = 0$$

$$s^2 (C^2 - 1) + S^2 (c^2 - 1) - 2Ss(Cc - 1) = 0$$

$$s^2 (S^2) + S^2 (-s^2) - 2Ss(Cc - 1) = 0$$

$$Ss(Cc - 1) = 0$$

The determinant of matrix (7.43) is zero, only if (Cc – 1) = 0 or S = or s = 0. (Cc-1) = 0 or S = 0, only if αL = 0, which once again leads to a trivial solution. However s = 0 if αL = nπ, n = all integers.

Substituting for αL from equation (7.33)

$$\alpha\, L = L \sqrt[4]{\frac{m\,\omega^2}{EI}} = n\pi\ , \ \omega = n^2\, \pi^2 \sqrt{\frac{EI}{m\, L^4}}\ , \ n = 1, 2, 3, \ldots$$

Note that mL = total mass on the beam and (EI/L^3) is proportional to the flexural stiffness of the beam. Thus

$$\omega = n^2\, \pi^2 \sqrt{\frac{Flexural\ stiffness}{Total\ mass}} \tag{7.44}$$

As can be seen, apart from the multiplying factor before the square root sign, equation (7.44) is similar to equation (1.2), which corresponds to the natural frequency of a single spring-mass system.

From equation (7.38), the displacement pattern (mode shape) is given by

$$V(x) = N_2(x)\, \theta_1 + N_4(x)\, \theta_2 \tag{7.45}$$

Substituting in equation (7.42a), αL = nπ, s = sinnπ = 0 and c = cosnπ = $(-1)^n$

$$S_1 = \frac{\{-S(-1)^n\}\alpha L}{\{1 - C(-1)^n\}}\ , \ S_2 = \frac{(S\ )\alpha L}{(1 - C(-1)^n)}\ , \ \therefore\ S_1 = -(-1)^n S_2$$

From the first row of the matrix (7.43)

$$\therefore \{S_1\, \theta_1 + S_2\, \theta_2\} = S_1 \{\theta_1 - (-1)^n \theta_2\} = 0$$

$$\therefore\ (-1)^n \theta_2 = \theta_1\ or\ (-1)^n \theta_1 = \theta_2$$

From equation (7.39), substituting for s = 0 and c = $(-1)^n$, the shape functions are

$$N_2(x) = \frac{\{C(-1)^n - 1\}(Sinh\alpha x + \sin\alpha x) - S(-1)^n (Cosh\alpha x - \cos\alpha x)}{2(1 - C(-1)^n)\alpha L}$$

$$N_4(x) = \frac{-\{C - (-1)^n\}(Sinh\alpha x - \sin\alpha x) + S\ (Cosh\alpha x - \cos\alpha x)}{2(1 - C(-1)^n)\alpha L}$$

Substituting for N_2 and N_4 in equation (7.45) and simplifying

$$V(x) = -[\frac{L}{n\pi} \sin n\pi \frac{x}{L}\ ]\ \theta_1$$

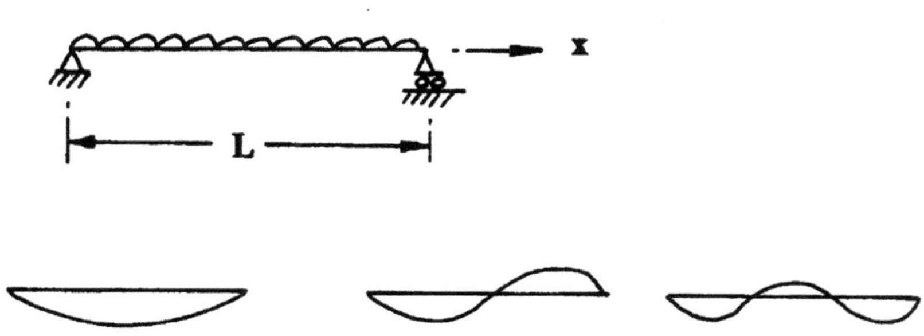

Fig. 7.6 Mode shapes of a simply supported beam

Figure 7.6 shows the first three natural frequencies and their corresponding mode shapes. The mode shapes are all sine waves. The first mode (n = 1) corresponds to a half sine wave with the maximum displacement at mid-span. However for the next frequency (n = 2) the mode shape is a full sine wave with the midspan becoming a node with zero deflection. Similarly, for the next frequency (n = 3), the mode shape consists of three half-sine waves with two nodes at third points. As in the case of the bar element discussed in Section 7.3, on either side of a node the beam deflects in opposite directions.

7.5.2 Cantilever Beam: Mode Shapes

As another example of how the dynamic stiffness matrix can be used to determine the natural frequencies and corresponding mode shapes, consider the cantilever beam shown in Fig. 7.7. The boundary conditions are $V_1 = \theta_1 = 0$ and $M_2 = F_2 = 0$ at the free end but V_2 and θ_2 are not zero. Substituting the boundary conditions in equation (7.41)

$$\begin{bmatrix} 0 \\ 0 \end{bmatrix} = \frac{EI}{L} \begin{bmatrix} \frac{T_1}{L^2} & \frac{Q_1}{L} \\ \frac{Q_1}{L} & S_1 \end{bmatrix} \begin{bmatrix} V_2 \\ \theta_2 \end{bmatrix} \tag{7.46}$$

A non-trivial solution to V_2 and θ_2 is possible only if the determinant is zero. Therefore $T_1\ S_1 - Q_1^2 = 0$. Substituting from equation (7.42) for S_1, T_1 and Q_1 and simplifying, the determinant becomes

$$Deter = (Cs + Sc)((Cs - Sc) - S^2 s^2 = C^2 s^2 - S^2 c^2 - S^2 s^2$$

$$= (C^2 - S^2)s^2 - S^2 c^2 = s^2 - (C^2 - 1)\, c^2 = s^2 + c^2 - C^2 c^2 = 1 - C^2 c^2$$

$$\therefore Deter = 1 - C^2\, c^2 = 0 \tag{7.47}$$

The natural frequencies are determined by solving the transcendental equation $C^2\, c^2 = 1$ or $C\, c = \pm 1$. The condition $C\, c = 1$ is satisfied only for $\alpha L = 0$, which is a trivial solution. The natural frequencies are determined by solving the equation $C\, c = -1$. First three roots are approximately given by $\alpha L = 1.88, 4.69, 7.85$. The corresponding natural frequencies are therefore

$$\alpha L = L\sqrt[4]{\frac{m\,\omega^2}{EI}} = (1.88,\ 4.69,\ 7.85) \text{ or } \omega = (\ 3.53,\ 21.996,\ 61.62)\sqrt{\frac{EI}{m\,L^4}} \tag{7.48}$$

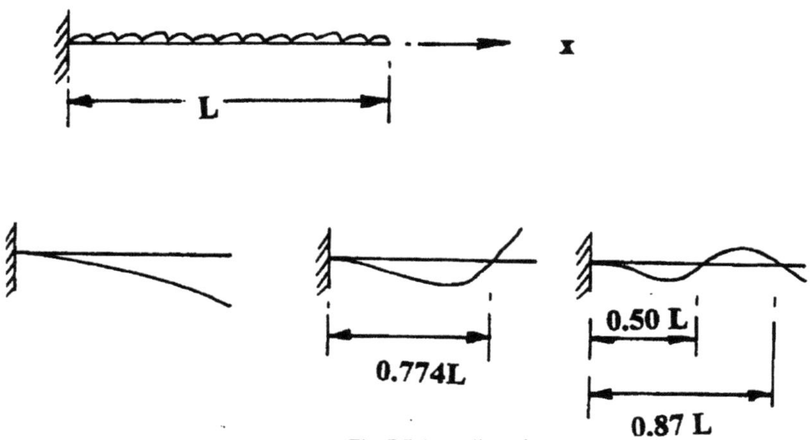

Fig. 7.7 A cantilever beam

Having obtained the natural frequency, the corresponding mode shape can be calculated from equation (7.38)

$$V(x) = N_3(x)\, V_2 + N_4(x)\, \theta_2 L$$

From the first row of matrix (7.46), $T_1\, V_2 + Q_1\, \theta_2 L = 0$

$$\therefore V(x) = \{N_3(x) - \frac{T_1}{Q_1} N_4(x)\}V_2$$

By plotting V(x), it can be shown that for the second natural frequency, there is a node at 0.774L from fixed end and for the third mode there are two nodes, the first at 0.50L and the second at 0.87L from the fixed end. Fig. 7.7 shows the modes of vibration of a cantilever beam.

7.5.3 Standard Results for Beam Vibration

The approach adopted in Sections 7.5.1 and 7.5.2, can be used to calculate the natural frequencies of beams with various support conditions. Table 7.1 summarises the results for the first three natural frequencies. The node positions given are measured from the left-hand support.

Table 7.1 Natural Frequencies of Beam Elements: $\omega = C\sqrt{EI/(mL^4)}$

Beam type	C	Node position
Simply supported beam	9.87	-
	39.48	0.5L
	88.83	0.33L, 0.67 L
Cantilever beam. Left end clamped	3.52	-
	22.04	0.77L
	61.70	0.50L, 0.87L
Clamped beam	22.37	-
	61.67	0.50 L
	120.90	0.36 L, 0.64 L
Propped cantilever. Left end clamped	15.41	-
	49.97	0.56 L
	104.25	0.38 L, 0.69 L

7.6 BEAM ELEMENT WITH SHEAR DEFORMATION INCLUDED

In the previous development, only deformation due to flexure was included. In most 'shallow' beams, that is beams whose span/depth ratio is large, flexural deformations do dominate. However, if the beam is comparatively 'stocky', with a 'small' span/depth ratio, then deformation due to shear can be important. For example, for a cantilever beam with a uniformly distributed load, if the span/depth ratio is less than about 10, then depending up the actual span/depth ratio, shear deformation can add anything up to 20% to the total tip deflection.

The main difference from the derivation in Section 7.4, comes from the fact that although plane sections remain plane before and after bending, they do not remain normal to the deformed neural axis.

Let v be the total deflection of the neutral axis. Let ψ be slope of the deformation due to bending and γ the shear strain. Like ψ, shear strain γ is also an angular deformation. As shown in Fig. 7.8, the total slope of the neutral axis is therefore

$$\frac{\partial v}{\partial x} = \psi + \gamma \tag{7.49}$$

From the moment–curvature relationship

$$M = EI \frac{\partial \psi}{\partial x} \tag{7.50}$$

Similarly, the shear force-average shear strain relationship is assumed to be

$$Q = GA_s \gamma = GA_s \{\frac{\partial v}{\partial x} - \psi\} \tag{7.51}$$

where G = shear or rigidity modulus, A_s = effective cross-sectional area resisting shear. Shear stress and shear strain in a cross section are not constant but in the interests of simplicity, shear force Q is related to average shear strain. It is because of this that A_s is much less than the actual cross-sectional area A. In the case of a rectangular section, A_s is about 0.83A. For an I–section or a T-section bending about an axis parallel to the flanges, A_s is approximately equal to 0.83 of the web area.

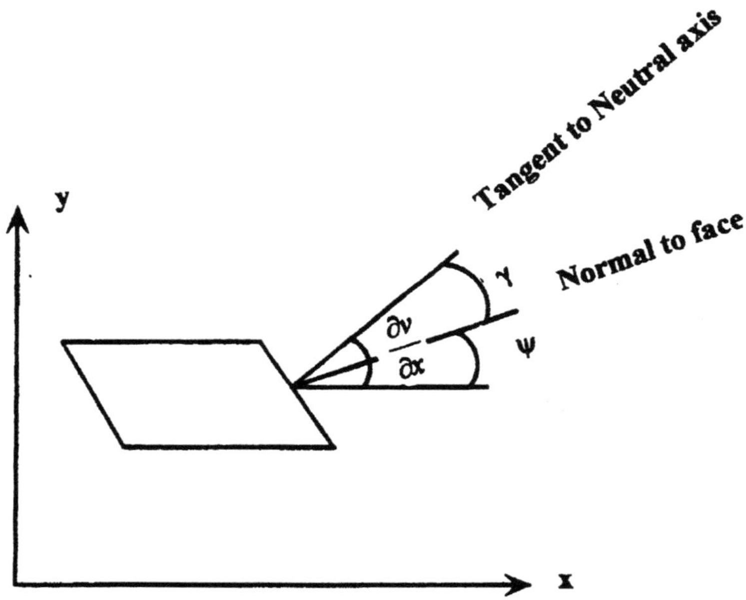

Fig. 7.8 Effect of shear deformation

Substituting for M and Q in terms of v and ψ in equations (7.26a) and (7.26b) and substituting for the distributed inertial force $q = -m(\partial^2 v / \partial t^2)$, the differential equations become

$$GA_s\{\frac{\partial^2 v}{\partial x^2} - \frac{\partial \psi}{\partial x}\} - m\frac{\partial^2 v}{\partial t^2} = 0 \qquad (7.52)$$

$$EI\frac{\partial^2 \psi}{\partial x^2} + GA_s\{\frac{\partial v}{\partial x} - \psi) = 0 \qquad (7.53)$$

The variable ψ can be eliminated from equations (7.52) and (7.53) as follows. Differentiating equation (7.52) twice with respect to x and the equation (7.53) once with respect to x

$$GA_s\{\frac{\partial^4 v}{\partial x^4} - \frac{\partial^3 \psi}{\partial x^3}\} - m\frac{\partial^4 v}{\partial x^2\,\partial t^2} = 0 \qquad (7.54a)$$

$$EI\frac{\partial^3 \psi}{\partial x^3} + GA_s\{\frac{\partial^2 v}{\partial x^2} - \frac{\partial \psi}{\partial x}) = 0 \qquad (7.54b)$$

Eliminating $\frac{\partial^3 \psi}{\partial x^3}$ from equations (7.54a) and (7.54b)

$$EI\frac{\partial^4 v}{\partial x^4} + GA_s\{\frac{\partial^2 v}{\partial x^2} - \frac{\partial \psi}{\partial x}\} - \frac{m\ EI}{GA_s}\frac{\partial^4 v}{\partial x^2\,\partial t^2} = 0 \qquad (7.54c)$$

Substituting in equation (7.54c)

$$GA_s\{\frac{\partial^2 v}{\partial x^2} - \frac{\partial \psi}{\partial x}\} = m\frac{\partial^2 v}{\partial t^2}$$

from equation (7.54a). This leads to the differential equation

$$EI\frac{\partial^4 v}{\partial x^4} + m\frac{\partial^2 v}{\partial t^2} - \frac{m\ EI}{GA_s}\frac{\partial^4 v}{\partial x^2\,\partial t^2} = 0 \qquad (7.55)$$

If GA_s is made very large, then equation (7.55) reverts to equation (7.30). Assuming undamped free vibration and substituting equation (7.31) in equation (7.55) and simplifying

$$EI\frac{d^4 V}{dx^4} + \omega^2\frac{m\,EI}{GA_s}\frac{d^2 V}{dx^2} - m\,\omega^2\ V = 0 \qquad (7.56)$$

As equation (7.56) is an ordinary differential equation in V with constant coefficients, the solution is obtained by assuming V(x) as $e^{\beta x}$. Substituting this expression in equation (7.56) and solving for β

$$\beta = \pm\alpha\sqrt{\{p - q\}}\ ,\ \pm j\alpha\sqrt{\{p + q\}}\ , j = \sqrt{-1}$$

$$\alpha = \sqrt[4]{\frac{m\omega^2}{EI}}\ ,\ p = \sqrt{[1 + q^2]}\ ,\ q = \frac{\alpha^2}{2}\frac{EI}{GA_s}$$

$$V(x) = C_1 \cos\alpha\sqrt{(p+q)}\, x + C_2 \sin\alpha\sqrt{(p+q)}\, x + C_3 \, Cosh\alpha\sqrt{(p-q)}\, x$$
$$+ C_4 \, Sinh\alpha\sqrt{(p-q)}\, x$$

If the shear rigidity GA_s is infinitely large, then q becomes zero and p becomes unity and the solution reverts to equation (7.33) derived ignoring shear deformation.

As in Section 7.3, for a simply supported beam

$$V(x) = C \sin n\pi \frac{x}{L} \tag{7.57}$$

where C is a constant.

Substituting equation (7.57) in equation (7.56)

$$\{EI\, n^4\, \pi^4 \frac{1}{L^4} + \omega^2 \frac{m\, EI}{G A_s} (-n^2\, \pi^2 \frac{1}{L^2}) - m\,\omega^2\} C \sin n\pi \frac{x}{L} = 0$$

Since C is not equal to zero, the terms inside the bracket must be zero. Therefore

$$\omega = n^2\, \pi^2 \sqrt{\frac{EI}{m\, L^4}} \frac{1}{\sqrt{\{1 + n^2\, \pi^2 \dfrac{EI}{G A_s\, L^2}\}}} \tag{7.58}$$

As is to be expected, inclusion of shear deformation reduces the stiffness of the beam and therefore reduces the frequency. As a rough guide to the influence of shear deformation on the natural frequency, in a simply supported rectangular steel beam b x d with a uniformly distributed mass, $EI = bd^3/12$, $A_s = bd/1.2$, $E = 200$ kN/mm^2 and $G = 77$ kN/mm^2, the parameter $EI/(G\, A_s\, L^2) = 0.26\, (d/L)^2$. If the span/depth ratios are 10, 5 and 2.5, then inclusion of shear deformation decreases the first or fundamental natural frequency (n = 1) by 1%, 5% and 16% respectively. For the same beam, the second natural frequency (n = 2) is reduced by 5%, 16% and 38% respectively.

7.7 EFFECT OF AXIAL LOAD ON NATURAL FREQUENCY OF BEAMS

It is well known that axial compressive loads decrease the translational and rotational stiffness of beams. Tensile axial load has the opposite effect. It is therefore important to be able to include this effect in the natural frequency calculations. Fig. 7.9 shows a beam element with an axial compressive force P.

Proceeding as in Section 7.4, the basic equations of equilibrium are

$$-Q + \{Q + \frac{\partial Q}{\partial x} \Delta x\} + q\, \Delta x = 0 \tag{7.59}$$

$$-M + \{M + \frac{\partial M}{\partial x} \Delta x\} + \{Q + \frac{\partial Q}{\partial x} \Delta x\}\Delta x + P \frac{\partial v}{\partial x} \Delta x + q\, \Delta x \frac{\Delta x}{2} = 0 \tag{7.60}$$

Comparing equations (7.59) with equation (7.24), and equation (7.60) with equation (7.25), the main difference in the latter case is the additional term due to the moment caused by P in deformed state of the beam. Simplifying and letting Δx tend to zero, equation (7.59) and (7.60) become

$$\frac{\partial Q}{\partial x} + q = 0 \tag{7.61}$$

$$\frac{\partial M}{\partial x} + Q + P\frac{\partial v}{\partial x} = 0 \tag{7.62}$$

Eliminating Q from equations (7.61) and (7.62), the differential equilibrium is

$$\frac{\partial^2 M}{\partial x^2} + P\frac{\partial^2 v}{\partial x^2} - q = 0 \tag{7.63}$$

Substituting for M in terms of curvature from equation (7.27) and q = $-m(\partial^2 v/\partial t^2)$, equation (7.63) can be written as

$$EI\,\frac{\partial^4 v}{\partial x^4} + P\frac{\partial^2 v}{\partial x^2} + m\frac{\partial^2 v}{\partial t^2} = 0 \tag{7.64}$$

Once again considering undamped free vibration, if V(x) is the amplitude of vibration and ω the frequency of vibration, then v = V(x) $\cos\omega t$. Substituting for v in equation (7.64)

$$EI\,\frac{d^4 V}{dx^4} + P\frac{d^2 V}{dx^2} - m\omega^2\, V = 0 \tag{7.65}$$

As equation (7.65) is an ordinary differential equation in V with constant coefficients, the solution is obtained by assuming V(x) as $e^{\beta x}$. Substituting this expression in equation (7.65) and solving for β

$$\beta^2 = -\frac{P}{2\,EI} \pm \sqrt{[(\frac{P}{2\,EI})^2 + \frac{m\omega^2}{EI}]}$$

$$\beta = \pm\sqrt{(p - q)} \quad , \ \pm j\sqrt{(p + q)} \ , \ j=\sqrt{-1}$$

$$p = \sqrt{[(\frac{P}{2\,EI})^2 + \frac{m\omega^2}{EI}]} \ , \ q = \frac{P}{2\,EI}$$

$$V(x) = C_1 \cos\alpha\sqrt{(p+q)}\,x + C_2 \sin\alpha\sqrt{(p+q)}\,x + C_3 Cosh\alpha\sqrt{(p-q)}\,x$$
$$+ C_4\, Sinh\alpha\sqrt{(p-q)}\,x$$

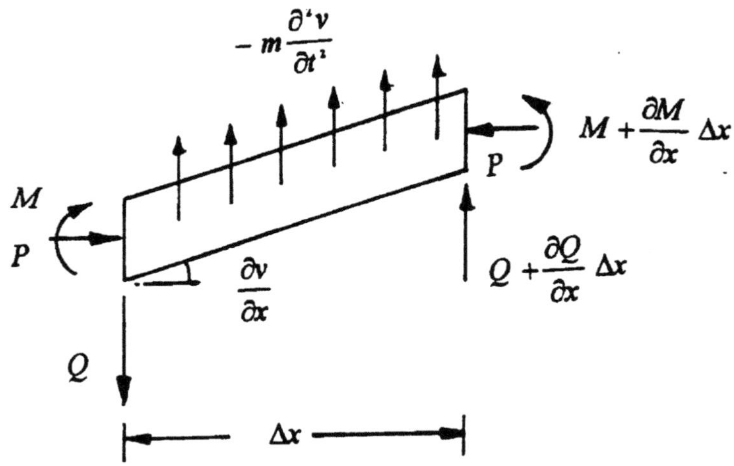

Fig. 7.9 Beam with axial load

As in Section 7.3, for a simply supported beam, $V(x) = C \sin n\pi\,(x/L)$, where C is a constant. Using this value in equation (7.65)

$$EI\,\frac{n^4\,\pi^4}{L^4} - P\frac{n^2\,\pi^2}{L^2} - m\,\omega^2 = 0 \qquad (7.66)$$

Solving equation (7.66) for ω

$$\omega^2 = \frac{EI}{m\,L^4}\{n^4\,\pi^4 - \rho\,n^2\,\pi^4\},\ \rho = \frac{P}{P_{Euler}},\ P_{Euler} = \pi^2\,\frac{EI}{L^2} \qquad (7.67)$$

Rearranging equation (7.67)

$$\omega = n^2\,\pi^2\sqrt{\frac{EI}{m\,L^4}}\sqrt{(1 - \frac{\rho}{n^2})} \qquad (7.68)$$

To get an idea of the effect of the axial compressive load on the first natural frequency (n = 1), if ρ is equal to 0.1, 0.3, 0.5, 0.7, then the natural frequency decreases by 5%, 14%, 29% and 45% respectively. As shown in Fig. 7.10, the reduction is almost linear up to a value of ρ = 0.6.

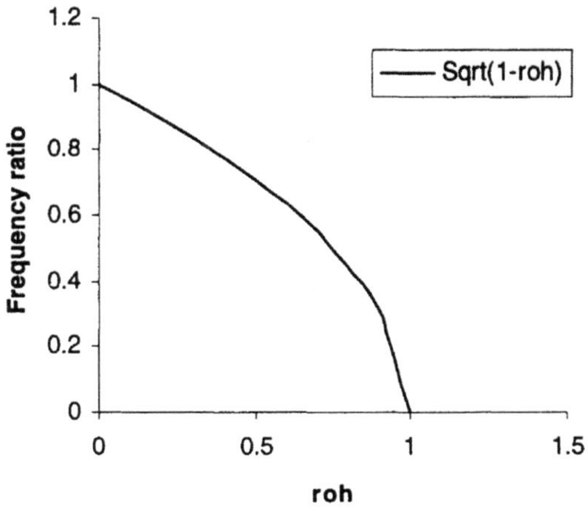

Fig. 7.10 Effect of axial load on natural frequency

7.8 SHAFT ELEMENT

Figure 7.11 shows a shaft element subjected to end torques and distributed inertial torque.

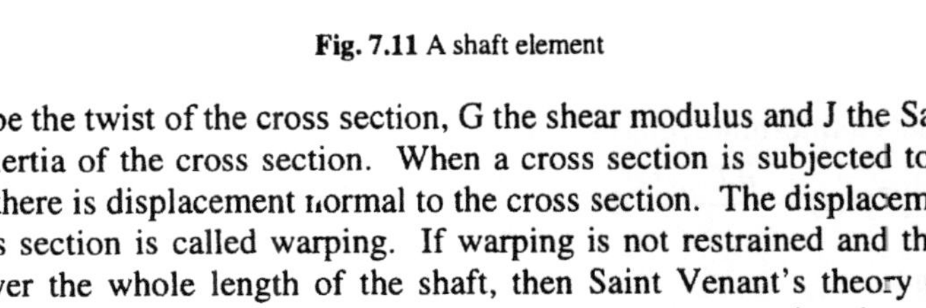

Fig. 7.11 A shaft element

Let $\Phi(x,t)$ be the twist of the cross section, G the shear modulus and J the Saint Venant torsional inertia of the cross section. When a cross section is subjected to torsion, it twists and there is displacement normal to the cross section. The displacement normal to the cross section is called warping. If warping is not restrained and the torque is constant over the whole length of the shaft, then Saint Venant's theory of uniform torsion is valid. The only 'force' that exists at a cross section is a torque, T.

Considering the equilibrium of an infinitesimal length, Δx, the torques at the two cut sections are

$$T(x,t) \text{ and } T(x,t) + \frac{\partial T(x,t)}{\partial x}\Delta x$$

As shown in Fig. 7.12, at any point at a radial distance r from the axis of rotation, the tangential acceleration a_t is given by

$$a_t = r\frac{\partial^2 \phi}{\partial t^2}$$

If m is the mass per unit length, then the mass of the element of area dA is m (dA/A). The tangential inertial force f on an infinitesimal element is equal to

$$f = -\frac{m}{A}\, dA\, r\frac{\partial^2 \phi}{\partial t^2}$$

The torque produced by this force is fr. The total inertial torque is thus given by

$$t = \int f\, r = -\int\frac{m}{A} dA\, r\frac{\partial^2 \phi}{\partial t^2}\, r = -\frac{m}{A}\frac{\partial^2 \phi}{\partial t^2}\int r^2\, dA = -\frac{m}{A} I_{Polar}\frac{\partial^2 \phi}{\partial t^2} \tag{7.69}$$

$$I_{Polar} = \int r^2\, dA$$

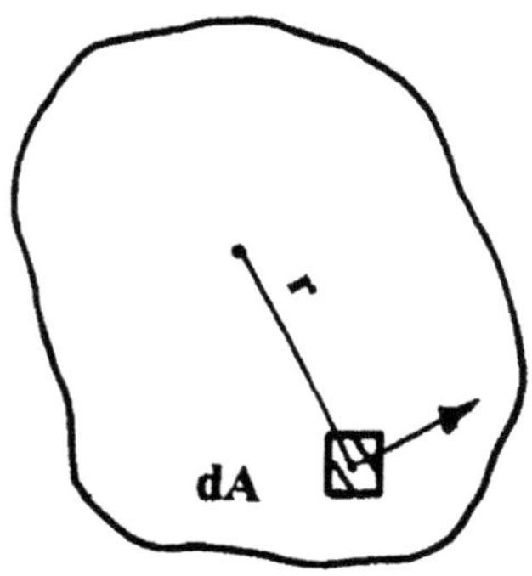

Fig. 7.12 Calculation of Inertial torque

The torsional equilibrium requires that

$$-T(x,t) + \{T(x,t) + \frac{\partial T(x,t)}{\partial x}\Delta x\} - \frac{m}{A} I_{Polar}\frac{\partial^2\phi(x,t)}{\partial t^2}\Delta x = 0 \tag{7.70}$$

Simplifying

$$\frac{\partial T(x,t)}{\partial x} - \frac{m}{A} I_{Polar} \frac{\partial^2 \phi(x,t)}{\partial t^2} = 0 \tag{7.71}$$

In Saint Venant's theory of uniform torsion, the relationship between the torque T and the corresponding twist Φ is given by

$$T(x,t) = GJ \frac{\partial \phi}{\partial x} \tag{7.72}$$

Substituting equation (7.72) in equation (7.71)

$$\frac{\partial}{\partial x}\{GJ \frac{\partial \phi(x,t)}{\partial x}\} - \frac{m}{A} I_{Polar} \frac{\partial^2 u(x,t)}{\partial t^2} = 0 \tag{7.73}$$

If GJ is constant, then

$$GJ \frac{\partial^2 \phi(x,t)}{\partial x^2} - \frac{m}{A} I_{Polar} \frac{\partial^2 \phi(x,t)}{\partial t^2} = 0 \tag{7.74}$$

This is the basic differential equation governing the shaft element. It is similar to the differential equation (7.5) for the bar element. Hence in the expressions corresponding to the bar element, if EA and m are replaced respectively by GJ and (m/A) I_{polar}, the corresponding expressions for the shaft element can be obtained. For the torque element shown in Fig. 7.13, the dynamic force-deformation relationship can be expressed in matrix form as

$$\begin{bmatrix} T_1 \\ T_2 \end{bmatrix} = \begin{bmatrix} k_{11} & k_{12} \\ k_{12} & k_{22} \end{bmatrix} \begin{bmatrix} \Phi_1 \\ \Phi_2 \end{bmatrix} \tag{7.75}$$

$$k_{11} = k_{22} = GJ \cot \alpha L, \quad k_{12} = -GJ \frac{1}{\sin \alpha L} \tag{7.76}$$

$$\alpha L = L \sqrt{\frac{m}{A} I_{Polar} \frac{\omega^2}{GJ}} \tag{7.77}$$

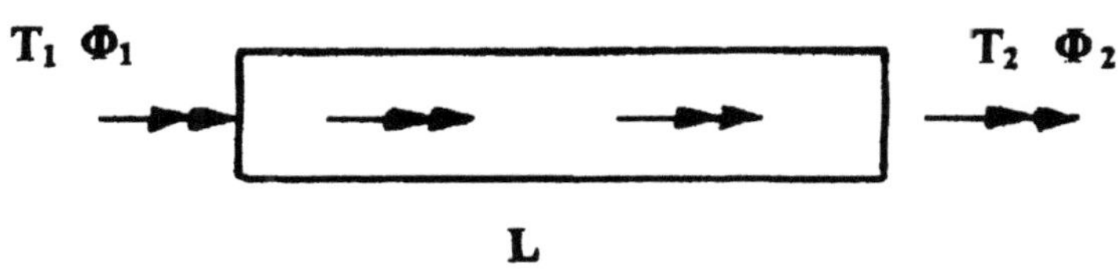

Fig. 7.13 A torsion element

7.9 APPROXIMATE SHAPE FUNCTIONS AND STIFFNESS COEFFICIENTS

In the previous sections, ordinary differential equations governing the undamped free vibration problem were solved exactly and the dynamic stiffness matrix relating the amplitudes of forces at the ends of an element and the corresponding deformation amplitude was obtained. However, as was shown, these 'dynamic stiffness coefficients' are non-linear functions of the stiffness, mass and frequency of vibration. In general these 'exact' dynamic stiffness matrices can be obtained in the case of problems governed by ordinary differential equations. This is an almost impossible task in the case of most problems governed by partial differential equations. Thus, although methods do exist for the determination of natural frequencies using these 'dynamic stiffness coefficients', in most cases in practice, the corresponding calculations are done by solving a general linear eigen-value problem given by

$$K\,X - \omega^2 M\,X = 0$$

where the dynamic stiffness matrix is approximately $[K - \omega^2 M]$.

The rest of the chapter is therefore devoted to deriving the static stiffness and mass matrices for line elements. However before doing so, it is useful to see how the approximate mass and stiffness matrices relate to the exact dynamic stiffness matrix. One result of such a comparison is that it gives an idea of the approximations involved and in addition what parameters need to be limited in order to obtain reasonably accurate solutions.

7.10 APPROXIMATE DYNAMIC STIFFNESS MATRIX FOR A BAR ELEMENT

The following results relating trigonometric functions and the corresponding convergent series are well known.

$$\sin x = x - \frac{x^3}{3!} + \frac{x^5}{5!} - \frac{x^7}{7!} + \ldots$$

$$\cos x = 1 - \frac{x^2}{2!} + \frac{x^4}{4!} - \frac{x^{67}}{6!} + \ldots$$

$$\cot x = \frac{1}{x} - \frac{x}{3} - \frac{x^5}{45} - \frac{2x^5}{945} - \ldots$$

$$\operatorname{cosec} x = \frac{1}{x} + \frac{x}{6} + \frac{7x^5}{360} + \frac{31x^5}{15120} - \ldots$$

Using these series, shape functions in equation (7.15) can be expanded in terms of powers of x. The shape functions become

$$N_1(x) = \cos\alpha x - \sin\alpha x\cot\alpha L = \{1 - \frac{(\alpha x)^2}{2!} + \frac{(\alpha x)^4}{4!} - \frac{(\alpha x)^6}{6!} + \ldots\}$$

$$- \{\alpha x - \frac{(\alpha x)^3}{3!} + \frac{(\alpha x)^5}{5!} - \frac{(\alpha x)^7}{7!} + \ldots\}[\frac{1}{\alpha L} - \frac{\alpha L}{3} - \frac{(\alpha L)^5}{45} - \frac{2(\alpha L)^5}{945} - \ldots]$$

$$N_2(x) = \frac{\sin\alpha x}{\sin\alpha L} = \frac{\alpha x - \frac{(\alpha x)^3}{3!} + \frac{(\alpha x)^5}{5!} - \frac{(\alpha x)^7}{7!} + \ldots}{(\alpha L) - \frac{(\alpha L)^3}{3!} + \frac{(\alpha L)^5}{5!} - \frac{(\alpha L)^7}{7!} + \ldots}$$

Retaining only terms in x, the above expressions for shape functions can be written as

$$N_1(x) = \{1 - \frac{x}{L}\}, \quad N_2(x) = \frac{x}{L} \tag{7.78}$$

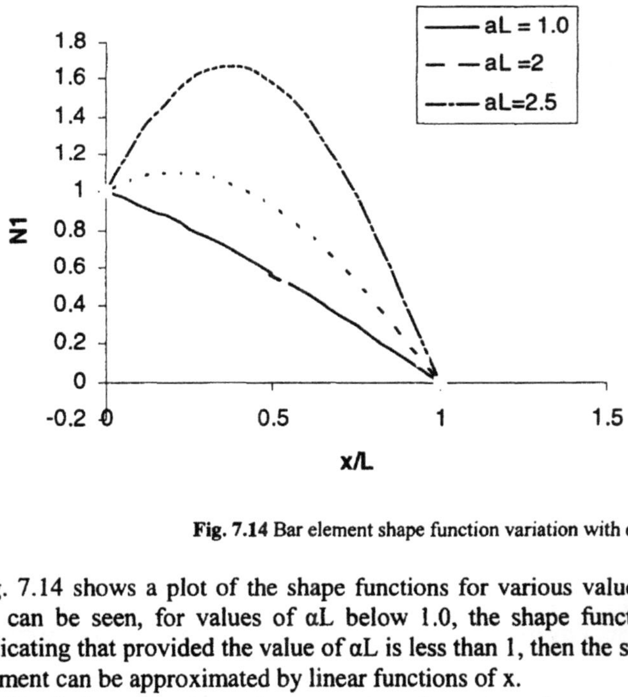

Fig. 7.14 Bar element shape function variation with αL

Fig. 7.14 shows a plot of the shape functions for various values of the parameter αL. As can be seen, for values of αL below 1.0, the shape functions are almost linear, indicating that provided the value of αL is less than 1, then the shape functions for a bar element can be approximated by linear functions of x.

Similarly, using the series expansions for cotx and cosecx, the dynamic stiffness coefficients in equation (7.20) can be expressed as

$$k_{11} = k_{22} = AE\alpha[\frac{1}{(\alpha L)} - \frac{(\alpha L)}{3} - \frac{(\alpha L)^5}{45} - \frac{2(\alpha L)^5}{945} - ...]$$

$$= \frac{AE}{L} - \frac{AE}{L}[\frac{(\alpha L)^2}{3} + \frac{(\alpha L)^4}{45} + \frac{2(\alpha L)^6}{945} - ...]$$

$$k_{12} = -AE\,\alpha[\frac{1}{(\alpha L)} + \frac{(\alpha L)}{6} + \frac{7(\alpha L)^3}{360} + \frac{31(\alpha L)^5}{15120} - ...]$$

$$= -\frac{AE}{L} - \frac{AE}{L}[\frac{(\alpha L)^2}{6} + \frac{7(\alpha L)4}{360} + \frac{31(\alpha L)^6}{15120} - ...]$$

Retaining only the two terms of the series, for 'small' values of αL, and substituting for αL from equation (7.10), the stiffness coefficients can be expressed as

$$k_{11} = k_{22} = \frac{AE}{L} - \frac{m\omega^2}{3}L \quad k_{12} = -\frac{AE}{L} - \frac{m\omega^2}{6}L \tag{7.79}$$

From equation (7.19), the dynamic stiffness relationship for a bar element becomes

$$\therefore \begin{bmatrix} F_1 \\ F_2 \end{bmatrix} = \frac{AE}{L}\begin{bmatrix} 1 & -1 \\ -1 & 1 \end{bmatrix}\begin{bmatrix} U_1 \\ U_2 \end{bmatrix} - \omega^2\frac{mL}{6}\begin{bmatrix} 2 & 1 \\ 1 & 2 \end{bmatrix}\begin{bmatrix} U_1 \\ U_2 \end{bmatrix} \tag{7.80}$$

If

$$K = \textit{Static stiffness matrix} = \frac{AE}{L}\begin{bmatrix} 1 & -1 \\ -1 & 1 \end{bmatrix} \tag{7.81}$$

$$M = \textit{Consistent mass matrix} = \frac{mL}{6}\begin{bmatrix} 2 & 1 \\ 1 & 2 \end{bmatrix} \tag{7.82}$$

$$\begin{bmatrix} F_1 \\ F_2 \end{bmatrix} = [K - \omega^2 M]\begin{bmatrix} U_1 \\ U_2 \end{bmatrix} \tag{7.83}$$

This shows that in a dynamic situation, the dynamic stiffness matrix can be expressed approximately as

Dynamic stiffness matrix = [static stiffness matrix – ω^2 mass matrix]

7.11 DERIVATION OF APPROXIMATE DYNAMIC STIFFNESS MATRIX

In the previous section, approximation to the dynamic stiffness matrix for a bar element was obtained by using series approximations to the exact dynamic stiffness matrix. Evidently this approach is possible only if the exact stiffness matrix is known. Unfortunately, for problems governed by partial differential equations, the exact stiffness matrix itself is not known. Therefore the above approach fails. Fortunately, there are many methods available in practice for the direct derivation of the approximate dynamic stiffness matrix. However, because of the fact that the exact dynamic stiffness matrix is not known, the approximation involved can only be established by conducting numerical experiments. The two most important and commonly used methods for the direct derivation of the approximate dynamic stiffness matrix are

- The method based on an approximate solution to the differential equation using Galerkin's weighted residual method
- The method based on calculus of variations such as the minimum total potential.

Provided, the assumptions involved are the same, both approaches lead to the same static stiffness and mass matrices.

7.12 METHOD OF MINIMUM TOTAL POTENTIAL

The most commonly used method based on variational calculus is the method of minimum total potential of structure. The total potential π is the sum of the strain energy and the loss of potential energy, by the external loads

$$\text{Strain energy} = \frac{1}{2}\int[\sum \mathit{Stress} \times \mathit{Strain}]\,dvol \tag{7.84}$$

It is understood that the product Stress x Strain includes all the relevant stress components and their corresponding strain components and the integration is carried out over the whole volume of the element.

The loss of potential by the external loads is simply the negative of the load times the corresponding displacement.

In any structure, which is an assemblage of a number of elements, the total potential π is the sum of π_e, the total potential for the individual elements.

7.12.1 Approximate Bar Element Dynamic Stiffness Matrix

In the case of an axial element, the only stress component σ is the axial stress and the corresponding axial strain ε, both of which are assumed to be constant over the cross section. Thus in the case of a vibrating bar element, from equation (7.4)

$$\varepsilon = \frac{\partial u(x,t)}{\partial x}, \quad \sigma = E\,\varepsilon, \quad \mathit{Strain\ energy} = \frac{1}{2}\int \{\frac{\partial u(x,t)}{\partial x}\}^2\, A\, dx \tag{7.85}$$

The 'external loads' acting are the applied forces at the ends of the element and the distributed inertial force $-m(\partial^2 u(x,t)/\partial t^2)$.

Considering the axial element shown in Fig. 7.2

$$\pi_e = \frac{1}{2} AE \int_{x=0}^{L} \frac{\partial u}{\partial x}^2 dx - \frac{1}{2}\int_0^L -m \frac{\partial^2 u}{\partial t^2} u \, dx - F_1(t)u(0,t) - F_2(0,t)u(L,t) \quad (7.86)$$

In the case of undamped free vibration, u(x,t) = U(x) cosωt. Therefore ignoring the time parameter and considering only the amplitudes of forces and displacements

$$\pi_e = \frac{1}{2} AE \int_{x=0}^{L} \{\frac{dU}{dx}\}^2 dx - \frac{1}{2}\int_0^L m\omega^2 U^2 \, dx - F_1 U_1 - F_2 U_2 \quad (7.87)$$

As an approximate solution, using the shape function from equation (7.78),

$$U(x) = N_1(x) U_1 + N_2(x) U_2 \quad (7.88)$$

Substituting equation (7.88) in equation (7.87) and taking the partial derivative of π_e with respect to U_1 and U_2

$$\frac{\partial \pi_e}{\partial U_j} = AE \int_{x=0}^{L} \frac{dN_j}{dx} \sum_{i=1}^{2} \frac{dN_i}{dx} U_i \, dx - m\omega^2 \int_0^L N_j \sum_{i=1} N_i U_i \, dx - F_j, j=1,2 \quad (7.89)$$

The integrals in equation (7.89) are shown in Table 7.2.

Table 7.2 Integrals of shape functions

i	$\int_0^L N_i N_j dx$		$\int_0^L \frac{dN_i}{dx}\frac{dN_j}{dx} dx$	
	j		j	
	1	2	1	2
1	$\frac{L}{3}$	$\frac{L}{6}$	$\frac{1}{L}$	$-\frac{1}{L}$
2	$\frac{L}{6}$	$\frac{L}{3}$	$-\frac{1}{L}$	$\frac{1}{L}$

Substituting for the integrals

$$\frac{\partial \pi_e}{\partial U_1} = \frac{AE}{L}\{U_1 - U_2\} - m\omega^2 L \left(\frac{U_1}{3} + \frac{U_2}{6}\right) - F_1$$

$$\frac{\partial \pi_e}{\partial U_2} = \frac{AE}{L}\{-U_1 + U_2\} - m\omega^2 L \left(\frac{U_1}{6} + \frac{U_2}{3}\right) - F_2$$

$$\begin{bmatrix} \frac{\partial \pi_e}{\partial U_1} \\ \frac{\partial \pi_e}{\partial U_2} \end{bmatrix} = [K - \omega^2 M]\begin{bmatrix} U_1 \\ U_2 \end{bmatrix} - \begin{bmatrix} F_1 \\ F_2 \end{bmatrix} \quad (7.90)$$

The static elastic stiffness matrix K and the mass matrix M for a bar element are given by equation (7.90) and are the same as equations (7.81) and (7.82), the latter derived by replacing the trigonometric terms in the exact stiffness relationship by the corresponding infinite series and as an approximation, retaining only the first term of the series.

7.12.2 Approximate Beam Element Dynamic Stiffness Matrix Including Flexural Deformation Only

In the case of beams, the two main 'stresses' are the bending moment M and the shear force Q. The corresponding 'strains' are curvature and an average shear strain. In the elastic state, the usual assumption made in the Engineer's Theory of Bending, is that plane sections remain plane before and after bending and that shear strain can be ignored. This enables the moment M to be related to the curvature by

$$M = EI\frac{\partial^2 v(x,t)}{\partial x^2} \tag{7.91}$$

where v(x,t) = the deflection of the neutral axis and EI = flexural rigidity.

Thus the strain energy can be expressed in terms of the displacement v(x,t) as

$$Strain\ energy = \frac{1}{2}\int_x M\ \frac{\partial^2 v(x,t)}{\partial x^2}\ dx\ = \frac{1}{2}\ EI\ \int\{\frac{\partial^2 v(x,t)}{\partial x^2}\}^2\ dx \tag{7.92}$$

The loss of potential by the external loads is simply the negative of the load times the corresponding displacement. The 'external loads' are the applied forces at the ends of the element and the distributed inertial force. In the case of a beam element, the normal inertial force per unit length is $-m(\partial^2 v(x,t)/\partial t^2)$.

Considering the bending element shown in Fig. 7.5

$$\begin{aligned}\pi_e = &\frac{1}{2}EI\int_{x=0}^{L}\frac{\partial^2 v}{\partial x^2}^2 dx - \frac{1}{2}\int_0^L -m\frac{\partial^2 v}{\partial t^2}v\,dx - F_1(t)\,v(0,t) - F_2(0,t)\,v(L,t)\\ &-M_1\frac{\partial v(0,t)}{\partial x} - M_2\frac{\partial v(L,t)}{\partial x}\end{aligned} \tag{7.93}$$

In the case of undamped free vibration, v(x,t) = V(x) cosωt. Therefore equation (7.93) simplifies to

$$\pi_e = \frac{1}{2}EI\int_{x=0}^{L}\{\frac{d^2V}{dx^2}\}^2 dx - m\omega^2\ \frac{1}{2}\int_0^L V^2\,dx - F_1\ V_1 - F_2V_2 - M_1\ \theta_1 - M_2\ \theta_2 \tag{7.94}$$

In order to evaluate the integrals, it is necessary to assume an approximate solution to V. An approximate solution can be obtained as the solution to the static problem

$$EI\ \frac{d^4\ V(x)}{d\ x^4} = 0 \tag{7.95}$$

$$V(x) = C_1 + C_2\, x + C_3\, x^2 + C_4\, x^3 \tag{7.96}$$

The constants C_1 to C_4 need to be obtained from the boundary conditions

$$x = 0,\ V = V_1 \ and\ \frac{dV}{dx} = -\theta_1 ,\ x = L,\ V = V_2 \ and\ \frac{dV}{dx} = -\theta_2 \tag{7.97}$$

From equations (7.96) and (7.97), the constants can be determined as

$$C_1 = V_1 ,\ C_2 = -\theta_1 ,\ C_3 = \frac{3(V_2 - V_1) + (2\theta_1 + \theta_2)L}{L^2},$$

$$C_4 = \frac{2(V_1 - V_2) - (\theta_1 + \theta_2)L}{L^3}$$

Substituting for the constants, equation (7.96) becomes

$$V = N_1(x)\, V_1 + N_2(x)\, \theta_1 + N_3(x)\, V_2 + N_4(x)\, \theta_2 \tag{7.98}$$

The shape functions $N_1(x)$ to $N_4(x)$ are given by

$$N_1(x) = 1 - 3(\frac{x}{L})^2 + 2(\frac{x}{L})^3,\ N_2(x) = -L[\frac{x}{L} - 2(\frac{x}{L})^2 + (\frac{x}{L})^3]$$

$$N_3(x) = 3(\frac{x}{L})^2 - 2(\frac{x}{L})^3,\ N_4(x) = L[(\frac{x}{L})^2 - (\frac{x}{L})^3] \tag{7.99}$$

Equation (7.99) gives approximate shape functions, while the exact shape functions are given by equation (7.39).

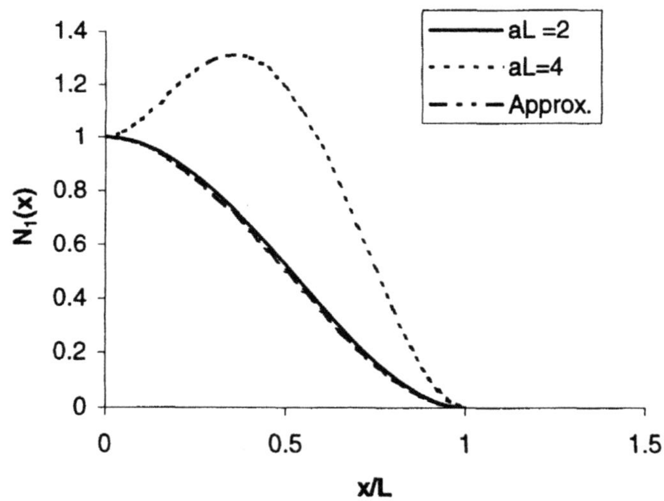

Fig. 7.15 Variation of beam shape function N_1 with αL

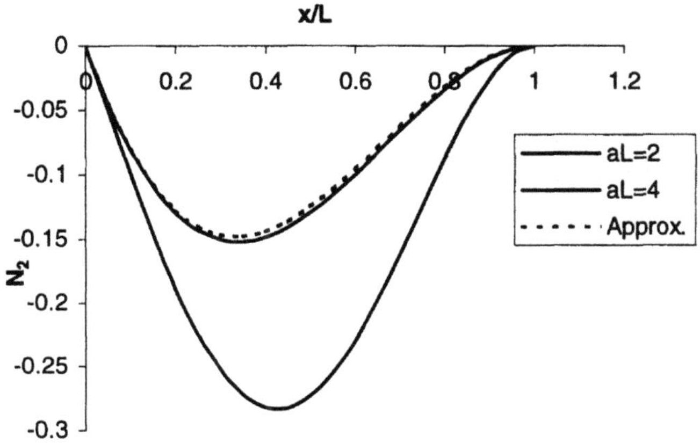

Fig. 7.16 Variation of beam shape function N_2 with αL

Figs 7.15 and 7.16 show a comparison between the exact and approximate shape functions. Once again it is clear that provided the value of $\alpha L < 2.0$, the two sets of shape functions are very close.

Substituting equations (7.98) and (7.99) in equation (7.94) and proceeding as in equation (7.89)

$$\frac{\partial \pi_e}{\partial \phi_i} = \int_0^L \{ \frac{d^2 N_j(x)}{d x^2} EI \sum_{i=1}^{4} \frac{d^2 N_i(x)}{d x^2} \phi_i - m\omega^2 N_j(x) \sum_{i=1}^{4} N_i(x)\phi_i \} dx - Force_j \qquad (7.100)$$

In equation (7.100), for simplicity in mathematical manipulation, a new variable Φ_i has been introduced, where

$\phi_1 = V_1$, $\phi_2 = \theta_1$, $\phi_3 = V_2$, $\phi_4 = \theta_2$

$Force_j = (F_1, j = 1), (M_1, j = 2), (F_2, j = 3)$ and $(M_2, j = 4)$.

Tables 7.3 and 7.4 show the integrals involved in equation (7.100).

Table 7.3 $\int_0^L \frac{d^2 N_i}{dx^2} \frac{d^2 N_j}{dx^2} dx$

i	j			
	1	2	3	4
1	$12/L^3$	$-6/L^2$	$-12/L^3$	$-6/L^2$
2	$-6/L^2$	$4/L$	$6/L^2$	$2/L$
3	$-12/L^3$	$6/L^2$	$12/L^3$	$6/L^2$
4	$-6/L^2$	$2/L$	$6/L^2$	$4/L$

Table 7.4 $\int_0^L N_i \; N_j \; dx$

i	j			
	1	2	3	4
1	156 L/420	-22L²/420	54 L/420	13L²/420
2	-22L²/420	4L³/420	-13L²/420	-3L³/420
3	54 L/420	-13L²/420	156 L/420	22L²/420
4	13L²/420	-3L³/420	22L²/420	4L³/420

Substituting for the integrals in equation (7.100), the static elastic stiffness matrix K and the consistent mass matrix M are given by

$$\begin{bmatrix} \dfrac{\partial \pi_e}{\partial V_1} \\ \dfrac{\partial \pi_e}{\partial \theta_1} \\ \dfrac{\partial \pi_e}{\partial V_2} \\ \dfrac{\partial \pi_e}{\partial \theta_2} \end{bmatrix} = [K - \omega^2 M] \begin{bmatrix} V_1 \\ \theta_1 \\ V_2 \\ \theta_2 \end{bmatrix} - \begin{bmatrix} F_1 \\ M_1 \\ F_2 \\ M_2 \end{bmatrix}$$

$$K = \frac{EI}{L} \begin{bmatrix} \dfrac{12}{L^2} & -\dfrac{6}{L} & -\dfrac{12}{L^2} & -\dfrac{6}{L} \\ -\dfrac{6}{L} & 4 & \dfrac{6}{L} & 2 \\ -\dfrac{12}{L^2} & \dfrac{6}{L} & \dfrac{12}{L^2} & \dfrac{6}{L} \\ -\dfrac{6}{L} & 2 & \dfrac{6}{L} & 4 \end{bmatrix}, \quad M = \frac{mL}{420} \begin{bmatrix} 156 & -22L & 54 & 13L \\ -22L & 4L^2 & -13L & -3L^2 \\ 54 & -13L & 156 & 22L \\ 13L & -3L^2 & 22L & 4L^2 \end{bmatrix} \tag{7.101}$$

7.12.3 Approximate Beam Element Dynamic Stiffness Matrix Including Flexural and Shear Deformations

In Section 7.12.2, only the strain energy due to bending was included. If shear deformation is to be included, then in the strain energy term, an additional term due to shear strain energy has to be included. This is given by $0.5 \int \tau \; \gamma \; dvol$, where τ = shear stress and γ = shear strain. The shear strain is not constant over the cross section.

However, in the interests of simplicity, an average shear stress and strain are considered for the whole cross section, $\tau = Q/A, \gamma = \tau/G$, then the shear strain energy becomes

$$\frac{1}{2}\int \tau\ \gamma\ dvol \ = \frac{1_s}{2}\ GA_s \int_{x=0}^{L} \gamma^2\, dx \tag{7.102}$$

where G = shear modulus and A_s = average cross-sectional area for shear stress calculation.

Proceeding as in Section 7.6, if v be the total deflection of the neutral axis, ψ the slope the deformation due to bending and γ the shear strain, then as shown in Fig. 7.8

$$\frac{\partial v}{\partial x} = \psi + \gamma \tag{7.103}$$

From the moment–curvature relationship

$$M = EI\ \frac{\partial \psi}{\partial x} \tag{7.104}$$

Similarly the shear force-average shear strain relationship is assumed to be

$$Q = GA_s\ \gamma = GA_s\ \{\frac{\partial v}{\partial x} - \psi\} \tag{7.105}$$

Substituting for M and Q, the total potential modified to include the strain energy due to shear is

$$\pi_e = \frac{1}{2} EI \int_{x=0}^{L} \{\frac{\partial \varphi}{\partial x}\}^2 dx + \frac{1}{2} GA_s \int_{x=0}^{L} \gamma^2\ dx - \frac{1}{2}\int_0^L - m\ \frac{\partial^2 v}{\partial t^2}\ v\, dx$$
$$- F_1(t)\, v(0,t) - F_2(0,t)\, v(L,t) - M_1 \frac{\partial v(0,t)}{\partial x} - M_2 \frac{\partial v(L,t)}{\partial x} \tag{7.106}$$

Because of undamped free vibration, only the amplitudes of forces and displacements need to be considered. Therefore equation (7.106) can be written as

$$\pi_e = \frac{1}{2} EI \int_{x=0}^{L} \{\frac{d\psi}{dx}\}^2 dx + \frac{1}{2} GA_s \int_{x=0}^{L} \gamma^2 dx - m\omega^2\ \frac{1}{2}\int_0^L V^2\, dx$$
$$- F_1\ V_1 - F_2 V_2 - M_1\ \theta_1 - M_2\ \theta_2 \tag{7.107}$$

An approximate solution to V(x) is assumed to be as given in equation (7.96). However the important difference is that the slope of the neutral axis is the sum of the slope caused by bending moment and the shear strain. Therefore

$$\frac{dV}{dx} = \psi + \gamma = C_2 + 2C_3\, x + 3C_4\, x^2 \tag{7.108}$$

Because of the distributed inertial forces, the shear force is not constant along the beam. However, in the interests of simplicity, assuming that the shear strain is constant and independent of x, the curvature of the neutral axis is given by

$$\frac{d^2V}{dx^2} = \frac{d\psi}{dx} = 2C_3 + 6C_4 x \tag{7.109}$$

The moment-curvature relationship is

$$M = EI \frac{\partial \psi}{\partial x} = EI\{2C_3 + 6\, C_4\, x\} \tag{7.110}$$

Shear strain can be calculated using the equation (7.26b).

$$\frac{\partial M}{\partial x} + Q = 0, \therefore EI\,(6\,C_4) + Q = 0,\ \gamma = \frac{Q}{G\,A_s} = -\frac{6\,EI}{G\,A_s} C_4 \tag{7.111}$$

For ease of mathematical manipulation, let

$$\beta = \frac{12\,EI}{G\,A_s\,L^2}, \quad \therefore \gamma = -0.5\,\beta\,L^2\,C_4 \tag{7.112}$$

$$\psi = \frac{dV}{dx} - \gamma = C_2 + 2C_3\,x + 3C_4\,x^2 + 0.5\,\beta\,L^2\,C_4 \tag{7.113}$$

The boundary conditions are

$$x = 0,\ V = V_1 \ and\ \psi = -\theta_1 \ ,\ x = L,\ V = V_2 \ and\ \psi = -\theta_2 \tag{7.114}$$

Using equations (7.112) to (7.114), and solving for the constants, C_1 to C_4

$$C_1 = V_1\ ,\ C_2 = \frac{\beta(V_2 - V_1) - (1 + 0.5\beta)\theta_1 L + 0.5\beta)\,\theta_2 L}{L^2}$$

$$C_3 = \frac{3(V_2 - V_1) + (2 + 0.5\beta)\theta_1 L + (1 - 0.5\beta)\,\theta_2 L}{L^2}$$

$$C_4 = \frac{2(V_1 - V_2) - (\theta_1 + \theta_2)L}{(1+\beta)L^3}$$

(7.115)

If shear deformation is ignored, then $\beta = 0$ and the above expressions become the same as the ones derived in Section 7.12.2. Substituting for the constants from equation (7.115)

$$V = C_1 + C_2\,x + C_3\,x^2 + C_4\,x^3 = N_1(x)V_1 + N_2(x)\theta_1 + N_3(x)V_2 + N_4(x)\theta_2 \tag{7.116}$$

The shape functions for displacement $N_1(x)$ to $N_4(x)$ are given by

$$N_1(x) = \frac{1}{(1+\beta)}\{1 + \beta - \beta\frac{x}{L} - 3(\frac{x}{L})^2 + 2(\frac{x}{L})^3\}$$

$$N_2(x) = -\frac{L}{(1+\beta)}[(1 + 0.5\beta)\frac{x}{L} - (2 + 0.5\beta)\,(\frac{x}{L})^2 + (\frac{x}{L})^3]$$

$$N_3(x) = \frac{1}{(1+\beta)}[\beta\frac{x}{L} + 3(\frac{x}{L})^2 - 2(\frac{x}{L})^3]$$

$$N_4(x) = \frac{L}{(1+\beta)}[0.5\,\beta\frac{x}{L} + (1 - 0.5\beta)(\frac{x}{L})^2 - (\frac{x}{L})^3\,]$$

(7.117)

Similarly, the shape functions for the slope caused by bending moment are

$$\psi = C_2 + 2C_3\,x + 3C_4\,x^2 + 0.5\,\beta\,L^2\,C_4 = N_{1\psi}(x)V_1 + N_{2\psi}(x)\,\theta_1 + N_{3\psi}(x)V_2 + N_{4\psi}(x)\theta_2$$

(7.118)

$$N_{1\psi}(x) = \frac{1}{(1+\beta)\,L}\,\{\,6(\frac{x}{L})^2 - 6\frac{x}{L} - \beta\}$$

$$N_{2\psi}(x) = -\frac{1}{(1+\beta)}\,[(4+\beta)\frac{x}{L} - 3\,(\frac{x}{L})^2 - (1+\beta)]$$

$$N_{3\psi}(x) = \frac{1}{(1+\beta)L}\,[6\frac{x}{L} - 6(\frac{x}{L})^2]$$

$$N_{4\psi}(x) = \frac{1}{(1+\beta)}[(2-\beta)(\frac{x}{L}) - 3(\frac{x}{L})^2]$$

(7.119)

Finally, the shape functions for the shear strain are given by

$$\gamma = -0.5\,\beta\,L^2\,C_4 = N_{1\gamma}(x)\,V_1 + N_{2\gamma}(x)\,\theta_1 + N_{3\gamma}(x)\,V_2 + N_{4\gamma}(x)\,\theta_2$$

(7.120)

$$N_{1\gamma}(x) = -N_{3\gamma}(x) = -\frac{\beta}{(1+\beta)}\frac{1}{L}\,,\quad N_{2\gamma}(x) = N_{4\gamma}(x) = \frac{0.5\,\beta}{(1+\beta)}$$

(7.121)

Substituting for the variables V(x), $\psi(x)$ and γ in terms of the shape functions in equation (7.106) and minimising π_e with respect to the 'displacements' at the ends of the element as in equation (7.100)

$$\frac{\partial\pi_e}{\partial\phi_i} = EI\int_0^L\frac{d\,N_{j\psi}(x)}{d\,x}\sum_{i=11}^{4}\frac{d\,N_{i\psi}(x)}{d\,x}\phi_i + GA_s\int_0^L N_{j\gamma}(x)\sum_{i=11}^{4}N_{i\gamma}(x)\phi_i$$

$$- m\omega^2\int_0^L N_j(x)\sum_{i=1}^{4}N_i(x)\,\phi_i\}\,dx - Force_j \qquad (7.122)$$

From equation (7.122), $(i,j)^{th}$ term of the static stiffness matrix K is given by

$$k_{ij} = EI\int_0^L\frac{dN_{j\psi}(x)}{dx}\frac{dN_{i\psi}(x)}{dx} + GA_s\int_0^L N_{j\gamma}(x)\int_0^L N_{j\gamma}(x)\,dx$$

$$= EI[\int_0^L\frac{dN_{j\psi}(x)}{dx}\frac{dN_{i\psi}(x)}{dx} + \frac{12}{\beta L^2}\int_0^L N_{j\gamma}(x)\int_0^L N_{j\gamma}(x)\,dx\;],\beta = \frac{12\,EI}{G\,A_s\,L^2}$$

(7.123)

Similarly, (i,j)[th] term of the consistent mass matrix is given by

$$m_{ij} = m \int_0^L N_j(x)\, N_i(x) \; dx \tag{7.124}$$

Tables 7.5 – 7.7 show the integrals $\int_0^L \frac{dN_{i\psi}}{dx}\frac{dN_{j\psi}}{dx}dx$, $\int_0^L N_{i\gamma}\, N_{j\gamma}\, dx$ and $\int_0^L N_i\, N_j\, dx$.

Substituting for the integrals, the static elastic stiffness matrix K and the mass matrix M are given by equation (7.126) and Table 7.8, respectively. It can be verified that if shear deformation is ignored, then β becomes zero and the static stiffness and mass matrices revert to the values shown in equation (7.101).

$$\begin{bmatrix} \frac{\partial \pi_e}{\partial V_1} \\ \frac{\partial \pi_e}{\partial \theta_1} \\ \frac{\partial \pi_e}{\partial V_2} \\ \frac{\partial \pi_e}{\partial \theta_2} \end{bmatrix} = [K \; - \; \omega^2 M] \begin{bmatrix} V_1 \\ \theta_1 \\ V_2 \\ \theta_2 \end{bmatrix} - \begin{bmatrix} F_1 \\ M_1 \\ F_2 \\ M_2 \end{bmatrix} \tag{7.125}$$

$$K = \frac{EI}{(1+\beta)\,L} \begin{bmatrix} \frac{12}{L^2} & -\frac{6}{L} & -\frac{12}{L^2} & -\frac{6}{L} \\ -\frac{6}{L} & 4+\beta & \frac{6}{L} & 2-\beta \\ -\frac{12}{L^2} & \frac{6}{L} & \frac{12}{L^2} & \frac{6}{L} \\ -\frac{6}{L} & 2-\beta & \frac{6}{L} & 4+\beta \end{bmatrix} \tag{7.126}$$

Table 7.5 $\int_0^L \frac{dN_{i\psi}}{dx}\frac{dN_{j\psi}}{dx}\,dx$

i	j			
	1	2	3	4
1	$\frac{12}{(1+\beta)^2 L^3}$	$-\frac{6}{(1+\beta)^2 L^2}$	$-\frac{12}{(1+\beta)^2 L^3}$	$-\frac{6}{(1+\beta)^2 L^2}$
2	$-\frac{6}{(1+\beta)^2 L^2}$	$\frac{(4+2\beta+\beta^2)}{(1+\beta)^2 L}$	$\frac{6}{(1+\beta)^2 L^2}$	$\frac{(2-2\beta-\beta^2)}{(1+\beta)^2 L}$
3	$-\frac{12}{(1+\beta)^2 L^3}$	$\frac{6}{(1+\beta)^2 L^2}$	$\frac{12}{(1+\beta)^2 L^3}$	$\frac{6}{(1+\beta)^2 L^2}$
4	$-\frac{6}{(1+\beta)^2 L^2}$	$\frac{(2-2\beta-\beta^2)}{(1+\beta)^2 L}$	$\frac{6}{(1+\beta)^2 L^2}$	$\frac{(4+2\beta+\beta^2)}{(1+\beta)^2 L}$

Table 7.6 $\int_0^L N_{i\gamma}N_{j\gamma}\,dx$

i	j			
	1	2	3	4
1	$\frac{\beta^2}{(1+\beta)^2 L}$	$-0.5\frac{\beta^2}{(1+\beta)^2}$	$-\frac{\beta^2}{(1+\beta)^2 L}$	$-0.5\frac{\beta^2}{(1+\beta)^2}$
2	$-0.5\frac{\beta^2}{(1+\beta)^2}$	$0.25\frac{\beta^2}{(1+\beta)^2}L$	$0.5\frac{\beta^2}{(1+\beta)^2}$	$0.25\frac{\beta^2}{(1+\beta)^2}L$
3	$-\frac{\beta^2}{(1+\beta)^2 L}$	$0.5\frac{\beta^2}{(1+\beta)^2}$	$\frac{\beta^2}{(1+\beta)^2 L}$	$0.5\frac{\beta^2}{(1+\beta)^2}$
4	$-0.5\frac{\beta^2}{(1+\beta)^2}$	$0.25\frac{\beta^2}{(1+\beta)^2}L$	$0.5\frac{\beta^2}{(1+\beta)^2}$	$0.25\frac{\beta^2}{(1+\beta)^2}L$

Table 7.7 $\int_0^L N_i \, N_j \, dx$

i	j = 1	j = 2	j = 3	j = 4
1	$\frac{L}{(1+\beta)^2}\{\frac{\beta^2}{3}+\frac{7\beta}{10}+\frac{13}{35}\}$	$-\frac{L^2}{(1+\beta)^2}\{\frac{\beta^2}{24}+\frac{11\beta}{120}+\frac{11}{210}\}$	$\frac{L}{(1+\beta)^2}\{\frac{\beta^2}{6}+\frac{3\beta}{10}+\frac{9}{70}\}$	$\frac{L^2}{(1+\beta)^2}\{\frac{\beta^2}{24}+\frac{3\beta}{40}+\frac{13}{420}\}$
2	$-\frac{L^2}{(1+\beta)^2}\{\frac{\beta^2}{24}+\frac{11\beta}{120}+\frac{11}{210}\}$	$\frac{L^3}{(1+\beta)^2}\{\frac{\beta^2}{120}+\frac{\beta}{60}+\frac{1}{105}\}$	$-\frac{L^2}{(1+\beta)^2}\{\frac{\beta^2}{24}+\frac{3\beta}{40}+\frac{13}{420}\}$	$-\frac{L^2}{(1+\beta)^2}\{\frac{\beta^2}{120}+\frac{\beta}{60}+\frac{1}{140}\}$
3	$\frac{L}{(1+\beta)^2}\{\frac{\beta^2}{6}+\frac{3\beta}{10}+\frac{9}{70}\}$	$-\frac{L^2}{(1+\beta)^2}\{\frac{\beta^2}{24}+\frac{3\beta}{40}+\frac{13}{420}\}$	$\frac{L}{(1+\beta)^2}\{\frac{\beta^2}{3}+\frac{7\beta}{10}+\frac{13}{35}\}$	$\frac{L^2}{(1+\beta)^2}\{\frac{\beta^2}{24}+\frac{11\beta}{120}+\frac{11}{210}\}$
4	$\frac{L^2}{(1+\beta)^2}\{\frac{\beta^2}{24}+\frac{3\beta}{40}+\frac{13}{420}\}$	$-\frac{L^2}{(1+\beta)^2}\{\frac{\beta^2}{24}+\frac{3\beta}{40}+\frac{13}{420}\}$	$\frac{L^2}{(1+\beta)^2}\{\frac{\beta^2}{24}+\frac{11\beta}{120}+\frac{11}{210}\}$	$\frac{L^3}{(1+\beta)^2}\{\frac{\beta^2}{120}+\frac{\beta}{60}+\frac{1}{105}\}$

Table 7.8 Mass matrix

$$M = \frac{mL}{(1+\beta)^2\,420}\begin{bmatrix} 140\beta^2+294\beta+156 & -(17.5\beta^2+38.5\beta+22)L & 70\beta^2+126\beta+54 & (17.5\beta^2+31.5\beta+13)L \\ -(17.5\beta^2+38.5\beta+22)L & (3.5\beta^2+7\beta+4)L^2 & -(17.5\beta^2+31.5\beta+13)L & -(3.5\beta^3+7\beta+3)L^2 \\ 70\beta^2+126\beta+54 & -(17.5\beta^2+31.5\beta+13)L & 140\beta^2+294\beta+156 & (17.5\beta^2+38.5\beta+22)L \\ (17.5\beta^2+31.5\beta+13)L & -(3.5\beta^3+7\beta+3)L^2 & (17.5\beta^2+38.5\beta+22)L & (3.5\beta^2+7\beta+4)L^2 \end{bmatrix}$$

7.12.4 Approximate Beam-column Element Dynamic Stiffness Matrix

When an axial force acts at the ends of the element, external work is done by the axial load and therefore there is a corresponding loss of potential. Ignoring the axial compressive strain due the axial load, although the neutral axis remains unaltered in length, but because of the fact that the beam bends, the bowing effect reduces distance between the ends of the element. The shortening Δ_p due to bowing is the difference between the straight length and the curved length.

$$\Delta_P = \int dx - \int ds, \quad ds^2 = dx^2 + dv^2, \quad ds = dx\sqrt{\{1 + (\frac{dv}{dx})^2\}} \tag{7.127}$$

Since the $\frac{dv}{dx}$ is small, $\sqrt{\{1 + (\frac{dv}{dx})^2\}} \approx 1 + \frac{1}{2}(\frac{dv}{dx})^2$. Therefore

$$\Delta_P = \int dx - \int \{1 + \frac{1}{2}(\frac{dv}{dx})^2\}dx = \frac{1}{2}\int_0^L\{\frac{dv}{dx}\}^2\, dx \tag{7.128}$$

The total potential modified to include the loss of potential by the axial load P is given by

$$\pi_e = \frac{1}{2}EI\int_{x=0}^{L}\{\frac{\partial\psi}{\partial x}\}^2 dx + \frac{1}{2}GA_s\int_{x=0}^{L}\gamma^2 dx - \frac{1}{2}\int_0^L -m\frac{\partial^2 v}{\partial t^2}v\, dx$$
$$-P\frac{1}{2}\int_0^L(\frac{\partial v}{\partial x})^2 dx - F_1(t)v(0,t) - F_2(0,t)v(L,t) - M_1\frac{\partial v(0,t)}{\partial x} - M_2\frac{\partial v(L,t)}{\partial x} \tag{7.129}$$

Because of undamped free vibration, only the amplitudes of forces and displacements need to be considered, and the total potential can be expressed as

$$\pi_e = \frac{1}{2}EI\int_{x=0}^{L}\{\frac{d\psi}{dx}\}^2 dx + \frac{1}{2}GA_s\int_{x=0}^{L}\gamma^2 dx - m\omega^2\frac{1}{2}\int_0^L V^2\, dx - P\frac{1}{2}\int_0^L(\frac{dV}{dx})^2 dx$$
$$-F_1 V_1 - F_2 V_2 - M_1\theta_1 - M_2\theta_2 \tag{7.130}$$

Proceeding as in equation (7.122)

$$\frac{\partial\pi_e}{\partial\phi_i} = EI\int_0^L\frac{dN_{j\psi}(x)}{dx}\sum_{i=11}^{4}\frac{dN_{i\psi}(x)}{dx}\phi_i\, dx + GA_s\int_0^L N_{j\gamma}(x)\sum_{i=11}^{4}N_{i\gamma}(x)\phi_i\, dx$$
$$-m\omega^2\int_0^L N_j(x)\sum_{i=1}^{4}N_i(x)\,\phi_i\}\, dx - P\int_0^L\frac{dN_j(x)}{dx}\sum_{i=1}^{4}\frac{dN_i(x)}{dx}\phi_i\, dx - Force_j \tag{7.131}$$

The new integral $\int_0^L \frac{dN_i}{dx}\frac{dN_j}{dx}dx$ is shown in Table 7.9.

Table 7.9 $\int_0^L \frac{dN_i}{dx}\frac{dN_j}{dx}dx$

i	j			
	1	2	3	4
1	C_1	$-C_4$	$-C_1$	$-C_4$
2	$-C_4$	C_2	C_4	$-C_3$
3	$-C_1$	C_4	C_1	C_4
4	$-C_4$	$-C_3$	C_4	C_2

$$C_1 = \frac{\beta^2 + 2\beta + 1.2}{(1+\beta)^2\ L},\ C_2 = \frac{\frac{\beta^2}{12} + \frac{\beta}{6} + \frac{2}{15}}{(1+\beta)^2} L,\ C_3 = \frac{\frac{\beta^2}{12} + \frac{\beta}{6} + \frac{1}{30}}{(1+\beta)^2} L\ ,\ C_4 = \frac{0.10}{(1+\beta)^2}$$

The matrix associated with P is called geometric stiffness matrix. The (i,j)[th] term g_{ij} of the geometric matrix G is given by

$$g_{ij} = P\int_0^L \frac{d\,N_j(x)}{d\,x}\frac{d\,N_i(x)}{d\,x}dx$$

The full stiffness relationship is given by

$$\begin{bmatrix} \frac{\partial \pi_e}{\partial V_1} \\ \frac{\partial \pi_e}{\partial \theta_1} \\ \frac{\partial \pi_e}{\partial V_2} \\ \frac{\partial \pi_e}{\partial \theta_2} \end{bmatrix} = [K\ -\ P\ G -\ \omega^2\ M\ ] \begin{bmatrix} V_1 \\ \theta_1 \\ V_2 \\ \theta_2 \end{bmatrix} - \begin{bmatrix} F_1 \\ M_1 \\ F_2 \\ M_2 \end{bmatrix} \tag{7.132}$$

7.12.5 Approximate Dynamic Stiffness Matrix of a Shaft Element

As indicated in Section 7.8, the differential equation governing the dynamic equilibrium equation for the bar element and shaft element are similar. Therefore for the shaft element shown in Fig. 7.11

$$\pi_e = \frac{1}{2} GJ \int_{x=0}^{L} \{\frac{\partial\phi(x,t)}{\partial x}\}^2 dx - \frac{m}{A} I_{Polar} \frac{1}{2}\int_{0}^{L} \frac{\partial^2\phi(x,t)}{\partial t^2}\phi \, dx - T_1(t)\phi(0,t) - T_2(0,t)\phi(L,t) \tag{7.133}$$

In the case of undamped free vibration, $\varphi(x, t) = \Phi(x) \cos\omega t$. Therefore considering only the amplitudes of forces and displacements

$$\pi_e = \frac{1}{2} GJ \int_{x=0}^{L} \{\frac{d\Phi}{dx}\}^2 dx - \frac{m}{A} I_{Polar} \omega^2 \frac{1}{2}\int_{0}^{L} \Phi^2 \, dx - T_1 \Phi_1 - T_2 \Phi_2 \tag{7.134}$$

Proceeding as in Section 7.12.1, in the case of the bar element, an approximate solution to $\Phi(x)$ is given by

$$\Phi(x) = N_1(x)\, \Phi_1 + N_2(x)\, \Phi_2,\ N_1(x) = 1 - x/L,\ N_2(x) = x/L \tag{7.135}$$

$$\begin{bmatrix} T_1 \\ T_2 \end{bmatrix} = [K - \omega^2 M]\begin{bmatrix} \Phi_1 \\ \Phi_2 \end{bmatrix} \tag{7.136}$$

The static elastic stiffness matrix K and mass matrix M are given by

$$K = \frac{GJ}{L}\begin{bmatrix} 1 & -1 \\ -1 & 1 \end{bmatrix}, \quad M = \frac{m}{A} I_{Polar} \frac{L}{6}\begin{bmatrix} 2 & 1 \\ 1 & 2 \end{bmatrix} \tag{7.137}$$

7.13 GALERKIN'S WEIGHTED RESIDUAL METHOD

Galerkin's weighted residual method is a method for solving approximately differential equations, both ordinary and partial. If an exact solution is substituted into the differential equation, then the error or 'residual' will be zero. However if an approximate solution is substituted, then the differential equation will not be satisfied over the region of interest and an error (residual) exists. In Galerkin's method, the approximate solution is written in terms of a set of 'trial' or shape functions and the aim is to make the weighted value of the residual zero. Normally the shape functions themselves are used as the weighting functions. The method is best illustrated by few simple examples.

7.13.1 Undamped Free Longitudinal Vibrations

The differential equation in terms of the amplitude of vibration U(x) and force F(x) is given by equation (7.3).

$$\frac{dF}{dx} + m\omega^2 U = 0 \tag{7.138}$$

Let an approximate solution to U(x) be as in equation (7.78). Because of the fact that the solution is approximate, when substituted into the differential equation (7.138), the approximate solution will lead to errors. In Galerkin's method, the weighted average error over the region of interest is made equal to zero. The weighting functions are simply chosen to be the shape functions themselves. In other words, the method requires that

$$\int_0^L \{\frac{dF}{dx} + m\omega^2 U\} N_j \, dx = 0 \, , \, j = 1{,}2 \tag{7.139}$$

One can see that because of the fact that the shape functions are linear functions of x, their second derivative needed in calculating the first derivative of force F will be zero. However, using integration by parts, evaluation of the second derivative can be avoided as follows

$$\int_0^L \frac{dF}{dx} N_j \, dx = \left[N_j \, F \right]_{x=0}^{L} - \int_0^L \frac{d \, N_j(x)}{d \, x} F \, dx \tag{7.140}$$

Note that

$$j = 1, N_1(x) = 1 - x/L, \therefore \; N_j \, F \Big|_{x=0}^{L} = - F = F_1$$

$$j = 2, \; N_2(x) = \; x/L, \therefore \; \; N_j \, F \Big|_{x=0}^{L} = F = F_2 \tag{7.141}$$

Substituting for F in terms of shape functions

$$F = AE \, \frac{dU}{dx} = AE \sum_{i=1}^{2} \frac{dN_i}{dx} U_i \tag{7.142}$$

Substituting equations (7.140) to (7.142) in equation (7.139)

$$F_j - AE \int_{x=0}^{L} [\frac{dN_1}{dx} U_1 + \frac{dN_2}{dx} U_2] \frac{dN_j}{dx} dx + m\omega^2 \int_{x=0}^{L} [N_1 U_1 + N_2 U_2] N_j \, dx = 0 \tag{7.143}$$

It is perhaps worth pointing out, that the process of integration by parts not only reduces the order of differentiation but even more important brings in to play the forces at the boundary of the region of interest. Equation (7.143) can be expressed in matrix form as

$$\begin{bmatrix} F_1 \\ F_2 \end{bmatrix} = AE \begin{bmatrix} \int_0^L \frac{dN_1}{dx}\frac{dN_1}{dx}dx & \int_0^L \frac{dN_1}{dx}\frac{dN_2}{dx}dx \\ \int_0^L \frac{dN_1}{dx}\frac{dN_2}{dx}dx & \int_0^L \frac{dN_2}{dx}\frac{dN_2}{dx}dx \end{bmatrix} \begin{bmatrix} U_1 \\ U_2 \end{bmatrix} - m\,\omega^2 \begin{bmatrix} \int_0^L N_1\,N_1\,dx & \int_0^L N_1\,N_2\,dx \\ \int_0^L N_2\,N_1\,dx & \int_0^L N_2\,N_2\,dx \end{bmatrix} \begin{bmatrix} U_1 \\ U_2 \end{bmatrix} \tag{7.144}$$

By comparing the equations (7.90) and (7.144), it can be seen that Galerkin's weighted residual method results in exactly the same static stiffness and consistent mass matrices as derived using the minimum potential approach.

7.13.2 Undamped Free Flexural Vibrations Including Bending Deformations Only

The differential equation of equilibrium in terms of the amplitude of vibration V(x) and M is given by equation (7.26c) as

$$\frac{d^2M}{dx^2} + m\omega^2\, V = 0 \tag{7.145}$$

Let an approximate solution to V(x) be as in equation (7.99). Galerkin's method requires that

$$\int_0^L \{\frac{d^2M}{dx^2} + m\omega^2 V\} N_j\, dx \;= 0\,,\; j = 1\; to\, 4 \tag{7.146}$$

Using integration by parts

$$\int_0^L \frac{d^2M}{dx^2} N_j\; dx = \left[N_j \frac{dM}{dx} \right]_{x=0}^{L} - \int_0^L \frac{d\,N_i(x)}{d\,x}\frac{dM}{dx}dx \tag{7.147}$$

Integrating by parts, the right-hand integral in equation (7.147)

$$\int_0^L \frac{dM}{dx}\frac{dN_j}{dx}\, dx = \left[\frac{dN_j}{dx} M \right]_{x=0}^{L} - \int_0^L \frac{d^2N_i(x)}{dx^2} M\, dx \tag{7.148}$$

Substituting equation (7.148) in equation (7.147)

$$\int_0^L \frac{d^2M}{dx^2} N_j\; dx = \left[N_j \frac{dM}{dx} \right]_{x=0}^{L} - \left[M \frac{dN_j}{dx} \right]_{x=0}^{L} + \int_0^L \frac{d^2N_i}{dx^2} M\, dx \tag{7.149}$$

Using equation (7.26b), the derivative of moment can be replace by -Q. Therefore

$$\int_0^L \frac{d^2M}{dx^2} N_j \, dx = -\left[N_j Q\right]_{x=0}^{L} - \left[M \frac{dN_j}{dx} \right]_{x=0}^{L} + \int_0^L \frac{d^2N_i}{dx^2} M \, dx \tag{7.150}$$

Table 7.10 shows the value of the shape functions and their first derivatives at x = 0 and x = L.

Table 7.10

	x = 0		x = L	
j	N_j	$\frac{dN_j}{dx}$	N_j	$\frac{dN_j}{dx}$
1	1	0	0	0
2	0	-1	0	0
3	0	0	1	0
4	0	0	0	-1

Using the data in Table 7.10, and noting that sagging bending moment and 'clockwise' shear are positive, from Fig. 7.5

$$\therefore \left[N_j(x) \; Q\right]_{x=0}^{L} + \left[\frac{dN_j(x)}{dx} M\right]_{x=0}^{L}$$

$$\begin{aligned} &= -Q_{x=0} = F_1, \; for \; j = 1 \\ &= M_{x=0} = M_1, \; for \; j = 2 \\ &= Q_{x=L} = F_2, \; for \; j = 3 \\ &= -M_{x=L} = M_2, \; for \; j = 4 \end{aligned} \tag{7.151}$$

Using equations (7.150) and (7.151) and substituting for M in terms of curvature from equation (7.27), equation (7.146) can be expressed as

$$\therefore F_1 = EI \int_0^L \frac{d^2 N_1(x)}{dx^2} \sum_{i=1}^{4} \frac{d^2 N_i(x)}{dx^2} \phi_i \; dx \; - m\omega^2 \int_{x=0}^{L} N_1 \sum_{i=1}^{4} N_i \phi_i \, dx$$

$$\therefore M_1 = EI \int_0^L \frac{d^2 N_2(x)}{dx^2} \sum_{i=1}^{4} \frac{d^2 N_i(x)}{dx^2} \phi_i \; dx \; - m\omega^2 \int_0^L N_2 \sum_{i=1}^{4} N_i \phi_i \, dx$$

$$\therefore F_2 = EI \int_0^L \frac{d^2 N_3(x)}{dx^2} \sum_{i=1}^{4} \frac{d^2 N_i(x)}{dx^2} \phi_i \; dx \; - m\omega^2 \int_0^L N_3 \sum_{i=1}^{4} N_i \; \phi_i \; dx$$

$$\therefore M_2 = EI \int_0^L \frac{d^2 N_4(x)}{dx^2} \sum_{i=1}^{4} \frac{d^2 N_i(x)}{dx^2} \phi_i \; dx \; - m\omega^2 \int_0^L N_4 \sum_{i=1}^{4} N_i \; \phi_i \; dx$$

(7.152)

The above four equations can be expressed in matrix form as shown in Table 7.11.

From the above equations, it is clear that the static stiffness matrix coefficient k_{ij} and the consistent mass matrix coefficient m_{ij} are given by

$$k_{ij} = EI\int_0^L \frac{d^2 N_j(x)}{d x^2}\frac{d^2 N_i(x)}{d x^2}\, dx \quad , m_{ij} = m\omega^2 \int_{x=0}^{L} N_j\, N_i\, dx$$

Above equations are identical to the equations derived in Section 7.12.2. Thus both Galerkin's weighted residual method and the method of minimum total potential lead to identical static stiffness and mass matrices.

Table 7.11

$$\begin{bmatrix} F_1 \\ M_1 \\ F_2 \\ M_2 \end{bmatrix} = EI\begin{bmatrix} \int_0^L \frac{d^2N_1}{dx^2}\frac{d^2N_1}{dx^2}dx & \int_0^L \frac{d^2N_1}{dx^2}\frac{d^2N_2}{dx^2}dx & \int_0^L \frac{d^2N_1}{dx^2}\frac{d^2N_3}{dx^2}dx & \int_0^L \frac{d^2N_1}{dx^2}\frac{d^2N_4}{dx^2}dx \\ & \int_0^L \frac{d^2N_2}{dx^2}\frac{d^2N_2}{dx^2}dx & \int_0^L \frac{d^2N_2}{dx^2}\frac{d^2N_3}{dx^2}dx & \int_0^L \frac{d^2N_2}{dx^2}\frac{d^2N_4}{dx^2}dx \\ & & \int_0^L \frac{d^2N_3}{dx^2}\frac{d^2N_3}{dx^2}dx & \int_0^L \frac{d^2N_3}{dx^2}\frac{d^2N_4}{dx^2}dx \\ \text{symmertic} & & & \int_0^L \frac{d^2N_4}{dx^2}\frac{d^2N_4}{dx^2}dx \end{bmatrix}$$

$$-m\omega^2\begin{bmatrix} \int_0^L N_1 N_1\, dx & \int_0^L N_1 N_2\, dx & \int_0^L N_1 N_3\, dx & \int_0^L N_1 N_4\, dx \\ & \int_0^L N_2 N_2\, dx & \int_0^L N_2 N_3\, dx & \int_0^L N_2 N_4\, dx \\ & & \int_0^L N_3 N_3\, dx & \int_0^L N_3 N_4\, dx \\ \text{symmetric} & & & \int_0^L N_4 N_4\, dx \end{bmatrix}\begin{bmatrix} V_1 \\ \theta_1 \\ V_2 \\ \theta_2 \end{bmatrix}$$

CHAPTER 8

FINITE ELEMENTS

8.1 INTRODUCTION

In Chapter 7, methods were discussed for deriving the static stiffness and consistent mass matrices for line elements such as beam, bar and shaft elements. Two fundamental assumptions made in the derivation of these matrices were

- what stresses are significant in any particular problem
- how the displacement in the member is related through shape functions to the displacements at the ends of the member.

A skeletal structure is assembled from clearly identifiable members (elements) connected at joints (nodes). A continuum such as a plate or a floor slab has no such identifiable members (elements) connected at joints (nodes). However, in order to extend the ideas developed in Chapter 7 to continuum structures, it is necessary to imagine that the structure is in fact assembled from a set of imaginary elements called finite elements connected only at imaginary joints called nodes. This process is known as spatial discretisation. Because of the fact that finite elements are imaginary, a structure could be discretised in various ways, using triangular, rectangular, quadrilateral finite elements with straight or curved boundaries. Some of the common shapes of elements used are shown in Fig. 8.1.

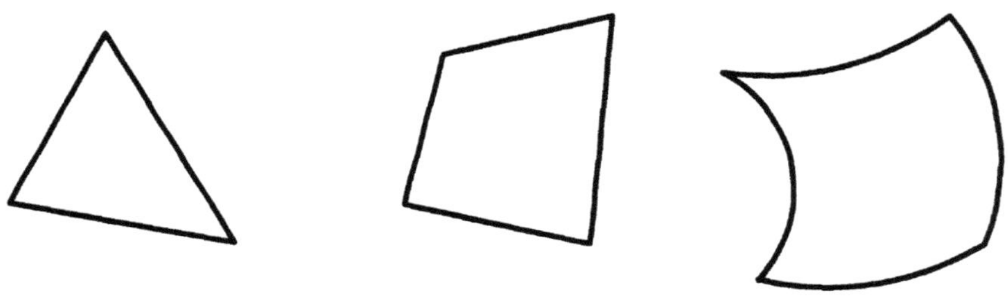

Fig. 8.1 Various shapes of finite elements

8.2 SUMMARY OF RESULTS FROM THE THEORY OF ELASTICITY

In order to develop the basic concepts in the theory of finite elements, consider a two-dimensional structure such as a wall subjected to forces acting in the plane of the structure. Let the structure lie in the vertical x-y plane. If the thickness is small and any out of plane deformation is ignored, then any point in the structure displaces by u in the x direction and by v, in the y direction. The displacements u and v are functions of the position of the point. The significant stresses are in-plane normal stresses σ_x, σ_y and shear stress τ_{xy}, as shown in Fig. 8.2.

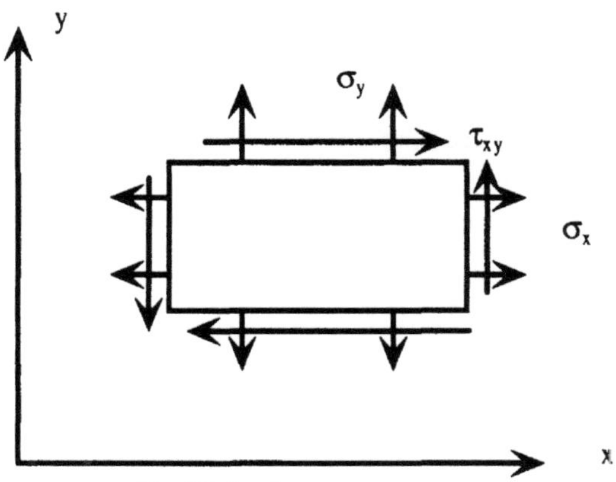

Fig. 8.2 In-plane stresses

The following results from the Theory of Elasticity will be needed for developing the properties of the finite element.

8.2.1 Equilibrium

It is well known from the Theory of Elasticity, that the two equations of equilibrium in the x and y directions are

$$\frac{\partial \sigma_x}{\partial x} + \frac{\partial \tau_{xy}}{\partial y} - m\frac{\partial^2 u}{\partial t^2} = 0\,, \quad \frac{\partial \sigma_y}{\partial y} + \frac{\partial \tau_{xy}}{\partial x} - m\frac{\partial^2 v}{\partial t^2} = 0 \tag{8.1}$$

In the above equations, m is the mass per unit volume and the partial differential of displacements u and v with respect to time t, correspond to accelerations.

8.2.2 Strain-Stress Relationship

A general orthotropic material has different material properties in the x and –y directions. Four material constants are required for specifying the strain-stress relationship. They are: E_x and E_y, Young's moduli in x and y directions, either the Poisson's ratio ν_x in the y direction or the Poisson's ratio ν_y in the x direction and G the shear modulus. The strain-stress relationships are given by

$$\varepsilon_x = \frac{\sigma_x}{E_x} - \upsilon_x \frac{\sigma_y}{E_y}\,, \ \varepsilon_y = \frac{\sigma_y}{E_y} - \upsilon_y \frac{\sigma_x}{E_x}\,, \ \gamma_{xy} = \frac{\tau_{xy}}{G}\,, \upsilon_x E_x = \upsilon_y E_y \tag{8.2}$$

If the material is isotropic, then $E_x = E_y = E$, Young's modulus, $\nu_x = \nu_y = \nu$, Poisson's ratio and $G = E/[2(1 + \nu)]$.

8.2.3 Stress-Strain Relationship

From the strain-stress relationships, the stress-strain relationships are given by

$$\sigma_x = \frac{E_x}{(1-\upsilon_x \upsilon_y)}\{\varepsilon_x + \upsilon_x \varepsilon_y\}\,,\ \sigma_y = \frac{E_y}{(1-\upsilon_x \upsilon_y)}\{\varepsilon_y + \upsilon_y \varepsilon_x\}\,,\ \tau_{xy} = G\gamma_{xy} \tag{8.3}$$

The above relationship in matrix form is given by

$$\sigma = D\,\varepsilon \tag{8.4}$$

$$\sigma = \begin{bmatrix}\sigma_x\\ \sigma_y\\ \tau_{xy}\end{bmatrix},\ D = \begin{bmatrix}\frac{E_x}{(1-\upsilon_x\upsilon_y)} & \frac{\upsilon_x E_x}{(1-\upsilon_x\upsilon_y)} & 0\\ \frac{\upsilon_y E_y}{(1-\upsilon_x\upsilon_y)} & \frac{E_y}{(1-\upsilon_x\upsilon_y)} & 0\\ 0 & 0 & G\end{bmatrix},\ \varepsilon = \begin{bmatrix}\varepsilon_x\\ \varepsilon_y\\ \gamma_{xy}\end{bmatrix} \tag{8.5}$$

σ and ε are known as stress and strain vectors respectively and D as elasticity matrix.

8.2.4 Strain energy

The expression for strain energy is given by

$$Strain\ energy = \frac{1}{2}\int_{x=-a}^{a}\int_{y=-b}^{b}\{\sigma_x \varepsilon_x + \sigma_y \varepsilon_y + \tau_{xy}\gamma_{xy}\}\,dy\,dx\,t \tag{8.6}$$

where t = thickness of the element.

Using the matrix notation of equation (8.5), equation (8.6) becomes

$$Strain\ energy = \frac{1}{2}\int_{x=-a}^{a}\int_{y=-b}^{b}\{\varepsilon^T \sigma\}\,dy\,dx\,t \tag{8.7}$$

8.2.5 Strain-Displacement Relationship

The strain-displacement relationships are given by

$$\varepsilon_x = \frac{\partial u}{\partial x}\,,\ \varepsilon_y = \frac{\partial v}{\partial y},\ \gamma_{xy} = \varepsilon_x = \frac{\partial u}{\partial y} + \frac{\partial v}{\partial x} \tag{8.8}$$

8.2.6 Equilibrium Equations in terms of Displacements

Substituting for strains in terms derivatives of displacements from equation (8.3), the equations of equilibrium given by equation (8.1) become

$$\frac{E_x}{(1-\upsilon_x\upsilon_y)}[\frac{\partial^2 u}{\partial x^2} + \upsilon_x \frac{\partial^2 v}{\partial x \partial y}] + G\{\frac{\partial^2 u}{\partial y^2} + \frac{\partial^2 v}{\partial x \partial y}\} - m\frac{\partial^2 u}{\partial t^2} = 0$$

$$\frac{E_y}{(1-\upsilon_x\upsilon_y)}[\frac{\partial^2 v}{\partial y^2}+\upsilon_y\frac{\partial^2 u}{\partial x\partial y}]+G\{\frac{\partial^2 v}{\partial x^2}+\frac{\partial^2 u}{\partial x\partial y}\}-m\frac{\partial^2 v}{\partial t^2}=0 \tag{8.9}$$

If damping is ignored, then for undamped, free vibration

$$u(x,y,t) = U(x,y)\cos\omega t\,,\quad v(x,y,t) = V(x,y)\cos\omega t$$

Equation (8.9), can be expressed in terms of displacement amplitudes as

$$\frac{E_x}{(1-\upsilon_x\upsilon_y)}[\frac{\partial^2 U}{\partial x^2}+\upsilon_x\frac{\partial^2 V}{\partial x\partial y}]+G\{\frac{\partial^2 U}{\partial y^2}+\frac{\partial^2 V}{\partial x\partial y}\}+m\omega^2 U=0$$

$$\frac{E_y}{(1-\upsilon_x\upsilon_y)}[\frac{\partial^2 V}{\partial y^2}+\upsilon_y\frac{\partial^2 U}{\partial x\partial y}]+G\{\frac{\partial^2 V}{\partial x^2}+\frac{\partial^2 U}{\partial x\partial y}\}+m\omega^2 V=0 \tag{8.10}$$

8.3 A PLANE STRESS RECTANGULAR ELEMENT: SHAPE FUNCTIONS

As an introduction to concepts in finite element method, the static stiffness and mass matrices will be developed for the rectangular element of size 2a x 2b as shown in Fig. 8.3. The four nodes are numbered in the anticlockwise direction. It is worth pointing out that in the finite element method, instead of approximating the unknown functions u and v over the entire region of interest as is done in the classical Ritz method, the approximation applies only over the individual finite elements. This procedure has three distinct advantages

- the error is minimised
- the solution method becomes general rather than being specific to a given problem
- the problem of finding an approximate expression for the unknown functions becomes, decidedly much simpler.

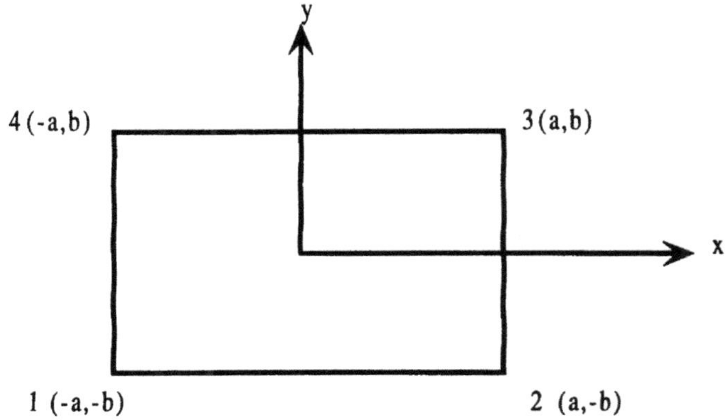

Fig. 8.3 A rectangular element

In the light of the above remarks, the first step is to assume displacement variation in the element in such a way that given the value of the displacements of the nodes corresponding to that element, displacement in the element can be determined. In this case since there are only four nodes, it is assumed that the displacement in the element is given by

$$U(x,y) = C_1 + C_2\,x + C_3\,y + C_4\,xy \tag{8.11}$$

The constants C_1 to C_4 can be determined from the boundary conditions as follows.

At node 1: $x = -a$, $y = -b$, $U(x,y) = u_1$
At node 2: $x = a$, $y = -b$, $U(x,y) = u_2$
At node 3: $x = a$, $y = b$, $U(x,y) = u_3$
At node 4: $x = -a$, $y = b$, $U(x,y) = u_4$

Solving for the constants, C_i,

$$C_1 = \frac{u_1 + u_2 + u_3 + u_4}{4}, \quad C_2 = \frac{-u_1 + u_2 + u_3 - u_4}{4a},$$

$$C_3 = \frac{-u_1 - u_2 + u_3 + u_4}{4b}, \quad C_4 = \frac{u_1 - u_2 + u_3 - u_4}{4ab}$$

Substituting for the constants in equation (8.11)

$$U(x,y) = N_1(x,y)\,u_1 + N_2(x,y)\,u_2 + N_3(x,y)\,u_3 + N_4(x,y)\,u_4 \tag{8.12a}$$

In a similar manner

$$V(x,y) = N_1(x,y)\,v_1 + N_2(x,y)\,v_2 + N_3(x,y)\,v_3 + N_4(x,y)\,v_4 \tag{8.12b}$$

The shape functions are given by

$$N_1(x,y) = \frac{1}{4}\{1 - \frac{x}{a} - \frac{y}{b} + \frac{x}{a}\frac{y}{b}\} = \frac{1}{4}(1 - \frac{x}{a})(1 - \frac{y}{b})$$

$$N_2(x,y) = \frac{1}{4}\{1 + \frac{x}{a} - \frac{y}{b} - \frac{x}{a}\frac{y}{b}\} = \frac{1}{4}(1 + \frac{x}{a})(1 - \frac{y}{b})$$

$$N_3(x,y) = \frac{1}{4}\{1 + \frac{x}{a} + \frac{y}{b} + \frac{x}{a}\frac{y}{b}\} = \frac{1}{4}(1 + \frac{x}{a})(1 + \frac{y}{b})$$

$$N_4(x,y) = \frac{1}{4}\{1 - \frac{x}{a} + \frac{y}{b} - \frac{x}{a}\frac{y}{b}\} = \frac{1}{4}(1 - \frac{x}{a})(1 + \frac{y}{b}) \tag{8.13}$$

It should be noted that along any edge, only the displacements of the nodes defining that edge govern the displacements of the edge. As an example, considering the edge 2-3 defined by $x = a$. The shape functions become

$$N_1(a,y) = 0,\ N_2(a,y) = 0.5(1 - y/b),\ N_3(a,y) = 0.5(1 + y/b),\ N_4(a,y) = 0$$

The displacement of the edge 2-3 in the x and y directions are

$$U(a,y) = 0.5\{(1 - y/b)u_2 + (1 + y/b)u_3\},\ V(a,y) = 0.5\{(1 - y/b)v_2 + (1 + y/b)v_3\}$$

This means that an adjacent element also has identical displacements along a common edge, thus ensuring that the resulting displacements do not leave gaps between elements.

It is worth noting that the shape functions are linear functions of x and y. Evidently, if the actual displacements in the element are strongly non-linear functions of x and y, then we need to make the elements in those areas as small as possible. It is the same phenomenon as trying to approximate a curve by a series of straight segments. How many segments are needed to reasonably replicate a curve depends on the curvature of the curve.

Having approximated the displacements using shape functions, the next stage is to calculate the strains using strain-displacement relationships from equation (8.8).

$$\begin{bmatrix}\varepsilon_x\\ \varepsilon_y\\ \gamma_{xy}\end{bmatrix}=\begin{bmatrix}\frac{\partial N_1}{\partial x} & 0 & \frac{\partial N_2}{\partial x} & 0 & \frac{\partial N_3}{\partial x} & 0 & \frac{\partial N_4}{\partial x} & 0\\ 0 & \frac{\partial N_1}{\partial y} & 0 & \frac{\partial N_2}{\partial y} & 0 & \frac{\partial N_3}{\partial y} & 0 & \frac{\partial N_4}{\partial y}\\ \frac{\partial N_1}{\partial y} & \frac{\partial N_1}{\partial x} & \frac{\partial N_2}{\partial y} & \frac{\partial N_2}{\partial x} & \frac{\partial N_3}{\partial y} & \frac{\partial N_3}{\partial x} & \frac{\partial N_4}{\partial y} & \frac{\partial N_4}{\partial x}\end{bmatrix}\begin{bmatrix}u_1\\ v_1\\ u_2\\ v_2\\ u_3\\ v_3\\ u_4\\ v_4\end{bmatrix} \tag{8.14}$$

$$\varepsilon \quad = \quad B \quad \delta$$

The above relationship is expressed as

$$\varepsilon = B\,\delta \tag{8.15}$$

B matrix contains the derivatives of displacements and the δ vector contains the nodal displacements.

Substituting for ε in terms of σ using equation (8.4)

$$\sigma = D\,\varepsilon = D\,B\,\delta \tag{8.16}$$

8.4 GALERKIN'S WEIGHTED RESIDUAL METHOD

The partial differential equation (8.10) can be solved using the Galerkin's weighted residual method. The procedure is identical to the approach adopted for solving ordinary differential equations in Chapter 7, Section 7.13.

Substituting in equation (8.1), the approximate solution to U(x,y) and V(x,y) given by equation (8.12), the weighted average residual over the whole element is made equal to zero.

$$\int_{x=-a}^{a}\int_{y=-b}^{b}\{\frac{\partial \sigma_x}{\partial x}+\frac{\partial \tau_{xy}}{\partial y}+m\omega^2\frac{\partial^2 U}{\partial t^2}\}N_j\,dy\,dx\,t=0 \tag{8.17}$$

$$\int_{x=-a}^{a}\int_{y=-b}^{b}\{\frac{\partial \sigma_y}{\partial y}+\frac{\partial \tau_{xy}}{\partial x}+m\omega^2\frac{\partial^2 V}{\partial t^2}\}N_j\,dy\,dx\,t=0 \tag{8.18}$$

Integrating partially with respect to x

$$\int_{y=-b}^{b}[\int_{x=-a}^{a}\frac{\partial \sigma_x}{\partial x}N_j\,dx]dy = \int_{y=-b}^{b}\sigma_x N_j\Big|_{x=-a}^{a}dy - \int_{y=-b}^{b}\int_{x=-a}^{a}\sigma_x\frac{\partial N_j}{\partial x}dx\,dy \tag{8.19a}$$

Integrating partially with respect to y

$$\int_{x=-a}^{a}[\int_{y=-b}^{b}\frac{\partial \tau_{xy}}{\partial y}N_j\,dy]dx = \int_{x=-a}^{a}\tau_{xy}N_j\Big|_{y=-b}^{b}dx - \int_{y=-b}^{b}\int_{x=-a}^{a}\tau_{xy}\frac{\partial N_j}{\partial y}dx\,dy \tag{8.19b}$$

Integrating by parts partially with respect to y

$$\int_{x=-a}^{a}[\int_{y=-b}^{b}\frac{\partial \sigma_y}{\partial y}N_j\,dy]dx = \int_{x=-a}^{a}\sigma_y N_j\Big|_{y=-b}^{b}dx - \int_{x=-a}^{a}\int_{y=-b}^{b}\sigma_y\frac{\partial N_j}{\partial y}dy\,dx \tag{8.20a}$$

Integrating partially with respect to x

$$\int_{y=-b}^{b}[\int_{x=-a}^{a}\frac{\partial\tau_{xy}}{\partial x}N_j\,dx]dy = \int_{y=-b}^{b}\tau_{xy}N_j\Big|_{x=-a}^{a}dy - \int_{x=-a}^{a}\int_{y=-b}^{b}\tau_{xy}\frac{\partial N_j}{\partial x}dy\,dx \tag{8.20b}$$

Substituting in equation (8.17), the integrals from equations (8.19a and b)

$$\int_{x=-a}^{a}\int_{y=-b}^{b}\{\sigma_x\frac{\partial N_j}{\partial x}+\tau_{xy}\frac{\partial N_j}{\partial y}-m\omega^2 U N_j\}dx\,dy = F_{xj} \tag{8.21}$$

Substituting in equation (8.18), the integrals from equations (8.20a and b)

$$\int_{x=-a}^{a}\int_{y=-b}^{b}\{\sigma_y\frac{\partial N_j}{\partial y}+\tau_{xy}\frac{\partial N_j}{\partial x}-m\omega^2 V N_j\}dx\,dy = F_{yj} \tag{8.22}$$

where

$$F_{xj} = \int_{y=-b}^{b}[\sigma_x]N_j t\Big|_{x=-a}^{a}dy + \int_{x=-a}^{a}[\tau_{xy}]N_j t\Big|_{y=-b}^{b}dx \tag{8.23}$$

$$F_{yj} = \int_{x=-a}^{a}[\sigma_y]N_j t\Big|_{y=-b}^{b}dx + \int_{y=-b}^{b}[\tau_{xy}]N_j t\Big|_{x=-a}^{a}dy \tag{8.24}$$

Using equation (8.3) and equation (8.8), replacing stresses in terms of displacements in equation (8.21) and equation (8.22)

$$\int_{x=-a}^{a}\int_{y=-b}^{b}\sum_{i=1}^{4}[\{\frac{E_x}{(1-\upsilon_x\upsilon_y)}\frac{\partial N_i}{\partial x}\frac{\partial N_j}{\partial x}+G\frac{\partial N_i}{\partial y}\frac{\partial N_j}{\partial y}\}u_i$$

$$+\{\frac{\upsilon_x E_x}{(1-\upsilon_x\upsilon_y)}\frac{\partial N_i}{\partial y}\frac{\partial N_j}{\partial x}+G\frac{\partial N_i}{\partial x}\frac{\partial N_j}{\partial y}\}v_i]t\,dy\,dx$$

$$-m\omega^2\int_{x=-a}^{a}\int_{y=-b}^{b}\sum_{i=1}^{4}N_i N_j\;u_i\,t\,dy\,dx = F_{xj}\;,\; j=1,2,3,4 \tag{8.25}$$

$$\int_{x=-a}^{a}\int_{y=-b}^{b}\sum_{i=1}^{4}[\{\frac{E_y}{(1-\upsilon_x\upsilon_y)}\frac{\partial N_i}{\partial y}\frac{\partial N_j}{\partial y}+G\frac{\partial N_i}{\partial x}\frac{\partial N_j}{\partial x}\}v_i$$

$$+\{\frac{\upsilon_y E_y}{(1-\upsilon_x\upsilon_y)}\frac{\partial N_i}{\partial x}\frac{\partial N_j}{\partial y}+G\frac{\partial N_i}{\partial y}\frac{\partial N_j}{\partial x}\}u_i]t\,dy\,dx$$

$$-m\omega^2\int_{x=-a}^{a}\int_{y=-b}^{b}\sum_{i=1}^{4}N_i N_j\;u_i\,t\,dy\,dx = F_{yj}\;,\; j=1,2,3,4 \tag{8.26}$$

Table 8.1, shows the value of shape function along the edges of the element.

Table 8.1 Values of shape functions along the edges

	X = -a	X = a	Y = -b	Y = b
N_1	0.5(1 - y/b)	0	0.5(1 - x/a)	0
N_2	0	0.5(1 - y/b)	0.5(1 + x/a)	0
N_3	0	0.5(1 + y/b)	0	0.5(1 + x/a)
N_4	0.5(1 + y/b)	0	0	0.5(1 - x/a)

Using the values from Table 8.1, the forces F_{xj} in equation (8.26) become

$$
\begin{aligned}
j=1, F_{x1} &= -\int_{y=-b}^{b}[\sigma_x at\, x=-a]0.5(1-\frac{y}{b})t\,dy - \int_{x=-a}^{a}[\tau_{xy} at\, y=-b]0.5(1-\frac{x}{a})t\,dx \\
j=2, F_{x2} &= \int_{y=-b}^{b}[\sigma_x at\, x=a]0.5(1-\frac{y}{b})t\,dy - \int_{x=-a}^{a}[\tau_{xy} at\, y=-b]0.5(1+\frac{x}{a})t\,dx \\
j=3, F_{x3} &= \int_{y=-b}^{b}[\sigma_x at\, x=a]0.5(1+\frac{y}{b})t\,dy + \int_{x=-a}^{a}[\tau_{xy} at\, y=b]0.5(1+\frac{x}{a})t\,dx \\
j=4, F_{x4} &= -\int_{y=-b}^{b}[\sigma_x at\, x=-a]0.5(1+\frac{y}{b})t\,dy + \int_{x=-a}^{a}[\tau_{xy} at\, y=b]0.5(1-\frac{x}{a})t\,dx
\end{aligned}
\tag{8.27}
$$

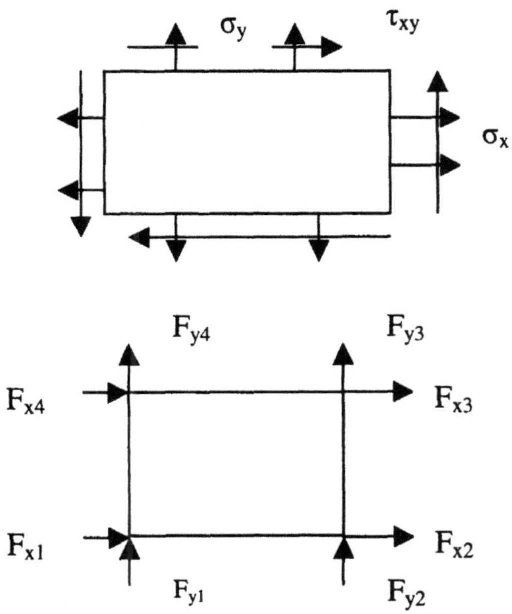

Fig. 8.4 Edge stresses and nodal forces

Similarly, the forces F_{yj} in equation (8.27), become

$$
\begin{aligned}
j=1, F_{y1} &= -\int_{x=-a}^{a}[\sigma_y at\, y=-b]0.5(1-\frac{x}{a})t\,dx - \int_{y=-b}^{b}[\tau_{xy}\, at\, x=-a]0.5(1-\frac{y}{b})t\,dy \\
j=2, F_{y2} &= -\int_{x=-a}^{a}[\sigma_y at\, y=-b]0.5(1+\frac{x}{a})t\,dx + \int_{y=-b}^{b}[\tau_{xy}\, at\, x=a]0.5(1-\frac{y}{b})t\,dy \\
j=3, F_{y3} &= \int_{x=-a}^{a}[\sigma_y at\, y=b]0.5(1+\frac{x}{a})t\,dx + \int_{y=-b}^{b}[\tau_{xy}\, at\, x=a]0.5(1+\frac{y}{b})t\,dy \\
j=4, F_{y4} &= \int_{x=-a}^{a}[\sigma_y at\, y=b]0.5(1-\frac{x}{a})t\,dx - \int_{y=-b}^{b}[\tau_{xy}\, at\, x=-a]0.5(1+\frac{y}{b})t\,dy
\end{aligned}
\tag{8.28}
$$

If the stresses on the edges are assumed to be constant, then from equation (8.27)

$$F_{x1} = \{-(\sigma_x at\, x=-a)bt - (\tau_{xy} at\, y=-b)at\}, \quad F_{x2} = \{(\sigma_x\, at\; x=a)bt - (\tau_{xy}\, at\; y=-b)at\}$$

$$F_{x3} = \{(\sigma_x at\, x=a)bt + (\tau_{xy} at\, y=b)at\}, \quad F_{x4} = \{-(\sigma_x at\, x=-a)bt + (\tau_{xy} at\, y=b)at\} \tag{8.29}$$

Similarly from equation (8.28)

$$F_{y1} = \{-(\sigma_y at\, y=-b)at - (\tau_{xy} at\, x=-a)bt\}, \quad F_{y2} = \{-(\sigma_y at\, y=-b)at + (\tau_{xy} at\, x=a)bt\}$$

$$F_{y3} = \{(\sigma_y at\, y=b)at + (\tau_{xy} at\, x=a)bt\}, \quad F_{y4} = \{(\sigma_x at\, y=b)at - (\tau_{xy} at\, x=-a)bt\} \tag{8.30}$$

As shown in Fig. 8.4, the forces F_{xj} and F_{yj} arise from the distributed stresses on the edges of the element. One consequence of the approximations used in the derivation of equations (8.25) and (8.26) is that when considering equilibrium, distributed stresses are 'replaced' by concentrated forces at the nodes. It is important to appreciate that when the elements are assembled to form a structure, at any node, these forces arising from internal stresses will sum to zero. Therefore there is no need to compute these forces.

8.4.1 Integrals of Shape Functions

Tables 8.2 and 8.3, show integrals of shape functions occurring in equations (8.21) and (8.22).

Table 8.2 Integrals of shape functions, in-plane case

i	$\frac{9}{ab}\int_{x=-a}^{a}\int_{y=-b}^{b} N_i N_j\, dy\, dx$				$6\frac{a}{b}\int_{x=-a}^{a}\int_{y=-b}^{b} \frac{\partial N_i}{\partial x}\frac{\partial N_j}{\partial x} dy\, dx$				$6\frac{b}{a}\int_{x=-a}^{a}\int_{y=-b}^{b} \frac{\partial N_i}{\partial y}\frac{\partial N_j}{\partial y} dy\, dx$			
	j				j				j			
	1	2	3	4	1	2	3	4	1	2	3	4
1	4	2	1	2	2	-2	-1	1	2	1	-1	-2
2	2	4	2	1	-2	2	1	-1	1	2	-2	-1
3	1	2	4	2	-1	1	2	-2	-1	-2	2	1
4	2	1	2	4	1	-1	-2	2	-2	-1	1	2

Table 8.3 Integrals of shape functions, in-plane case

i	$4\int_{x=-a}^{a}\int_{y=-b}^{b} \frac{\partial N_i}{\partial x}\frac{\partial N_j}{\partial y} dy\, dx$				$4\int_{x=-a}^{a}\int_{y=-b}^{b} \frac{\partial N_i}{\partial y}\frac{\partial N_j}{\partial x} dy\, dx$			
	j				j			
	1	2	3	4	1	2	3	4
1	1	1	-1	-1	1	-1	-1	1
2	-1	-1	1	1	1	-1	-1	1
3	-1	-1	1	1	-1	1	1	-1
4	1	1	-1	-1	-1	1	1	-1

8.4.2 Explicit K and M Matrices

Substituting in equations (8.21) and (8.22), the values of integrals shown in Tables 8.2 and 8.3, explicit expressions for the coefficients of static stiffness and consistent mass matrices can be derived. Tables 8.4 and 8.5 show, respectively, the static stiffness matrix K and the consistent mass matrix M

8.5 DERIVATION OF K AND M MATRICES BY MINIMISATION OF TOTAL POTENTIAL

As explained in Section 7.12, the total potential consists of three parts. They are

- strain energy stored
- loss of potential by external loads
- loss of potential by inertial forces.

8.5.1 Strain energy

Substituting for σ = D B δ from equation (8.16) and for ε = Bδ from equation (8.15), the expression for strain energy in equation (8.7) becomes

$$Strain\,energy = \frac{1}{2}\delta^T \left[\int_{x=-a}^{a} \int_{y=-b}^{b} \{B^T D B\} dy\, dx\, t \right] \delta \tag{8.31}$$

where, since δ is not a function of x or y, it has been taken out of the integral sign.

The coefficients of the matrix B^T D B consist of products of the derivatives of shape functions and elastic constants. For example, the first three coefficients are

$$coeff(1,1) = \frac{E_x}{(1-\upsilon_x\upsilon_y)} \frac{\partial N_1}{\partial x} \frac{\partial N_1}{\partial x} + G \frac{\partial N_1}{\partial y} \frac{\partial N_1}{\partial y}$$

$$coeff(1,2) = \frac{\upsilon_x E_x}{(1-\upsilon_x\upsilon_y)} \frac{\partial N_1}{\partial x} \frac{\partial N_1}{\partial y} + G \frac{\partial N_1}{\partial y} \frac{\partial N_1}{\partial x}$$

$$coeff(1,2) = \frac{\upsilon_x E_x}{(1-\upsilon_x\upsilon_y)} \frac{\partial N_1}{\partial x} \frac{\partial N_2}{\partial y} + G \frac{\partial N_1}{\partial y} \frac{\partial N_2}{\partial x}$$

8.5.2 Loss of Potential by External Loads

If the forces acting in x and y direction at the ith node are, F_{xi} and F_{yi} as shown in Fig. 8.4, the loss of potential by these loads acting at the joints is equal to

$$Loss\ of\ potential\ by\, external\ loads = -\left[\sum_{i=1}^{4} F_{xi}\, u_i + \sum_{i=1}^{4} F_{yi}\, v_i\right] = -\delta^T\ F \tag{8.32}$$

where

$$F^T = \begin{bmatrix} F_{x1} & F_{y1} & F_{x2} & F_{y2} & F_{x3} & F_{y3} & F_{x4} & F_{y4} \end{bmatrix} \tag{8.33}$$

The vector F is known as the nodal force vector.

8.5.3 Loss of Potential by Inertial Forces

The loss of potential by the distributed inertial forces is calculated as follows. The inertial forces are given in x and y directions respectively by mω^2 U and mω^2 V respectively. Substituting for U and V in equation (8.12), in terms of shape functions, the inertial forces in the x and y directions become, respectively, m$\omega^2 \Sigma N_i u_i$ and m $\omega^2 \Sigma N_i v_i$. The loss of potential by the distributed inertial forces is therefore given by

$$Loss\ of\ potential = -\frac{1}{2} m\omega^2 [\int_{x=-a}^{a} \int_{y=-b}^{b} \{U\sum_{i=1}^{4} N_i u_i + V\sum_{i=1}^{4} N_i v_i\} dy\, dx\, t] \tag{8.34}$$

Replacing U and V in terms of the approximate solution, equation (8.34) becomes

$$Loss\ of\ potential = -\frac{1}{2} m\omega^2 [\int_{x=-a}^{a} \int_{y=-b}^{b} \{\sum_{j=1}^{4} N_j u_j \sum_{i=1}^{4} N_i u_i + \sum_{j=1}^{4} N_j v_j \sum_{i=1}^{4} N_i v_i\} dy\, dx\ t] \tag{8.35}$$

Equation (8.35) can be more conveniently written as

$$Loss\ of\ potential = -\frac{1}{2} m\omega^2 [\int_{x=-a}^{a} \int_{y=-b}^{b} \{\sum_{i=1}^{4} \sum_{j=1}^{4} N_i N_j u_i u_j\} dy\, dx\, t + \int_{x=-a}^{a} \int_{y=-b}^{b} \{\sum_{i=1}^{4} \sum_{j=1}^{4} N_i N_j v_i v_j\} dy\, dx\, t] \tag{8.36}$$

Or in matrix form equation (8.36) can be expressed as

$$Loss\ of\ potential\ by\ inertial\ forces = -\frac{1}{2} m\omega^2 \delta^T [\int_{x=-a}^{a} \int_{y=-b}^{b} N\ dy\, dx\ t] \delta \tag{8.37}$$

where the matrix N is given by

$$N = \begin{bmatrix} N_1N_1 & 0 & N_1N_2 & 0 & N_1N_3 & 0 & N_1N_4 & 0 \\ & N_1N_1 & 0 & N_1N_2 & 0 & N_1N_3 & 0 & N_1N_4 \\ & & N_2N_2 & 0 & N_2N_3 & 0 & N_2N_4 & 0 \\ & & & N_2N_2 & 0 & N_2N_3 & 0 & N_2N_4 \\ & & & & N_3N_3 & 0 & N_3N_4 & 0 \\ symm. & & & & & N_3N_3 & 0 & N_3N_4 \\ & & & & & & N_4N_4 & 0 \\ & & & & & & & N_4N_4 \end{bmatrix} \tag{8.38}$$

8.5.4 Matrices K and M

From equations (8.31), (8.34) and (8.37), the total potential for the element is given by

$$\pi_e = \frac{1}{2}\delta^T [\int_{x=-a}^{a} \int_{y=-b}^{b} \{B^T D B\} dy\, dx\, t]\delta - \frac{1}{2} m\omega^2 \delta^T [\int_{x=-a}^{a} \int_{y=-b}^{b} \{N\} dy\, dx\, t]\delta - \delta^T F \tag{8.39}$$

Differentiating equation (8.39) with respect to δ, the matrix relationship can be expressed as

$$\frac{\partial \pi_e}{\partial \delta} = [\int_{x=-a}^{a} \int_{y=-b}^{b} \{B^T D B\} dy\, dx\, t]\delta - m\omega^2 [\int_{x=-a}^{a} \int_{y=-b}^{b} \{N\} dy\, dx\, t] - F \tag{8.40}$$

The static stiffness matrix K and consistent mass matrix M are given by

$$K = [\int_{x=-a}^{a} \int_{y=-b}^{b} \{B^T D B\} dy\, dx\, t]\, ,\; M = m [\int_{x=-a}^{a} \int_{y=-b}^{b} \{N\} dy\, dx\, t] \tag{8.41}$$

The force–deformation relationship is therefore

$$\frac{\partial \pi_e}{\partial \delta} = [K - m\omega^2 M]\delta - F \tag{8.42}$$

The explicit expressions for K and M are given in Tables 8.4 and 8.5.

Table 8.5 Mass matrix

u_1	v_1	u_2	v_2	u_3	v_3	u_4	v_4
$4\alpha_7$	0	$2\alpha_7$	0	α_7	0	$2\alpha_7$	0
0	$4\alpha_7$	0	$2\alpha_7$	0	α_7	0	$2\alpha_7$
$2\alpha_7$	0	$4\alpha_7$	0	$2\alpha_7$	0	α_7	0
0	$2\alpha_7$	0	$4\alpha_7$	0	$2\alpha_7$	0	α_7
α_7	0	$2\alpha_7$	0	$4\alpha_7$	0	$2\alpha_7$	0
0	α_7	0	$2\alpha_7$	0	$4\alpha_7$	0	$2\alpha_7$
$2\alpha_7$	0	α_7	0	$2\alpha_7$	0	$4\alpha_7$	0
0	$2\alpha_7$	0	α_7	0	$2\alpha_7$	0	$4\alpha_7$

$$\alpha_7 = m\frac{abt}{9}$$

8.6 CONCEPT OF MAPPING AND ISOPARAMETRIC ELEMENTS

In the previous sections, the static stiffness and mass matrices were developed for a plane stress rectangular element. Rectangular elements restrict the range of problems that can be solved because

- they cannot model inclined or curved edges

Table 8.4 Static stiffness matrix

u_1	v_1	u_2	v_2	u_3	v_3	u_4	v_4
$2(\alpha_1+\alpha_2)$	$\alpha_3+\alpha_4$	$-2\alpha_1+\alpha_2$	$\alpha_3-\alpha_4$	$-(\alpha_1+\alpha_2)$	$-(\alpha_3+\alpha_4)$	$\alpha_1-2\alpha_2$	$-\alpha_3+\alpha_4$
$\alpha_3+\alpha_4$	$2(\alpha_5+\alpha_6)$	$-\alpha_3+\alpha_4$	$\alpha_5-2\alpha_6$	$-(\alpha_3+\alpha_4)$	$-(\alpha_5+\alpha_6)$	$\alpha_3-\alpha_4$	$-2\alpha_5+\alpha_6$
$-2\alpha_1+\alpha_2$	$-\alpha_3+\alpha_4$	$2(\alpha_1+\alpha_2)$	$-(\alpha_3+\alpha_4)$	$\alpha_1-2\alpha_2$	$\alpha_3-\alpha_4$	$-(\alpha_1+\alpha_2)$	$(\alpha_3+\alpha_4)$
$\alpha_3-\alpha_4$	$\alpha_5-2\alpha_6$	$-(\alpha_3+\alpha_4)$	$2(\alpha_5+\alpha_6)$	$-\alpha_3+\alpha_4$	$-2\alpha_5+\alpha_6$	$\alpha_3+\alpha_4$	$-(\alpha_5+\alpha_6)$
$-(\alpha_1+\alpha_2)$	$-(\alpha_3+\alpha_4)$	$\alpha_1-2\alpha_2$	$-\alpha_3+\alpha_4$	$2(\alpha_1+\alpha_2)$	$(\alpha_3+\alpha_4)$	$-2\alpha_1+\alpha_2$	$\alpha_3-\alpha_4$
$-(\alpha_3+\alpha_4)$	$-(\alpha_5+\alpha_6)$	$\alpha_3-\alpha_4$	$-2\alpha_5+\alpha_6$	$(\alpha_3+\alpha_4)$	$2(\alpha_5+\alpha_6)$	$-\alpha_3+\alpha_4$	$\alpha_5-2\alpha_6$
$\alpha_1-2\alpha_2$	$\alpha_3-\alpha_4$	$-(\alpha_1+\alpha_2)$	$(\alpha_3+\alpha_4)$	$-2\alpha_1+\alpha_2$	$-\alpha_3+\alpha_4$	$2(\alpha_1+\alpha_2)$	$-(\alpha_3+\alpha_4)$
$-\alpha_3+\alpha_4$	$-2\alpha_5+\alpha_6$	$(\alpha_3+\alpha_4)$	$-(\alpha_5+\alpha_6)$	$\alpha_3-\alpha_4$	$\alpha_5-2\alpha_6$	$-(\alpha_3+\alpha_4)$	$2(\alpha_5+\alpha_6)$

$$\alpha_1=\frac{E_x}{(1-\upsilon_x\upsilon_y)}\frac{bt}{6a},\alpha_2=G\frac{at}{6b}\ ,\alpha_3=\frac{\upsilon_x E_x}{(1-\upsilon_x\upsilon_y)}\frac{t}{4},\alpha_4=G\frac{t}{4},$$

$$\alpha_5=\frac{E_y}{(1-\upsilon_x\upsilon_y)}\frac{at}{6b},\ \alpha_6=G\frac{bt}{6a}$$

- it is not possible to use finer elements in areas where strains change rapidly and to use coarser elements elsewhere. This process is known as mesh refining.

A quadrilateral element would certainly answer the first requirement and, to a certain extent, the second requirement as well.

Consider the quadrilateral element shown in Fig. 8.5. Equation (8.11) is an approximate expression for u and a similar expression for v.

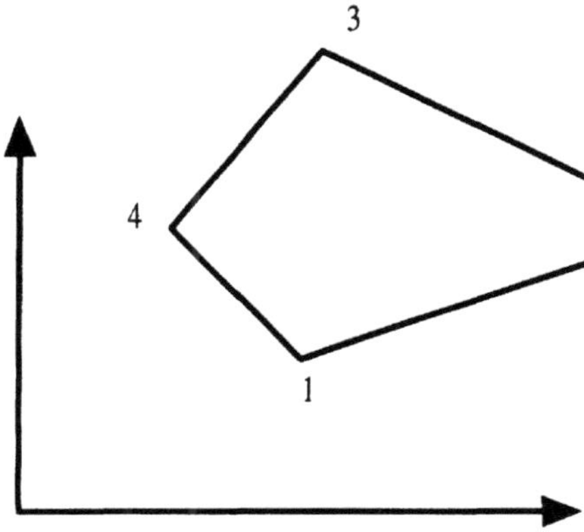

Fig. 8.5 A quadrilateral element

The reader will very quickly realise that even to solve constants C_1 to C_4, in order to develop the shape functions, becomes algebraically very complex. The complexity increases when one has to integrate the shape functions over a general quadrilateral. To carry out this task algebraically is extremely complex. However the entire complexity can be avoided by using the concept of mapping.

Consider a square element in the (ξ,η) system axis, as shown in Fig. 8.6. The origin is at (0,0) and the four edges are defined by $\xi = \pm 1$ and $\eta = \pm 1$. If

$$u(\xi,\eta) = C_1 + C_2\,\xi + C_3\,\eta + C_4\,\xi\eta$$

Proceeding as in Section 8.3, the shape functions for this element are given by

$$n_1(\xi,\eta) = \frac{1}{4}(1-\xi)(1-\eta),\ n_2(\xi,\eta) = \frac{1}{4}(1+\xi)(1-\eta)$$

$$n_3(\xi,\eta) = \frac{1}{4}(1+\xi)(1+\eta),\ n_4(\xi,\eta) = \frac{1}{4}(1-\xi)(1+\eta) \tag{8.43}$$

Consider now the mapping functions

$$x(\xi,\eta) = n_1(\xi,\eta)x_1 + n_2(\xi,\eta)x_2 + n_3(\xi,\eta)x_3 + n_4(\xi,\eta)x_4 \tag{8.44a}$$

$$y(\xi,\eta) = n_1(\xi,\eta)y_1 + n_2(\xi,\eta)y_2 + n_3(\xi,\eta)y_3 + n_4(\xi,\eta)y_4 \tag{8.44b}$$

where (x_i,y_i) are the coordinates of the ith node of the quadrilateral element shown in Fig. 8.5.

$$x(-1,-1) = x_1,\ x(1,-1) = x_2,\ x(1,1) = x_3,\ x(-1,1) = x_4$$
$$y(-1,-1) = y_1,\ y(1,-1) = y_2,\ y(1,1) = y_3,\ y(-1,1) = y_4$$

Thus the four corners of the quadrilateral, simply map on to the four corners of the square element.

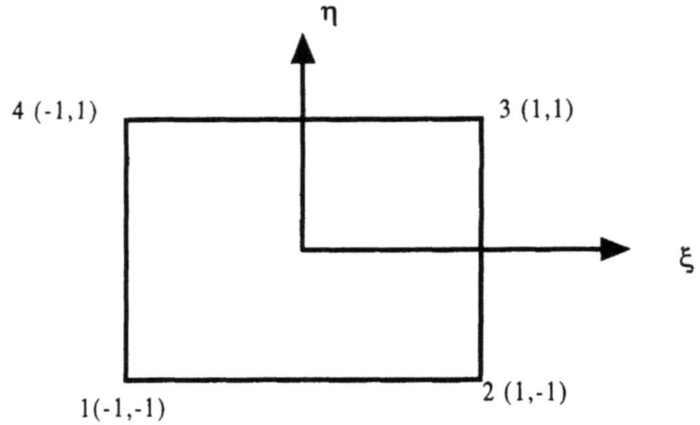

Fig. 8.6 A unit square element

Consider the four edges

$\xi = 1$: $x(1,\eta) = 0.5(1-\eta)\,x_2 + 0.5(1+\eta)x_3$

$\xi = -1$: $x(-1,\eta) = 0.5(1-\eta)x_1 + 0.5(1+\eta)x_4$

$\eta = 1$: $y(\xi,1) = 0.5(1+\xi)y_3 + 0.5(1-\xi)y_4$

$\eta = -1$: $y(\xi,-1) = 0.5(1-\xi)y_1 + 0.5(1+\xi)y_2$

One can see that edges 2-3, 1-4, 1-2 and 4-3 map linearly on to the edges $\xi = 1$, $\xi = -1$, $\eta = -1$ and $\eta = 1$ respectively.

Figure 8.7 shows the contours corresponding to equal values of (ξ,η) on the quadrilateral element. When the functions used for approximating the displacements are used for mapping from the (ξ,η) to (x,y) system, the mapping is called an isoparametric transformation and the corresponding element as an isoparametric element.

Having established the mapping, the same shape functions can be used to calculate the displacements in the element. Thus

$$u(\xi,\eta) = \sum_{i=1}^{4} n_i(\xi,\eta)\, u_i \;,\; v(\xi,\eta) = \sum_{i=1}^{4} n_i(\xi,\eta)\, v_i \tag{8.45}$$

The fact that u and v are expressed in the (ξ,η) system causes no difficulty, because, for a given value of (ξ,η), the corresponding value of (x,y) is known from equation (8.44). In order to derive expressions for the static stiffness and mass matrices, we need

$$N_i(x,y),\ \frac{\partial N_i(x,y)}{\partial x}\ and\ \frac{\partial N_i(x,y)}{\partial y}\ ,\ i = 1,\ 2, 3\ and\ 4.$$

$n_i(\xi,\eta)$ becomes $N_i(x,y)$ as a result of a change of variables. $N_i(x,y)$ can be explicitly derived, provided one can find explicit expressions for ξ and η in terms of x and y. Unfortunately this is not always possible. However, this is not really required when we understand the concept of change of variables in partial differentiation.

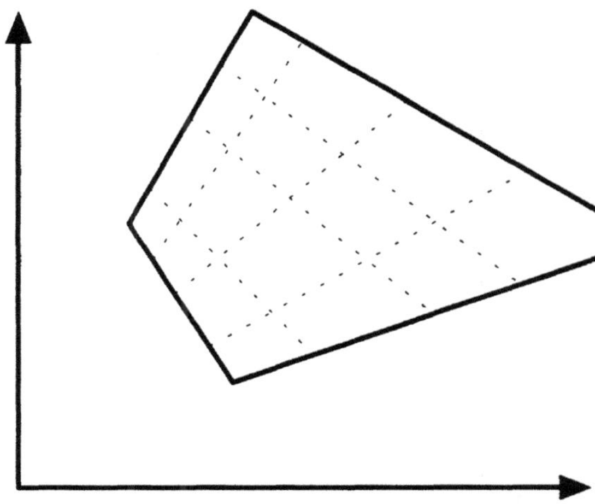

Fig. 8.7 Contours of ξ and η in the x-y plane

8.6.1 Change of Variables and Implicit Functions

Let F(x,y,z,...) be a real valued function of variables x, y, z, ... whose first order derivatives exist and are continuous. Further let $x = x(\xi,\eta,\zeta, \ldots)$, $y = y(\xi,\eta,\zeta, \ldots)$, $z = z(\xi,\eta,\zeta, \ldots)$, ...be differentiable functions of the real variables, ξ, η, ζ, ...

Then if F(x,y,z, ...) becomes the function $f(\xi,\eta,\zeta, \ldots)$ as a result of this change of variables, then

$$\begin{aligned}
\frac{\partial f}{\partial \xi} &= \frac{\partial F}{\partial x}\frac{\partial x}{\partial \xi} + \frac{\partial F}{\partial y}\frac{\partial y}{\partial \xi} + \frac{\partial F}{\partial z}\frac{\partial z}{\partial \xi} + \ldots \\
\frac{\partial f}{\partial \eta} &= \frac{\partial F}{\partial x}\frac{\partial x}{\partial \eta} + \frac{\partial F}{\partial y}\frac{\partial y}{\partial \eta} + \frac{\partial F}{\partial z}\frac{\partial z}{\partial \eta} + \ldots \\
\frac{\partial f}{\partial \zeta} &= \frac{\partial F}{\partial x}\frac{\partial x}{\partial \zeta} + \frac{\partial F}{\partial y}\frac{\partial y}{\partial \zeta} + \frac{\partial F}{\partial z}\frac{\partial z}{\partial \zeta} + \ldots
\end{aligned} \tag{8.46}$$

The above equations can be expressed in matrix form as

$$\begin{bmatrix} \frac{\partial f}{\partial \xi} \\ \frac{\partial f}{\partial \eta} \\ \frac{\partial f}{\partial \zeta} \\ \cdot \\ \cdot \end{bmatrix} = J \begin{bmatrix} \frac{\partial F}{\partial x} \\ \frac{\partial F}{\partial y} \\ \frac{\partial F}{\partial z} \\ \cdot \\ \cdot \end{bmatrix}, \quad J = \begin{bmatrix} \frac{\partial x}{\partial \xi} & \frac{\partial x}{\partial \xi} & \frac{\partial x}{\partial \xi} & \cdot & \cdot \\ \frac{\partial x}{\partial \eta} & \frac{\partial x}{\partial \eta} & \frac{\partial x}{\partial \eta} & \cdot & \cdot \\ \frac{\partial x}{\partial \zeta} & \frac{\partial x}{\partial \zeta} & \frac{\partial x}{\partial \zeta} & \cdot & \cdot \\ \cdot & \cdot & \cdot & \cdot & \cdot \\ \cdot & \cdot & \cdot & \cdot & \cdot \end{bmatrix} \tag{8.47}$$

The matrix which connects the partial differentials of $f(\xi,\eta,\zeta, \ldots)$ with respect to its variables $(\xi,\eta,\zeta\ \ldots)$ and the partial differentials of F(x,y,z) with respect to its variables (x,y,z, ...) is called the Jacobian matrix J. If the number of variables of f and F are equal, then the Jacobian matrix J is square and its inverse exists. If the inverse of the Jacobian matrix does not exist, then mapping cannot be done.

When integrating, the following rule applies

$$\int F(x,y,z,\ldots)dx\,dy\,dz\ldots = \int f(\xi,\eta,\zeta,\ldots)d\xi\,d\eta\,d\zeta \times Det\,of\ J$$

8.6.1.1 Change of Variables: Example

The ideas in the above section are illustrated by a simple example involving the change of variables from polar to Cartesian system.

$$f(r,\theta) = r^2(1+3\cos\theta\sin\theta)\ ,\quad x = r\cos\theta\ ,\quad y = r\sin\theta$$

F(x,y) is obtained from f(r,θ) by replacing r and θ by corresponding expressions in terms of x and y. In this example this can be done and

$$F(x,y) = x^2 + y^2 + 3xy$$

However, the partial derivatives of F with respect to x and y can be calculated without knowing this explicit expression.

$$f(r,\theta) = r^2(1+3\cos\theta\sin\theta)$$

$$\frac{\partial f}{\partial r} = 2r(1+3\cos\theta\sin\theta),\quad \frac{\partial f}{\partial\theta} = 3r^2(\cos^2\theta - \sin^2\theta)$$

$$\frac{\partial x}{\partial r} = \cos\theta,\ \frac{\partial y}{\partial r} = \sin\theta,\quad \frac{\partial x}{\partial\theta} = -r\sin\theta,\ \frac{\partial y}{\partial\theta} = r\cos\theta$$

$$\begin{bmatrix} \dfrac{\partial f}{\partial r} = 2r(1+3\cos\theta\sin\theta) \\ \dfrac{\partial f}{\partial\theta} = 3r^2(\cos^2\theta - \sin^2\theta) \end{bmatrix} = \begin{bmatrix} \dfrac{\partial x}{\partial r} = \cos\theta & \dfrac{\partial y}{\partial r} = \sin\theta \\ \dfrac{\partial x}{\partial\theta} = -r\sin\theta & \dfrac{\partial y}{\partial\theta} = r\cos\theta \end{bmatrix} \begin{bmatrix} \dfrac{\partial F}{\partial x} \\ \dfrac{\partial F}{\partial y} \end{bmatrix}$$

$$J = \begin{bmatrix} \dfrac{\partial x}{\partial r} = \cos\theta & \dfrac{\partial y}{\partial r} = \sin\theta \\ \dfrac{\partial x}{\partial\theta} = -r\sin\theta & \dfrac{\partial y}{\partial\theta} = r\cos\theta \end{bmatrix}$$

The determinant of the Jacobian matrix is given by

$$Det\,J = \frac{\partial x}{\partial r}\frac{\partial y}{\partial\theta} - \frac{\partial x}{\partial\theta}\frac{\partial y}{\partial r} = r(\cos^2\theta + \sin^2\theta) = r$$

It can be shown that the inverse of the Jacobian matrix is

$$J^{-1} = \begin{bmatrix} \cos\theta & -\dfrac{\sin\theta}{r} \\ \sin\theta & \dfrac{\cos\theta}{r} \end{bmatrix}$$

The partial derivatives of F with respect to x and y can be calculated as

$$\begin{bmatrix} \dfrac{\partial F}{\partial x} \\ \dfrac{\partial F}{\partial y} \end{bmatrix} = J^{-1} \begin{bmatrix} \dfrac{\partial f}{\partial r} \\ \dfrac{\partial f}{\partial \theta} \end{bmatrix} = \begin{bmatrix} \cos\theta & -\dfrac{\sin\theta}{r} \\ \sin\theta & \dfrac{\cos\theta}{r} \end{bmatrix} \begin{bmatrix} 2r(1+3\cos\theta\sin\theta) \\ 3r^2(\cos^2\theta - \sin^2\theta) \end{bmatrix}$$

Carrying out the multiplication

$$\begin{bmatrix} \dfrac{\partial F}{\partial x} \\ \dfrac{\partial F}{\partial y} \end{bmatrix} = \begin{bmatrix} r(2\cos\theta + 3\sin\theta) \\ r(3\cos\theta + \sin\theta) \end{bmatrix}$$

Substituting for x = r cosθ, y = r sinθ, it can be verified from the explicit expression for F(x,y), that the above calculated values of the derivatives are correct.

$$\begin{bmatrix} \dfrac{\partial F}{\partial x} \\ \dfrac{\partial F}{\partial y} \end{bmatrix} = \begin{bmatrix} 2x+3y \\ 3x+2y \end{bmatrix}$$

$$\int F(x,y)dx\,dy = \int f(r,\theta)\,dr\,d\theta\,[Det\,of\;\; J = r] = \int f(r,\theta)\; r\,dr\,d\theta$$

Note that because the partial derivatives of F with respect to x and y are in terms of r and θ, integration will need to be done in the (r,θ) system.

8.6.2 Cartesian Derivatives of Shape Functions

Using equation (8.46), the relationship between the derivatives of shape functions $N_i(x,y)$ with respect to (x,y) and of shape functions $n_i(\xi,\eta)$ with respect to (ξ,η) are given by

$$\frac{\partial n_i(\xi,\eta)}{\partial \xi} = \frac{\partial N_i(x,y)}{\partial x}\frac{\partial x}{\partial \xi} + \frac{\partial N_i(x,y)}{\partial y}\frac{\partial y}{\partial \xi}$$

$$\frac{\partial n_i(\xi,\eta)}{\partial \eta} = \frac{\partial N_i(x,y)}{\partial x}\frac{\partial x}{\partial \eta} + \frac{\partial N_i(x,y)}{\partial y}\frac{\partial y}{\partial \eta} \tag{8.48}$$

From equation (8.44)

$$\frac{\partial x}{\partial \xi} = \sum_{i=1}^{4} \frac{\partial n_i(\xi,\eta)}{\partial \xi} x_i\,, \quad \frac{\partial x}{\partial \eta} = \sum_{i=1}^{4} \frac{\partial n_i(\xi,\eta)}{\partial \eta} x_i \tag{8.49a}$$

$$\frac{\partial y}{\partial \xi} = \sum_{i=1}^{4} \frac{\partial n_i(\xi,\eta)}{\partial \xi} y_i\,, \quad \frac{\partial y}{\partial \eta} = \sum_{i=1}^{4} \frac{\partial n_i(\xi,\eta)}{\partial \eta} y_i \tag{8.49b}$$

$$J = \begin{bmatrix} \dfrac{\partial x}{\partial \xi} & \dfrac{\partial y}{\partial \xi} \\ \dfrac{\partial x}{\partial \eta} & \dfrac{\partial y}{\partial \eta} \end{bmatrix}, \quad J^{-1} = \frac{1}{Det\,J} \begin{bmatrix} \dfrac{\partial y}{\partial \eta} & -\dfrac{\partial y}{\partial \xi} \\ -\dfrac{\partial x}{\partial \eta} & \dfrac{\partial x}{\partial \xi} \end{bmatrix} \tag{8.50}$$

$$Det\,J = \frac{\partial x}{\partial \xi}\frac{\partial y}{\partial \eta} - \frac{\partial x}{\partial \eta}\frac{\partial y}{\partial \xi} \tag{8.51}$$

$$\begin{bmatrix} \dfrac{\partial N_i(x,y)}{\partial x} \\ \dfrac{\partial N_i(x,y)}{\partial y} \end{bmatrix} = J^{-1} \begin{bmatrix} \dfrac{\partial n_i(\xi,\eta)}{\partial \xi} \\ \dfrac{\partial n_i(\xi,\eta)}{\partial \eta} \end{bmatrix} \tag{8.52}$$

Note that $\partial N_i(x,y)/\partial x$ and $\partial N_i(x,y)/\partial y$ will be in terms of (ξ,η). This aids the evaluation of the integral involving these partial differentials

$$\iint F(x,y)\, dy\, dx = \int_{\xi=-1}^{1} \int_{\eta=-1}^{1} f(\xi,\eta)\, d\eta\, d\xi\, Det\, J \tag{8.53}$$

8.7 NUMERICAL INTEGRATION BY GAUSSIAN QUADRATURE

Evaluation of static stiffness and consistent mass matrices requires the evaluation of integrals of the form

$$\int_{\xi=-1}^{1} \int_{\eta=-1}^{1} f(\xi,\eta)\, d\eta\, d\xi\, Det\, J \tag{8.54}$$

The analytical evaluation of these double integrals is too complex. Evaluation of these integrals by numerical means is almost obligatory. There are a large number of different rules available for numerical integration. The one commonly favoured in finite element programs is the Gauss quadrature. Most numerical integration methods, such as Simpson's rule, require the evaluation of the function at equal intervals. However in Gauss quadrature, the evaluation is done at unequal intervals. The method is designed to integrate polynomials, and the number of points at which the function needs to be calculated depends on the degree of the polynomial. Consider the following function, which is a polynomial of order (2n - 1) and it contains 2 n terms

$$f(\xi) = a_1 + a_2\xi + a_3\xi^2 + \ldots + a_{2n-1}\xi^{2n-2} + a_{2n}\xi^{2n-1} \tag{8.55}$$

Integrating this expression between the limits (-1 and +1), all the terms in even powers of ξ contribute nothing to the integral. Therefore

$$I = \int_{\xi=-1}^{1} f(\xi)\, d\xi = 2a_1 + \frac{2}{3}a_3 + \ldots + \frac{2}{(2n-1)}a_{2n-1} \tag{8.56}$$

If we assume that the function is evaluated at n points only and that the weighted sum is exactly equal to the value of the integral, then it is implied that

$$I = \sum_{i=1}^{n} f(\xi_i)w_i = a_1\sum_{i=1}^{n} w_i + a_2\sum_{i=1}^{n} w_i\xi_i + a_3\sum_{i=1}^{n} w_i\xi_i^2 \ldots + a_{2n-1}\sum_{i=1}^{n} w_i\xi_i^{2n-1} + a_{2n}\sum_{i=1}^{n} w_i\xi_i^{2n} \tag{8.57}$$

where w_i = i[th] weight.

Comparing equations (8.56) and (8.57)

$$\sum_{i=1}^{n} w_i = 2,\ \sum_{i=1}^{n} w_i\xi_i = 0,\ \sum_{i=1}^{n} w_i\xi_i^2 = \frac{2}{3}, \ldots \sum_{i=1}^{n} w_i\xi_i^{2n-1} = \frac{2}{(2n-1)},\ \sum_{i=1}^{n} w_i\xi_i^{2n-1} = 0 \tag{8.58}$$

These are 2n equations in n values of ξ_i and w_i. Therefore, a polynomial of degree (2n −1) requires n points at which the function needs to be evaluated in order to obtain the exact value of the integral. It is usual to have the values of ξ_i symmetrically distributed about the axis.

Table 8.6 Abscissas and weight coefficients

n	a	w	Integral exact up to degree
2	$\pm 1/\sqrt{3}$	1.0	3
3	$\pm\sqrt{(3/5)}$	5/9	5
	0	8/9	
4	±0.86113	0.34786	7
	±0.33998	0.65215	
5	±0.90618	0.23693	9
	±0.53847	0.47863	
	0	0.56889	

For example, if n = 2, then assume without any loss of generality $w_1 = w_2 = w$, and $\xi_1 = -a$ and $\xi_2 = +a$. Solving equation (8.58), $a = 1/\sqrt{3}$, $w = 1$.

Similarly if n = 3, then assume without any loss of generality $w_1 = w_3$, and $\xi_1 = -a$ and $\xi_3 = +a$., $\xi_2 = 0$. Solving equation (8.58), $a = \sqrt{(3/5)}$, $w_1 = 5/9$, $w_2 = 8/9$.

The general formula for Gaussian integration is

$$\int_{\xi=-1}^{+1} f(\xi)\,d\xi = \sum_{i=1}^{n} f(a_i)\,w_i \tag{8.59}$$

where w_i is called a weight coefficient and a_i the abscissa. These are shown in Table 8.6 for different values of n.

The evaluation of a double integral requires repeated evaluation of the function at n^2 points. The integral is given by

$$\int_{\xi=-1}^{1}\int_{\eta=-1}^{1} f(\xi,\eta)\,d\eta\,d\xi = \sum_{i=1}^{n}\sum_{j=1}^{n} f(a_i, a_j)\,w_i w_j \tag{8.60}$$

For example if

$$f(\xi,\eta) = \xi^5 + 3\xi^3\eta^2 + 3\xi^2\eta^3 + \eta^5$$

then using only n = 3, we need nine evaluations of the function to be integrated.

$$\begin{aligned}\int_{\xi=-1}^{1}\int_{\eta=-1}^{1} f(\xi,\eta)\,d\eta\,d\xi = &\{f(-a,-a) + f(a,-a)\}(5/9) + f(0,-a)(8/9)\\ &+\{f(-a,0) + f(a,0)\}(5/9) \;\; + f(0,0)(8/9)\\ &+\{f(-a,a) + f(a,a)\}(5/9) \;\; + f(0,a)(8/9)\end{aligned}$$

where $a = \sqrt{(3/5)}$

8.8 EVALUATION OF K AND M MATRICES OF ISOPARAMETRIC ELEMENT

The steps involved in evaluating the K and M matrices can be summarised as follows.

- Choose the element.
- Calculate the D matrix.
- Store the coordinates of its nodes in matrix ELCOD

$$ELCOD = \begin{bmatrix} x_1 & x_2 & x_3 & x_4 \\ y_1 & y_2 & y_3 & y_4 \end{bmatrix}$$

- Zero the KMAT and MASS matrices. In FORTRAN this is accomplished by KMAT = 0, MASS = 0
- Start cycle on Gaussian integration. Choose k integration points.
- Choose the coordinates ξ_i and weight w_i of the ith Gauss point, i = 1 to k.
- Choose the coordinates η_j and weight w_j of the jth Gauss point, j = 1 to k.
- Store the value of shape functions in a vector SHAPE

$$SHAPE = \begin{bmatrix} n_1(\xi,\eta) \\ n_2(\xi,\eta) \\ n_3(\xi,\eta) \\ n_4(\xi,\eta) \end{bmatrix}$$

- Store the partial derivates of $n_i(\xi_i, \eta_j)$ in a matrix DERIV

$$DERIV = \begin{bmatrix} \dfrac{\partial n_1}{\partial \xi} & \dfrac{\partial n_2}{\partial \xi} & \dfrac{\partial n_3}{\partial \xi} & \dfrac{\partial n_4}{\partial \xi} \\ \dfrac{\partial n_1}{\partial \eta} & \dfrac{\partial n_2}{\partial \eta} & \dfrac{\partial n_3}{\partial \eta} & \dfrac{\partial n_4}{\partial \eta} \end{bmatrix}$$

- Calculate the Jacobian matrix, J

$$J = \begin{bmatrix} \sum_{i=1}^{4} \dfrac{\partial n_i}{\partial \xi} x_i & \sum_{i=1}^{4} \dfrac{\partial n_i}{\partial \xi} y_i \\ \sum_{i=1}^{4} \dfrac{\partial n_i}{\partial \eta} x_i & \sum_{i=1}^{4} \dfrac{\partial n_i}{\partial \eta} y_i \end{bmatrix} = DERIV \times ELCOD^T$$

This is accomplished in FORTRAN 90 by a simple statement

J = matmul(DERIV,transpose(ELCOD))

- Calculate the determinant of J and its inverse

$$DetJ = j(1,1) \times j(2,2) - j(1,2) \times j(2,1)$$

$$JINVESRSE = \frac{1}{DET\ J} \begin{bmatrix} j(2,2) & -j(1,2) \\ -j(2,1) & j(1,1) \end{bmatrix}$$

- Calculate the Cartesian derivatives

$$CARTD = \begin{bmatrix} \frac{\partial N_1}{\partial x} & \frac{\partial N_2}{\partial x} & \frac{\partial N_3}{\partial x} & \frac{\partial N_4}{\partial x} \\ \frac{\partial N_1}{\partial y} & \frac{\partial N_2}{\partial y} & \frac{\partial N_3}{\partial y} & \frac{\partial N_4}{\partial y} \end{bmatrix} = JINVERSE \times DERIV$$

This is accomplished in FORTRAN 90 by a simple statement

CARTD = matmul(JINVERSE,DERIV)

- From CARTD, B matrix can be constructed

$$B = \begin{bmatrix} \frac{\partial N_1}{\partial x} & 0 & \frac{\partial N_2}{\partial x} & 0 & \frac{\partial N_3}{\partial x} & 0 & \frac{\partial N_4}{\partial x} & 0 \\ 0 & \frac{\partial N_1}{\partial y} & 0 & \frac{\partial N_2}{\partial y} & 0 & \frac{\partial N_3}{\partial y} & 0 & \frac{\partial N_4}{\partial y} \\ \frac{\partial N_1}{\partial y} & \frac{\partial N_1}{\partial x} & \frac{\partial N_2}{\partial y} & \frac{\partial N_2}{\partial x} & \frac{\partial N_3}{\partial y} & \frac{\partial N_3}{\partial x} & \frac{\partial N_4}{\partial y} & \frac{\partial N_4}{\partial x} \end{bmatrix}$$

- Form $[B^T D B] w_i w_j$ Det J.

This is accomplished in FORTRAN 90 by a simple statement

KMAT = KMAT + $w_i \times w_j \times$ DET J × matmul(transpose(B),matmul(D,B))

- Form matrix TEMPORARY= SHAPE x SHAPE^T

$$TEMPORARY = \begin{bmatrix} n_1n_1 & n_1n_2 & n_1n_3 & n_1n_4 \\ n_2n_1 & n_2n_2 & n_2n_3 & n_2n_4 \\ n_3n_1 & n_3n_2 & n_3n_3 & n_3n_4 \\ n_4n_1 & n_4n_2 & n_4n_3 & n_4n_4 \end{bmatrix}$$

In FORTRAN 90

TEMPORARY = matmul(SHAPE, transpose(SHAPE))

- From this the N matrix can be formed

$$N = \begin{bmatrix} N_1N_1 & 0 & N_1N_2 & 0 & N_1N_3 & 0 & N_1N_4 & 0 \\ & N_1N_1 & 0 & N_1N_2 & 0 & N_1N_3 & 0 & N_1N_4 \\ & & N_2N_2 & 0 & N_2N_3 & 0 & N_2N_4 & 0 \\ & & & N_2N_2 & 0 & N_2N_3 & 0 & N_2N_4 \\ & & & & N_3N_3 & 0 & N_3N_4 & 0 \\ symm. & & & & & N_3N_3 & 0 & N_3N_4 \\ & & & & & & N_4N_4 & 0 \\ & & & & & & & N_4N_4 \end{bmatrix}$$

- MASS = MASS + w_i x w_j x DET J x N
- Repeat the calculations for all values of j Gauss points.
- Repeat the calculations for all values of i Gauss points.

At the end of the calculations, KMAT and MASS matrices will contain the element static stiffness and mass matrices respectively.

8.9 IN-PLANE QUADRATIC ISOPARAMETRIC ELEMENT

In Section 8.6, attention was concentrated on developing a quadrilateral element as it was more general than a rectangular element. The strain distribution in the quadrilateral element is linear. More complex elements can be developed where the strain distribution is even more complex. Generally it has been found that it is more economical for an accurate analysis to use less number of complex elements rather than a large number of simple elements. One of the most commonly used elements in two-dimensional analysis is an eight node quadratic element as shown in Fig. 8.8. Apart from the nodes at the corners, it has nodes at the middle of the edges as well. Considering the square element shown in the (ξ,η) axes, with the edges defined by $\xi = \pm 1$, $\eta = \pm 1$, the shape function can be developed by assuming that

$$u(\xi,\eta) = C_1 + C_2\xi + C_3\eta + C_4\xi^2 + C_5\xi\eta + C_6\eta^2 + C_7\xi^2\eta + C_8\xi\eta^2 \tag{8.61}$$

It can be noticed that since there are only eight nodes, only eight constants can be used in defining u. Using the boundary conditions

$u(-1,-1) = u_1$, $u(0,-1) = u_2$, $u(1,-1) = u_3$, $u(-1,0) = u_8$, $u(1,0) = u_4$
$u(-1,1) = u_7$, $u(0,1) = u_6$, $u(1,1) = u_5$

the constants can be determined and the shape functions are given as follows

$$\begin{aligned}
n_1(\xi,\eta) &= -\frac{1}{4}(1-\xi)(1-\eta)(1+\xi+\eta), \quad n_2(\xi,\eta) = \frac{1}{2}(1-\xi^2)(1-\eta)\\
n_3(\xi,\eta) &= -\frac{1}{4}(1+\xi)(1-\eta)(1-\xi+\eta), \quad n_4(\xi,\eta) = \frac{1}{2}(1+\xi)(1-\eta^2)\\
n_5(\xi,\eta) &= -\frac{1}{4}(1+\xi)(1+\eta)(1-\xi-\eta), \quad n_6(\xi,\eta) = \frac{1}{2}(1-\xi^2)(1+\eta)\\
n_7(\xi,\eta) &= -\frac{1}{4}(1-\xi)(1+\eta)(1+\xi-\eta), \quad n_1(\xi,\eta) = \frac{1}{2}(1-\xi)(1-\eta^2)
\end{aligned} \tag{8.62}$$

Once the shape functions are known, the rest of the development follows the same steps as in Section 8.6.

8.10 SUMMARY OF RESULTS FROM THE THEORY OF THIN PLATES

Kirchoff's thin plate theory is the two-dimensional equivalent of the Engineer's Theory of Bending, and the usual assumption that plane sections remain plane and shear deformation is negligible is made. Because plate is the continuum equivalent of a plane grid, it bends about the x and y axes and, in addition, also experiences a twist. The 'stresses' to be considered are

- M_x, bending moment about the x axis and the corresponding shear force Q_x
- M_y, bending moment about the x axis and the corresponding shear force Q_y
- M_{xy}, the twisting moment.

The following results from the classical Theory of Plates will be needed for developing the properties of the element. The deflection of the plate in the z direction is w(x,y).

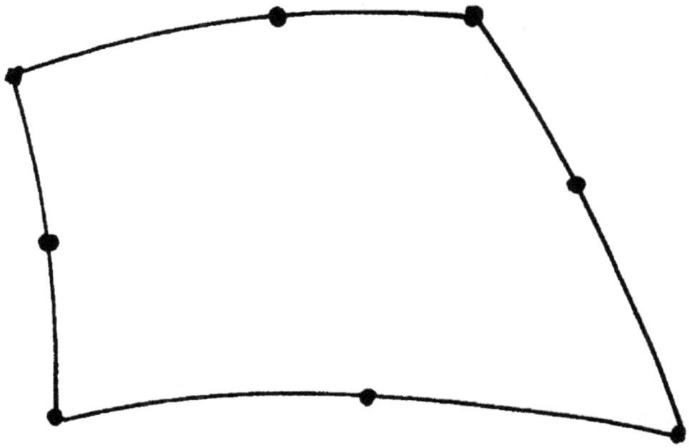

Fig. 8.8 Eight node quadrilateral element

8.10.1 Equilibrium

Using the forces acting on an element, as shown in Fig.8.9, the equations of equilibrium are

$$\frac{\partial M_x}{\partial x} + Q_x - \frac{\partial M_{xy}}{\partial y} = 0, \frac{\partial M_y}{\partial y} + Q_y - \frac{\partial M_{xy}}{\partial x} = 0, \frac{\partial Q_x}{\partial x} + \frac{\partial Q_y}{\partial y} - m\frac{\partial^2 w}{\partial t^2} = 0 \qquad (8.63)$$

Eliminating Q_x and Q_y from equation(8.63), the equation of equilibrium is

$$\frac{\partial^2 M_x}{\partial x^2} + \frac{\partial^2 M_y}{\partial y^2} - 2\frac{\partial^2 M_{xy}}{\partial x \partial y} + m\frac{\partial^2 w}{\partial t^2} = 0 \qquad (8.64)$$

8.10.2 Moment-Curvature Relationship

The 'moment-curvature' relationships are given by

$$M_x = D_x[\frac{\partial^2 w}{\partial x^2} + \upsilon_x \frac{\partial^2 w}{\partial y^2}],\ M_y = D_y[\frac{\partial^2 w}{\partial y^2} + \upsilon_y \frac{\partial^2 w}{\partial x^2}],\ M_{xy} = -D_{xy} 2\frac{\partial^2 w}{\partial x \partial y} \qquad (8.65)$$

$$D_x = \frac{E_x}{(1-\upsilon_x \upsilon_y)}\frac{t^3}{12}, D_y = \frac{E_y}{(1-\upsilon_x \upsilon_y)}\frac{t^3}{12}, D_{xy} = G\frac{t^3}{12} \qquad (8.66)$$

The above relationships can be expressed in matrix form as

$$\sigma = D\,\varepsilon \qquad (8.67)$$

$$\sigma = \begin{bmatrix} M_x \\ M_y \\ M_{xy} \end{bmatrix}, \ \varepsilon = \begin{bmatrix} \frac{\partial^2 w}{\partial x^2} \\ \frac{\partial^2 w}{\partial y^2} \\ -2\frac{\partial^2 w}{\partial x \partial y} \end{bmatrix}, \ D = \begin{bmatrix} D_x & \upsilon_x D_x & 0 \\ \upsilon_y D_y & D_y & 0 \\ 0 & 0 & D_{xy} \end{bmatrix} \tag{8.68}$$

8.10.3 Strain energy

The expression for strain energy is given by

$$Strain\,energy = \frac{1}{2}\int_{x=-a}^{a}\int_{y=-b}^{b}\{M_x \frac{\partial^2 w}{\partial x^2} + M_y \frac{\partial^2 w}{\partial y^2} + M_{xy}(-2\frac{\partial^2 w}{\partial x \partial y}\}\,dy\,dx \tag{8.69}$$

Using the matrix notation of equation (8.68), equation (8.69), becomes

$$Strain\,energy = \frac{1}{2}\int_{x=-a}^{a}\int_{y=-b}^{b}\{\varepsilon^T \sigma\}\,dy\,dx\,t \tag{8.70}$$

8.10.4 Equations of Equilibrium in terms of Displacements

Substituting equations (8.65) in equation (8.64)

$$D_x \frac{\partial^4 w}{\partial x^4} + 2(\upsilon_x D_x + 2D_{xy})\frac{\partial^4 w}{\partial x^2 \partial y^2} + D_y \frac{\partial^4 w}{\partial y^4} + m\frac{\partial^2 w}{\partial t^2} = 0 \tag{8.71}$$

If damping is ignored and the vibration is free, then we need to consider only the amplitudes. Substituting for w = W(x,y) cosωt, equation (8.71) can be modified in to

$$D_x \frac{\partial^4 W}{\partial x^4} + 2(\upsilon_x D_x + 2D_{xy})\frac{\partial^4 W}{\partial x^2 \partial y^2} + D_y \frac{\partial^4 W}{\partial y^4} - m\omega^2 W = 0 \tag{8.72}$$

8.10.5 Frequencies of vibration of a simply supported plate

Equation (8.72) can be used to calculate the frequency of vibration of a simply supported plate of dimension 2a x 2b. Taking the origin of coordinates at the centre of the plate and the axes parallel to the edges, displacement amplitude W can be expressed as a double Fourier series

$$W = \sum_{p=1}^{\infty}\sum_{q=1}^{\infty} c_{pq} \sin\frac{p\pi x}{a} \sin\frac{q\pi y}{b}$$

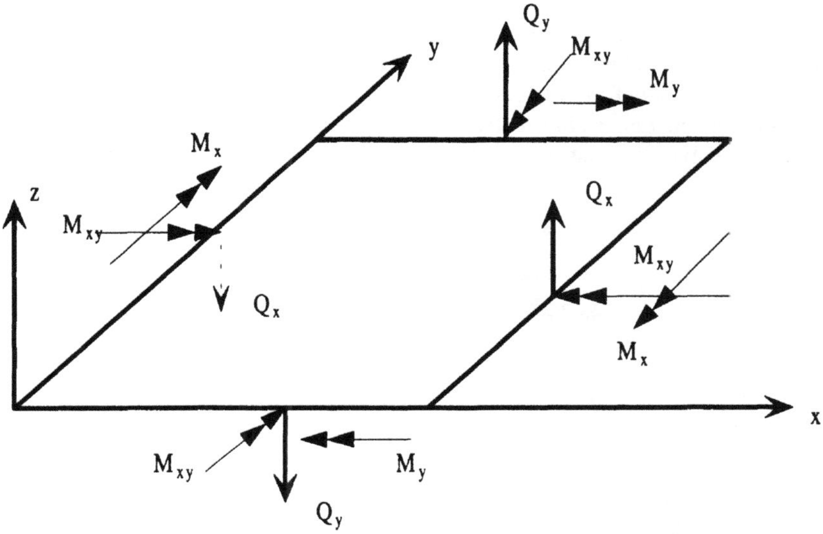

Fig. 8.9 Forces acting on an element of a plate

This expression satisfies the boundary conditions of zero displacement and zero curvature at the edges. Substituting W into equation (8.72) and simplifying

$$\omega^2 = \frac{\pi^4}{m}\{D_x\frac{p^4}{a^4} + 2(v_x D_x + 2D_{xy})\frac{p^2 q^2}{a^2 b^2} + D_y\frac{q^4}{b^4}\}$$

For an isotropic plate, $D = Et^3/[12(1\text{-}v^2)]$ and ω is given by

$$\omega = \pi^2\{\frac{p^2}{a^2}+\frac{q^2}{b^2}\}\sqrt{\frac{D}{m}}$$

8.11 A KIRCHOFF RECTANGULAR PLATE-BENDING ELEMENT: SHAPE FUNCTIONS

Figure 8.10 shows a rectangular plate-bending element. As in the case of a beam element, we need to consider at each node a translation and because the plate is a two-dimensional structure, a rotation about the x axis and a rotation about the y axis. At each node there are therefore three 'displacements' to consider and the element has therefore twelve degrees of freedom. Assume, therefore, a polynomial expression for W consisting of twelve arbitrary constants

$$W = C_1 + C_2x + C_3y + C_4x^2 + C_5xy + C_6y^2 + C_7x^3 + C_8x^2y + C_9xy^2 + C_{10}y^3 \\ + C_{11}x^3y + C_{12}xy^3 \tag{8.73}$$

Note that the above polynomial includes all linear, quadratic and cubic polynomial terms but only two terms out of five of the quartic polynomial. The terms excluded are x^4, x^2y^2 and y^4. This was an arbitrary choice except that it is important that the displacement pattern chosen must be independent of the direction of the co-ordinate axes chosen. This is known as 'geometric invariance' criterion. Although the x^2 y^2 term does satisfy the necessary criteria, it could not be included because of the restriction on being able to have only twelve independent constants. The next step is to solve the constants in terms of the 'displacements' at the nodes of the element.

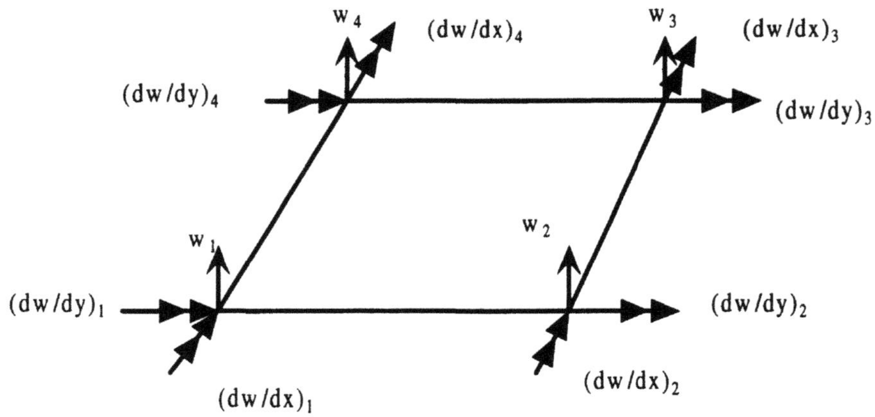

Fig. 8.10 A rectangular plate-bending element

$$At\ node\ i, W = w_i\ ,\ \frac{\partial W}{\partial y} = \theta_{xi},\ \frac{\partial W}{\partial x} = -\theta_{yi}\ ,\ \mathrm{i = 1, 2, 3\ and\ 4} \tag{8.74}$$

$$W = \sum_{i=1}^{4}\{N_i(x,y)\,w_i + N_{i+1}(x,y)\theta_{xi} + N_{i+2}(x,y)\theta_{yi} = \sum_{i=1}^{12} N_i(x,y)\delta_i$$

where for ease of mathematical manipulations

$$\delta^T = [\delta_1\quad \delta_2\quad \delta_3\quad \delta_4\quad \delta_5\quad \delta_6\quad \delta_7\quad \delta_8\quad \delta_9\quad \delta_{10}\quad \delta_{11}\quad \delta_{12}]$$

$$= \lfloor w_1\quad \theta_{x1}\quad \theta_{y1}\quad w_2\quad \theta_{x2}\quad \theta_{y2}\quad w_3\quad \theta_{x3}\quad \theta_{y3}\quad w_4\quad \theta_{x4}\quad \theta_{y4} \rfloor \tag{8.75}$$

The shape functions and their derivatives are, setting, $\xi = x/a$ and $\eta = y/b$,

$$N_1 = 1 - \frac{1}{4}(1+\xi)(1+\eta) - \frac{1}{8}\{(1+\xi)^2(2-\xi)(1-\eta) + (1-\xi)(1+\eta)^2(2-\eta)\}$$

$$\frac{\partial N_1}{\partial \xi} = -\frac{1}{4}\frac{1}{a}(1+\eta) + \frac{1}{8}\frac{1}{a}\{3(1-\xi^2)(1-\eta) + (1+\eta)^2(2-\eta)\}$$

$$\frac{\partial N_1}{\partial \eta} = -\frac{1}{4}\frac{1}{b}(1+\xi) + \frac{1}{8}\frac{1}{b}\{(1+\xi)^2(2-\xi) + 3(1-\xi)(1-\eta^2)\}$$

$$N_2 = \frac{b}{8}(1-\xi)(1-\eta)^2(1+\eta)$$

$$\frac{\partial N_2}{\partial \xi} = -\frac{1}{8}\frac{b}{a}(1-\eta)^2(1+\eta)\,,\ \frac{\partial N_2}{\partial \eta} = -\frac{1}{8}(1-\xi)(1-\eta)(1+3\eta)$$

$$N_3 = -\frac{a}{8}(1-\xi)^2(1+\xi)(1-\eta)$$

$$\frac{\partial N_3}{\partial \xi} = \frac{1}{8}(1-\xi)(1+3\xi)(1-\eta)\,,\ \frac{\partial N_3}{\partial \eta} = \frac{1}{8}\frac{a}{b}(1-\xi)^2(1+\xi)$$

$$N_4 = -\frac{1}{8}(1+\xi)(1-\eta)\{(1+\xi)(\xi-2)+(1+\eta)\eta\}$$

$$\frac{\partial N_4}{\partial \xi} = \frac{1}{8}\frac{1}{a}(1-\eta)\{3(1-\xi^2)+(1+\eta)\eta\}$$

$$\frac{\partial N_4}{\partial \eta} = \frac{1}{8}\frac{1}{b}(1+\xi)\{(1+\xi)(\xi-2)-(1-3\eta^2)\}$$

$$N_5 = \frac{b}{8}(1+\xi)(1-\eta)^2(1+\eta)$$

$$\frac{\partial N_5}{\partial \xi} = \frac{1}{8}\frac{b}{a}(1-\eta)^2(1+\eta)\,,\ \frac{\partial N_5}{\partial \eta} = -\frac{1}{8}(1+\xi)(1-\eta)(1+3\eta)$$

$$N_6 = \frac{a}{8}(1+\xi)^2(1-\xi)(1-\eta)$$

$$\frac{\partial N_6}{\partial \xi} = \frac{1}{8}(1+\xi)(1-3\xi)(1-\eta)\,,\ \frac{\partial N_6}{\partial \eta} = -\frac{1}{8}\frac{a}{b}(1+\xi)^2(1-\xi)$$

$$N_7 = -\frac{1}{8}(1+\xi)(1+\eta)\{2+(1+\xi)(\xi-2)+(1+\eta)(\eta-2)\}$$

$$\frac{\partial N_7}{\partial \xi} = -\frac{1}{8}\frac{1}{a}(1+\eta)\{3\xi^2-1+(1+\eta)(\eta-2)\}$$

$$\frac{\partial N_7}{\partial \eta} = -\frac{1}{8}\frac{1}{b}(1+\xi)\{(1+\xi)(\xi-2)-1+3\eta^2\}$$

$$N_8 = -\frac{b}{8}(1+\xi)(1+\eta)^2(1-\eta)$$

$$\frac{\partial N_8}{\partial \xi} = -\frac{1}{8}\frac{b}{a}(1+\eta)^2(1-\eta)\,,\ \frac{\partial N_8}{\partial \eta} = -\frac{1}{8}(1+\xi)(1+\eta)(1-3\eta)$$

$$N_9 = \frac{a}{8}(1+\xi)^2(1-\xi)(1+\eta)$$

$$\frac{\partial N_9}{\partial \xi} = \frac{1}{8}(1+\xi)(1-3\xi)(1+\eta)\,,\ \frac{\partial N_9}{\partial \eta} = \frac{1}{8}\frac{a}{b}(1+\xi)^2(1-\xi)$$

$$N_{10} = -\frac{1}{8}(1-\xi)(1+\eta)\{(1+\xi)\xi+(1+\eta)(\eta-2)\}$$

$$\frac{\partial N_{10}}{\partial \xi} = -\frac{1}{8}\frac{1}{a}(1+\eta)\{(1-3\xi^2)-(1+\eta)(\eta-2)\}$$

$$\frac{\partial N_{10}}{\partial \eta} = -\frac{1}{8}\frac{1}{b}(1-\xi)\{(1+\xi)\xi-3(1-\eta^2)\}$$

$$N_{11} = -\frac{b}{8}(1-\xi)(1+\eta)^2(1-\eta)$$

$$\frac{\partial N_{11}}{\partial \xi} = \frac{1}{8}\frac{b}{a}(1-\eta)(1+\eta)^2, \quad \frac{\partial N_{11}}{\partial \eta} = -\frac{1}{8}(1-\xi)(1+\eta)(1-3\eta)$$

$$N_{12} = -\frac{a}{8}(1+\xi)(1-\xi)^2(1+\eta)$$

$$\frac{\partial N_{12}}{\partial \xi} = \frac{1}{8}(1+3\xi)(1-\xi)(1+\eta), \quad \frac{\partial N_{12}}{\partial \eta} = -\frac{1}{8}\frac{a}{b}(1+\xi)(1-\xi)^2 \qquad (8.76)$$

In order to check the continuity of displacements across element boundaries consider the edge 2-3 defined by x = a

$$w(a, y) = N_4 w_2 + N_5\theta_{x2} + N_7 w_3 + N_8\theta_{x3}$$

$$\frac{\partial w}{\partial y} = \frac{\partial N_4}{\partial y} w_2 + \frac{\partial N_5}{\partial y}\theta_{x2} + \frac{\partial N_7}{\partial y} w_3 + \frac{\partial N_8}{\partial y}\theta_{x3}$$

All other shape functions are zero.

This shows that the displacement w and the slope $\partial w/\partial y$ along that edge are governed by the displacements of the nodes governing that edge. This means that an adjacent element defined by the same two nodes also has identical displacements along a common edge, thus ensuring continuity of these two 'displacements'.

When checking the continuity of slope $\partial w/\partial x$ along the edge 2-3, it is found that not only the derivatives of shape functions at the nodes defining the edge 2-3, but also the derivatives of shape functions at nodes 1 and 2, are also involved. Since across any edge only two nodes are common, clearly there is no continuity of the slope

$$\frac{\partial w}{\partial x} = \frac{\partial N_4}{\partial x} w_2 + \frac{\partial N_5}{\partial x}\theta_{x2} + \frac{\partial N_6}{\partial x}\theta_{y2} + \frac{\partial N_7}{\partial x} w_3 + \frac{\partial N_8}{\partial x}\theta_{x3} + \frac{\partial N_9}{\partial x}\theta_{y3}$$

$$+ \frac{\partial N_1}{\partial x} w_1 + \frac{\partial N_2}{\partial x}\theta_{x1} + \frac{\partial N_{10}}{\partial x} w_4 + \frac{\partial N_{11}}{\partial x}\theta_{x4}$$

Such elements which violate continuity conditions are called non-conforming elements. However, such elements are extensively used in practice and appear to behave well.

Once the shape functions are known, the stiffness matrix for the element can be determined using the Galerkin's weighted residual method or by the minimum total potential approach.

8.11.1 Conditions That Displacement Functions Must Satisfy

The basic premise of the finite element technique rests up on the assumption that as the number of elements increases, solution converges to the exact solution in the limit. It can be shown that for a finite element solution to produce convergent results, the element displacement functions must be continuous and have continuous derivatives up to an order (n - 1), where n is the highest order of the derivative in the expression for strain energy. In the case of plate vibration problems, only second order derivatives are involved. Therefore, the function w and its first order derivatives must be continuous within and across the elements. If these conditions are satisfied, then the element is said to be a conforming element. When a conforming element is used in the analysis, the predicted frequency will always be lower than the exact value, because discretisation introduces constraints on a continuous structure. On the other hand, if the element is non-conforming, then the predicted frequency can be higher or lower than the exact frequency. When polynomial functions are used to generate shape functions, they must be complete polynomials of at least degree n. However, terms from polynomials of degree higher than n can be used without their being complete but they must satisfy 'geometric invariance' criterion. In other words, if x and y terms are interchanged, there must be no difference in the resulting polynomial.

8.12 GALERKIN'S WEIGHTED RESIDUAL METHOD

When using the approximate solution given by equation (8.71) to solve the differential equation (8.60), the average weighted residual is zero, when

$$\int_{x=-a}^{a}\int_{y=-b}^{b}\{\frac{\partial^2 M_x}{\partial x^2}+\frac{\partial^2 M_y}{\partial y^2}-2\frac{\partial^2 M_{xy}}{\partial x\partial y}+m\frac{\partial^2 w}{\partial t^2}\}N_j\,dx\,dy=0 \tag{8.77}$$

Integrating partially with respect to x

$$\int_{y=-b}^{b}[\int_{x=-a}^{a}\frac{\partial^2 M_x}{\partial x^2}N_j\,dx]dy=\int_{y=-b}^{b}\frac{\partial M_x}{\partial x}N_j\Big|_{x=-a}^{a}dy-\int_{y=-b}^{b}\int_{x=-a}^{a}\frac{\partial M_x}{\partial x}\frac{\partial N_j}{\partial x}dx\,dy \tag{8.78}$$

Integrating once again with respect to x, the integral on the right-hand side of equation (8.78)

$$\int_{y=-b}^{b}[\int_{x=-a}^{a}\frac{\partial M_x}{\partial x}\frac{\partial N_j}{\partial x}dx]dy=\int_{y=-b}^{b}M_x\frac{\partial N_j}{\partial x}\Big|_{x=-a}^{a}dy-\int_{y=-b}^{b}\int_{x=-a}^{a}M_x\frac{\partial^2 N_j}{\partial x^2}dx\,dy \tag{8.79}$$

Similarly integrating partially with respect to y

$$\int_{x=-a}^{a}[\int_{y=-b}^{b}\frac{\partial^2 M_y}{\partial y^2}N_j\,dy]dx=\int_{x=-a}^{a}\frac{\partial M_y}{\partial y}N_j\Big|_{y=-b}^{b}dx-\int_{x=-a}^{a}\int_{y=-b}^{b}\frac{\partial M_y}{\partial y}\frac{\partial N_j}{\partial y}dx\,dy \tag{8.80}$$

Integrating once again with respect to y, the integral on the right-hand side of equation (8.80)

$$\int_{x=-a}^{a}\left[\int_{y=-b}^{b}\frac{\partial M_y}{\partial y}\frac{\partial N_j}{\partial y}dy\right]dx=\int_{x=-a}^{a}M_y\frac{\partial N_j}{\partial y}\Big|_{y=-b}^{b}dx-\int_{y=-b}^{b}\int_{x=-a}^{a}M_y\frac{\partial^2 N_j}{\partial y^2}dx\,dy \tag{8.81}$$

Integrating partially with respect to x

$$\int_{y=-b}^{b}\left[\int_{x=-a}^{a}\frac{\partial^2 M_{xy}}{\partial x\partial y}N_j\,dx\right]dy=\int_{y=-b}^{b}\frac{\partial M_{xy}}{\partial y}N_j\Big|_{x=-a}^{a}dy-\int_{y=-b}^{b}\int_{x=-a}^{a}\frac{\partial M_{xy}}{\partial y}\frac{\partial N_j}{\partial x}dx\,dy \tag{8.82}$$

Integrating once again with respect to y, the integral on the right-hand side of equation (8.82)

$$\int_{x=-a}^{a}\left[\int_{y=-b}^{b}\frac{\partial M_{xy}}{\partial y}\frac{\partial N_j}{\partial x}dx\right]dy=\int_{x=-a}^{a}M_{xy}\frac{\partial N_j}{\partial x}\Big|_{y=-b}^{b}dx-\int_{y=-b}^{b}\int_{x=-a}^{a}M_{xy}\frac{\partial^2 N_j}{\partial x\partial y}dx\,dy \tag{8.83}$$

Integrating partially with respect to y

$$\int_{x=-a}^{a}\left[\int_{y=-b}^{b}\frac{\partial^2 M_{xy}}{\partial x\partial y}N_j\,dy\right]dx=\int_{x=-a}^{a}\frac{\partial M_{xy}}{\partial x}N_j\Big|_{y=-b}^{b}dx-\int_{y=-b}^{b}\int_{x=-a}^{a}\frac{\partial M_{xy}}{\partial x}\frac{\partial N_j}{\partial y}dx\,dy \tag{8.84}$$

Integrating once again with respect to x, the integral on the right-hand side of equation (8.84)

$$\int_{y=-b}^{b}\left[\int_{x=-a}^{a}\frac{\partial M_{xy}}{\partial x}\frac{\partial N_j}{\partial y}dy\right]dx=\int_{y=-b}^{b}M_{xy}\frac{\partial N_j}{\partial y}\Big|_{x=-a}^{a}dy-\int_{y=-b}^{b}\int_{x=-a}^{a}M_{xy}\frac{\partial^2 N_j}{\partial x\partial y}dx\,dy \tag{8.85}$$

Note that from equation (8.63)

$$\int_{y=-b}^{b}\left\{\frac{\partial M_x}{\partial x}-\frac{\partial M_{xy}}{\partial y}\right\}N_j\Bigg|_{x=-a}^{a}dy=\int_{y=-b}^{b}\{-Q_x\}N_j\Big|_{x=-a}^{a}dy \tag{8.86}$$

$$\int_{x=-a}^{a}\left\{\frac{\partial M_y}{\partial y}-\frac{\partial M_{xy}}{\partial x}\right\}N_j\Bigg|_{y=-b}^{b}dx=\int_{x=-a}^{a}\{-Q_y\}N_j\Big|_{y=-b}^{b}dy \tag{8.87}$$

Substituting equations (8.78) to (8.87) in equation (8.73) and simplifying

$$\int_{x=-a}^{a}\int_{y=-b}^{b}\left\{M_x\frac{\partial^2 N_j}{\partial x^2}-2M_{xy}\frac{\partial^2 N_j}{\partial x\partial y}+M_y\frac{\partial^2 N_j}{\partial y^2}-m\omega^2\,wN_j\right\}dx\,dy=F_j \tag{8.88}$$

where

$$F_j=\int_{y=-b}^{b}\left\{M_{xy}\frac{\partial N_j}{\partial y}-M_x\frac{\partial N_j}{\partial x}-Q_xN_j\right\}\Bigg|_{x=-a}^{a}dy$$
$$+\int_{x=-a}^{a}\left\{M_{xy}\frac{\partial N_j}{\partial x}-M_y\frac{\partial N_j}{\partial y}-Q_yN_j\right\}\Bigg|_{y=-b}^{b}dx \tag{8.89}$$

Substituting for moments in terms of curvatures and twists from equation (8.68), in equation (8.89)

$$\int_{x=-a}^{a}\int_{y=-b}^{b}\{D_x\frac{\partial^2 w}{\partial x^2}\frac{\partial^2 N_j}{\partial x^2}+2(\upsilon_x D_x+2D_{xy})\frac{\partial^2 w}{\partial x\partial y}\frac{\partial^2 N_j}{\partial x\partial y}+D_y\frac{\partial^2 w}{\partial y^2}\frac{\partial^2 N_j}{\partial y^2}$$
$$-m\omega^2\ wN_j\}dx\,dy=F_j \tag{8.90}$$

Substituting equation (8.75) in equation (8.90)

$$\int_{x=-a}^{a}\int_{y=-b}^{b}\sum_{i=1}^{12}\{D_x\frac{\partial^2 N_i}{\partial x^2}\frac{\partial^2 N_j}{\partial x^2}+2(\upsilon_x D_x+2D_{xy})\frac{\partial^2 N_i}{\partial x\partial y}\frac{\partial^2 N_j}{\partial x\partial y}+D_y\frac{\partial^2 N_i}{\partial y^2}\frac{\partial^2 N_j}{\partial y^2}$$
$$-m\omega^2\ N_iN_j\}\delta_i\,dx\,dy=F_j,\ j=1-12 \tag{8.91}$$

From equation (8.87), the general term for the static stiffness and consistent mass matrices are given by

$$k_{ij}=\int_{x=-a}^{a}\int_{y=-b}^{b}\{D_x\frac{\partial^2 N_i}{\partial x^2}\frac{\partial^2 N_j}{\partial x^2}+2(\upsilon_x D_x+2D_{xy})\frac{\partial^2 N_i}{\partial x\partial y}\frac{\partial^2 N_j}{\partial x\partial y}$$
$$+D_y\frac{\partial^2 N_i}{\partial y^2}\frac{\partial^2 N_j}{\partial y^2}\}dxdy \tag{8.92}$$

$$m_{ij}=m\omega^2\int_{x=-a}^{a}\int_{y=-b}^{b}N_iN_j\,dx\,dy \tag{8.93}$$

The coefficients of K and M matrices can be evaluated using Tables 8.7 to 8.10. The forces F_j arise from the distributed 'stresses' on the edges of the element. It is important to appreciate that when the elements are assembled to form a structure, at any node, these forces arising from internal stresses will sum to zero. Therefore there is no need to compute these forces.

Table 8.7 Integral: $6\frac{a}{b}\int_{x=-a}^{a}\int_{y=-b}^{b}\frac{\partial^2 N_i}{\partial x^2}\frac{\partial^2 N_j}{\partial x^2}dy\,dx$

i	j											
	1	2	3	4	5	6	7	8	9	10	11	12
1	$6/a^2$	0	6/a	$6/a^2$	0	6/a	$3/a^2$	0	3/a	$-3/a^2$	0	3/a
2	0	0	0	0	0	0	0	0	0	0	0	0
3	6/a	0	8	6/a	0	8	3/a	0	2	-3/a	0	4
4	$6/a^2$	0	6/a	$6/a^2$	0	6/a	$3/a^2$	0	3/a	$-3/a^2$	0	3/a
5	0	0	0	0	0	0	0	0	0	0	0	0
6	6/a	0	8	6/a	0	8	3/a	0	4	-3/a	0	2
7	$3/a^2$	0	3/a	$3/a^2$	0	3/a	$6/a^2$	0	6/a	$-6/a^2$	0	6/a
8	0	0	0	0	0	0	0	0	0	0	0	0
9	3/a	0	2	3/a	0	4	6/a	0	8	-6/a	0	4
10	$-3/a^2$	0	-3/a	$-3/a^2$	0	-3/a	$-6/a^2$	0	-6/a	$6/a^2$	0	-6/a
11	0	0	0	0	0	0	0	0	0	0	0	0
12	3/a	0	4	3/a	0	2	6/a	0	4	-6/a	0	8

Table 8.8 Integral $6\frac{b}{a}\int_{x=-a}^{a}\int_{y=-b}^{b}\frac{\partial^2 N_i}{\partial y^2}\frac{\partial^2 N_j}{\partial x^2}dy\,dx$

i	j											
	1	2	3	4	5	6	7	8	9	10	11	12
1	$6/b^2$	-6/b	0	$-3/b^2$	-3/b	0	$-3/b^2$	-3/b	0	$6/b^2$	-6/b	0
2	-6/b	8	0	3/b	4	0	3/b	2	0	-6/b	4	0
3	0	0	0	0	0	0	0	0	0	0	0	0
4	$-3/b^2$	3/b	0	$6/b^2$	6/b	0	$6/b^2$	6/b	0	$-3/b^2$	3/b	0
5	-3/b	4	0	6/b	8	0	6/b	4	0	-3/b	2	0
6	0	0	0	0	0	0	0	0	0	0	0	0
7	$-3/b^2$	3/b	0	$6/b^2$	6/b	0	$6/b^2$	6/b	0	$-3/b^2$	3/b	0
8	-3/b	2	0	6/b	4	0	6/b	8	0	-3/b	4	0
9	0	0	0	0	0	0	0	0	0	0	0	0
10	$6/b^2$	-6/b	0	$-3/b^2$	-3/b	0	$-3/b^2$	-3/b	0	$6/b^2$	-6/b	0
1	-6/b	4	0	3/b	2	0	3/b	4	0	-6/b	8	0
12	0	0	0	0	0	0	0	0	0	0	0	0

Table 8.9 Integral $12ab\int_{x=-a}^{a}\int_{y=-b}^{b}\frac{\partial^2 N_i}{\partial x\partial y}\frac{\partial^2 N_j}{\partial x\partial y}dy\,dx$

i	j											
	1	2	3	4	5	6	7	8	9	10	11	12
1	3	0	0	3	0	0	3	0	0	3	0	0
2	0	b^2	0	0	$-b^2$	0	0	b^2	0	0	$-b^2$	0
3	0	0	a^2	0	0	$-a^2$	0	0	a^2	0	0	$-a^2$
4	3	0	0	3	0	0	3	0	0	3	0	0
5	0	$-b^2$	0	0	b^2	0	0	$-b^2$	0	0	b^2	0
6	0	0	$-a^2$	0	0	a^2	0	0	$-a^2$	0	0	a^2
7	3	0	0	3	0	0	3	0	0	3	0	0
8	0	b^2	0	0	$-b^2$	0	0	b^2	0	0	$-b^2$	0
9	0	0	a^2	0	0	$-a^2$	0	0	a^2	0	0	$-a^2$
10	3	0	0	3	0	0	3	0	0	3	0	0
11	0	$-b^2$	0	0	b^2	0	0	$-b^2$	0	0	b^2	0
12	0	0	$-a^2$	0	0	a^2	0	0	$-a^2$	0	0	a^2

Table 8.10 Integral $\frac{6300}{ab}\int_{x=-a}^{a}\int_{y=-b}^{b} N_i N_j \, dy \, dx$

i	j											
	1	2	3	4	5	6	7	8	9	10	11	12
1	3454	922b	-922a	1226	398b	548a	394	-232b	232a	1226	-548b	-398a
2	922b	$320b^2$	-252ab	398b	$160b^2$	168ab	232b	$-120b^2$	112ab	548b	$-240b^2$	-168ab
3	-922a	-252ab	$320a^2$	-548a	-168ab	$-240a^2$	-232a	112ab	$-120a^2$	-398a	168ab	$160a^2$
4	1226	398b	-548a	3454	922b	922a	1226	-548b	398a	394	-232b	-232a
5	398b	$160b^2$	-168ab	922b	$320b^2$	252ab	548b	$-240b^2$	168ab	232b	$-120b^2$	-112ab
6	548a	168ab	$-240a^2$	922a	252ab	$320a^2$	398a	-168ab	$160a^2$	232a	-112ab	$-120a^2$
7	394	232b	-232a	1226	548b	398a	3454	-922b	922a	1226	-398b	-548a
8	-232b	232b	112ab	-548b	$-240b^2$	-168ab	-922b	$320b^2$	-252ab	-398b	$160b^2$	168ab
9	232a	112ab	$-120a^2$	398a	168ab	$160a^2$	922a	-252ab	$320a^2$	548a	-168ab	$-240a^2$
10	1226	548b	-398a	394	232b	232a	1226	-398b	548a	3454	-922b	-922a
11	-548b	$-240b^2$	168ab	-232b	$-120b^2$	-112ab	-398b	$160b^2$	-168ab	-922b	$320b^2$	252ab
12	-398a	-168ab	$160a^2$	-232a	-112ab	$-120a^2$	-548a	168ab	$-240a^2$	-922a	252ab	$320a^2$

8.13 DERIVATION OF K AND M MATRICES BY MINIMISATION OF TOTAL POTENTIAL

Following the procedure explained in Section 8.4 for the derivation of K and M matrices for a plane stress element, the three components making up the total potential are calculated as follows.

8.13 1. Strain energy

Substituting for w from equation (8.75), for strain ε in equation (8.68) and carrying out the differentiation, ε = B δ, where

$$B = \begin{bmatrix} \frac{\partial^2 N_1}{\partial x^2} & \frac{\partial^2 N_2}{\partial x^2} & \frac{\partial^2 N_3}{\partial x^2} & \cdots\cdots & \frac{\partial^2 N_{11}}{\partial x^2} & \frac{\partial^2 N_{12}}{\partial x^2} \\ \frac{\partial^2 N_1}{\partial y^2} & \frac{\partial^2 N_2}{\partial y^2} & \frac{\partial^2 N_3}{\partial y^2} & \cdots\cdots & \frac{\partial^2 N_{11}}{\partial y^2} & \frac{\partial^2 N_{12}}{\partial y^2} \\ -2\frac{\partial^2 N_1}{\partial x \partial y} & -2\frac{\partial^2 N_2}{\partial x \partial y} & -2\frac{\partial^2 N_3}{\partial x \partial y} & \cdots\cdots & -2\frac{\partial^2 N_{11}}{\partial x \partial y} & -2\frac{\partial^2 N_{12}}{\partial x \partial y} \end{bmatrix} \tag{8.94}$$

Substituting for σ = D ε from equation (8.64), the expression for strain energy in equation (8.66) becomes

$$Strain\ energy = \frac{1}{2}\delta^T \left[\int_{x=-a}^{a} \int_{y=-b}^{b} \{ B^T D B \}\, dy\, dx \right] \delta \tag{8.95}$$

8.13.2 Loss of Potential by External Loads

At any node i, as shown in Fig. 8.11, the nodal force are a force F_{zi}, in the vertical z direction, a couple M_{xi} about the x-axis and another couple M_{yi} about the y axis. The loss of potential by these loads is equal to

$$Loss\ of\ potential\ by\ external\ loads = -\delta^T\ F \tag{8.96}$$

where the nodal force vector F is given by

$$F = \begin{bmatrix} F_{z1} & M_{x1} & M_{y1} & \cdots\cdots & F_{z4} & M_{x4} & M_{y4} \end{bmatrix}^T \tag{8.97}$$

8.13.3 Loss of Potential by Inertial Forces

The loss of potential by the distributed inertial forces is calculated as follows. The inertial force in the z directions is m ω^2 w. Substituting for w in terms of the shape functions, the inertial forces in the z direction is $m\omega^2\ \Sigma N_i\ \delta_i$. The loss of potential by the distributed inertial forces is therefore given by

$$Loss\ of\ potential = -\frac{1}{2} m\omega^2 [\int_{x=-a}^{a} \int_{y=-b}^{b} \{w \sum_{i=1}^{4} N_i \delta_i\} dy\, dx] \tag{8.98}$$

$$Loss\ of\ potential = -\frac{1}{2} m\omega^2 [\int_{x=-a}^{a} \int_{y=-b}^{b} \{\sum_{j=1}^{4} N_j \delta_j \sum_{i=1}^{4} N_i \delta_i\} dy\, dx] \tag{8.99}$$

Equation (8.99) can be more conveniently written as

$$Loss\ of\ potential = -\frac{1}{2} m\omega^2 [\int_{x=-a}^{a} \int_{y=-b}^{b} \{\sum_{i=1}^{4} \sum_{j=1}^{4} N_i N_j \delta_i \delta_j\} dy\, dx] \tag{8.100}$$

Or in matrix form equation (8.100) can be expressed as

$$Loss\ of\ potential\ by\ inertial\ forces = -\frac{1}{2} \omega^2 \delta^T M\, \delta \tag{8.101}$$

where the $(i,j)^{th}$ coefficient of the matrix M is given by

$$(i, j)^{th}\ element = [m \int_{x=-a}^{a} \int_{y=-b}^{b} \{N_i\ N_j\} dy\, dx] \tag{8.102}$$

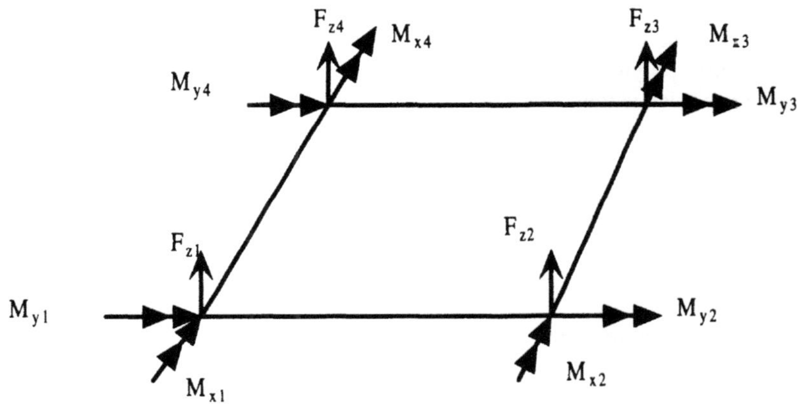

Fig. 8.11 Nodal forces on a rectangular plate-bending element

8.13.4 Matrices K and M

From equations (8.95), (8.96) and (8.101), the total potential for the element is given by

$$\pi_e = \frac{1}{2} \delta^T [\int_{x=-a}^{a} \int_{y=-b}^{b} \{B^T D B\} dy\, dx\, t] \delta - \frac{1}{2} \omega^2 \delta^T M\, \delta - \delta^T F \tag{8.103}$$

Differentiating equation (8.103) with respect to δ, the matrix relationship can be expressed as

$$\frac{\partial \pi_e}{\partial \delta} = [\int_{x=-a}^{a} \int_{y=-b}^{b} \{B^T D B\} dy\, dx\, t] \delta - \omega^2 M\, \delta - F \tag{8.104}$$

The static stiffness matrix K and consistent mass matrix M are given by

$$K = [\int_{x=-a}^{a} \int_{y=-b}^{b} \{B^T D B\} dy\, dx\, t] \text{ , mass matrix} = M \tag{8.105}$$

The force–deformation relationship is therefore the same as equation (8.42).

8.14 A CONFORMING RECTANGULAR THIN PLATE ELEMENT

It was shown in Section 8.11, that the twelve degree freedom rectangular plate-bending element is a non-conforming element. It is possible to develop a corresponding conforming element with 16 degrees of freedom.

Therefore assume a polynomial expression for W consisting of 16 arbitrary constants.

$$W = C_1 + C_2 x + C_3 y + C_4 x^2 + C_5 xy + C_6 y^2 + C_7 x^3 + C_8 x^2 y + C_9 xy^2 + C_{10} y^3$$
$$+ C_{11} x^3 y + C_{12} x^2 y^2 + C_{13} xy^3 + C_{14} x^3 y^2 + C_{15} x^2 y^3 + C_{16} x^3 y^3 \tag{8.106}$$

Note that the above polynomial includes all linear, quadratic and cubic polynomial terms but only three terms out of five of the quartic polynomial, two terms out of six of the quintic polynomial and only one term from the six degree polynomial. The reason for choice is the same as that in Section 8.11.

The next step is to solve the constants in terms of the 'displacements' at the nodes of the element

$$At\ node\ i, W = w_i\ ,\ \frac{\partial W}{\partial y} = \theta_{xi},\ \frac{\partial W}{\partial x} = -\theta_{yi},\ \frac{\partial^2 W}{\partial x \partial y} = \theta_{xyi}\ ,\ \text{i} = 1, 2, 3 \text{ and } 4$$

$$W = \sum_{i=1}^{4} \{N_i(x,y) w_i + N_{i+1}(x,y)\theta_{xi} + N_{i+2}(x,y)\theta_{yi} + N_{i+3}(x,y)\theta_{xyi}\}$$

W can be expressed more compactly as

$$W = \sum_{i=1}^{16} N_i(x,y)\delta_i$$

where for ease of algebraic manipulation

$$\delta^T = \lfloor \delta_1 \quad \delta_2 \quad \delta_3 \quad \delta_4 \quad . \quad . \quad . \quad . \quad \delta_{13} \quad \delta_{14} \quad \delta_{15} \quad \delta_{16} \rfloor$$
$$= \lfloor w_1 \quad \theta_{x1} \quad \theta_{y1} \quad \theta_{xy1} \quad . \quad . \quad . \quad . \quad w_4 \quad \theta_{x3} \quad \theta_{y4} \quad \theta_{xy4} \rfloor$$

Once the shape functions are known, the static stiffness and mass matrices can be developed using the procedure followed in Section 8.13.

8.15 MINDLIN-REISSNER ISOPARAMETRIC QUADRATIC THICK PLATE-BENDING ELEMENT

Just as the Kirchoff plate-bending element was the two-dimensional equivalent of a beam element with shear deformation effects ignored, in a similar manner the Mindlin-Reissner element is equivalent to the beam element where shear deformation effects have been included. Shear deformation effects become important when modelling cellular and sandwich plates. The assumptions made are similar to the ones made in developing the beam static stiffness and consistent mass matrices for a beam element in Section 7.6. The total rotation of the cross section is the sum of the slope due to bending and the average strain due to shear force

$$\frac{\partial w}{\partial x} = \psi_x + \gamma_x, \ \frac{\partial w}{\partial y} = \psi_y + \gamma_y \tag{8.107}$$

The moment-curvature relationships are

$$M_x = D_x \frac{\partial \psi_x}{\partial x} + \upsilon_x D_x \frac{\partial \psi_y}{\partial y}, \quad M_y = D_y \frac{\partial \psi_y}{\partial y} + \upsilon_x D_x \frac{\partial \psi_x}{\partial x}$$

$$M_{xy} = -D_{xy}\{\frac{\partial \psi_x}{\partial y} + \frac{\partial \psi_y}{\partial x}\} \tag{8.108}$$

The above equations for flexure can be expressed in matrix form as

$$\sigma_f = D_f \ \varepsilon_f \tag{8.109}$$

$$\sigma_f = \begin{bmatrix} M_x \\ M_y \\ M_{xy} \end{bmatrix}, \ \varepsilon_f = \begin{bmatrix} \frac{\partial \psi_x}{\partial x} \\ \frac{\partial \psi_y}{\partial y} \\ -\{\frac{\partial \psi_x}{\partial y} + \frac{\partial \psi_y}{\partial x}\} \end{bmatrix}, \ D_f = \begin{bmatrix} D_x & \upsilon_x D_x & 0 \\ \upsilon_y D_y & D_y & 0 \\ 0 & 0 & D_{xy} \end{bmatrix} \tag{8.110}$$

The shear force-shear strain relationships are

$$Q_x = G A_s \ \gamma_x = G A_s \{\frac{\partial w}{\partial x} - \psi_x\}, \ Q_y = G A_s \ \gamma_y = G A_s \{\frac{\partial w}{\partial y} - \psi_y\} \tag{8.111}$$

The above equations for flexure can be expressed in matrix form as

$$\sigma_s = D_s \ \varepsilon_s \tag{8.112}$$

where

$$\sigma_s = \begin{bmatrix} Q_x \\ Q_y \end{bmatrix}, \ D_s = \begin{bmatrix} GA_s & 0 \\ 0 & GA_s \end{bmatrix}, \ \varepsilon_s = \begin{bmatrix} \frac{\partial w}{\partial x} - \psi_x \\ \frac{\partial w}{\partial y} - \psi_y \end{bmatrix} \tag{8.113}$$

Referring to the element shown in Fig. 8.9, at each node i, the three degrees of freedoms are a translation in the z direction, a rotation ψ_x about the y axis and ψ_y a rotation about the x axis. Therefore the 'displacements' in the element are

$$w(\xi,\eta) = \sum_{i=1}^{8} n_i(\xi,\eta)\, w_i, \ \psi_x(\xi,\eta) = \sum_{i=1}^{8} n_i(\xi,\eta)\psi_{xi}, \ \psi_y(\xi,\eta) = \sum_{i=1}^{8} n_i(\xi,\eta)\psi_{yi} \tag{8.114}$$

Note that ψ_x and ψ_y are the slopes of the deflected curve due to bending, and are treated as independent variables and are not expressed as the first derivative of displacement, w.

The coordinate 'mapping' is given by

$$x(\xi,\eta) = \sum_{i=1}^{8} n_i(\xi,\eta)\, x_i, \quad y(\xi,\eta) = \sum_{i=1}^{8} n_i(\xi,\eta)\, y_i \tag{8.115}$$

The shape functions n_i are given by equation (8.62) and the nodal displacement vector δ is given by

$$\delta = \{w_1\,,\psi_{x1}\,,\psi_{y1}\,,w_2\,,\psi_{x2}\,,\psi_{y2}\,,\ldots,w_8\,,\psi_{x8}\,,\psi_{y8}\}^T \tag{8.116}$$

The flexural and shear strain vectors can be expressed as

$$\varepsilon_f = B_f\ \delta \tag{8.117}$$

$$B_f(x,y) = \begin{bmatrix} 0 & \frac{\partial N_1}{\partial x} & 0 & 0 & \frac{\partial N_2}{\partial x} & 0 & \cdots & 0 & \frac{\partial N_8}{\partial x} & 0 \\ 0 & 0 & \frac{\partial N_1}{\partial y} & 0 & 0 & \frac{\partial N_2}{\partial y} & \cdots & 0 & 0 & \frac{\partial N_8}{\partial y} \\ 0 & -\frac{\partial N_1}{\partial y} & -\frac{\partial N_1}{\partial x} & 0 & -\frac{\partial N_2}{\partial y} & -\frac{\partial N_2}{\partial x} & \cdots & 0 & -\frac{\partial N_8}{\partial y} & -\frac{\partial N_8}{\partial x} \end{bmatrix} \tag{8.118}$$

$$\varepsilon_s = B_s\ \delta \tag{8.119}$$

$$B_s(x,y) = \begin{bmatrix} \frac{\partial N_1}{\partial x} & -N_1 & 0 & \frac{\partial N_2}{\partial x} & -N_2 & 0 & \cdots & \frac{\partial N_8}{\partial x} & -N_8 & 0 \\ \frac{\partial N_1}{\partial y} & 0 & -N_1 & \frac{\partial N_2}{\partial y} & 0 & -N_2 & \cdots & \frac{\partial N_4}{\partial y} & 0 & -N_4 \end{bmatrix} \tag{8.120}$$

The derivatives of shape functions N_i with respect to (x,y) are related to the derivatives shape functions n_i with respect to (ξ,η) through the inverse of the Jacobian as shown in equations (8.49) to (8.52). The strain energy can be calculated from

$$Strain\ energy = \frac{1}{2}\int_{x=-a}^{a}\int_{y=-b}^{b}\{\varepsilon_f^T\,\sigma_f + \varepsilon_s^T\,\sigma_s\}dydx \tag{8.121}$$

Substituting equations (8.109), (8.112), (8.117) and (8.119), equation (8.121) becomes

$$Strain\ energy = \frac{1}{2}\delta^T[\int_{x=-a}^{a}\int_{y=-b}^{b}\{B_f^T\,D_f\,B_f + B_s^T\,D_s\,B_s\}dydx]\delta \tag{8.122}$$

Substituting

$$K = [\int_{x=-a}^{a}\int_{y=-b}^{b}\{B_f^T\,D_f\,B_f + B_s^T\,D_s\,B_s\}dydx] \tag{8.123}$$

The strain energy stored is given by

$$Strain\ energy = \frac{1}{2}\delta^T\,K\,\delta \tag{8.124}$$

The loss of potential by the inertial force is given by

$$Loss\ of\ potential = -\int_{x=-a}^{a}\int_{y=-b}^{b}\{-m\,\frac{\partial^2 w}{\partial t^2}\,w\}\,dy\,dy \tag{8.125}$$

Substituting for w in terms of amplitudes as w = W cosωt, and expressing the displacement W in terms of the shape functions as $W = \sum_{i=1}^{8} n_i W_i$ and simplifying,

$$Loss\ of\ potential = -\frac{1}{2} m\omega^2 \int_{x=-a}^{a} \int_{y=-b}^{b} [\sum_{i=1}^{8}\sum_{j=1}^{8} n_i n_j W_i W_j] dy\, dx \tag{8.126}$$

The equation (8.126) can be expressed in matrix form as

$$Loss\ of\ potential = \frac{1}{2}\omega^2 \delta^T M \delta \tag{8.127}$$

Where (i,j)th coefficient of the matrix M is given by

$$m_{ij} = m \int_{x=-a}^{a} \int_{y=-b}^{b} N_i N_j\, dx\, dy = m \int_{\xi=-1}^{1} \int_{\eta=-1}^{1} n_i n_j\, d\eta\, d\xi\ DetJ \tag{8.128}$$

Finally the loss of potential by nodal forces is given by

$$Loss\ of\ potential = -\frac{1}{2}\delta^T F \tag{8.129}$$

Where

$$F = \begin{bmatrix} F_{z1} & M_{x1} & M_{y1} & F_{z2} & M_{x2} & M_{y2} & . & . & . & F_{z8} & M_{x8} & M_{y8} \end{bmatrix}^T \tag{8.130}$$

The total potential for the element is therefore

$$\pi_e = \frac{1}{2}\delta^T K \delta - \omega^2 \frac{1}{2}\delta^T M \delta - \delta^T F \tag{8.131}$$

Differentiating equation (8.131) with respect to δ

$$\frac{\partial \pi_e}{\partial \delta} = K\delta - \omega^2 M \delta - F \tag{8.132}$$

The integrals in calculating the coefficients of K and M matrices can be evaluated using Gaussian integration.

8.16 ISOPARAMETRIC QUADRATIC FLAT SHELL ELEMENT

A flat thin shell element can be developed which can analyse structures where both in-plane and out of plane effects are important. At each node there are five degrees of freedom $[u_i, v_i, w_i, (\partial w/\partial x)_i, (\partial w/\partial y)_i]$. This element is simply 'addition' of the in-plane and plate-bending elements developed in Sections 8.9 and 8.15 respectively.

Such an element can be used for the analysis of folded plate and box girder structures. Although the plate-bending element used is a non-conforming element, this is preferred to a conforming element, purely because it is difficult to interpret the compatibility of the twist term $\partial^2 w/\partial x \partial y$ at the nodes where several inclined plates normal to the x-z plane intersect.

8.17 REDUCED INTEGRATION

When the appropriate number of integration points is used to calculate the exact integral, it is called full integration. When the number of integration points used are less than the optimum, the calculated integral will be in error and this is called reduced integration. It might be puzzling at first as to why anybody should want reduced integration. To illustrate the need for this, consider the rectangular plane stress element considered in Section 8.3. Imagine that the nodes are subjected to the following displacements.

$u_1 = -u_4 = \Delta,\ u_3 = -u_2 = \Delta,\ v_1 = v_2 = v_3 = v_4 = 0$

From equations (8.11) and (8.12), displacements are

$u(x,y) = \Delta\,(x/a)\,(y/b),\ v(x,y) = 0$

The strains are

$$\varepsilon_x = \frac{\partial u}{\partial x} = \frac{\Delta}{a}\frac{y}{b},\ \varepsilon_y = \frac{\partial v}{\partial y} = 0,\ \gamma_{xy} = \frac{\partial u}{\partial y} + \frac{\partial v}{\partial x} = \frac{\Delta}{b}\frac{x}{a}$$

Figure 8.12 shows the intended displacement pattern, which is a pure bending displacement, but from the shape functions the calculated displacement pattern leads to shear strains.

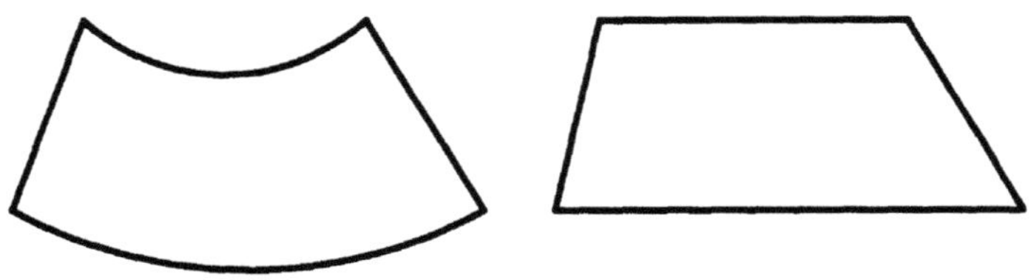

Fig. 8.12 Spurious shear strains

The normal strain in the x direction varies linearly over the depth of the element as is the case under pure bending deformation. The corresponding strain energy is given by

$$\frac{1}{2}\int \varepsilon_x \sigma_x\, dvol = \frac{1}{2}E\int \varepsilon_x^2\ dA\ dx = \frac{1}{2}E\frac{\Delta^2}{a^2}\frac{1}{b^2}\int y^2\ dA\ dx = \frac{1}{2}E\frac{\Delta^2}{a^2}\frac{1}{b^2}I\int dx$$

Note that by definition the second moment of area I = $\int y^2\ dA$. The strain at the top face is $\varepsilon = \Delta / a$ and the curvature is $\varepsilon / b = \Delta / ab$. Therefore the strain energy is given by

$$\frac{1}{2}\int \varepsilon_x \sigma_x\, dvol = \frac{1}{2}E\,I\ Curvature^2 \int dx$$

Thus integration using two Gauss points, which integrates exactly polynomials up to third degree, leads to the correct strain energy. However the undesirable shear strain is zero only at the centre of the element. Therefore when calculating the strain energy due to normal strains if one point integration rule is used, then the strain energy due to this spurious shear strain can be eliminated. In other words, in this case using reduced integration is beneficial. Reduced integration has the effect of 'softening' the contribution to stiffness matrix from shear strain energy.

A somewhat similar situation arises when very thin plates are analysed using isoparametric plate-bending elements. The contribution to stiffness matrix from shear strain energy is directly proportional to thickness. On the other hand, contribution from 'flexural' strain energy is directly proportional to the cube of the thickness. Because of this, in very thin elements, off diagonal terms due to shear strain energy could be much larger than the diagonal terms due to flexural strain energy. This leads to the serious problem of 'locking', where as the thickness of the element is reduced, one gets erroneous results. The problem can be overcome by using reduced integration when calculating shear strain energy terms. However, exact integration should be used for flexural strain energy terms.

CHAPTER 9

FINITE STRIP METHOD

9.1 INTRODUCTION

In Chapter 8, the finite element method was described. This is certainly a very versatile method. Unfortunately, its very versatility makes it quite expensive to use. In this chapter, the finite strip method, which has many similarities with the finite element method but perhaps is not as versatile, will be described. It is particularly suited for the analysis of structures with constant cross-section and the end boundary conditions that do not change transversely.

9.2 BENDING FINITE STRIP WITH SIMPLY SUPPORTED ENDS

Consider a plate as shown in Fig. 9.1, which is simply supported at the ends parallel to the x axis. The edges parallel to the y axis can have any boundary condition. The only restriction is that these conditions are constant along the entire span. The plate can be imagined to be divided into a series of strips of constant width parallel to the y axis. These imaginary strips are called finite strips and the edges of the strip are called nodal lines. In contrast to the finite element, we have a finite strip and instead of nodes we have nodal lines.

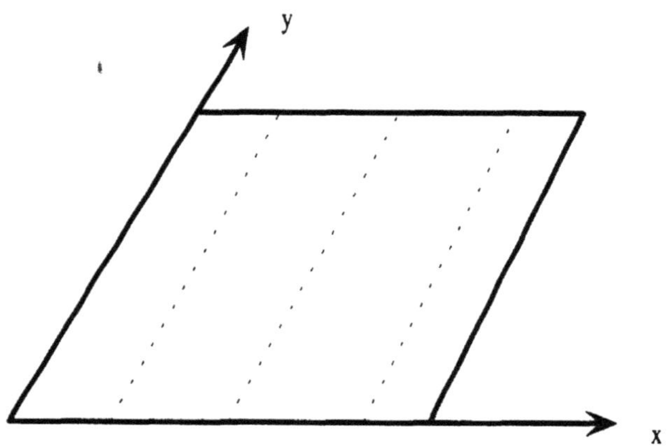

Fig. 9.1 A simply supported plate

Consider a strip simply supported on the edges parallel to the x axis as shown in Fig. 9.2.

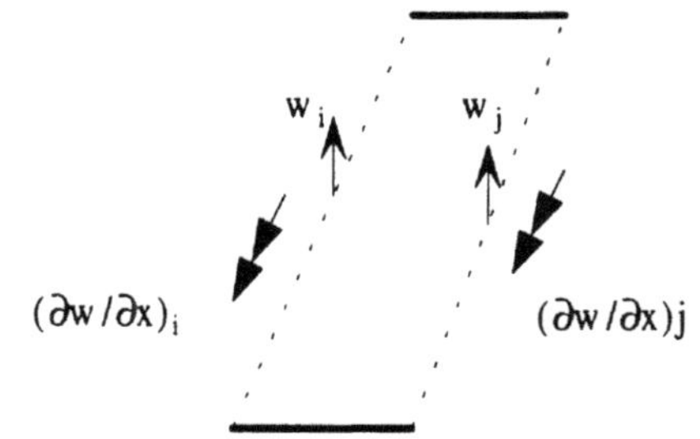

Fig. 9.2 A general bending strip

The displacement in the z direction at the ith and j[th] nodal lines can be expressed as a Fourier series.

$$w_i(y) = \sum_{n=1}^{\infty} w_{in} \sin\frac{n\pi y}{L}, \quad w_j(y) = \sum_{n=1}^{\infty} w_{jn} \sin\frac{n\pi y}{L} \tag{9.1}$$

w_{in} and w_{jn} are coefficients of the Fourier sine series.

Equation (9.1) satisfies the simply supported boundary condition, $w_i(y) = w_j(y) = 0$ and $\frac{d^2 w_i(y)}{dy^2} = 0$ and $\frac{d^2 w_j(y)}{dy^2} = 0$ at y = 0 and L.

Having described the variation of displacement along a nodal line, a similar series can be used to describe the rotation about the y axis as well. Thus

$$(\frac{\partial w}{\partial x})_i = \theta_{yi}(y) = \sum_{n=1}^{\infty} \theta_{yin} \sin\frac{n\pi y}{L}, \quad (\frac{\partial w}{\partial x})_j = \theta_{yj}(y) = \sum_{n=1}^{\infty} \theta_{yjn} \sin\frac{n\pi y}{L} \tag{9.2}$$

In the transverse or x-direction, the displacement variation can be described using the shape functions used for the beam in Section 7.12.2. Therefore

$$w(x, y) = N_1(x)\, w_i(y) + N_2(x)\theta_{yi}(y) + N_3(x)\, w_j(y) + N_4(x)\theta_{yj}(y)$$

$$= \sum_{n=1}^{\infty} \{N_1(x)\, w_{in} + N_2(x)\theta_{yin} + N_3(x)\, w_{jn} + N_4(x)\theta_{yjn}\} \sin\frac{n\pi y}{L} \tag{9.3}$$

where

$$N_1 = 1 - 3(\frac{x}{b})^2 + 2(\frac{x}{b})^3, \quad N_2 = -b\{\frac{x}{b} - 2(\frac{x}{b})^2 + (\frac{x}{b})^3\}$$

$$N_3 = 3(\frac{x}{b})^2 - 2(\frac{x}{b})^3, \quad N_4 = b\{(\frac{x}{b})^2 - (\frac{x}{b})^3\} \tag{9.4}$$

For a thin Kirchoff plate, the moment-curvature relationship is given by equation (8.64) and the expression for strain energy is by equation (8.65).

Replacing the moments in equation (8.65) by curvatures from equation (8.64)

$$Strain\ energy = \frac{1}{2}\int_{x=0}^{b}\int_{y=0}^{L}\{D_x(\frac{\partial^2 w}{\partial x^2})^2 + D_y(\frac{\partial^2 w}{\partial y^2})^2$$

$$+(\upsilon_x D_x + \upsilon_y D_y)\frac{\partial^2 w}{\partial x^2}\frac{\partial^2 w}{\partial y^2} + 4D_{xy}(\frac{\partial^2 w}{\partial x\partial y})^2\}\,dy\,dx \tag{9.5}$$

where

$$\frac{\partial^2 w(x,y)}{\partial x^2} = \sum_{n=1}^{\infty}\{\frac{d^2N_1(x)}{dx^2}w_{in} + \frac{d^2N_2(x)}{dx^2}\theta_{yin}$$

$$+\frac{d^2N_3(x)}{dx^2}w\theta_{jn} + \frac{d^2N_4(x)}{dx^2}\theta_{yjn}\}\sin\frac{n\pi y}{L} \tag{9.6}$$

$$\frac{\partial^2 w(x,y)}{\partial y^2} = -\sum_{n=1}^{\infty}\{N_1(x)\,w_{in} + N_2(x)\theta_{yin} + N_3(x)\,w_{jn}$$

$$+N_4(x)\theta_{yjn}\}(\frac{n\pi}{L})^2\sin\frac{n\pi y}{L} \tag{9.7}$$

$$\frac{\partial^2 w(x,y)}{\partial x\partial y} = \sum_{n=1}^{\infty}\{\frac{dN_1(x)}{dx}w_{in} + \frac{dN_2(x)}{dx}\theta_{yin} + \frac{d^2N_3(x)}{dx})\,w_{jn}$$

$$+\frac{d^2N_4(x)}{dx}\theta_{yjn}\}(\frac{n\pi}{L})\cos\frac{n\pi y}{L} \tag{9.8}$$

From Section 8.13.3, the loss of potential by the inertial force is given by

$$Loss\ of\ potential = -\frac{1}{2}m\omega^2\int_{x=0}^{b}\int_{y=0}^{L}w^2\,dy\,dx \tag{9.9}$$

The total potential for the strip is
π_e = strain energy + loss of potential by inertial force + external forces.
Following the steps in Section 8.13.4

$$\frac{\partial\pi_e}{\partial\delta_{jm}} = \int_{x=0}^{b}\int_{y=0}^{L}\{D_x(\frac{\partial^2 w}{\partial x^2})\frac{d^2N_j}{dx^2}\sin\frac{m\pi y}{L} - D_y(\frac{\partial^2 w}{\partial y^2})N_j(\frac{m\pi}{L})^2\sin\frac{m\pi y}{L}$$

$$-0.5(\upsilon_x D_x + \upsilon_y D_y)\frac{\partial^2 w}{\partial x^2}N_j(\frac{m\pi}{L})^2\sin\frac{m\pi y}{L} + 0.5(\upsilon_x D_x$$

$$+\upsilon_y D_y)\frac{\partial^2 w}{\partial y^2}\frac{d^2N_j}{dx^2}\sin\frac{m\pi y}{L} + 4D_{xy}\frac{\partial^2 w}{\partial x\partial y}\frac{dN_j}{dx}(\frac{m\pi}{L})\cos\frac{m\pi y}{L}\,dx\,dy$$

$$-m\omega^2\int_{x=0}^{b}\int_{y=0}^{L}wN_j\sin\frac{m\pi y}{L}\,dy\,dx - F_{jm} \tag{9.10}$$

where $\delta_{1m} = w_{im}$, $\delta_{2m} = \theta_{yim}$, $\delta_{3m} = w_{jm}$, $\delta_{4m} = w_{yjm}$.
Using the well-known orthogonal property

$$\int_0^L \sin\frac{n\pi y}{L}\sin\frac{m\pi y}{L}dy = \int_0^L \cos\frac{n\pi y}{L}\cos\frac{m\pi y}{L}dy = \frac{L}{2}, m=n \qquad (9.11)$$
$$=0, m\neq n$$

equation (9.10) becomes

$$\frac{\partial\pi_e}{\partial\delta_{jm}} = \frac{L}{2}\int_{x=0}^{b} \{D_x\sum_{i=1}^{4}(\frac{d^2N_i}{dx^2}\delta_{im})\frac{d^2N_j}{dx^2} + D_y\sum_{i=1}^{4}(N_i\delta_{im}(\frac{m\pi}{L})^2)N_j(\frac{m\pi}{L})^2$$
$$-0.5(\upsilon_x D_x + \upsilon_y D_y)\sum_{i=1}^{4}(\frac{d^2N_i}{dx^2}\delta_{im})N_j(\frac{m\pi}{L})^2 - 0.5(\upsilon_x D_x$$
$$+\upsilon_y D_y)\sum_{i=1}^{4}(N_i\delta_{im})\frac{d^2N_j}{dx^2} + 4D_{xy}\sum_{i=1}^{4}[\frac{dN_i}{dx}\delta_{im}(\frac{m\pi}{L})]\frac{dN_j}{dx}(\frac{m\pi}{L})\}dy\,dx$$
$$-m\omega^2\frac{L}{2}\int_{x=0}^{b}\sum_{i=1}^{4}(N_i\delta_{im})N_j\,dy\,dx - F_{jm} \qquad (9.12)$$

Equation (9.12) can be expressed more compactly as

$$\frac{\partial\pi_e}{\partial\delta_{jm}} = \sum_{i=1}^{4}k_{ijm}\,\delta_{im} - \omega^2\sum_{i=1}^{4}m_{ijm}\,\delta_{im} - F_{jm} \qquad (9.13)$$

where

$$k_{ijm} = D_x L\frac{1}{2}\int_{x=0}^{b} D_x\frac{d^2N_i}{dx^2}\frac{d^2N_j}{dx^2}dx + D_y L(\frac{m\pi}{L})^4\frac{1}{2}\int_{x=0}^{b} N_iN_j dx$$
$$-D_1(\frac{m\pi}{L})^2 L\int_{x=0}^{b}\frac{1}{2}[\frac{d^2N_i}{dx^2}N_j + N_i\frac{d^2N_j}{dx^2}]dx + D_{xy}L(\frac{m\pi}{L})^2 2\int_{x=0}^{b}\frac{dN_i}{dx}\frac{dN_j}{dx}dx \qquad (9.14)$$
$$D_1 = \upsilon_x D_x = \upsilon_y D_y\ ,\ m_{ijm} = \frac{L}{2}m\int_{x=0}^{b}\{N_j\ N_i\}dx$$

The integrals $\int_0^b \frac{d^2N_i}{dx^2}\frac{d^2N_j}{dx^2}dx$ and $\int_0^b N_i\,N_j\,dx$ in equation (9.14) can be evaluated from Tables 7.2 and 7.3 by replacing L by b and the other integral is given in Table 9.1.

If $\alpha_m = \dfrac{m\pi}{L}$, and the displacements are in the order

$$\delta_m^T = \{w_{im}, \theta_{yim}, w_{jm}, \theta_{yjm}\}$$

the symmetric coefficients of the static stiffness and mass matrices are given by

$$k_{11} = k_{33} = D_x\frac{6L}{b^3} + D_y\alpha_m^4\frac{13bL}{70} + \alpha_m^2\frac{L}{b}(1.2D_1 + 2.4D_{xy})$$
$$k_{12} = -[D_x\frac{3L}{b^2} + D_y\alpha_m^4\frac{b^2L11}{210} + \alpha_m^2 L(D_1 1.1 + D_{xy}0.2)]$$
$$k_{13} = -[D_x\frac{6L}{b^3} - D_y\alpha_m^4\frac{bL9}{140} + \alpha_m^2\frac{L}{b}(D_1 1.2 + D_{xy}2.4)]$$

Table 9.1 Integrals of shape functions

i	$b\int_0^b \frac{1}{2}\{N_i \frac{d^2N_j}{dx^2} + \frac{d^2N_i}{dx^2} N_j\}dx$			
	j			
	1	2	3	4
1	-1.2	0.6b	1.2	0.1b
2	0.6b	-0.1333	-0.1b	0.03333 b^2
3	1.2	-0.1b	-1.2	-0.6b
4	0.1b	0.03333 b^2	-0.6b	-0.13333

$$k_{14} = k_{13}\,,\; k_{23} = -k_{13}$$

$$k_{22} = k_{44} = D_x \frac{2L}{b} + D_y \alpha_m^4 \frac{b^3 L}{210} + \alpha_m^2 \frac{L}{b}(D_1 \frac{2}{15} + D_{xy} \frac{4}{15})$$

$$k_{24} = -[-D_x \frac{L}{b} + D_y \alpha_m^4 \frac{b^3 L1}{280} + \alpha_m^2 \frac{L}{b}(D_1 + D_{xy}\, 2)\frac{1}{30}]$$

$$k_{34} = [D_x \frac{3L}{b^2} + D_y \alpha_m^4 \frac{b^2 L11}{420} + \alpha_m^2 L(D_1 0.6 + D_{xy}\, 0.2)] \tag{9.15}$$

$$m_{11} = m_{33} = mLb\frac{13}{70},\; m_{12} = -\,mLb^2 \frac{11}{420},\; m_{13} = mLb\frac{9}{140},\; m_{14} = mLb^2 \frac{13}{840}$$

$$m_{22} = m_{44} = mLb^3 \frac{1}{210},\; m_{23} = -\,mLb^2 \frac{13}{840},\; m_{24} = -\,mLb^3 \frac{1}{280}$$

$$m_{34} = mLb^2 \frac{11}{420} \tag{9.16}$$

9.3 IN-PLANE FINITE STRIP WITH SIMPLY SUPPORTED ENDS

The displacement in the y direction at the ith and jth nodal lines can be expressed as a Fourier series

$$u_i(y) = \sum_{n=1}^{\infty} u_{in} \sin\frac{n\pi y}{L},\; u_j(y) = \sum_{n=1}^{\infty} u_{jn} \sin\frac{n\pi y}{L}$$

$$u(x,y) = N_1(x)u_i(y) + N_2(x)u_j(y) \tag{9.17}$$

$$N_1 = (1 - \frac{x}{b})\,,\; N_2 = (\frac{x}{b}) \tag{9.18}$$

Note that the assumed displacement pattern ensures that the displacement in the y direction is zero at the ends.

Similarly the displacement in the y direction at the ith and j[th] nodal lines can be expressed as a Fourier series

$$v_i(y) = \sum_{n=1}^{\infty} v_{in} \cos\frac{n\pi y}{L}, \quad v_j(y) = \sum_{n=1}^{\infty} v_{jn} \cos\frac{n\pi y}{L}$$

$$v(x, y) = N_1(x) v_i(y) + N_2(x) v_j(y) \tag{9.19}$$

where, u_{in}, u_{jn}, v_{in} and v_{jn} are coefficients of the Fourier series.

The reason for choosing cosine series instead of sine series for representing the displacement in the y direction is to ensure that at the ends of the strip $\sigma_y = 0$.

Using equations (8.5), (8.6) and (8.8)

$$Strain\ energy = \frac{1}{2} t \int_{x=0}^{b} \int_{y=0}^{L} \{D_x (\frac{\partial u}{\partial x})^2 + D_y (\frac{\partial v}{\partial y})^2 + (\upsilon_x D_x + \upsilon_y D_y) \frac{\partial u}{\partial x}\frac{\partial v}{\partial y}$$

$$+ G(\frac{\partial u}{\partial y} + \frac{\partial u}{\partial y})^2 \} \, dy \, dx \tag{9.20}$$

where

$$D_x = \frac{E_x}{(1-\upsilon_x \upsilon_y)} \; ; \; D_y = \frac{E_y}{(1-\upsilon_x \upsilon_y)} \; ;$$

$$\frac{\partial u(x, y)}{\partial x} = \sum_{n=1}^{\infty} \{ \frac{dN_1(x)}{dx} u_{in} + \frac{dN_2(x)}{dx}) u_{jn} \} \sin\frac{n\pi y}{L} \tag{9.21}$$

$$\frac{\partial u(x, y)}{\partial y} = \sum_{n=1}^{\infty} \{ N_1(x) u_{in} + N_2(x) \, u_{jn} \} \frac{n\pi}{L} \cos\frac{n\pi y}{L} \tag{9.22}$$

$$\frac{\partial v(x, y)}{\partial y} = - \sum_{n=1}^{\infty} \{ N_1(x) v_{in} + N_2(x) v_{in} \} (\frac{n\pi}{L}) \sin\frac{n\pi y}{L} \tag{9.23}$$

$$\frac{\partial v(x, y)}{\partial x} = \sum_{n=1}^{\infty} \{ \frac{dN_1(x)}{dx} v_{in} + \frac{dN_2(x)}{dx} v_{in} \} \cos\frac{n\pi y}{L} \tag{9.24}$$

From Section 8.5.3, the loss of potential by the inertial force is given by

$$Loss\ of\ potential = -\frac{1}{2} m \omega^2 \int_{x=0}^{b} \int_{y=0}^{L} [u^2 + v^2] \, dy \, dx \tag{9.25}$$

The total potential for the strip is

π_e = strain energy + loss of potential by inertial force + external forces.

Following the steps in Section 8.5.4

$$\frac{\partial \pi_e}{\partial u_{jm}}=t\int_{x=0}^{b}\int_{y=0}^{L}\{D_x(\frac{\partial u}{\partial x})\frac{dN_j}{dx}\sin\frac{m\pi y}{L}+$$

$$0.5(\upsilon_x D_x+\upsilon_y D_y)\frac{\partial v}{\partial y}\frac{dN_j}{dx}\sin\frac{m\pi y}{L}+G(\frac{\partial u}{\partial y}+\frac{\partial v}{\partial x})N_j(\frac{m\pi}{L})\cos\frac{m\pi y}{L}\}dy\,dx$$

$$-m\omega^2\int_{x=0}^{b}\int_{y=0}^{L}u N_j\sin\frac{m\pi y}{L}dy\,dx-F_{xjm} \tag{9.26}$$

$$\frac{\partial \pi_e}{\partial v_{jm}}=t\int_{x=0}^{b}\int_{y=0}^{L}\{-D_y(\frac{\partial v}{\partial y})\frac{m\pi}{L}\sin\frac{m\pi y}{L}-$$

$$0.5(\upsilon_x D_x+\upsilon_y D_y)\frac{\partial u}{\partial x}N_j(\frac{m\pi}{L})\sin\frac{m\pi y}{L}+G(\frac{\partial u}{\partial y}+\frac{\partial v}{\partial x})\frac{dN_j}{dx}\cos\frac{m\pi y}{L}\}dy\,dx$$

$$-m\omega^2\int_{x=0}^{b}\int_{y=0}^{L}v N_j\cos\frac{m\pi y}{L}dy\,dx-F_{yjm} \tag{9.27}$$

Using the well-known orthogonal property given in equation (9.11), equation (9.26) and (9.27) become

$$\frac{\partial \pi_e}{\partial u_{jm}}=\frac{L}{2}t\int_{x=0}^{b}\{D_x\sum_{i=1}^{2}(\frac{dN_i}{dx}u_{im})\frac{dN_j}{dx}-0.5(\upsilon_x D_x+\upsilon_y D_y)\sum_{i=1}^{2}(N_i v_{im})(\frac{m\pi}{L})\frac{dN_j}{dx}$$

$$+G\sum_{i=1}^{2}[N_i u_{im}(\frac{m\pi}{L})+\frac{dN_i}{dx}v_{im}]N_j(\frac{m\pi}{L})\}dx$$

$$-m\omega^2\frac{L}{2}\int_{x=0}^{b}\sum_{i=1}^{2}(N_i u_{im})N_j\,dy\,dx-F_{xjm} \tag{9.28}$$

$$\frac{\partial \pi_e}{\partial v_{jm}}=\frac{L}{2}t\int_{x=0}^{b}\{D_y\sum_{i=1}^{2}(N_i v_{im}(\frac{m\pi}{L}))(\frac{m\pi}{L})N_j-0.5(\upsilon_x D_x$$

$$+\upsilon_x D_x)\sum_{i=1}^{2}(\frac{dN_i}{dx}u_{im}(\frac{m\pi}{L})N_j+G\sum_{i=1}^{2}[N_i u_{im}(\frac{m\pi}{L})+\frac{dN_i}{dx}v_{im}]\frac{dN_j}{dx}\}dx$$

$$-m\omega^2\frac{L}{2}\int_{x=0}^{b}\sum_{i=1}^{2}(N_i v_{im})N_j\,dy\,dx-F_{yjm} \tag{9.29}$$

Regrouping, equations (9.28) and (9.29) simplify to

$$\frac{\partial \pi_e}{\partial u_{jm}}=\frac{L}{2}t\sum_{i=1}^{2}\int_{x=0}^{b}[D_x\frac{dN_i}{dx}\frac{dN_j}{dx}+G(\frac{m\pi}{L})^2N_iN_j]dx\,u_{im}$$

$$+\frac{L}{2}t\sum_{i=1}^{2}\int_{x=0}^{b}[-D_1\frac{m\pi}{L}N_i\frac{dN_j}{dx}+G\frac{m\pi}{L}\frac{dN_i}{dx}N_j]dx\,v_{ij}$$

$$-m\omega^2\frac{L}{2}\int_{x=0}^{b}\sum_{i=1}^{2}(N_i u_{im})N_j\,dx-F_{xjm} \tag{9.30}$$

$$\frac{\partial \pi_e}{\partial v_{jm}} = \frac{L}{2} t \sum_{i=1}^{2} \int_{x=0}^{b} [D_y (\frac{m\pi}{L})^2 N_i N_j + G (\frac{m\pi}{L}) \frac{dN_i}{dx} \frac{dN_j}{dx}] dx\, v_{im}$$

$$+ \frac{L}{2} t \sum_{i=1}^{2} \int_{x=0}^{b} [-0.5 D_1 (\frac{m\pi}{L}) \frac{dN_i}{dx} N_j + G (\frac{m\pi}{L}) N_i \frac{dN_j}{dx}] dx\; u_{im}$$

$$- m\omega^2 \frac{L}{2} \int_{x=0}^{b} \sum_{i=1}^{2} (N_i v_{im}) N_j\, dy\, dx - F_{yjm} \tag{9.31}$$

The integrals in equations (9.30) and (9.31) can be evaluated from Table 9.2.

Table 9.2

i	$b\int_0^b \frac{dN_i}{dx}\frac{dN_j}{dx}dx$		$\frac{6}{b}\int_0^b N_i N_j\, dx$		$\int_0^b N_i \frac{dN_j}{dx}dx$		$\int_0^b \frac{dN_i}{dx} N_j\, dx$	
	j		j		j		j	
	1	2	1	2	1	2	1	2
1	1	-1	2	1	-0.5	0.5	-0.5	-0.5
2	-1	1	1	2	-0.5	0.5	0.5	0.5

If $\alpha_m = m\pi / L$, $D_1 = \upsilon_x D_x = \upsilon_y D_y$ and the displacements are in the order $\delta_m^T = \{u_{im}, v_{im}, u_{jm}, v_{jm}\}$, the symmetric coefficients of the static stiffness and mass matrices are given by

$$k_{11} = k_{33} = t[D_x \frac{L}{2b} + G\frac{Lb}{12}\alpha_m^2],\; k_{12} = t[D_1 \frac{L}{4}\alpha_m - G\frac{L}{4}\alpha_m]$$

$$k_{13} = t[-D_x \frac{L}{2b} + G\frac{Lb}{12}\alpha_m^2],\; k_{14} = t[D_1 \frac{L}{4}\alpha_m + G\frac{L}{4}\alpha_m]$$

$$k_{22} = k_{44} = t[G\frac{L}{b} + D_y \frac{Lb}{6}\alpha_m^2],\; k_{23} = t[-D_1 \frac{L}{4}\alpha_m + G\frac{L}{4}\alpha_m]$$

$$k_{24} = t[-G\frac{L}{2b} + D_y \frac{Lb}{12}\alpha_m^2],\; k_{34} = k_{23} \tag{9.32}$$

The only non-zero symmetric coefficients of the mass matrix are

$$m_{11} = m_{22} = m_{33} = m_{44} = \frac{mLb}{6},\; m_{12} = m_{34} = \frac{mLb}{12} \tag{9.33}$$

9.4 AUXILIARY NODAL LINE TECHNIQUE

In finite element analysis, there are two ways of improving computational accuracy. One is to increase the number of simple elements, for example simple rectangular or triangular elements, and another is to use elements with many more nodes than simple element, such as quadratic elements with mid-side nodes. Generally the second option is found to be more economical.

A similar approach can be adopted in the case of the finite strip method. One can generate strips with more than just two nodal lines.

9.4 1 Thin Plate-bending Strip With an Auxiliary Nodal Line

Proceeding as in Section 9.2, assuming an additional nodal line at the middle of the strip, as shown in Fig. 9.3

$$w(x,y)=\sum_{n=1}^{\infty}\{N_1(x)w_{in}+N_2(x)\theta_{yin}+N_3(x)w_{kn}+N_4(x)\theta_{ykn}$$

$$+N_5(x)w_{Jn}+N_6(x)\theta_{yJn}\}\sin\frac{n\pi y}{L} \tag{9.34}$$

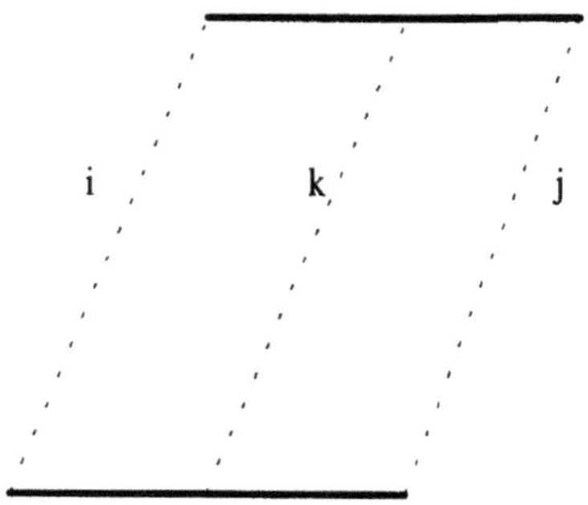

Fig. 9.3 A finite strip with a nodal line in the middle

The shape functions are determined by assuming that in the x direction, the displacement varies as a quintic polynomial

$$w(x,y)=\sum_{n=1}^{\infty}\{f_n(x)\}\sin\frac{n\pi y}{L} \tag{9.35}$$

$$f_n(x)=C_{1n}+C_{2n}x+C_{3n}x^2+C_{4n}x^3+C_{5n}x^4+C_{6n}x^5 \tag{9.36}$$

The constants C_{1n} to C_{6n} are determined from the condition

$$f_n(0)=w_{in},\ f_n(0.5b)=w_{kn},\ f_n(b)=w_{jn}$$

$$\frac{df_n(0)}{dx}=\theta_{yin}\ ,\ \frac{df_n(0.5b)}{dx}=\theta_{ykn}\ ,\ \frac{df_n(b)}{dx}=\theta_{yjn} \tag{9.37}$$

Solving the constants, the shape functions are given by

$$N_1=1-23(\frac{x}{b})^2+66(\frac{x}{b})^3-68(\frac{x}{b})^4+24(\frac{x}{b})^5,$$

$$N_2 = b\{\frac{x}{b} - 6(\frac{x}{b})^2 + 13(\frac{x}{b})^3 - 12(\frac{x}{b})^4 + 4(\frac{x}{b})^5\},$$

$$N_3 = 16(\frac{x}{b})^2 - 32(\frac{x}{b})^3 + 16(\frac{x}{b})^4,$$

$$N_4 = -b\{-8(\frac{x}{b})^2 + 32(\frac{x}{b})^3 - 40(\frac{x}{b})^4 + 16(\frac{x}{b})^5\},$$

$$N_5 = 7(\frac{x}{b})^2 - 34(\frac{x}{b})^3 + 52(\frac{x}{b})^4 - 24(\frac{x}{b})^5,$$

$$N_6 = -b\{-(\frac{x}{b})^2 + 5(\frac{x}{b})^3 - 8(\frac{x}{b})^4 + 4(\frac{x}{b})^5\} \qquad (9.38)$$

Tables 9.3 to 9.6 give the necessary integrals for the derivation of stiffness and mass matrices using equation (9.14).

Table 9.3 $b^3 \int_0^b \frac{d^2N_i}{dx^2} \frac{d^2N_j}{dx^2} dx$

i	j					
	1	2	3	4	5	6
1	145.4857	-32.5143b	-102.4	-54.8571b	-43.0857	-6.9143b
2	-32.5143b	9.4857b^2	25.60b	9.1429 b^2	6.9143b	1.0857 b^2
3	-102.4	25.60b	204.8	0	-102.4	-25.6b
4	-54.8571b	9.1429 b^2	0	36.5714 b^2	54.8571b	9.1429 b^2
5	-43.0857	6.9143b	-102.4	54.8571b	145.4857	32.5143b
6	-6.9143b	1.0857 b^2	-25.6b	9.1429 b^2	32.5143b	9.4857 b^2

Table 9.4 $\frac{100}{b} \int_0^b N_i N_j dx$

i	j					
	1	2	3	4	5	6
1	15.0938	-0.8225b	6.3492	1.1544b	1.8903	0.2092b
2	-0.8225b	0.0577 b^2	-0.6349b	-0.0866 b^2	-0.2092b	-0.0217 b^2
3	6.3492	-0.6349b	40.6349	0	6.3492	0.6349b
4	1.1544b	-0.0866 b^2	0	0.9235 b^2	-1.1544b	-0.0866 b^2
5	1.8903	-0.2092b	6.3492	-1.1544b	15.0938	0.8225b
6	0.2092b	-0.0217 b^2	0.6349b	-0.0866 b^2	0.8225b	0.0577 b^2

Table 9.5 $100b\int_0^b \frac{dN_i}{dx}\frac{dN_j}{dx}\,dx$

i	j					
	1	2	3	4	5	6
1	264.7619	-6.1905b	-243.8095	-38.0953b	-20.9524	1.4286b
2	-6.1905b	4.4444 b^2	7.6191b	-1.2698 b^2	-1.4286b	-0.7937 b^2
3	-243.8095	7.6191b	487.6191	0	-243.8095	-7.6191b
4	-38.0953b	-1.2698 b^2	0	40.6349 b^2	38.0953b	-1.2698 b^2
5	-20.9524	-1.4286b	-243.8095	38.0952b	264.7619	6.1905b
6	1.4286b	-0.7937 b^2	-7.6191b	-1.2698 b^2	6.1905b	4.4444 b^2

Table 9.6 $100b\int_0^b \frac{1}{2}\{\frac{d^2N_i}{dx^2}N_j + N_i\frac{d^2N_j}{dx^2}\}\,dx$

i	j					
	1	2	3	4	5	6
1	-264.7619	56.1905b	243.8095	38.0952b	20.9524	-1.4286b
2	56.1905b	-4.4445 b^2	-7.6191b	1.2698 b^2	1.4286b	0.7937 b^2
3	243.8095	-7.6190b	-487.6190	0	243.8095	7.6191b
4	38.0952b	1.2698 b^2	0	-40.6349 b^2	-38.0952b	1.2698 b^2
5	20.9524	1.4286b	243.8095	-38.0952b	-264.7620	-56.1905b
6	-1.4286b	0.7937 b^2	7.6191b	1.2698 b^2	-56.1905b	-4.4445 b^2

9.4.2 In-plane Strip with an Auxiliary nodal line

Proceeding as in Section 9.3, assuming an additional nodal line at the middle of the strip as shown in Fig. 9.3

$$u_i(y) = \sum_{n=1}^{\infty} u_{in}\sin\frac{n\pi y}{L},\quad u_K(y) = \sum_{n=1}^{\infty} u_{Kn}\sin\frac{n\pi y}{L},\quad u_j(y) = \sum_{n=1}^{\infty} u_{jn}\sin\frac{n\pi y}{L}$$

$$u(x,y) = N_1(x)u_i(y) + N_2(x)u_k(y) + N_3(x)u_j(y) \tag{9.39}$$

Similarly

$$v_i(y) = \sum_{n=1}^{\infty} v_{in}\cos\frac{n\pi y}{L},\quad v_k(y) = \sum_{n=1}^{\infty} v_{kn}\cos\frac{n\pi y}{L},\quad v_j(y) = \sum_{n=1}^{\infty} v_{jn}\cos\frac{n\pi y}{L}$$

$$v(x,y) = N_1(x)v_i(y) + N_2(x)v_k(y) + N_3(x)v_j(y) \tag{9.40}$$

The shape functions are determined by assuming that in the x direction, the displacement varies as a cubic polynomial

$$f_n(x) = C_{1n} + C_{2n}x + C_{3n}x^2 \tag{9.41}$$

The constants C_{1n} to C_{6n} are determined from the condition

$$f_n(0) = u_{in},\ f_n(0.5b) = u_{kn},\ f_n(b) = u_{jn}$$

Solving the constants, the shape functions are given by

$$N_1 = 1 - 3(\frac{x}{b}) + 2(\frac{x}{b})^2,\quad N_2 = 4\{\frac{x}{b} - (\frac{x}{b})^2\},\quad N_3 = -(\frac{x}{b}) + 2(\frac{x}{b})^2 \tag{9.42}$$

The necessary integrals for the derivation of stiffness and mass matrices using equations (9.29) and (9.31) are given in Tables 9.7 and 9.8.

If $\alpha_m = m\pi / L$, $D_1 = \upsilon_x D_x = \upsilon_y D_y$ and the displacements are in the order $\delta_m^T = \{u_{im}, v_{im}, u_{km}, v_{km}, u_{jm}, v_{jm}\}$, symmetric coefficients of the static stiffness and mass matrices are given by

Table 9.7 Integrals of Shape functions

i	$\frac{1}{b}\int_0^b N_i N_j\,dx$			$b\int_0^b \frac{dN_i}{dx}\frac{dN_j}{dx}\,dx$		
	j			j		
	1	2	3	1	2	3
1	0.1333	0.06667	-0.03333	2.3333	-2.6667	0.3333
2	0.06667	0.53333	0.06667	-2.6667	5.3333	-2.6667
3	-0.03333	0.06667	0.1333	0.3333	-2.6667	2.3333

Table 9.8 Integrals of Shape functions

i	$b\int_0^b N_i \frac{dN_j}{dx}\,dx$			$b\int_0^b \frac{dN_i}{dx} N_j\,dx$		
	j			j		
	1	2	3	1	2	3
1	-0.5	0.6667	-0.16667	-0.5	-0.6667	0.16667
2	-0.6667	0	0.6667	0.6667	0	-0.6667
3	0.16667	-0.6667	0.5	-0.16667	0.6667	0.5

$$k_{11} = k_{55} = t[D_x \frac{L}{b}\frac{7}{6} + G\frac{Lb}{15}\alpha_m^2]$$

$$k_{13} = t[-D_x \frac{L}{b}\frac{4}{3} + G\frac{Lb}{30}\alpha_m^2],\ k_{15} = t[D_x \frac{L}{b}\frac{1}{6} - G\frac{Lb}{60}\alpha_m^2]$$

$$k_{12} = t[D_1 \frac{L}{4} \alpha_m - G\frac{L}{4}\alpha_m],\ k_{14} = t[-D_1 \frac{L}{3} \alpha_m - G\frac{L}{3}\alpha_m]$$
$$k_{16} = t[D_1 \frac{L}{12} \alpha_m + G\frac{L}{12}\alpha_m],\ k_{33} = t[D_x \frac{L}{b}\frac{8}{3} + G\,Lb\frac{4}{15}\alpha_m^2]$$
$$k_{35} = k_{13},\ k_{23} = -k_{14},\ k_{34} = 0,\ k_{36} = k_{14}$$
$$k_{22} = k_{66} = t[G\frac{L}{b}\frac{7}{6} + D_y \frac{Lb}{15}\alpha_m^2],\ k_{24} = k_{66} = t[-G\frac{L}{b}\frac{4}{3} + D_y \frac{Lb}{30}\alpha_m^2]$$
$$k_{26} = t[G\frac{L}{b}\frac{1}{6} - D_y \frac{Lb}{60}\alpha_m^2],\ k_{44} = t[G\frac{L}{b}\frac{8}{3} + D_y\,Lb\frac{4}{15}\alpha_m^2]$$
$$k_{46} = t[-G\frac{L}{b}\frac{4}{3} + D_y\,Lb\frac{1}{30}\alpha_m^2],\ k_{45} = k_{14},\ k_{56} = -k_{12} \tag{9.43}$$

The only non-zero symmetric coefficients of the mass matrix are

$$m_{11} = m_{22} = m_{44} = m_{66} = \frac{2mLb}{15},\ m_{33} = m_{44} = \frac{8mLb}{15},$$
$$m_{13} = m_{24} = m_{35} = m_{46} = \frac{mLb}{15},\ m_{15} = m_{26} = -\frac{mLb}{30} \tag{9.44}$$

9.5 MINDLIN-REISSNER PLATE-BENDING FINITE STRIP

Mindlin-Reissner plate-bending strip can be developed in a manner similar to Kirchoff strip as in Section 9.2. The main difference lies in the fact that as in Section 8.15, the rotations of the cross section due to flexure are treated as independent variables

$$w(x, y) = \sum_{n=1}^{\infty} \{N_1(x)\, w_{in} + N_2(x)\, w_{jn}\} \sin\frac{n\pi y}{L}$$
$$\Psi_x(x, y) = \sum_{n=1}^{\infty} \{N_1(x)\, \Psi_{x\,in} + N_2(x)\, \Psi_{x\,jn}\} \sin\frac{n\pi y}{L}$$
$$\Psi_y(x, y) = \sum_{n=1}^{\infty} \{N_1(x)\, \Psi_{y\,in} + N_2(x)\, \Psi_{y\,jn}\} \cos\frac{n\pi y}{L} \tag{9.45}$$

The shape functions are given by equation (9.18). The strain energy is given by

$$Strain\ energy = \frac{1}{2}\int_{y=0}^{L}\int_{x=0}^{b} \{D_x (\frac{\partial \Psi_x}{\partial x})^2 + D_y (\frac{\partial \Psi_y}{\partial y})^2$$
$$+ D_1 \frac{\partial \Psi_x}{\partial x}\frac{\partial \Psi_y}{\partial y} + D_{xy} (\frac{\partial \Psi_x}{\partial y} + \frac{\partial \Psi_y}{\partial x})^2$$
$$+ GA_s(\frac{\partial w}{\partial x} - \Psi_x)^2 + GA_s(\frac{\partial w}{\partial y} - \Psi_y)^2 \}dy\,dx \tag{9.46}$$

Differentiating equation (9.46) successively with respect to w_{in}, Ψ_{xin} and Ψ_{yin}, and taking advantage of the orthogonal property of sine and cosine functions

$$\frac{\partial}{\partial w_{1n}}(strain\ energy) = G\,A_s \frac{L}{2}\int_{x=0}^{b}\sum_{m=1}^{2}[\frac{dN_m}{dx}w_{mn} - N_m\Psi_{xmn}]\frac{dN_l}{dx}dx$$

$$+G\,A_s\frac{L}{2}\int_{x=0}^{b}\sum_{m=1}^{2}[N_m(\frac{n\pi}{L})w_{mn} - N_m\Psi_{ymn}]N_l(\frac{n\pi}{L})]dx \qquad (9.47)$$

Regrouping equation (9.47)

$$\frac{\partial}{\partial w_{1n}}(strain\ energy) = G\,A_s \frac{L}{2}\int_{x=0}^{b}\sum_{m=1}^{2}[\frac{dN_m}{dx}\frac{dN_l}{dx} + (\frac{n\pi}{L})^2 N_m N_l]\,dx\,w_{mn}$$

$$-G\,A_s\frac{L}{2}\int_{x=0}^{b}\sum_{m=1}^{2}[N_m\frac{dN_l}{dx}]dx\,\Psi_{xmn} - G\,A_s\frac{L}{2}\int_{x=0}^{b}\sum_{m=1}^{2}[(\frac{n\pi}{L})N_m N_l]dx\Psi_{ymn} \qquad (9.48)$$

$$\frac{\partial}{\partial \Psi_{x1n}}(strain\ energy) = D_x\frac{L}{2}\int_{x=0}^{b}\sum_{m=1}^{2}[\frac{dN_m}{dx}\Psi_{xmn}]\frac{dN_l}{dx}dx$$

$$+D_1\frac{L}{2}\int_{x=0}^{b}\sum_{m=1}^{2}[-N_m(\frac{n\pi}{L})\Psi_{ymn}]\frac{dN_l}{dx}dx$$

$$+D_{xy}\frac{L}{2}\int_{x=0}^{b}\sum_{m=1}^{2}[N_m(\frac{n\pi}{L})\Psi_{xmn} + \frac{dN_m}{dx}\Psi_{ymn}]N_l(\frac{n\pi}{L})dx$$

$$+G\,A_s\frac{L}{2}\int_{x=0}^{b}\sum_{m=1}^{2}[\frac{dN_m}{dx}w_{mn} - N_m\Psi_{xmn}](-N_l)]\}dx \qquad (9.49)$$

Regrouping equation (9.49)

$$\frac{\partial}{\partial \Psi_{x1n}}(strain\ energy) = -G\,A_s\frac{L}{2}\int_{x=0}^{b}\sum_{m=1}^{2}[\frac{dN_m}{dx}(N_l)]dx\,w_{mn}$$

$$\frac{L}{2}\int_{x=0}^{b}\sum_{m=1}^{2}[D_x\frac{dN_m}{dx}\frac{dN_l}{dx} + \{D_{xy}(\frac{n\pi}{L})^2 + G\,A_s\}N_m N_l]\Psi_{xmn}dx$$

$$+\frac{L}{2}\int_{x=0}^{b}\sum_{m=1}^{2}[-D_1 N_m\frac{dN_l}{dx} + D_{xy}\frac{dN_m}{dx}N_l](\frac{n\pi}{L})\Psi_{ymn}dx \qquad (9.50)$$

$$\frac{\partial}{\partial \Psi_{y1n}}(strain\ energy) = D_y\frac{L}{2}\int_{x=0}^{b}\sum_{m=1}^{2}[N_m(-\frac{n\pi}{L})\Psi_{ymn}]N_l(-\frac{n\pi}{L})dx$$

$$+D_1\frac{L}{2}\int_{x=0}^{b}\sum_{m=1}^{2}[(\frac{dN_m}{dx})\Psi_{xmn}]N_l(-\frac{n\pi}{L})dx$$

$$+D_{xy}\frac{L}{2}\int_{x=0}^{b}\sum_{m=1}^{2}[N_m(\frac{n\pi}{L})\Psi_{xmn} + \frac{dN_m}{dx}\Psi_{ymn}]\frac{dN_l}{dx}dx$$

$$+G\,A_s\frac{L}{2}\int_{x=0}^{b}\sum_{m=1}^{2}[N_m(\frac{n\pi}{L})w_{mn} - N_m\Psi_{ymn}](-N_l)]\}dx \qquad (9.51)$$

Regrouping equation (9.51)

$$\frac{\partial}{\partial\Psi_{y\ln}}(strain\ energy) = -G A_s \frac{L}{2}\int_{x=0}^{b}\sum_{m=1}^{2}[N_m N_l(\frac{n\pi}{L})]w_{mn}\,dx$$
$$+\frac{L}{2}\int_{x=0}^{b}\sum_{m=1}^{2}[-D_1\frac{dN_m}{dx}N_l + D_{xy}N_m\frac{dN_l}{dx}]\Psi_{xmn}(\frac{n\pi}{L})dx$$
$$+\frac{L}{2}\int_{x=0}^{b}\sum_{m=1}^{2}[D_y N_m N_l(\frac{n\pi}{L})^2 + D_{xy}\frac{dN_m}{dx}\frac{dN_l}{dx} - +G A_s N_m N_l)]\Psi_{ymn}dx$$

(9.52)

Numerical studies indicate that the 'best value' is obtained by using only one point Gaussian integration. The necessary integrals are shown in Table 9.9.

Table 9.9 Integrals of shape functions

i	$b\int_0^b \frac{dN_i}{dx}\frac{dN_j}{dx}dx$		$\frac{1}{b}\int_0^b N_i N_j\,dx$		$\int_0^b N_i \frac{dN_j}{dx}dx$		$\int_0^b \frac{dN_i}{dx}N_j\,dx$	
	j		j		j		j	
	1	2	1	2	1	2	1	2
1	0.5	-0.5	0.125	0.125	-0.25	0.25	-0.25	-0.25
2	-0.5	0.5	0.125	0.125	-0.25	0.25	0.25	0.25

Taking A_s = t/1.2, where t is the thickness of the strip and 1.2 is the area shear factor for a rectangular section, the coefficients of the stiffness matrix for the strip is given by

$$k_{11} = k_{33} = GA_s\frac{L}{8b}[4 + b^2\alpha_m^2],\ k_{12} = GA_s\frac{L}{4},\ k_{15} = -GA_s\frac{L}{4}$$

$$k_{13} = k_{16} = -G A_s\frac{Lb}{8}\alpha_m,\ k_{14} = GA_s\frac{L}{8b}[-4 + b^2\alpha_m^2]$$

$$k_{22} = k_{44} = \frac{Lb}{8}[D_x\frac{4}{b^2} + (D_{xy}\ \alpha_m^2 + GA_s)]$$

$$k_{23} = \frac{L}{4}[D_1 - D_{xy}]\alpha_m,\ k_{24} = GA_s\frac{L}{4}$$

$$k_{25} = \frac{Lb}{8}[-D_x\frac{4}{b^2} + (D_{xy}\ \alpha_m^2 + GA_s)],\ k_{26} = \frac{L}{4}[D_1 + D_{xy}]\alpha_m$$

$$k_{33} = k_{66} = \frac{Lb}{8}[D_x\alpha_m^2 + 4D_{xy}\ + GA_s]$$

$$k_{34} = -GA_s\frac{L}{8} \quad , k_{35} = \frac{L}{4}[D_1 - D_{xy}], \; k_{36} = \frac{Lb}{8}[D_x\alpha_m^2 - 4D_{xy} + GA_s]$$

$$k_{45} = -GA_s\frac{L}{4} \quad , k_{46} = -GA_s\frac{Lb}{8}\alpha_m, \; k_{56} = \frac{L}{4}[-D_1 + D_{xy}]\alpha_m \tag{9.53}$$

9.6 CUBIC SPLINE STRIP METHOD

The main attraction of the finite strip method for the case of strips with simply supported ends is that the orthogonal property of the sine and cosine functions given by equation (9.11) decouples the stiffness matrix for each harmonic. In other words, if the structure is simply supported at its ends and has constant properties along the y axis, then the structure can be analysed for each harmonic and the results summed. This generally leads to a much more efficient analysis than the finite element method. Unfortunately, all these advantages are lost if the structure does not have simply supported ends or has properties which are not constant along its length.

Cubic splines instead of Fourier sine or cosine series have been used for representing the variation of a function in the y direction along a nodal line. The main advantage is that it is more economic than the finite element method, when curvature continuity is required.

9.6.1 CUBIC SPLINES

Figure 9.4 shows a cubic spline φ.
It is a symmetrical curve consisting of four cubic segments f_i, with continuity up to second derivative at the junctions. Taking the origin at the axis of symmetry, the four equally spaced segments at a distance h apart are

$$f_1(y) = \frac{1}{6h^3}(y+2h)^3 \; , \; -2h \le y \le -h$$

$$f_2(y) = \frac{1}{6h^3}[h^3 + 3h^2(y+h) + 3h(y+h)^2 - 3(y+h)^3], \; -h \le y \le 0$$

$$f_3(y) = \frac{1}{6h^3}[h^3 + 3h^2(h-y) + 3h(h-y)^2 - 3(h-y)^3], \; 0 \le y \le h$$

$$f_4(y) = \frac{1}{6h^3}(2h-y)^3 \; , \; h \le y \le 2h \tag{9.54}$$

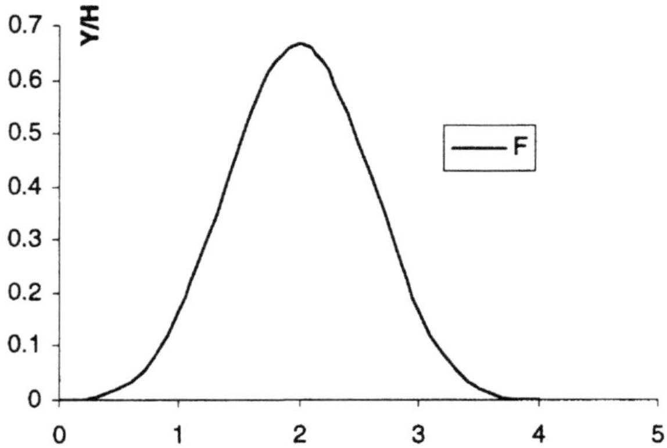

Fig. 9.4 Cubic spline

The first derivatives are

$$\frac{df_1(y)}{dy}=\frac{1}{2h^3}(y+2h)^2\ ,\ -2h\le y\le -h$$

$$\frac{df_2(y)}{dy}=\frac{1}{2h^3}[h^2\ +2h(y+h)\ -3(y+h)^2],\ -h\le y\le 0$$

$$\frac{df_3(y)}{dy}=\frac{1}{2h^3}[-h^2\ -2h(h-y)\ +3(h-y)^2],\ 0\le y\le h$$

$$\frac{df_4(y)}{dy}=-\frac{1}{2h^3}(2h-y)^2\ ,\ h\le y\le 2h \tag{9.55}$$

The second derivatives are

$$\frac{d^2f_1(y)}{dy^2}=\frac{1}{h^3}(y+2h)\ ,\ -2h\le y\le -h$$

$$\frac{d^2f_2(y)}{dy^2}=\frac{1}{h^3}[h\ -3(y+h)],\ -h\le y\le 0$$

$$\frac{d^2f_3(y)}{dy^2}=\frac{1}{h^3}[h\ -3(h-y)],\ 0\le y\le h$$

$$\frac{d^2f_4(y)}{dy^2}=\frac{1}{h^3}(2h-y)\ ,\ h\le y\le 2h \tag{9.56}$$

The integrals of the four functions are

$$\int_{-2h}^{-h} f_1(y)\,dy = \frac{h}{24},\ \int_{-2h}^{0} f_2(y)\,dy = \frac{11h}{24},\int_{0}^{h} f_3(y)\,dy = \frac{11h}{24},\int_{h}^{2h} f_1(y)\,dy = \frac{h}{24}$$

Table 9.10 summarises the value of the spline and its derivatives at intersection points. As shown in Fig. 9.5, at any node i, contributions come from three splines centred respectively at (i-1), i and (i+1). However over the region between nodes i and (i+1), contributions come from four splines centred at (i-1), i, (i+1) and (i+2). The splines φ_{i-1}, φ_i, φ_{i+1} and φ_{i+2} contribute respectively functions f_4, f_3, f_2 and f_1. Using cubic splines, any function F(y) in the interval $a \le y \le b$, can be represented by a series of (m+2) splines centred at equal distances of h = (b-a)/m.

$$F(y) = \sum_{i=-1}^{m+1} C_i\ \phi_i \tag{9.57}$$

where, C_i are constants and the spline φ_i is centred at y = a + i (b-a)/m.

Table 9.10

Function	y				
	-2h	-h	0	h	2h
φ	0	$\frac{1}{6}$	$\frac{4}{6}$	$\frac{1}{6}$	0
$\frac{d\phi}{dy}$	0	$\frac{1}{2h}$	0	$-\frac{1}{2h}$	0
$\frac{d^2\phi}{dy^2}$	0	$\frac{1}{h^2}$	$-\frac{2}{h^2}$	$\frac{1}{h^2}$	0

9.6 2 Example

As a simple example, consider representing the following function F(y) in the region $0 \le y \le 6$ and given that the values of the function at intervals of 2 are

F(0) = 20, F(2) = 68, F(4) = 988, F(6) = 5084

The slopes at the ends are prescribed as F'(0) = -2, F'(6) = 394

It is required to use a cubic spline interpolation to represent the function.

Solution: Since the region of interest is 6 units long and there are nodes at intervals of 2, therefore m = 6/2 = 3, m+2 = 5. We need six cubic splines to approximate the function. Representing F(y) as

$$F(y) = \sum_{i=-1}^{4} C_i\ \phi_i$$

By equating the value of the function at the nodes

$$\frac{1}{6}C_{-1}+\frac{4}{6}C_0+\frac{1}{6}C_1=20$$

$$\frac{1}{6}C_0+\frac{4}{6}C_1+\frac{1}{6}C_2=68$$

$$\frac{1}{6}C_1+\frac{4}{6}C_2+\frac{1}{6}C_3=988$$

$$\frac{1}{6}C_2+\frac{4}{6}C_3+\frac{1}{6}C_4=5084$$

The slopes at the ends are given by

$$-\frac{1}{2h}C_{-1}+\frac{1}{2h}C_1=-2$$

$$-\frac{1}{2h}C_2+\frac{1}{2h}C_4=394$$

The six simultaneous equations can be solved for the six unknowns.
C_0 = -89.0667, C_1 = 234.1333, C_2= -439.4666, C_3 = 7451.7333
C_{-1} = -81.0667, C_4 = 1136.5334

9.6.3 Incorporation of Boundary Conditions

The main advantage of the spline function interpolation is that any boundary condition can be represented by using modified rather than basic cubic splines. This is shown for two commonly occurring cases in practice.

9.6.2.1 Clamped End

Consider the case of a fixed end condition, requiring that

$$F(y)=\frac{dF}{dy}=0 \quad at\ y=0$$

From the values of the function and their derivatives shown in Table 9.10, this requires that

$$\frac{1}{6}C_{-1}+\frac{4}{6}C_0+\frac{1}{6}C_1=0\ ,\quad -\frac{1}{2h}C_{-1}+\frac{1}{2h}C_1=0$$

Solving

$C_{-1} = C_1$, C_0 = -0.5 C_1

Therefore

$$F(y)=\sum_{i=-1}^{m+1}C_i\ \phi_i=C_1\{\phi_{-1}-0.5\phi_0+\phi_1\}+\sum_{i=2}^{m+1}C_i\ \phi_i$$

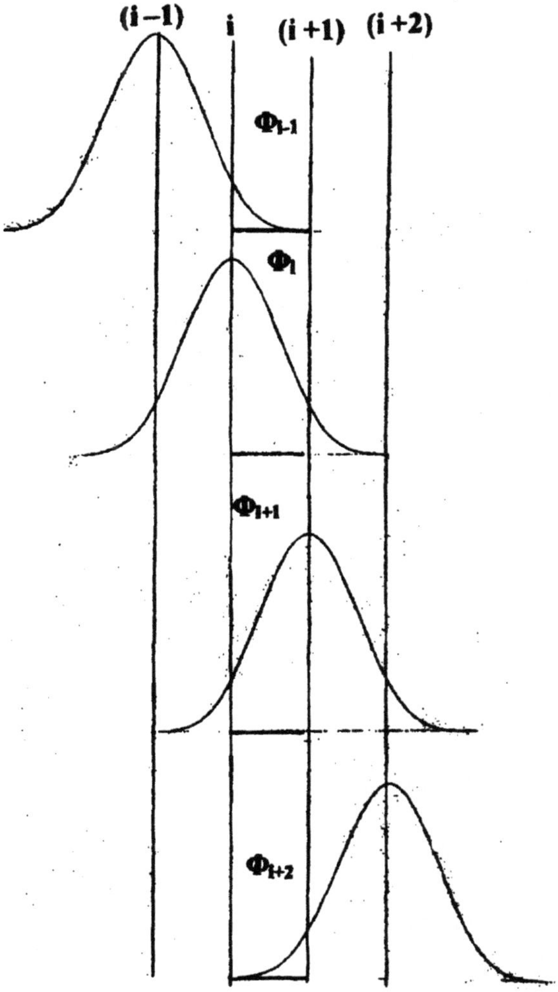

Fig. 9.5 Four splines which contribute to the displacement in a segment

$$\therefore F(y) = C_{-1}\bar{\phi}_{-1} + C_0\bar{\phi}_0 + C_1\bar{\phi}_1 + \sum_{i=2}^{m+1} C_i\,\phi_i \tag{9.58}$$

where modified spline functions $\bar{\phi}_{-1}, \bar{\phi}_0, \bar{\phi}_1$ are used instead of φ_{-1}, φ_0 and φ_1. The modified splines are

$$\bar{\phi}_{-1} = 0\,,\ \bar{\phi}_0 = 0\,,\ \bar{\phi}_1 = \{\phi_{-1} - 0.5\phi_0 + \phi_1\} \tag{9.59}$$

The modified spline function $\bar{\phi}_1$ is constructed from four cubic functions, which are, using equation (9.54)

$$\bar{f}_1(y) = f_1(y) + f_3(y+2h) - 0.5\,f_2(y+h),$$

$$\bar{f}_2(y) = f_2(y) + f_4(y+2h) - 0.5 f_3(y+h),$$

$$\bar{f}_3(y) = f_3(y) - 0.5\,f_4(y+h),\ \bar{f}_4(y) = f_4(y)$$

9.6.2.2 Simply Supported End

Similarly at a simply supported end

$$F(y) = 0 \ at\ y = 0$$

This requires that

$$\frac{1}{6}C_{-1} + \frac{4}{6}C_0 + \frac{1}{6}C_1 = 0$$

Solving for C_{-1}

$$C_{-1} = -C_1 - 4C_0$$

Therefore

$$F(y) = \sum_{i=-1}^{m+1} C_i\,\phi_i = C_0\{-4\phi_{-1} + \phi_0\} + C_1\{-\phi_{-1} + \phi_1\} + \sum_{i=2}^{m+1} C_i\,\phi_i$$

$$\therefore F(y) = C_{-1}\bar{\phi}_{-1} + C_0\bar{\phi}_0 + C_1\{\bar{\phi}_1\} + \sum_{i=2}^{m+1} C_i\,\phi_i \tag{9.60}$$

where modified spline functions $\bar{\phi}_{-1}, \bar{\phi}_0, \bar{\phi}_1$ are used instead of φ_{-1}, φ_0 and φ_1. The modified splines are

$$\bar{\phi}_{-1} = 0\,,\ \bar{\phi}_0 = \{-4\phi_{-1} + \phi_0\}\,,\ \bar{\phi}_1 = \{-\phi_{-1} + \phi_1\} \tag{9.61}$$

The modified spline function $\bar{\phi}_0$ is constructed from four cubic functions, which are, using equation (9.54)

$$\bar{f}_1(y) = f_1(y) - 4\,f_2(y+h),\ \bar{f}_2(y) = f_2(y) - 4\,f_3(y+h),$$

$$\bar{f_3}(y) = f_3(y) - 4 f_4(y+h),\ \bar{f_4}(y) = f_4(y)$$

Similarly, the modified spline function $\bar{\phi}_1$ is constructed from four cubic functions, which are, using equation (9.48),

$$\bar{f_1}(y) = f_1(y) - f_3(y+2h),\ \bar{f_2}(y) = f_2(y) - f_4(y+2h),$$

$$\bar{f_3}(y) = f_3(y),\ \bar{f_4}(y) = f_4(y)$$

Thus by using modified spline functions to suit the prescribed boundary conditions, any type of boundary condition can be simulated.

9.7 K AND M MATRICES FOR CUBIC SPLINE FINITE STRIP

The approach adopted for deriving the necessary matrices using the minimum total potential is identical to that used in Sections 9.2 and 9.3. The difference arises from the fact that in the previous sections, only trigonometric sine or cosine functions were used for interpolation but here the cubic splines are used.

9.7.1 In-plane Cubic Spline Finite Strip

The displacements are defined by

$$u(x, y) = \sum_{k=-1}^{m+1} N_1(x)\ \phi^u_{im}\ u_{im} + N_2(x)\ \phi^u_{jm}\ u_{jm}$$

$$v(x, y) = \sum_{k=-1}^{m+1} N_1(x)\ \phi^v_{im}\ v_{im} + N_2(x)\ \phi^v_{jm}\ v_{jm}$$

The shape functions are given in equation (9.18).

Note that the cubic splines φ have a superscript of u or v indicating that these splines are for interpolating respectively u or v. Depending on the boundary conditions, the splines will have to be modified as in indicated in Section 9.6.2.

9.7.2 Cubic Spline Plate-bending Finite Strip

The displacements are defined by

$$W(x, y) = \sum_{k=-1}^{m+1} N_1(x)\ \phi^w_{im}\ w_{im} + N_2(x)\ \phi^\theta_{im}\ \theta_{im}$$

$$+ N_3(x)\ \phi^w_{jm}\ w_{jm} + N_4(x)\ \phi^\theta_{jm}\ \theta_{jm}$$

The shape functions are given in equation (9.4).

Note that the cubic splines φ have a superscript of w or θ indicating that these splines are for interpolating respectively w or θ. Once again, depending on the boundary conditions, the splines will have to be modified as in indicated in Section 9.6.2.

CHAPTER 10

DIRECT STIFFNESS METHOD

10.1 INTRODUCTION

This chapter is a short introduction to the stiffness method. Since the readers are likely to be familiar with the stiffness method, the objective is to give a brief introduction to the method so that the reader will be familiar with the notation, expressions for stiffness and mass matrices used in the programs and the meaning of certain terms used in the programs.

For the 'exact' solution to the forces and deformations set up in a structure, it is necessary to satisfy the following fundamental requirements.

- Equilibrium as required by equations of statics. In the case of dynamic problems, inertial forces and damping forces have to be included.
- Compatibility of displacements not only along an element but also at the joints where elements intersect.
- Prescribed force or displacement constraints at supports.
- Forces and deformations being compatible with the assumed element and material properties.

10.2 MEMBER AXIS STIFFNESS MATRIX OF LINE ELEMENTS

The derivation of the static stiffness and mass matrices for line elements was discussed in Chapter 7. In the following the x axis is orientated along the length of the member with the positive axis directed from end 1 to end 2. The three main types of line elements considered are

Bar element: The static stiffness matrix for a bar element shown in Fig. 7.2 is given by

$$\begin{bmatrix} F_{a1} \\ F_{a2} \end{bmatrix} = \frac{AE}{L} \begin{bmatrix} 1 & -1 \\ -1 & 1 \end{bmatrix} \begin{bmatrix} \Delta_{a1} \\ \Delta_{a2} \end{bmatrix} \tag{10.1}$$

Beam element: The stiffness relationship for a beam element shown in Fig. 7.5, taking into consideration bending and shear deformations, is given by

$$\begin{bmatrix} F_{n1} \\ M_1 \\ F_{n2} \\ M_2 \end{bmatrix} = \frac{EI}{L(1+\beta)} \begin{bmatrix} \frac{12}{L^2} & -\frac{6}{L} & -\frac{12}{L^2} & -\frac{6}{L} \\ -\frac{6}{L} & 4+\beta & \frac{6}{L} & 2-\beta \\ -\frac{12}{L^2} & \frac{6}{L} & \frac{12}{L^2} & \frac{6}{L} \\ -\frac{6}{L} & 2-\beta & \frac{6}{L} & 4+\beta \end{bmatrix} \begin{bmatrix} \Delta_{n1} \\ \theta_1 \\ \Delta_{n2} \\ \theta_2 \end{bmatrix}, \beta = \frac{12EI}{GA_sL^2} \tag{10.2}$$

Two-dimensional rigid-jointed frame element: This is a combination of bar and beam elements. The element stiffness matrix is given by

$$\begin{bmatrix} F_{a1} \\ F_{n1} \\ M_1 \\ F_{a2} \\ F_{n2} \\ M_2 \end{bmatrix} = \frac{EI}{L(1+\beta)} \begin{bmatrix} \alpha & 0 & 0 & -\alpha & 0 & 0 \\ & \frac{12}{L^2} & -\frac{6}{L} & 0 & -\frac{12}{L^2} & -\frac{6}{L} \\ & & 4+\beta & 0 & \frac{6}{L} & 2-\beta \\ & & & \alpha & 0 & 0 \\ & symm. & & & \frac{12}{L^2} & \frac{6}{L} \\ & & & & & 4+\beta \end{bmatrix} \begin{bmatrix} \Delta_{a1} \\ \Delta_{n1} \\ \theta_1 \\ \Delta_{a2} \\ \Delta_{n2} \\ \theta_2 \end{bmatrix}, \alpha = \frac{A}{I}(1+\beta) \tag{10.3}$$

Shaft element: The stiffness relationship for a shaft element resisting uniform torque shown in Fig.7.11 is given by

$$\begin{bmatrix} T_1 \\ T_2 \end{bmatrix} = \frac{GJ}{L} \begin{bmatrix} 1 & -1 \\ -1 & 1 \end{bmatrix} \begin{bmatrix} \Phi_1 \\ \Phi_2 \end{bmatrix} \tag{10.4}$$

Plane grid member: This is a combination of shaft and beam elements. The element stiffness matrix is given by

$$\begin{bmatrix} F_{n1} \\ T_1 \\ M_1 \\ F_{n2} \\ T_2 \\ M_2 \end{bmatrix} = \frac{EI}{L(1+\beta)} \begin{bmatrix} \frac{12}{L^2} & 0 & -\frac{6}{L} & -\frac{12}{L^2} & 0 & -\frac{6}{L} \\ & \alpha & 0 & 0 & -\alpha & 0 \\ & & 4+\beta & \frac{6}{L} & 0 & 2-\beta \\ & & & \frac{12}{L^2} & 0 & \frac{6}{L} \\ & symm. & & & \alpha & 0 \\ & & & & & 4+\beta \end{bmatrix} \begin{bmatrix} w_1 \\ \Phi_1 \\ \theta_1 \\ w_2 \\ \Phi_2 \\ \theta_2 \end{bmatrix}, \alpha = \frac{GJ}{EI}(1+\beta) \tag{10.5}$$

10.3 MEMBER AXIS STIFFNESS MATRIX OF FINITE ELEMENTS

The derivation of the static stiffness and mass matrices for finite elements was discussed in Chapter 8. The two main types of finite elements considered were

- in-plane element
- plate-bending element.

In all cases it is assumed that the element lies in the x-y plane and the z axis is positive upwards.

10.3.1 In-plane Finite Elements

The displacements are arranged in the order

$$\delta = [u_1\ v_1\ u_2\ v_2 \ldots u_n\ v_n\ ]^T$$

where n = number of nodes in the element, n = 4 for a quadrilateral element with corner nodes only and n = 8 for a quadratic isoparametric element with corner and mid-side nodes.

Explicit stiffness and mass matrices are given in Tables 8.4 and 8.5 respectively.

10.3.2 Plate-bending Finite Elements

For non-conforming Kirchoff thin plate bending rectangular element, the displacements are arranged in the order

$$\delta = [w_1\ \theta_{x1}\ \ \theta_{y1}\ w_2\ \theta_{x2}\ \ \theta_{y2}\ w_3\ \theta_{x3}\ \theta_{y3}\ w_4\ \theta_{x4}\ \ \theta_{y4}\ ]^T$$

For a conforming Mindlin-Reissner isoparametric quadratic thick plate bending element

$$\delta = [w_1\ \varphi_{x1}\ \ \varphi_{y1}\ w_2\ \varphi_{x2}\ \ \varphi_{y2} \ldots w_8\ \varphi_{x8}\ \ \varphi_{y8}\ ]^T$$

Expressions for the coefficients of stiffness and mass matrices are given by equations (8.88) and (8.89) respectively.

10.3.3 Flat Shell finite element

Just as the two-dimensional rigid-jointed frame member is a combination of bar and beam elements, the flat shell element is a combination of in-plane and plate-bending elements. In this case the displacements are arranged as follows.

Using for the bending part non-conforming Kirchoff thin plate-bending rectangular element, the displacements are arranged in the order

$$\delta = [u_1\ v_1\ w_1\ \theta_{x1}\ \ \theta_{y1}\ u_2\ v_2\ w_2\ \theta_{x2}\ \ \theta_{y2}\ u_3\ v_3\ w_3\ \theta_{x3}\ \theta_{y3}\ u_4\ v_4\ w_4\ \theta_{x4}\ \ \theta_{y4}\ ]^T$$

For a conforming Mindlin-Reissner isoparametric quadratic thick plate-bending element

$$\delta = [u_1\ v_1\ w_1\ \varphi_{x1}\ \ \varphi_{y1}\ u_2\ v_2\ w_2\ \varphi_{x2}\ \ \varphi_{y2}\ ...u_8\ v_8\ w_8\ \varphi_{x8}\ \ \varphi_{y8}\ ]^T$$

10.4 MEMBER AXIS STIFFNESS MATRIX OF FINITE STRIP ELEMENTS

The derivation of the static stiffness and mass matrices for finite strips was discussed in Chapter 9. The ordering of displacements is the same as in the case of corresponding finite elements. In the case of simply supported strips, because the orthogonal property of sine and cosine functions decouples the stiffness matrix for each harmonic, the displacements correspond to a particular harmonic.
For the in-plane strip, the displacements are arranged in the order

$$\delta = [u_{1m}\ v_{1m}\ u_{2m}\ v_{2m}\ ...u_{nm}\ v_{nm}\ ]^T$$

For the Kirchoff thin plate-bending strip

$$\delta = [w_{1m}\quad \theta_{y1m}\ w_{2m}\quad \theta_{y2m}\ ...w_{nm}\ \theta_{ynm}\ ]^T$$

For the Mindlin-Reissner thick plate-bending strip

$$\delta = [w_{1m}\ \varphi_{x1m}\ \ \varphi_{y1m}\ w_{2m}\ \varphi_{x2m}\ \ \varphi_{y2m}\ ...w_{nm}\ \varphi_{xnm}\ \ \varphi_{ynm}]^T$$

For a flat shell element using Kirchoff thin plate-bending strip

$$\delta = [u_{1m}\ v_{1m}\ w_{1m}\ \theta_{y1m}\ \ u_{2m}\ v_{2m}\ w_{2m}\ \theta_{y2m}\ ...u_{nm}\ v_{nm}\ w_{nm}\ \theta_{ynm}]^T$$

For a flat shell element using Mindlin-Reissner thick plate-bending strip

$$\delta = [\ u_{1m}\ v_{1m} w_{1m}\ \varphi_{x1m}\ \ \varphi_{y1m}\ u_{2m}\ v_{2m} w_{2m}\ \varphi_{x2m}\ \ \varphi_{y2m}\ ...u_{nm}\ v_{nm} w_{nm}\ \varphi_{xnm}\ \ \varphi_{ynm}]^T$$

where n = number of nodal lines in the strip and m = number of the harmonic.

10.5 MEMBER AXIS TO GLOBAL AXIS STIFFNESS MATRIX

The member axis stiffness matrices given in the previous sections are not particularly useful because, in practice, at a joint (or node) members (or elements) meet at different angles. In order to be able to establish equations of equilibrium at a joint, it is necessary to assume that the member is inclined to the chosen coordinate system and that forces and displacements at the ends of the member act along the coordinate axes. Therefore it is necessary to convert the member axis stiffness matrices to the new general system.

10.5.1 Two-dimensional Pin-jointed Member

In this case, both the global and the local axes lie in the vertical x-y plane. Fig. 10.1 shows the orientation of the member.

The relationship between forces (displacements) in the two systems are as follows.

Displacements: $\Delta_a = \quad u_{Global}\,\ell + v_{Global}\,m$

Forces: $F_{x\,Global} = F_a \ell$, $F_{y\,Global} = F_a m$

The direction cosines are given by

$$\ell = \frac{(x_2 - x_1)}{L}, \quad m = \frac{(y_2 - y_1)}{L}$$

where L = length of the member and (x_i, y_i) are the coordinates of the ith end of the member.

Making use of the above relationships, the stiffness matrix in the global axes can be calculated and is given in equation (10.6).

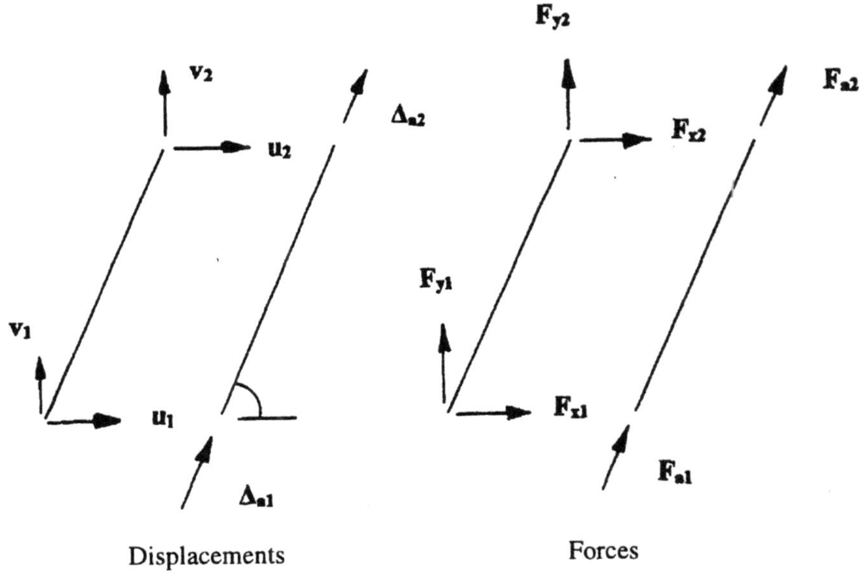

Fig. 10.1 Local and global displacements and forces: pin-jointed member

$$\begin{bmatrix} F_{x1} \\ F_{y1} \\ F_{x2} \\ F_{y2} \end{bmatrix} = \frac{AE}{L} \begin{bmatrix} \ell^2 & \ell m & -\ell^2 & -\ell m \\ & m^2 & -\ell m & -m^2 \\ & & \ell^2 & \ell m \\ symm. & & & m^2 \end{bmatrix} \begin{bmatrix} u_1 \\ v_1 \\ u_2 \\ v_2 \end{bmatrix} \tag{10.6}$$

10.5.2 Two-dimensional Rigid-jointed Member

In this case, both the global and the local axes lie in the vertical x-y plane. Fig. 10.2 shows the orientation of the member.

The relationship between the forces (displacements) in the two systems are as follows.

Displacements:

$$\Delta_a = u_{Global}\,\ell + v_{Global}\,m\ ,\ \Delta_n = -u_{Global}\,m + v_{Global}\,\ell\ ,\ \theta_{Local} = \theta_{Global}$$

Forces:

$$F_{x\,Global} = F_a\,\ell - F_n\,m\ ,\ F_{y\,Global} = F_a\,m + F_n\,\ell\ ,\ M_{Global} = M_{Local}$$

The direction cosines are given by

$$\ell = \frac{(x_2 - x_1)}{L}\ ,\ m = \frac{(y_2 - y_1)}{L}$$

Making use of the above relationships, the stiffness matrix in the global axes can be calculated and is given in equation (10.7).

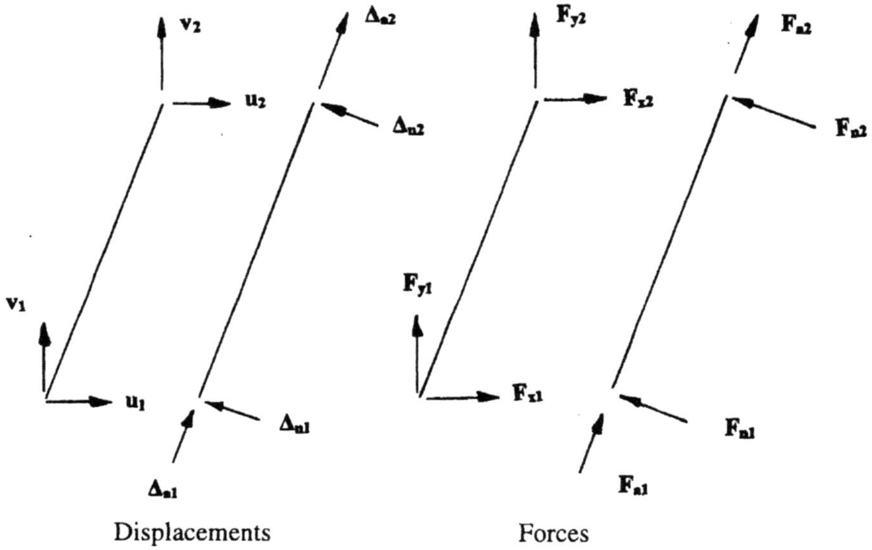

Fig. 10.2 Local and global forces and displacements: two-dimensional rigid-joined member

$$\begin{bmatrix} F_{x1} \\ F_{y1} \\ M_1 \\ F_{x2} \\ F_{y2} \\ M_2 \end{bmatrix} = \begin{bmatrix} k_{11} & k_{12} & k_{13} & -k_{11} & -k_{12} & k_{13} \\ & k_{22} & k_{23} & -k_{12} & -k_{22} & k_{23} \\ & & k_{33} & -k_{13} & -k_{23} & k_{36} \\ & & & k_{11} & k_{12} & -k_{13} \\ & & & & k_{22} & -k_{23} \\ & & & & & k_{33} \end{bmatrix} \begin{bmatrix} u_1 \\ v_1 \\ \theta_1 \\ u_2 \\ v_2 \\ \theta_2 \end{bmatrix} \tag{10.7}$$

$$k_{11} = \frac{12EI}{L^3(1+\beta)}m^2 + \frac{AE}{L}\ell^2\ ,\ k_{12} = -\frac{12EI}{L^3(1+\beta)}\ell m + \frac{AE}{L}\ell m\ ,$$

$$k_{13} = \frac{6EI}{L^2(1+\beta)}m\ ,\ k_{22} = \frac{12EI}{L^3(1+\beta)}\ell^2 + \frac{AE}{L}m^2\ ,k_{23} = \frac{6EI}{L^2(1+\beta)}\ell$$

$$k_{33} = \frac{EI(4+\beta)}{L(1+\beta)}\ ,k_{34} = \frac{EI(2-\beta)}{L(1+\beta)}$$

10.5.3 Plane Grid Member

Figure 10.3 shows the orientation of the member. The relationship between forces and displacements in the two systems are as follows.
Displacements:

$$\Phi = \theta_{x\,Global}\,\ell + \theta_{y\,Global}\,m,\quad \theta = -\theta_{x\,Global}\,m + \theta_{y\,Global}\,\ell\ ,\ w_{Local} = w_{Global}$$

Forces:

$$M_{x\,Global} = T\,\ell - M\,m\ ,\ M_{y\,Global} = Tm + M\,\ell\ ,\ F_{Local} = F_{Global}$$

$$\ell = \frac{(x_2 - x_1)}{L}\ ,\ m = \frac{(y_2 - y_1)}{L}$$

Making use of the above relationships, the stiffness matrix in the global axes can be calculated and is given in equation (10.8).

$$k_{11} = \frac{12\,EI}{L^3\,(1+\beta)}\ ,\ k_{12} = \frac{6\,EI}{L^2\,(1+\beta)}\,m,\ k_{13} = -\frac{6\,EI}{L^2\,(1+\beta)}\,\ell$$

$$k_{22} = \frac{(4+\beta)\,EI}{L(1+\beta)}\,m^2 + \frac{GJ}{L}\,\ell^2\ ,\ k_{23} = -\frac{(4+\beta)\,EI}{L(1+\beta)}\,\ell m + \frac{GJ}{L}\,\ell m$$

$$k_{25} = \frac{(2-\beta)\,EI}{L(1+\beta)}\,m^2 - \frac{GJ}{L}\,\ell^2\ ,\ k_{26} = -\frac{(2-\beta)\,EI}{L(1+\beta)}\,\ell m - \frac{GJ}{L}\,\ell m$$

$$k_{33} = \frac{(4+\beta)\,EI}{L(1+\beta)}\,\ell^2 + \frac{GJ}{L}\,m^2$$

$$\ell = \frac{(x_2 - x_1)}{L}\ ,\ m = \frac{(y_2 - y_1)}{L}$$

$$\begin{bmatrix} F_{z1} \\ M_{x1} \\ M_{y1} \\ F_{z2} \\ M_{x2} \\ M_{y2} \end{bmatrix} = \begin{bmatrix} k_{11} & k_{12} & k_{13} & -k_{11} & k_{12} & k_{13} \\ & k_{22} & k_{23} & -k_{12} & k_{25} & k_{26} \\ & & k_{33} & -k_{13} & k_{26} & k_{25} \\ & & & k_{11} & -k_{12} & -k_{13} \\ & & & & k_{22} & k_{23} \\ & & & & & k_{33} \end{bmatrix} \begin{bmatrix} w_1 \\ \theta_{x1} \\ \theta_{y1} \\ w_2 \\ \theta_{x2} \\ \theta_{y2} \end{bmatrix} \qquad (10.8)$$

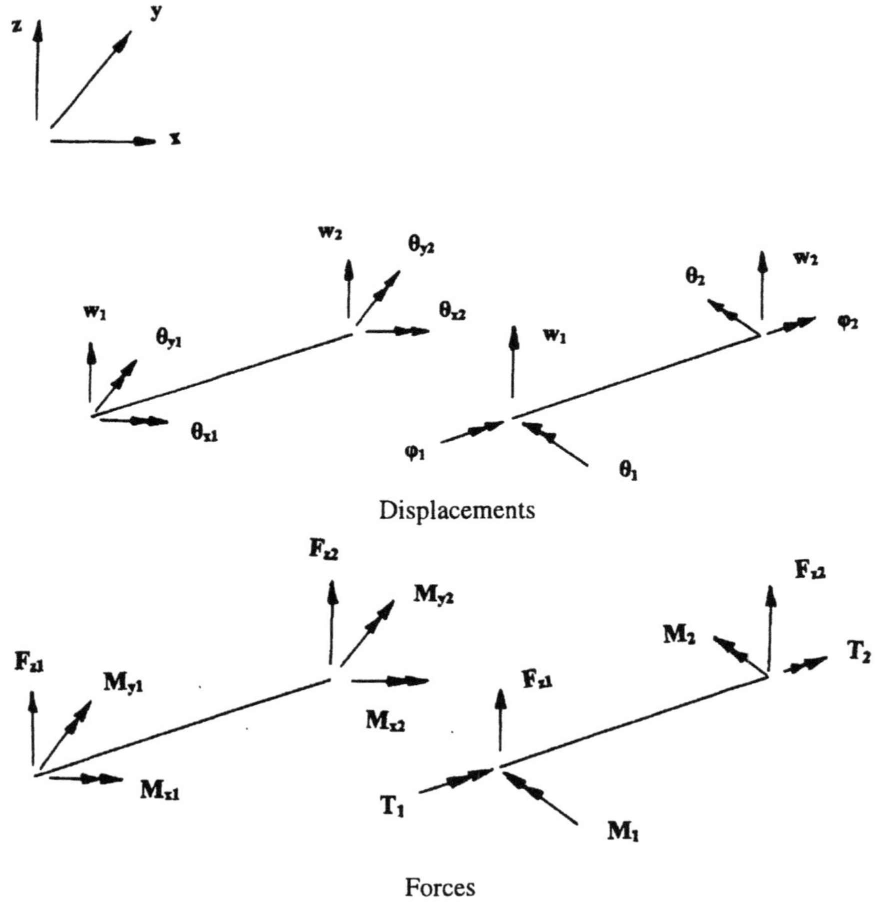

Fig. 10.3 Local and global forces and displacements: plane grid member

10.5.4 Flat Shell Finite Strip

Figure 10.4 shows a thin strip in the local coordinate system. Fig. 10.5 shows a strip orientated such that the length of the strip is normal to the vertical x-z plane. For the strip which uses thin plate-bending formulation, the relationship between the forces (displacements) in the two systems are as follows.

Displacements:

$$u_{local} = u_{Global}\,\ell + w_{Global}\,n\ ,\ w_{local} = -u_{Global}\,m + w_{Global}\,\ell$$

$$v_{Local} = v_{Global}\ ,\ \theta_{y\,Local} = \theta_{y\,Global}$$

Forces:

$$F_{x\,Global} = F_{x\,Local}\,\ell - F_{z\,Local}\,n\ ,\ F_{z\,Global} = F_{x\,Local}\,n + F_{z\,Local}\,\ell$$

$$F_{y\,Global} = F_{y\,Local}\ ,\ M_{y\,Global} = M_{y\,Local}$$

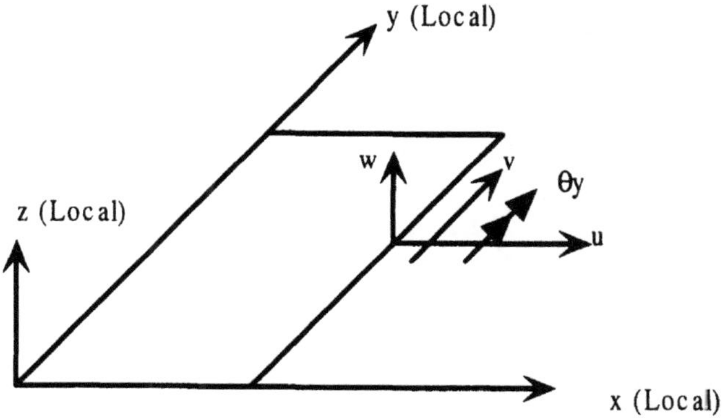

Fig. 10.4 Nodal displacements in a thin flat shell strip

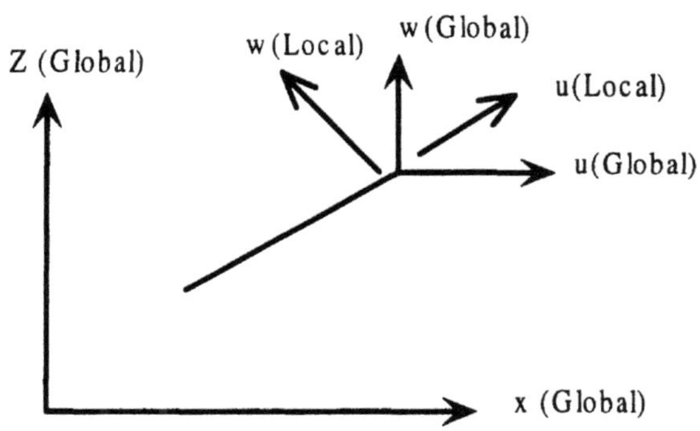

Fig. 10.5 Local and global displacements in a thin shell strip

For the strip which uses thick plate-bending formulation, Fig. 10.6 shows a strip in the local coordinate system and Fig. 10.7 shows a strip orientated such that the length of the strip is normal to the vertical x-z plane.

Displacements:

$$u_{local} = u_{Global}\,\ell + w_{Global}\,n\ ,\ w_{local} = -u_{Global}\,m + w_{Global}\,\ell$$

$$\Psi_{y\,local} = \Psi_{y\,Global}\,\ell + \Psi_{z\,Global}\,n\ ,\ v_{Local} = v_{Global}\ ,\ \Psi_{x\,Local} = \Psi_{x\,Global}$$

Note the introduction of an additional degree of freedom $\Psi_{z\,Global}$.

Forces:

$$F_{x\,Global} = F_{x\,Local}\,\ell - F_{z\,Local}\,n\ ,\ F_{z\,Global} = F_{x\,Local}\,n + F_{z\,Local}\,\ell$$

$$F_{y\,Global} = F_{y\,Local}\ ,\ M_{y\,Global} = M_{y\,Local}$$

$$M_{x\,Global} = M_{x\,Local}\,\ell\ ,\ M_{z\,Global} = M_{x\,Local}\,n$$

$$\ell = \frac{(x_2 - x_1)}{L}\ ,\ n = \frac{(z_2 - z_1)}{L}$$

Note that the global stiffness matrix needs six degrees of freedom per nodal line but in the member axes system, there are only five degrees of freedom per nodal line.

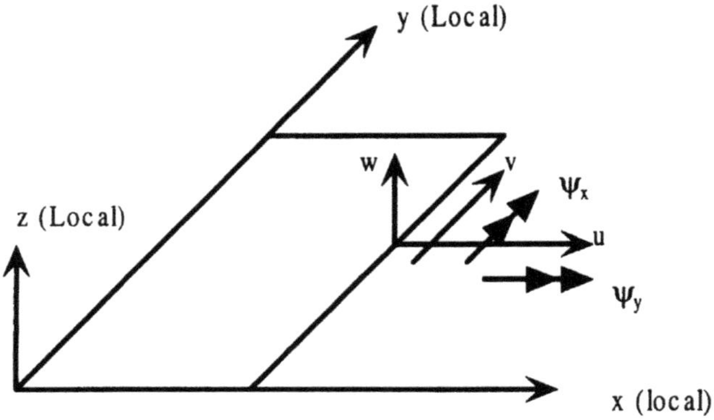

Fig. 10.6 Nodal displacements in a thick flat shell strip

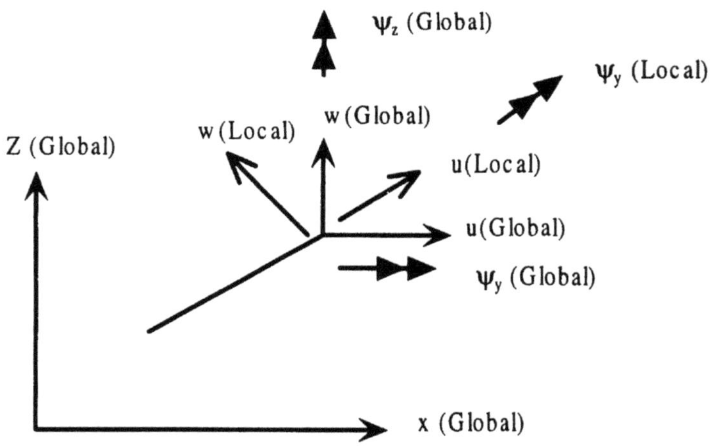

Fig. 10.7 Local and global displacements in a thick flat shell strip

10.6 STRUCTURAL STIFFNESS MATRIX

As a simple example to develop the concept of the structural stiffness matrix, consider the two-dimensional pin-jointed truss shown in Fig. 10.8.

Joints 1 and 3 are fully restrained. Joints 2 and 4 can displace both horizontally and vertically. Let the unknown joint displacements at joint 2 be $u = d_1$, and $v = d_2$.

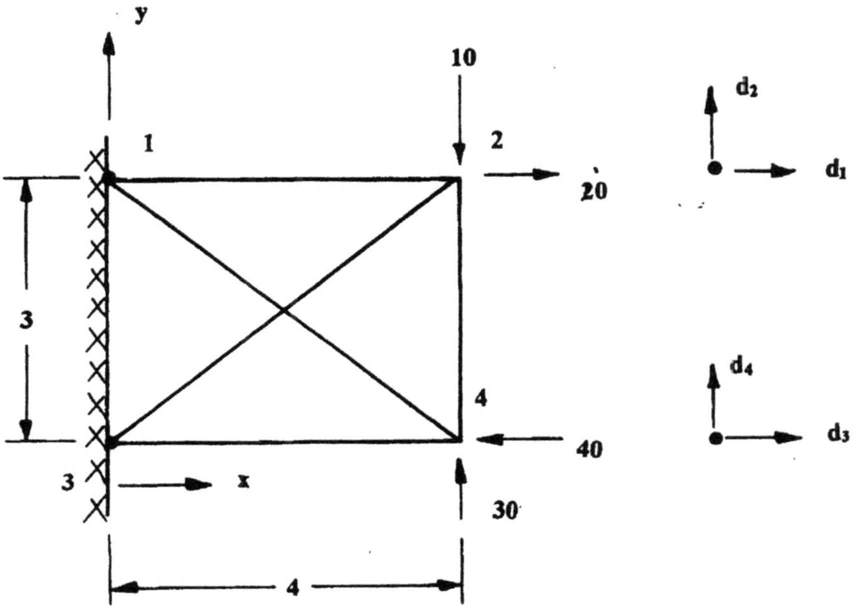

Fig. 10.8 A pin-jointed structure

Similarly at joint 4, u = d_3 and v = d_4. Thus there are four unknown joint displacements to be determined. This information is shown in Table 10.1.

At joints 2 and 4, which can translate, the equations of equilibrium which have to be satisfied are

Σ horizontal forces = 0 and Σ vertical forces = 0

It should be noted that there is no need to consider equilibrium at the support joints. In a stable structure, supports will provide the necessary forces at the supported end of the members to ensure their equilibrium.

The equations of equilibrium at each joint are written in terms of the forces in the bars meeting at that joint and the external load acting at that joint. The contribution from the members to the equations of equilibrium are best expressed in terms of displacements as given by the stiffness matrix for the member. Table 10.2 can be used to calculate the direction cosines for the five members of the truss.

Table 10.1

Joint	Joint displacements	
	u	v
1	0	0
2	d_1	d_2
3	0	0
4	d_3	d_4

Table 10.2

Member	End 1	End 2	x_1	y_1	x_2	y_2	L	ℓ	m
1-2	1	2	0	3	4	3	4	1.0	0
1-4	1	4	0	3	4	0	5	0.8	-0.6
2-3	2	3	4	3	0	0	5	-0.8	-0.6
3-4	3	4	0	0	4	0	4	1.0	0
4-2	4	2	4	0	4	3	3	0	1.0

Note that the designation of the members follows the convention joint number at end 1 followed by joint number at end 2. The designation is purely arbitrary but once a particular choice is made, it must be adhered to rigidly.

It is also necessary to identify for each member the end displacements (u_1, v_1) and (u_2, v_2) in terms of the known and unknown joint displacements as shown in Table 10.3.

Table 10.3

Member	Joint number at		Displacements			
	End 1	End 2	u_1	v_1	u_2	v_2
1-2	1	2	0	0	d_1	d_2
1-4	1	4	0	0	d_3	d_4
2-3	2	3	d_1	d_2	0	0
3-4	3	4	0	0	d_3	d_4
4-2	4	2	d_3	d_4	d_1	d_2

From the information in Tables 10.1 to 10.3, the stiffness matrix for each element can be calculated using equation (10.6).

<u>Member 1-2:</u>

$$\begin{bmatrix} F_{x2} & = & \textit{reaction at node} \, 1 \\ F_{y1} & = & \textit{reaction at node} \, 1 \\ F_{x2} & = & F_1 \\ F_{y1} & = & F_2 \end{bmatrix} = \begin{bmatrix} a_{11} & a_{12} & a_{13} & a_{14} \\ a_{21} & a_{22} & a_{23} & a_{24} \\ a_{31} & a_{32} & a_{33} & a_{34} \\ a_{41} & a_{42} & a_{43} & a_{44} \end{bmatrix} \begin{bmatrix} u_1 = 0 \\ v_1 = 0 \\ u_2 = d_1 \\ v_1 = d_2 \end{bmatrix}$$

Note that F_{x2} and F_{y2} act in the directions of displacements d_1 and d_2 respectively. Hence they are designated F_1 and F_2 respectively. The subscripts 1 and 2 simply indicate that these forces act in the direction of displacements d_1 and d_2 respectively. The coefficients a_{ij} are numerical values depending on the values of A, L, E, ℓ and m of the element. They are obtained by substituting the appropriate values in the stiffness matrix shown in equation (10.6).

From the above matrix we have

$$\begin{bmatrix} F_1 \\ F_2 \end{bmatrix} = \begin{bmatrix} a_{33} & a_{34} \\ a_{43} & a_{44} \end{bmatrix} \begin{bmatrix} d_1 \\ d_2 \end{bmatrix}$$

Member 1-4:

$$\begin{bmatrix} F_{x1} & = & reaction\ at\ node\ 1 \\ F_{y1} & = & reaction\ at\ node\ 1 \\ F_{x2} & = & F_3 \\ F_{y1} & = & F_4 \end{bmatrix} = \begin{bmatrix} b_{11} & b_{12} & b_{13} & b_{14} \\ b_{21} & b_{22} & b_{23} & b_{24} \\ b_{31} & b_{32} & b_{33} & b_{34} \\ b_{41} & b_{42} & b_{43} & b_{44} \end{bmatrix} \begin{bmatrix} u_1 = 0 \\ v_1 = 0 \\ u_2 = d_3 \\ v_1 = d_2 \end{bmatrix}$$

Note the designation of forces as F_3 and F_4 to indicate these forces act in the direction of displacement d_3 and d_4 respectively. From the above matrix we have

$$\begin{bmatrix} F_3 \\ F_4 \end{bmatrix} = \begin{bmatrix} b_{33} & b_{34} \\ b_{43} & b_{44} \end{bmatrix} \begin{bmatrix} d_3 \\ d_4 \end{bmatrix}$$

Members 2-3:

$$\begin{bmatrix} F_{x1} & = & F_1 \\ F_{y1} & = & F_2 \\ F_{x2} & = & reaction\ at\ node\ 3 \\ F_{y1} & = & reaction\ at\ node\ 3 \end{bmatrix} = \begin{bmatrix} c_{11} & c_{12} & c_{13} & c_{14} \\ c_{21} & c_{22} & c_{23} & c_{24} \\ c_{31} & c_{32} & c_{33} & c_{34} \\ c_{41} & c_{42} & c_{43} & c_{44} \end{bmatrix} \begin{bmatrix} u_1 = d_1 \\ v_1 = d_2 \\ u_2 = 0 \\ v_1 = 0 \end{bmatrix}$$

From the above matrix we have

$$\begin{bmatrix} F_1 \\ F_2 \end{bmatrix} = \begin{bmatrix} c_{33} & c_{34} \\ c_{43} & c_{44} \end{bmatrix} \begin{bmatrix} d_1 \\ d_2 \end{bmatrix}$$

Member 3-4:

$$\begin{bmatrix} F_{x1} & = & reaction\ at\ node\ 3 \\ F_{y1} & = & reaction\ at\ node\ 3 \\ F_{x2} & = & F_3 \\ F_{y1} & = & F_4 \end{bmatrix} = \begin{bmatrix} e_{11} & e_{12} & e_{13} & e_{14} \\ e_{21} & e_{22} & e_{23} & e_{24} \\ e_{31} & e_{32} & e_{33} & e_{34} \\ e_{41} & e_{42} & e_{43} & e_{44} \end{bmatrix} \begin{bmatrix} u_1 = 0 \\ v_1 = 0 \\ u_2 = d_3 \\ v_1 = d_4 \end{bmatrix}$$

From the above matrix we have

$$\begin{bmatrix} F_3 \\ F_4 \end{bmatrix} = \begin{bmatrix} e_{33} & e_{34} \\ e_{43} & e_{44} \end{bmatrix} \begin{bmatrix} d_3 \\ d_4 \end{bmatrix}$$

Member 4-2:

$$\begin{bmatrix} F_{x1} & = & F_3 \\ F_{y1} & = & F_4 \\ F_{x2} & = & F_1 \\ F_{y1} & = & F_2 \end{bmatrix} = \begin{bmatrix} f_{11} & f_{12} & f_{13} & f_{14} \\ f_{21} & f_{22} & f_{23} & f_{24} \\ f_{31} & f_{32} & f_{33} & f_{34} \\ f_{41} & f_{42} & f_{43} & f_{44} \end{bmatrix} \begin{bmatrix} u_1 = d_3 \\ v_1 = d_4 \\ u_2 = d_1 \\ v_1 = d_2 \end{bmatrix}$$

From the above matrix we have

$$\begin{bmatrix} F_3 \\ F_4 \\ F_1 \\ F_2 \end{bmatrix} = \begin{bmatrix} f_{11} & f_{12} & f_{13} & f_{14} \\ f_{21} & f_{22} & f_{23} & f_{24} \\ f_{31} & f_{32} & f_{33} & f_{34} \\ f_{41} & f_{42} & f_{43} & f_{44} \end{bmatrix} \begin{bmatrix} d_3 \\ d_4 \\ d_1 \\ d_2 \end{bmatrix}$$

The equations of equilibrium to be satisfied are

$$\begin{aligned} 20 = \Sigma F_1 = &\{a_{33}\, d_1 + a_{34}\, d_2\} \text{ from member 1-2} \\ &+ \{c_{11}\, d_1 + c_{12}\, d2\} \text{ from member 2-3} \\ &+ \{f_{31}\, d_3 + f_{32}\, d_4 + f_{33}\, d_1 + f_{34}\, d_2\} \text{ from member 4-2} \end{aligned}$$

$$\begin{aligned} -10 = \Sigma F_2 = &\{a_{43}\, d_1 + a_{44}\, d_2\} \text{ from member 1-2} \\ &+ \{c_{21}\, d_1 + c_{22}\, d_2\} \text{ from member 2-3} \\ &+ \{f_{41}\, d_3 + f_{42}\, d_4 + f_{43}\, d_1 + f_{44}\, d_2\} \text{ from member 4-2} \end{aligned}$$

$$\begin{aligned} -40 = \Sigma F_3 = &\{b_{33}\, d_3 + b_{34}\, d_4\} \text{ from member 1-4} \\ &+ \{e_{33}\, d_3 + e_{34}\, d_4\} \text{ from member 3-4} \\ &+ \{f_{11}\, d_3 + f_{12}\, d_4 + f_{13}\, d_1 + f_{14}\, d_2\} \text{ from member 4-2} \end{aligned}$$

$$\begin{aligned} 30 = \Sigma F_4 = &\{b_{43}\, d_3 + b_{44}\, d_4\} \text{ from member 1-4} \\ &+ \{e_{43}\, d_3 + e_{44}\, d_4\} \text{ from member 3-4} \\ &+ \{f_{21}\, d_3 + f_{22}\, d_4 + f_{23}\, d_1 + f_{24}\, d_2\} \text{ from member 4-2} \end{aligned}$$

The above four equations can be expressed in matrix form as

$$\begin{bmatrix} 20 \\ -10 \\ -40 \\ 30 \end{bmatrix} = \begin{bmatrix} a_{33} + c_{11} + f_{33} & a_{34} + c_{12} + f_{34} & f_{31} & f_{32} \\ & a_{44} + c_{22} + f_{44} & f_{41} & f_{42} \\ & & b_{33} + e_{33} + f_{11} & b_{34} + e_{34} + f_{12} \\ \textit{symmetric} & & & b_{44} + e_{44} + f_{22} \end{bmatrix} \begin{bmatrix} d_1 \\ d_2 \\ d_3 \\ d_4 \end{bmatrix}$$

The matrix connecting the vector of applied forces at the joints and the vector of the corresponding joint displacements is known as the structural stiffness matrix. Like the element stiffness matrices from which it is assembled, the structural stiffness matrix is also symmetric.

10.7 AUTOMATIC ASSEMBLY OF STRUCTURAL STIFFNESS MATRIX

It is clear from the way the equations of equilibrium at the joints are established, that the procedure consists simply of adding the appropriate contribution from each member to the particular equation of equilibrium. Since the method is designed for automatic computation, it is necessary to establish a systematic procedure for establishing the equations of equilibrium.

Consider the stiffness matrix for the member 4-2. The displacements at the ends are given by $u_1 = d_3$, $v_1 = d_4$, $u_2 = d_1$ and $v_2 = d_2$. Let the freedom number of the displacements at the ends be stored in the vector MEMDIS. The name of the vector MEMDIS standards for MEMber end DISplacements. For 4-2 the elements of MEMDIS are given by

MEMDIS(1) = 3 since $u_1 = d_3$
MEMDIS(2) = 4 since $v_1 = d_4$
MEMDIS(3) = 1 since $u_2 = d_1$
MEMDIS(4) = 2 since $v_2 = d_2$

The stiffness relationship for the element 4-2 is given by

$$\begin{bmatrix} F_3 \\ F_4 \\ F_1 \\ F_2 \end{bmatrix} = \begin{bmatrix} f_{11} & f_{12} & f_{13} & f_{14} \\ f_{21} & f_{22} & f_{23} & f_{34} \\ f_{31} & f_{32} & f_{33} & f_{34} \\ f_{41} & f_{42} & f_{42} & f_{44} \end{bmatrix} \begin{bmatrix} d_3 = \text{MEMDIS (1)} \\ d_4 = \text{MEMDIS (2)} \\ d_1 = \text{MEMDIS (3)} \\ d_2 = \text{MEMDIS (4)} \end{bmatrix}$$

Let the Element STIFfness matrix be matrix ESTIF and the structural or Global stiffness matrix be GSTIF. Evidently element f_{11} contributes to the location (3,3) of the structural stiffness matrix as can be verified from an examination of the structural stiffness matrix. Location (3,3) is the same as location {MEMDIS(1), MEMDIS(1)}. Similarly element f_{32} contributes to the location (1,4) of the structural stiffness matrix. Location (1,4) is the same as location {MEMDIS(3), MEMDIS(2)}. It can be easily verified by taking further elements that the rule
Element ESTIF(i,j) of the member stiffness matrix is added to the element
GSTIF{MEMDIS(i), MEMDIS(j)} of the structural stiffness matrix GSTIF.

As another example of the use of MEMDIS vector, consider the stiffness matrix for member 1-2. For this element $u_1 = 0$, $v_1 = 0$, $u_2 = d_1$ and $v_2 = d_2$. Therefore

MEMDIS(1) = 0 since $u_1 = 0$
MEMDIS(2) = 0 since $v_1 = 0$
MEMDIS(3) = 1 since $u_2 = d_1$
MEMDIS(4) = 2 since $v_2 = d_2$

The stiffness relationship for the element is given by

$$\begin{bmatrix} F_{x1} = \textit{reaction at node } 1 \\ F_{y1} = \textit{reaction at node } 1 \\ F_{x2} = F_1 \\ F_{y1} = F_2 \end{bmatrix} = \begin{bmatrix} a_{11} & a_{12} & a_{13} & a_{14} \\ a_{21} & a_{22} & a_{23} & a_{24} \\ a_{31} & a_{32} & a_{33} & a_{34} \\ a_{41} & a_{42} & a_{43} & a_{44} \end{bmatrix} \begin{bmatrix} u_1 = 0 & = MEMDIS(1) \\ v_1 = 0 & = MEMDIS(2) \\ u_2 = 0 & = MEMDIS(3) \\ v_1 = d_2 & = MEMDIS\ (4) \end{bmatrix}$$

Consider element a_{31}, since $F_1 = a_{31}(0) + a_{32}(0) + a_{33}d_1 + a_{34}\,d_2 = (a_{33}\,d_1 + a_{34}\,d_2)$, evidently the element a_{31} does not figure in the structural stiffness matrix. According to the rule, element a_{31} is added to the structural stiffness matrix at the location {MEMDIS(3), MEMDIS(1)} = (1,0). Since the location (1,0) has no meaning, element a_{31} is simply ignored.

As was indicated in Section 4.7.3, more efficient storage schemes like the vector format is used in practice for storing the matrix GSTIF.

10.8 MASS MATRICES

The element mass matrices derived in Chapters 7, 8 and 9 are called consistent mass matrices because they are calculated from the inertial forces arising from the assumed displacement. The element mass matrices can be assembled in the same way as element stiffness matrices to obtain the global mass matrix.

In many cases it is not necessary to use consistent mass matrices. Instead it is assumed that the mass is concentrated only at the nodes (joints). When this is done, the resulting mass matrices become diagonal matrices. Such mass matrices are called lumped mass matrices. One consequence of using the lumped mass matrix is that acceleration force with respect to rotation is generally taken as zero. In many cases, use of the lumped mass matrices can result in considerable savings both in terms of computational effort and also in computer memory requirements.

CHAPTER 11

DIRECT INTEGRATION

11.1 INTRODUCTION

It was shown in Chapter 3 that the determination of the dynamic response of multi-degree of freedom systems requires the solution of simultaneous linear differential equations expressed in matrix form as

$$M\, X''(t) + C\, X'(t) + K\, X(t) = F(t) \tag{11.1}$$

The damping matrix is generally assumed to follow Rayleigh damping given by

$$C = \alpha M + \beta K \tag{11.2}$$

where α and β are constants.

The general eigen-value Problem for determining the eigen-vectors Φ and the eigen-values Δ corresponding to equation (11.1) is given by

$$K\, \Phi = M\, \Phi\, \Lambda \tag{11.3}$$

It was shown in Section 3.5, that if the eigen-vectors Φ are known, then using the mode superposition method, the n simultaneous equations in equation (11.1) are decoupled in to a set of n ordinary differential equations in single unknowns. These equations are conveniently solved using the Duhamel integral approach explained in Section 2.8 and as shown in Section 3.5.4, the final response is combined from responses at individual frequencies.

The mode superposition method assumes that the matrices M, K and C remain constant. Unfortunately when structures are subjected to large forces such as might occur during an earthquake or a blast loading, most practical structures do not remain in the elastic stage. Sections of the structures yield, leading to reduction of the stiffness matrix K and hence in the damping matrix too. In such a case the method of mode superposition cannot be used. Equation (11.1) has to be integrated taking into account changes in K and C matrices.

11. 2 DIRECT NUMERICAL INTEGRATION

Direct numerical integration aims to calculate the value of X and its derivatives at discrete intervals of time Δt. The aim is to start from known initial conditions at say $t = t_0$, and calculate the values of the unknown at $t = t_0 + \Delta t$. In order to be able to do this, it is necessary to assume the variation of the variables over the interval Δt. Many different types of assumptions are made in practice. Some of these are discussed in the following sections.

It is perhaps worth stating that if the problem does not involve any non-linearities in the stiffness matrix, for large-scale problems, it is worth using the mode superposition method (Section 3.5), to uncouple the simultaneous equations before embarking on numerical integration.

11.2.1 Linear Acceleration Method

In the linear acceleration method, it is assumed that over the time interval Δt, acceleration varies linearly. Therefore

$$X^{''}(t+\tau) = X^{''}(t)+\frac{\tau}{\Delta t}\,[X^{''}(t+\Delta t) - X^{''}(t)] \tag{11.4}$$

Integrating equation (11.4) with respect to τ

$$X^{'}(t+\tau) = X^{'}(t)+X^{''}(t)\tau+\frac{\tau^2}{2\Delta t}\,[X^{''}(t+\Delta t) - X^{''}(t)] \tag{11.5}$$

Integrating equation (11.5) with respect to τ

$$X(t+\tau) = X(t) + X^{'}(t)\tau + X^{''}(t)\frac{\tau^2}{2}+\frac{\tau^3}{6\Delta t}\,[X^{''}(t+\Delta t) - X^{''}(t)] \tag{11.6}$$

Substituting τ = Δt, in equations (11.5) and (11.6) and simplifying

$$X^{'}(t+\Delta t) = X^{'}(t) + \frac{\Delta t}{2}\,[X^{''}(t+\Delta t) + X^{''}(t)] \tag{11.7}$$

$$X(t+\Delta t) = X(t) + X^{'}(t)\Delta t+\frac{\Delta t^2}{6}\,[X^{''}(t+\Delta t) + 2X^{''}(t)] \tag{11.8}$$

Solving equations (11.7) and (11.8) for X'(t + Δt) and X''(t + Δt)

$$X^{'}(t+\Delta t) = \frac{3}{\Delta t}[X(t+\Delta t) - X(t)] - 2X^{'}(t) - \frac{\Delta t}{2}\,X^{''}(t) \tag{11.9}$$

$$X^{''}(t+\Delta t) = \frac{6}{\Delta t^2}[X(t+\Delta t) - X(t)] - \frac{6}{\Delta t}\,X^{'}(t) - 2X^{''}(t) \tag{11.10}$$

At time (t + Δt), dynamic equilibrium equation (11.1) becomes

$$M\,X^{''}(t+\Delta t) + C\,X^{'}(t+\Delta t) + K\,X(t+\Delta t) = F(t+\Delta t) \tag{11.11}$$

Substituting equations (11.9) and (11.10) in equation (11.11) and simplifying

$$[\frac{6}{(\Delta t)^2}M + \frac{3}{\Delta t}C + K]\,X(t+\Delta t) = F(t+\Delta t) + C[\frac{3}{\Delta t}X(t) + 2X^{'}(t)$$

$$+\frac{\Delta t}{2}X^{''}(t)] + M[\frac{6}{(\Delta t)^2}X(t) + \frac{6}{\Delta t}X^{'}(t) + 2X^{''}(t)] \tag{11.12}$$

Since all the quantities on the right-hand side of equation (11.12) are known, the displacement X (t+ Δt) at the end of time step Δt can be calculated. Using equation (11.9), the velocity X'(t + Δt) at the end of time step Δt can be calculated. Although using equation (11.10), the acceleration at the end of time step can be calculated, in order to prevent errors in displacement and velocity calculations affecting equilibrium, acceleration is calculated directly from the equilibrium equation (11.11).

Note that the solution of equation (11.12) requires the 'inversion' of the matrix

$[6M(\Delta t)^{-2} + 3C(\Delta t)^{-1} + K]$. In order to reduce costs, it is clearly advisable to keep the time increment Δt constant, so that factorisation of the matrix by Cholesky or Crout decomposition is done only once. Of course, this observation is valid only if there is no change in the stiffness and damping matrices.

11.2.2 The Wilson-θ Method

The Wilson-θ method is a variation on the linear acceleration method. As opposed to the linear acceleration method, it is assumed that over the time interval θ Δt, acceleration varies linearly, where θ ≥ 1. Therefore

$$\ddot{X}(t+\tau) = \ddot{X}(t) + \frac{\tau}{\theta \Delta t}[\ddot{X}(t+\theta\Delta t) - \ddot{X}(t)] \tag{11.13}$$

Integrating equation (11.13) with respect to τ

$$\dot{X}(t+\tau) = \dot{X}(t) + \ddot{X}(t)\tau + \frac{\tau^2}{2\theta\Delta t}[\ddot{X}(t+\theta\Delta t) - \ddot{X}(t)] \tag{11.14}$$

Integrating equation (11.14) with respect to τ

$$X(t+\tau) = X(t) + \dot{X}(t)\tau + \ddot{X}(t)\frac{\tau^2}{2} + \frac{\tau^3}{6\theta\Delta t}[\ddot{X}(t+\theta\Delta t) - \ddot{X}(t)] \tag{11.15}$$

Substituting τ = θΔt in equations (11.14) and (11.15) and simplifying

$$\dot{X}(t+\theta\Delta t) = \dot{X}(t) + \frac{\theta\Delta t}{2}[\ddot{X}(t+\theta\Delta t) + \ddot{X}(t)] \tag{11.16}$$

$$X(t+\theta\Delta t) = X(t) + \theta\Delta t\,\dot{X}(t) + \frac{\theta^2\Delta t^2}{6}[\ddot{X}(t+\theta\Delta t) + 2\ddot{X}(t)] \tag{11.17}$$

Solving equations (11.16) and (11.17) for X'(t +θ Δt) and X"(t + θΔt)

$$\dot{X}(t+\theta\Delta t) = \frac{3}{\theta\Delta t}[X(t+\theta\Delta t) - X(t)] - 2\dot{X}(t) - \frac{\theta\Delta t}{2}\ddot{X}(t) \tag{11.18}$$

$$\ddot{X}(t+\theta\Delta t) = \frac{6}{\theta^2\Delta t^2}[X(t+\theta\Delta t) - X(t)] - \frac{6}{\theta\Delta t}\dot{X}(t) - 2\ddot{X}(t) \tag{11.19}$$

At time (t + θΔt), dynamic equilibrium equation (11.1) becomes

$$M\ddot{X}(t+\theta\Delta t) + C\dot{X}(t+\theta\Delta t) + KX(t+\theta\Delta t) = F(t+\theta\Delta t) \tag{11.20}$$

where

$$F(t+\theta\Delta t) = F(t) + \theta[F(t+\Delta t) - F(t)] \tag{11.21}$$

Substituting equations (11.18) and (11.19) in equation (11.21) and simplifying

$$\left[\frac{6}{(\Delta t)^2}M + \frac{3}{\Delta t}C + K\right]X(t+\Delta t) = F(t+\Delta t) + C\left[\frac{3}{\Delta t}X(t) + 2\dot{X}(t) + \frac{\Delta t}{2}\ddot{X}(t)\right] + M\left[\frac{6}{(\Delta t)^2}X(t) + \frac{6}{\Delta t}\dot{X}(t) + 2\ddot{X}(t)\right] \tag{11.22}$$

Since all the quantities on the right-hand side of equation (11.22), which represents the equation of equilibrium at time (t+ θΔt), are known, the displacement X(t+θΔt) can be calculated.

From equation (11.19), solving for X"(t+ θΔt)

$$X^{\cdot\cdot}(t+\theta\,\Delta t)=\frac{6}{\theta^2\Delta t^2}[X(t+\theta\Delta t)-X(t)]-\frac{6}{\theta\Delta t}X^{\cdot}(t)-2X^{\cdot\cdot}(t) \tag{11.23}$$

Knowing the value of X"(t+ θΔt), the next step is to calculate the displacement, velocity and acceleration vector at time (t + Δt). Note the difference from the linear accelerator method, where equilibrium was established at time (t + Δt) and all other quantities were also calculated for time (t + Δt).

Substituting τ = Δt, in equations (11.13) to (11.15)

$$X^{\cdot\cdot}(t+\Delta t)=\frac{1}{\theta}X^{\cdot\cdot}(t+\theta\,\Delta t)+\frac{(\theta-1)}{\theta}X^{\cdot\cdot}(t) \tag{11.24}$$

Similarly from equations (11.14) and (11.15)

$$X^{\cdot}(t+\Delta t)=X^{\cdot}(t)+X^{\cdot\cdot}(t)\Delta t+\frac{\Delta t}{2\theta}[X^{\cdot\cdot}(t+\theta\,\Delta t)-X^{\cdot\cdot}(t)] \tag{11.25}$$

$$X(t+\Delta t)=X(t)+X^{\cdot}(t)\Delta t+X^{\cdot\cdot}(t)\frac{\Delta t^2}{2}+\frac{\Delta t^2}{6\theta}[X^{\cdot\cdot}(t+\theta\,\Delta t)-X^{\cdot\cdot}(t)] \tag{11.26}$$

Note that if Rayleigh damping is used, then C = α M + β K. Substituting this value in equation (11.22)

$$[\{\frac{6}{(\theta\Delta t)^2}+\frac{3}{\theta\Delta t}\alpha\}M+(1+\frac{3}{\theta\Delta t}\beta)K]X(t+\theta\,\Delta t)=F(t+\theta\Delta t)$$

$$+[(2+\frac{\theta\Delta t}{2}\alpha)M+\frac{\theta\Delta t}{2}\beta\,K\,]X^{\cdot\cdot}(t)+[(\frac{6}{\theta\Delta t}+2\alpha)M+2\beta K]X^{\cdot}(t)$$

$$+[\{\frac{6}{(\theta\Delta t)^2}+\frac{3}{\theta\Delta t}\alpha\}M+\frac{3}{\theta\Delta t}\beta K\,]X(t) \tag{11.27}$$

11.2.2.1 Example on the Wilson-θ Method

The use of equations derived in the previous section is illustrated by a specific example. The three-storey frame shown in Fig. 3.1 and discussed in Section 3.2.4 is used. The stiffness and mass matrices are

$$K=\begin{bmatrix}14&-6&0\\-6&10&-4\\0&-4&4\end{bmatrix},\ M=\begin{bmatrix}4&0&0\\0&4&0\\0&0&2\end{bmatrix}$$

Rayleigh damping is assumed with C = α M + β K, α = 0.048 and β = 0.043.

It is assumed that the base of the frame is subjected to acceleration as shown in Fig. 11.1.

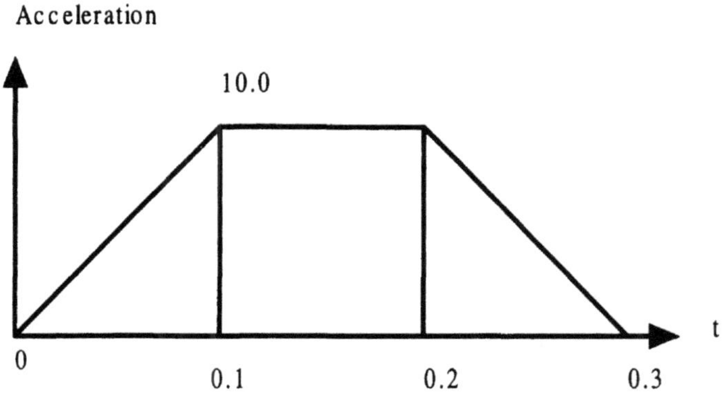

Fig. 11.1 Base acceleration versus time

The differential equations to be solved are expressed in matrix form as shown in Section 3.4 by

$$M\frac{d^2X}{dt^2} + (\alpha M + \beta K)\frac{dX}{dt} + K X = -\frac{d^2X_{Base}}{dt^2} M I$$

where I = a column vector with unit values.

In the following calculations, θ = 1.4 and Δt = 0.005 seconds. Using these values, the following matrices are calculated.

$$MBAR = [(2 + \frac{\theta\Delta t}{2}\alpha)M + \frac{\theta\Delta t}{2}\beta K] = 2.000168M + 1.505x10^{-4} K$$

$$MBAR = \begin{bmatrix} 8.0028 & -0.0009 & 0 \\ -0.0009 & 8.0022 & -0.0006 \\ 0 & -0.0006 & 4.0009 \end{bmatrix}$$

$$CBAR = [(\frac{6}{\theta\Delta t} + 2\alpha)M + 2\beta K] = 857.238857\,M + 0.086\,K$$

$$CBAR = \begin{bmatrix} 3430.16 & -0.5160 & 0 \\ -0.5160 & 3429.82 & -0.3440 \\ 0 & -0.3440 & 1714.82 \end{bmatrix}$$

$$KBAR = [\{\frac{6}{(\theta\Delta t)^2} + \frac{3}{\theta\Delta t}\alpha\}M + \frac{3}{\theta\Delta t}\beta K] = 122.4696x10^3\,M + 18.428574K$$

$$KBAR = 10^3 \begin{bmatrix} 490.136 & -0.1106 & 0 \\ -0.1106 & 490.063 & -0.0737 \\ 0 & -0.0737 & 245.013 \end{bmatrix}$$

$$STIFBAR = [\{\frac{6}{(\theta\Delta t)^2} + \frac{3}{\theta\Delta t}\alpha\}M + (1+\frac{3}{\theta\Delta t}\beta)K] = 122.4696x10^3\,M + 19.428574\,K$$

$$STIFBAR = 10^3\begin{bmatrix} 490.150 & -0.1166 & 0 \\ -0.1166 & 490.073 & -0.0777 \\ 0 & -0.0777 & 245.017 \end{bmatrix}$$

Cholesky factorisation of STIFFBAR = U^TU, gives

$$U = \begin{bmatrix} 700.107 & -0.1665 & 0 \\ 0 & 700.052 & -0.1110 \\ 0 & 0 & 494.992 \end{bmatrix}$$

Initial conditions: The storey displacement and velocity at the start are taken as zero. Since the base acceleration is also zero at the start, the acceleration at storey levels are also zero. Therefore

$$X(0) = \begin{bmatrix} 0 \\ 0 \\ 0 \end{bmatrix}, \quad \frac{dX(0)}{dt} = \begin{bmatrix} 0 \\ 0 \\ 0 \end{bmatrix}, \quad -\frac{d^2X_{Base}(0)}{dt^2}M\ I = \begin{bmatrix} 0 \\ 0 \\ 0 \end{bmatrix}, \quad \frac{d^2X(0)}{dt^2} = \begin{bmatrix} 0 \\ 0 \\ 0 \end{bmatrix}$$

Increment time by $\Delta t = 0.005$. Acceleration of the base = 0.5 and acceleration at a time increment of $\theta\ \Delta t = 0.007$ is equal to 0.7.

$$-\frac{d^2X_{Base}(0+\theta\ \Delta t)}{dt^2}M\ I = \begin{bmatrix} -2.8 \\ -2.8 \\ -1.4 \end{bmatrix}$$

From equation (11.27)

$$STIFBAR\times X(0+\theta\ \Delta t) = -\frac{d^2X_{Base}(0+\theta\ \Delta t)}{dt^2}M\ I + MBAR\times X^{''}(0)$$

$$+ CBAR \times X^{'}(0) + KBAR\times X(0)$$

$$STIFBAR\times X(0+\theta\ \Delta t) = \begin{bmatrix} -2.8 \\ -2.8 \\ -1.4 \end{bmatrix}$$

Solving for X(0 + θ Δt), $X(0+\theta\ \Delta t) = 10^{-5}\begin{bmatrix} -0.571389 \\ -0.571571 \\ -0.571570 \end{bmatrix}$

From equation (11.23), $X^{''}(0+\theta\ \Delta t) = \begin{bmatrix} -0.699660 \\ -0.699882 \\ -0.699882 \end{bmatrix}$

From equation (11.24), $\ddot{X}(0+\Delta t) = \begin{bmatrix} -0.499757 \\ -0.499916 \\ -0.499916 \end{bmatrix}$

From equation (11.25), $\dot{X}(0+\Delta t) = 10^{-2}\begin{bmatrix} -0.124939 \\ -0.124979 \\ -0.124979 \end{bmatrix}$

From equation (11.26), $X(0+\Delta t) = 10^{-5}\begin{bmatrix} -0.208232 \\ -0.208298 \\ -0.208298 \end{bmatrix}$

This completes one cycle of calculations. The next stage is to obtain values at time (Δt+Δt), using the values at (0+Δt) as starting values.

Acceleration of the base at a time increment of (Δt +θ Δt) = 0.012 is equal to 1.2.

$$X(\Delta t) = 10^{-5}\begin{bmatrix} -0.208232 \\ -0.208298 \\ -0.208298 \end{bmatrix}, \quad \frac{dX(\Delta t)}{dt} = 10^{-2}\begin{bmatrix} -0.124939 \\ -0.124979 \\ -0.124979 \end{bmatrix}$$

$$-\frac{d^2X_{Base}(\Delta t+\theta\,\Delta t)}{dt^2} M\,I = \begin{bmatrix} -4.8 \\ -4.8 \\ -2.4 \end{bmatrix}, \quad \frac{d^2X(\Delta t)}{dt^2} = \begin{bmatrix} -0.499757 \\ -0.499916 \\ -0.499916 \end{bmatrix}$$

From equation (11.27)

$$STIFBAR \times X(\Delta t+\theta\,\Delta t) = -\frac{d^2X_{Base}(\Delta t+\theta\,\Delta t)}{dt^2} M\,I + MBAR \times \ddot{X}(\Delta t)$$
$$+ CBAR \times \dot{X}(\Delta t) + KBAR \times X(\Delta t)$$

$$STIFBAR \times X(\Delta t+\theta\,\Delta t) = \begin{bmatrix} -14.1044 \\ -14.1055 \\ -7.05277 \end{bmatrix}$$

Solving for X(Δt + θ Δt), $X(\Delta t+\theta\,\Delta t) = 10^{-4}\begin{bmatrix} -0.287824 \\ -0.287940 \\ -0.287940 \end{bmatrix}$

From equation (11.23), $\ddot{X}(\Delta t+\theta\,\Delta t) = \begin{bmatrix} -1.19898 \\ -1.19965 \\ -1.19965 \end{bmatrix}$

From equation (11.24), $\ddot{X}(\Delta t+\Delta t) = \begin{bmatrix} -0.999201 \\ -0.999729 \\ -0.999729 \end{bmatrix}$

From equation (11.25), $\dot{X}(\Delta t+\Delta t) = 10^{-2}\begin{bmatrix} -0.499679 \\ -0.499890 \\ -0.499890 \end{bmatrix}$

From equation (11.26), $X(\Delta t+\Delta t) = 10^{-4}\begin{bmatrix} -0.166573 \\ -0.166634 \\ -0.166634 \end{bmatrix}$

This completes two cycles of calculations. The next stage is to obtain values at time (2Δt+Δt), using the values at (2Δt) as starting values. Calculations can be continued to the desired time.

11.2.3 The Newmark Method

One simple assumption that can be made is to assume that over the period Δt, the acceleration is the average of the acceleration at time t and t + Δt. Thus

$$\ddot{X}(t+\tau) = \frac{1}{2}[\ddot{X}(t+\Delta t) + \ddot{X}(t)] \tag{11.28}$$

Integrating equation (11.28) with respect to τ

$$\dot{X}(t+\tau) = \dot{X}(t) + \frac{\tau}{2}[\ddot{X}(t+\Delta t) + \ddot{X}(t)] \tag{11.29}$$

$$X(t+\tau) = X(t) + \dot{X}(t)\tau + \frac{1}{4}\tau^2[\ddot{X}(t+\Delta t) + \ddot{X}(t)] \tag{11.30}$$

Substituting τ = Δt

$$\dot{X}(t+\Delta t) = \dot{X}(t) + \frac{1}{2}[\ddot{X}(t+\Delta t) + \ddot{X}(t)] \tag{11.31}$$

$$X(t+\Delta t) = X(t) + \dot{X}(t)\Delta t + \frac{1}{4}(\Delta t)^2[\ddot{X}(t+\Delta t) + \ddot{X}(t)] \tag{11.32}$$

Newmark made a more general assumption that the acceleration and velocity and displacement are given by

$$\dot{X}(t+\Delta t) = \dot{X}(t) + [\delta\,\ddot{X}(t+\Delta t) + (1-\delta)\,\ddot{X}(t)]\Delta t \tag{11.33}$$

$$X(t+\Delta t) = X(t) + \dot{X}(t)\Delta t + (\Delta t)^2\,[(0.5-\alpha)\ddot{X}(t) + \alpha\,\ddot{X}(t+\Delta t)] \tag{11.34}$$

Note that if δ = 0.50 and α = 0.25 leads to the average acceleration assumption while δ = 0.50 and α = 0.1667 corresponds to the linear acceleration method.

Recommended values are $\delta \geq 0.50$, $\alpha \geq 0.25(0.5+\delta)^2$

Solving equations (11.33) and (11.34) for velocity and acceleration at time (t + Δt)

$$\ddot{X}(t+\Delta t) = \frac{1}{(\Delta t)^2\,\alpha}\{X(t+\Delta t) - X(t)\} - \frac{1}{\Delta t\,\alpha}\dot{X}(t) - \frac{(0.5-\alpha)}{\alpha}\ddot{X}(t) \tag{11.35}$$

$$\dot{X}(t+\Delta t) = \frac{\delta}{\alpha}\frac{1}{\Delta t}\{X(t+\Delta t) - X(t)\} + (1-\frac{\delta}{\alpha})\dot{X}(t) + \frac{(\alpha-0.5\delta)}{\alpha}\Delta t\,\ddot{X}(t) \tag{11.36}$$

Using equations (11.35) and (11.36), the equilibrium equation (11.11) at time (t+Δt) becomes

$$[M\frac{1}{(\Delta t)^2\alpha}+C\frac{\delta}{\alpha\,\Delta t}+K]X(t+\Delta t)=[M\frac{1}{(\Delta t)^2\alpha}+C\frac{\delta}{\alpha\,\Delta t}]X(t)$$

$$+[M\frac{1}{\Delta t\,\alpha}+C(1-\frac{\delta}{\alpha})]X^{'}(t)+[M\frac{(0.5-\alpha)}{\alpha}-C\frac{(\alpha-0.5\delta)}{\alpha}]X^{''}(t)+F(t+\Delta t)$$

(11.37)

Having calculated X(t+Δt), $X^{''}$(t+Δt) can be calculated from equation (11.35) and $X^{'}$(t+Δt) from equation (11.36).

11.2.4 The Central Difference Method

In the central difference method, acceleration and velocities are expressed as follows

$$X^{''}(t)=\frac{1}{(\Delta t)^2}\,[X(t-\Delta t)-\;2X(t)+X(t+\Delta t)] \tag{11.38}$$

$$X^{'}(t)=\frac{1}{2\Delta t}\,[X(t+\Delta t)-\;X(t-\Delta t)] \tag{11.39}$$

At time (t), dynamic equilibrium equation (11.1) becomes

$$M\,X^{''}(t)+C\,X^{'}(t)+K\,X(t)=F(t) \tag{11.40}$$

Substituting equations (11.38) and (11.39) in equation (11.40) and simplifying

$$[\frac{1}{(\Delta t)^2}M+\frac{1}{\Delta t}C]\;X(t+\Delta t)=F(t)-[\frac{1}{(\Delta t)^2}M-\frac{1}{\Delta t}C]\;X(t-\Delta t)$$

$$+[\frac{2}{(\Delta t)^2}M+K]\;X(t) \tag{11.41}$$

Note that the displacement at time (t+Δt) is obtained from equilibrium consideration at time t.

In general one knows the starting values of displacement and velocity at time t = 0. The acceleration at time t = 0, can be found from equilibrium equation at t = 0, from equation (11.40) as

$$M\,X^{''}(0)+C\,X^{'}(0)+K\,X(0)=F(0) \tag{11.42}$$

However in order to calculate the displacement at time (0 + Δt), we need displacement at time (0 - Δt). Using equations (11.38) and (11.39)

$$X^{''}(0)=\frac{1}{(\Delta t)^2}\,[X(0-\Delta t)-\;2X(0)+X(0+\Delta t)] \tag{11.43}$$

$$X^{'}(0)=\frac{1}{2\Delta t}\,[X(0+\Delta t)-\;X(0-\Delta t)] \tag{11.44}$$

Solving for X(0 - Δt) from equations (11.43) and (11.44)

$$X(0-\Delta t)=\;X(0)-\Delta t X^{'}(0)\;+\frac{(\Delta t)^2}{2}\;X^{''}(0) \tag{11.45}$$

Apart from this preliminary work, the rest of the calculations are fairly standard.

11.3 EXPLICIT AND IMPLICIT INTEGRATION SCHEMES

It was noticed in the previous sections, that except for the central difference method, other methods determined the displacement X(t + Δt) by considering equilibrium at time (t + Δt). Such methods are called implicit integration methods. These methods require the 'inversion' of the effective stiffness matrix of the form

$$K_{Effective} = a_1 M + a_2 C + K$$

where a_1 and a_2 are 'constants' for a fixed value of Δt. This can make the method quite expensive, although the 'inversion' or factorisation needs to be done only once provided Δt is kept constant throughout the calculations.

In the case of the central difference method, the displacement X(t + Δt) is determined by considering equilibrium at time t. Such methods are called explicit integration methods. This requires the 'inversion' of the effective mass matrix of the form

$$M_{Effective} = a_1 M + a_2 C$$

where a_1 and a_2 are 'constants' for a fixed value of Δt. If damping is ignored and the mass matrix is diagonal, then the 'inversion' of effective mass matrix is a trivial process. This makes the method inexpensive. However if the mass matrix is not diagonal and damping needs to be considered, then this advantage is lost.

11.4 STABILITY OF INTEGRATION SCHEMES

An integration scheme is said to be stable if there exists a finite time step Δt for which finite change in the values of displacement, velocity etc., does not lead to an ever-increasing change in subsequent values of displacement, etc. Numerical analysts have studied the stability of the schemes and their conclusions are as follows.

Wilson-θ Method: The method is unconditionally stable for $\theta \geq 1.37$.

The Newmark Average Acceleration Method: The method is unconditionally stable for $\alpha = 0.25$ and $\delta = 0.5$. However it has also been suggested that the following values of the parameters will also lead to unconditional stability.

$$\delta \geq 0.5, \ \alpha \geq 0.25(0.5 + \delta)^2$$

The Central Difference Method: Stable solution can be obtained only if

$\Delta t \leq 2/\omega_n$, ω_n = highest natural frequency of the system.

It has to be stressed that a value of time step, which leads to a stable solution, does not necessarily guarantee an accurate solution. It is possible that for an accurate solution, a smaller time step than that required for stability of the solution is needed. Unconditionally stable methods are preferred in practice except in cases of non-linear situations, where considerations of accurately accounting for non-linearity might dictate a very small time step, which can guarantee a stable solution.

11.5 ACCURACY OF INTEGRATION SCHEMES

Errors invariably occur in any integration scheme where continuous functions are discretised at intervals of time Δt. In general the smaller the value of Δt, the greater the accuracy, but it is achieved at the expense of many calculations. Two types of error normally occur. They are

Amplitude decay: It has been found that if the undamped free vibration of a single degree of freedom system is analysed using quite a small time step Δt, the resulting displacement displays amplitude decay as though damping was present. This damping results from numerical dissipation.

It is worth pointing out that in finite element modelling of a structure, the modelling generally leads to poor modelling for higher modes. Numerical damping can be quite useful in damping out the inaccurate response due to higher modes.

Period elongation: Another type of error is where the time between successive peaks (troughs) keeps increasing. This is called period elongation.

In the final analysis the only way of ensuring that an accurate solution has been obtained is to redo the analysis using a smaller time step.

11.6 NON-LINEAR ANALYSIS

In linear analysis considered in the previous sections, stiffness, mass and damping matrices are assumed to remain unchanged throughout the loading history. For economy in design, it is not possible to design structures which remain in the linear range for all possible design loading. Certain amount of yielding in parts of the structure must be accepted. In the case of frame structures, such yielding is concentrated at 'plastic hinges' while in the case of general structures, yielding can spread over fairly large areas. Such yielding naturally affects the stiffness and damping of structures. Two major consequences of such changes are

- the mode superposition method is inapplicable
- the equilibrium has to treated on an incremental basis.

The incremental equilibrium equation becomes

$$M\,\Delta X^{\cdot\cdot}(t) + C(t)\Delta X^{\cdot}(t) + K_{Tangent}(t)\Delta X(t) = \Delta F(t) \tag{11.46}$$

Note that both C and $K_{Tangent}$ matrices are now functions of time. $K_{Tangent}$ is the tangent stiffness matrix, generally taken to be the value at time t.

11.6.1 Non-linear Analysis: The Wilson-θ Method

As an example of developing formulae for calculations, consider the Wilson-θ Method. From equation (11.17)

$$X(t+\theta\,\Delta t) = X(t) + \theta\,\Delta t\,X^{\cdot}(t) + \frac{\theta^2\Delta t^2}{6}\,[X^{\cdot\cdot}(t+\theta\,\Delta t) + 2X^{\cdot\cdot}(t)] \tag{11.47}$$

Assuming that approximately

$$X(t+\theta\,\Delta t) - X(t) = \theta\,\Delta X \tag{11.48}$$

$$X^{''}(t+\theta\,\Delta t) - X^{''}(t) = \theta\,\Delta X^{''} \tag{11.49}$$

Substituting equations (11.48) and (11.49) in equation (11.47)

$$\theta\,\Delta X = \theta\,\Delta t\,X^{'}(t)+\frac{\theta^2\Delta t^2}{6}\,\theta\,\Delta X^{''} + \frac{\theta^2\Delta t^2}{2}\,X^{''}(t) \tag{11.50}$$

From which

$$\theta\,\Delta X^{''} = \frac{6}{\theta^2\Delta t^2}\{\theta\,\Delta X - \theta\,\Delta t\,X^{'}(t) - \frac{\theta^2\Delta t^2}{2}\,X^{''}(t)\} \tag{11.51}$$

Similarly from equation (11.16)

$$X^{'}(t+\theta\,\Delta t) = X^{'}(t) + \frac{\theta\,\Delta t}{2}\,[X^{''}(t+\theta\,\Delta t) + X^{''}(t)] \tag{11.52}$$

Assuming that approximately

$$X^{'}(t+\theta\,\Delta t) - X^{'}(t) = \theta\,\Delta X^{'} \tag{11.53}$$

From which

$$\theta\,\Delta X^{'} = \frac{\theta\,\Delta t}{2}(\theta\,\Delta\,X^{''})+\theta\,\Delta t\,X^{''}(t) \tag{11.54}$$

Substituting for $\theta\ \Delta X^{''}$ from equation (11.51) and simplifying

$$\theta\,\Delta X^{'} = \frac{3}{\theta\,\Delta t}\theta\,\Delta X - 3\,X^{'}(t) - \frac{\theta\,\Delta t}{2}\,X^{''}(t) \tag{11.55}$$

The incremental equilibrium equation at time $(t + \theta\Delta t)$ is

$$M\,\theta\,\Delta X^{''}(t) + C(t)\theta\,\Delta X^{'}(t) + K_{Tangent}(t)\theta\,\Delta X(t) = \theta\,\Delta F(t) \tag{11.56}$$

Substituting equations (11.51) and (11.55) in equation (11.56) and simplifying

$$\{\frac{6}{(\theta\,\Delta t)^2}M + \frac{3}{(\theta\,\Delta t)}C(t)+K_{Tangent}(t)\}\theta\,\Delta X$$

$$=\theta\,\Delta F + \{\frac{6}{(\theta\,\Delta t)}M + 3C(t)\}X^{'} + \{3M+\frac{\theta\,\Delta t}{2}C(t)\}X^{''} \tag{11.57}$$

Solve equation (11.57) for $\theta\ \Delta X$ and determine $\theta\ \Delta X^{'}$ and $\theta\ \Delta X^{''}$ from equations (11.55) and (11.51) respectively.

The displacement, velocity and acceleration at time $(t + \Delta t)$ can be obtained as follows

$$X(t+\Delta t) = X(t) + \frac{1}{\theta}(\theta\,\Delta X) \tag{11.58}$$

$$X^{'}(t+\Delta t) = X^{'}(t) + \frac{1}{\theta}(\theta\,\Delta X^{'}) \tag{11.59}$$

$$X^{''}(t+\Delta t) = X^{''}(t) + \frac{1}{\theta}(\theta\,\Delta X^{''}) \tag{11.60}$$

11.6.2 Non-linear Analysis: Newmar method

Taking $\theta = 1$ and substituting equations (11.48) and (11.49) in equation (11.35)

$$\Delta X^{''} = \frac{1}{(\Delta t)^2\ \alpha}\Delta X - \frac{1}{\Delta t\ \alpha}X^{'}(t) - \frac{0.5}{\alpha}X^{''}(t) \tag{11.61}$$

Similarly substituting equation (11.53) in equation (11.36) and simplifying

$$\Delta X^{'} = \frac{\delta}{\alpha}\frac{1}{\Delta t}\Delta X - \frac{\delta}{\alpha}X^{'}(t) + \frac{(\alpha - 0.5\delta)}{\alpha}\Delta t\, X^{''}(t) \tag{11.62}$$

Using equations (11.61) and (11.62), the incremental equilibrium equation (11.46) at time (t + Δt) becomes

$$[M\frac{1}{(\Delta t)^2\alpha} + C(t)\frac{\delta}{\alpha\,\Delta t} + K_{Tangent}(t)]\Delta X =$$

$$+[M\frac{1}{\Delta t\,\alpha} + C\frac{\delta}{\alpha}]X^{'}(t) + [M\frac{0.5}{\alpha} - C\frac{(\alpha - 0.5\delta)}{\alpha}\Delta t]X^{''}(t) + \Delta F \tag{11.63}$$

11.6.3 Equilibrium Iterations

In equations (11.57) and (11.63), it is assumed that $K_{Tangent}$ remains constant over the time interval Δt. However as the displacement increases, strains increase leading to perhaps more yielding and a change of $K_{Tangent}$. It is therefore necessary to iteratively check on equilibrium and the state of stress within each time increment. Generally Newton-Rhapson or a modified version of it is used. The reader should consult books on non-linear finite element analysis for more information on this important aspect.

11.7 PROGRAMS

A series of elementary programs using the theory developed in this chapter are given. The reader is well advised to run these programs on small sets of data to acquire experience.

11.7.1 Program WILSON

```
!***************************************************************************
! DIRECT INTEGRATION BY WILSON'S THETA METHOD
! LUMPED MASS MATRIX ASSUMED TO BE DIAGONAL
! STRUCTURE SUBJECTED TO BASE ACCELERATION
! SOLUTION OF SIMULTANEOUS EQUATIONS USING CHOLESKY FACTORISATION
!***************************************************************************
        PROGRAM WILSON
!******VARIABLE DECLARATION**************************************************
        IMPLICIT NONE
        REAL:: SUM,FACTOR,STEP,T,C1,C2,C3,C4,C5,C6,C7,C8,C9,&
        &C10,C11,ALPHA,BETA,THETA,THETA1,THETA2,DELTAT,&
        &THETADT,THETADT2,DELTAT2,TMAX,T1,T2,A1,A2
        INTEGER::I,J,IM1,IP1,K,N,NM1,JP1,NDPTS,NMAX,NMAXM2,&
        &I1,I1M1
```

```
        REAL,ALLOCATABLE:: STIF(:,:),MASS(:,:),MBAR(:,:),CBAR(:,:),&
        &KBAR(:,:),STIFBAR(:,:),ACLR0(:),VEL0(:),DIS0(:),DIS1(:),&
        &ACLR1(:),VEL1(:),U(:,:), F(:),Y(:),X(:)
        REAL,ALLOCATABLE:: AINPUT(:),TINPUT(:),ACLR(:),TIME(:)
!****** FILE MANAGEMENT*****************************************************
        OPEN(UNIT=1, FILE='D:\USER\FORTRAN\WILSONIN.TXT',&
        &STATUS='OLD',ACTION='READ')
        OPEN(UNIT=2, FILE ='D:\USER\FORTRAN\WILSONOUT.TXT',&
        &STATUS='REPLACE',ACTION='WRITE')
!******* INPUT SIZE OF MATRICES*********************************************
        READ(1,*) N
        ALLOCATE(STIF(N,N),MASS(N,N),MBAR(N,N),CBAR(N,N),&
        &KBAR(N,N),STIFBAR(N,N),ACLR0(N),VEL0(N),DIS0(N),&
        &DIS1(N),ACLR1(N),VEL1(N),U(N,N), F(N),Y(N),X(N))
!*******INPUT UPPER TRIANGULAR PART OF STIFFNESS AND***MASS MATRICES***
!******** INPUT THE DATA ROW BY ROW***********************
        READ(1,*) ((STIF(I,J),J=I,N),I=1,N)
        READ(1,*) ((MASS(I,J),J=I,N),I=1,N)
!******FILL UP THE LOWER SYMMETRIC HALF*******************
        DO I = 2,N
        DO J = 1,(I-1)
        STIF(I,J)=STIF(J,I)
        MASS(I,J)=MASS(J,I)
        END DO
        END DO
!*******PRINT THE INPUT MATRICES****************************
        WRITE(2,*)' STIFFNESS MATRIX'
        DO I = 1,N
        WRITE(2,'(6(4X,F10.4))')(STIF(I,J), J = 1,N)
        END DO
        WRITE(2,*)' MASS MATRIX'
        DO I = 1,N
        WRITE(2,'(6(4X,F10.4))')(MASS(I,J), J = 1,N)
        END DO
!*******READ RAYLEIGH DAMPING CONSTANTS*****************
!       C = ALPHA M + BETA K*********************************
        READ(1,*)ALPHA,BETA
        WRITE(2,'(A,E14.6,A,E14.6)')'ALPHA = ',ALPHA,'   BETA = ',BETA
!*******READ THETA AND TIME INCREMENT*******************
        READ(1,*)THETA,DELTAT
        WRITE(2,'(A,F10.4,A,E14.6)')'THETA = ',THETA,'   &
        &DELTA T = ',DELTAT
        THETA1=1.0/THETA
        THETA2=1-THETA1
        DELTAT2=DELTAT**2
        THETADT=THETA*DELTAT
        THETADT2=THETADT**2
        C1=6/THETADT2
        C2=3/THETADT
        C3=C1+C2*ALPHA
        C4=1.0+C2*BETA
        STIFBAR=C3*MASS+C4*STIF
        WRITE(2,*)' STIFBAR MATRIX'
        DO I = 1,N
        WRITE(2,'(6(4X,E14.6))')(STIFBAR(I,J), J = 1,N)
```

```
      END DO
      C5=ALPHA*0.5*THETADT+2
      C6=BETA*0.5*THETADT
      MBAR=C5*MASS+C6*STIF
      WRITE(2,*)' MBAR MATRIX'
      DO I = 1,N
      WRITE(2,'(6(4X,E14.6))')(MBAR(I,J), J = 1,N)
      END DO
      C7=2*ALPHA+6/THETADT
      C8=2*BETA
      CBAR=C7*MASS+C8*STIF
      WRITE(2,*)' CBAR MATRIX'
      DO I = 1,N
      WRITE(2,'(6(4X,E14.6))')(CBAR(I,J), J = 1,N)
      END DO
      KBAR=C3*MASS+(C4-1)*STIF
      C9=0.5*DELTAT*THETA1
      C10=0.5*DELTAT2
      C11=DELTAT2*THETA1/6.0
      WRITE(2,*)' KBAR MATRIX'
      DO I = 1,N
      WRITE(2,'(6(4X,E14.6))')(KBAR(I,J), J = 1,N)
      END DO
!*******CHOLESKY FACTORISATION OF STIFBAR ************
!*******CALCULATE THE FIRST ROW OF U*********************
      U(1,1) = SQRT(STIFBAR(1,1))
      FACTOR=1/U(1,1)
      DO J=2,N
      U(1,J)=STIFBAR(1,J)*FACTOR
      END DO
!*******CALCULATE THE ELEMENTS OF ROWS 2 TO N***********
      DO I = 2,N
      IP1=I+1
!*******CALCULATE THE DIAGONAL ELEMENT*****************
      IM1=I-1
      SUM=0
      DO K= 1,IM1
      SUM = SUM + U(K,I)*U(K,I)
      END DO
      U(I,I)=SQRT(STIFBAR(I,I)-SUM)
      IF(I==N)EXIT
!*******CALCULATE THE OFF DIAGONAL ELEMENTS************
      FACTOR=1.0/U(I,I)
      DO J= IP1,N
      SUM = 0
      DO K = 1,IM1
      SUM = SUM + U(K,I)*U(K,J)
      END DO
      U(I,J)=(STIFBAR(I,J)-SUM)*FACTOR
      END DO
      END DO
!*******PRINT U MATRIX****************************************
      WRITE(2,*) 'U MATRIX'
      DO I = 1,N
      WRITE(2,'(6(4X,E14.6))')(U(I,J), J=I,N)
```

```
        END DO
!*******INPUT DATA FOR BASE ACCELERATION-TIME*****
!      RELATIONSHIP ASSUMED PIECEWISE LINEAR******
        READ(1,*)NDPTS
        WRITE(2,'(A,I4)')'NUMBER OF DATA POINTS DEFINING&
        &ACCELERATION-TIME VARIATION =  ',NDPTS
        READ(1,*)TMAX
        WRITE(2,'(A,F10.4)')'MAXIMUM TIME UP TO WHICH &
        &RESPONSE IS REQUIRED = ',TMAX
        NMAX = NINT(TMAX/DELTAT) + 1
        ALLOCATE(AINPUT(NDPTS),TINPUT(NDPTS),ACLR(NMAX),&
        &TIME(NMAX))
        WRITE(2,*)'BASE ACCELERATION-TIME RELATIONSHIP'
        WRITE(2,*)'TIME        ACCLR'
        DO  I = 1,NDPTS
        READ(1,*)AINPUT(I),TINPUT(I)
         WRITE(2,*)TINPUT(I),AINPUT(I)
        END DO
!*******INTERPOLATE THE ACCLR-TIME INPUT TO OBTAIN *****
!      ACCELERATION AND TIME  SPACED AT EQUAL *******
!      INTERVALS OF DELTAT********************************
        ACLR(1)=AINPUT(1)
        TIME(1)=TINPUT(1)
        T=0
        DO I=2,NMAX
        T=T+DELTAT
        TIME(I)=T
!********CHECK IN WHICH SEGMENT OF INPUT DATA THE POINT LIES
        DO J = 1,NDPTS
        JP1=J+1
        IF(T>TINPUT(J) .AND. T<TINPUT(JP1)) THEN
        A1=AINPUT(J)
        A2=AINPUT(JP1)
        T1=TINPUT(J)
        T2=TINPUT(JP1)
        STEP=T-T1
        ACLR(I)=A1+(A2-A1)*STEP/(T2-T1)
        EXIT
        END IF
        END DO
        IF ((T+DELTAT)> TINPUT(NDPTS) )EXIT
        END DO
        WRITE(2,*)'INTERPOLATED BASE ACCELERATION-TIME &
        &RELATIONSHIP'
        DO I= 1,NMAX
        WRITE(2,'(2(4X,E14.6))')TIME(I),ACLR(I)
        END DO
!*******START OF CALCULATIONS****************************
!*******SET INITIAL DISPLACEMENT AND VELOCITY VECTORS***
!      AS ZERO*************************************************
        VEL0=0.0
        DIS0=0.0
!******INITIAL ACCELERATION OF ALL MASSES AS EQUAL******
!      TO BASE ACCELERATION********************************
        ACLR0=ACLR(1)
```

```
!*******PRINT DISPLACEMENT, VELOCITY AND ACCELERATION**
        WRITE(2,*)'TIME'
        WRITE(2,*)'DISPLACEMENTS  VELOCITY   ACCLERATION'
        WRITE(2,'(F10.4)')TIME(1)
        DO I = 1,N
        WRITE(2,'(3(4X,E14.6))')DIS0(I),VEL0(I),ACLR0(I)
        END DO
!*******START OF ITERATIONS ON TIME**********************
        NMAXM2=NMAX-2
        DO I1 = 2,NMAXM2
        I1M1=I1-1
!*******CALCULATE THE MASS TIMES ACCELERATION AT*****
!       EACH LEVEL AT A ***TIME INCREMENT OF THETA * DELTAT*
        DO J= 1,N
        F(J)=-(ACLR(I1M1)+THETA*(ACLR(I1)-ACLR(I1M1)))*MASS(J,J)
        END DO
!*******CALCULATE THE NET FORCE*************************
        F=F+MATMUL(KBAR,DIS0)+MATMUL(CBAR,VEL0)+&
        &MATMUL(MBAR,ACLR0)
!*******USE FORWARD SUBSTITUTION TO SOLVE ***********
!       U TRANSFORM ×Y = F*******************************
        Y(1)=F(1)/U(1,1)
        DO I = 2,N
        SUM = 0
        IM1=I-1
        DO J= 1,IM1
        SUM = SUM + U(J,I)*Y(J)
        END DO
        Y(I)= (F(I)-SUM)/U(I,I)
        END DO
!*******USE BACK SUBSTITUTION TO SOLVE U ×X = Y**********
!*******VECTOR X HAS  THE DISPLACEMENT AT TIME *******
!       (T+THETA DT)**************************************
        X(N)=Y(N)/U(N,N)
        NM1=N-1
        DO I= NM1,1,-1
        SUM = 0
        IP1=I+1
        DO J= N,IP1,-1
        SUM=SUM + U(I,J)*X(J)
        END DO
        X(I)=(Y(I)-SUM)/U(I,I)
        END DO
!*******CALCULATE ACCELERATION AT TIME (T+THETA DT)****
        ACLR1=C1*(X-DIS0)-2*C2*VEL0-2*ACLR0
!*******CALCULATE DISPLACEMENT AND VELOCITY AT TIME (T+ DT)
        DIS1=DIS0+VEL0*DELTAT+C10*ACLR0+C11*(ACLR1-ACLR0)
        VEL1=VEL0+ACLR0*DELTAT+C9*(ACLR1-ACLR0)
!*******CALCULATE ACCELERATION AT TIME (T+ DT)***********
        ACLR1=THETA1*ACLR1+THETA2*ACLR0
        WRITE(2,'(F10.4)')TIME(I1)
        DO J = 1,N
        WRITE(2,'(3(4X,E14.6))')DIS1(J),VEL1(J),ACLR1(J)
        END DO
!*******RESET THE VALUES FOR NEXT ITERATION************
```

```
          ACLR0=ACLR1
          VEL0=VEL1
          DIS0=DIS1
          END DO
          END PROGRAM WILSON
```

11.7.2 Program NEWMARK

```
!*****************************************************************
! DIRECT INTEGRATION BY NEWMARK'S AVERAGE ACCELERATION
! METHOD. LUMPED MASS MATRIX ASSUMED TO BE DIAGONAL
! STRUCTURE SUBJECTED TO BASE ACCELERATION
! SOLUTION OF SIMULTANEOUS EQUATIONS USING CHOLESKY
! FACTORISATION
!*****************************************************************
          PROGRAM NEWMARK
!*******VARIABLE DECLARATION*********************************
          IMPLICIT NONE
          REAL:: SUM,FACTOR,STEP,T,C1,C2,C3,C4,C5,C6,C7,C8,C9,&
          &ALPHA,BETA,NALPHA,NDELTA,DELTAT,DELTAT2,TMAX,&
          &T1,T2,A1,A2
          INTEGER::I,J,IM1,IP1,K,N,NM1,JP1,NDPTS,NMAX,NMAXM2,&
          &I1,I1M1
          REAL,ALLOCATABLE:: STIF(:,:),MASS(:,:),MBAR(:,:),CBAR(:,:),&
          &KBAR(:,:),STIFBAR(:,:),ACLR0(:),VEL0(:),DIS0(:),DIS1(:),&
          &ACLR1(:),VEL1(:),U(:,:), F(:),Y(:),X(:)
          REAL,ALLOCATABLE:: AINPUT(:),TINPUT(:),ACLR(:),TIME(:)
!*******  FILE MANAGEMENT***********************************
          OPEN(UNIT=1, FILE='D:\USER\FORTRAN\NEWMARKIN.TXT',&
          STATUS='OLD',ACTION='READ')
          OPEN(UNIT=2, FILE ='D:\USER\FORTRAN\NEWMARKOUT.TXT',&
          &STATUS='REPLACE',ACTION='WRITE')
!******  INPUT SIZE OF MATRICES*********************************
          READ(1,*) N
          ALLOCATE(STIF(N,N),MASS(N,N),MBAR(N,N),CBAR(N,N),&
          &KBAR(N,N),STIFBAR(N,N),ACLR0(N),VEL0(N),DIS0(N),&
          &DIS1(N),ACLR1(N),VEL1(N),U(N,N), F(N),Y(N),X(N))
!******INPUT UPPER TRIANGULAR PART OF STIFFNESS AND ***MASS MATRICES*
!*******INPUT THE DATA ROW BY ROW*************************
          READ(1,*) ((STIF(I,J),J=I,N),I=1,N)
          READ(1,*) ((MASS(I,J),J=I,N),I=1,N)
!*******FILL UP THE LOWER SYMMETRIC HALF******************
          DO I = 2,N
          DO J = 1,(I-1)
          STIF(I,J)=STIF(J,I)
          MASS(I,J)=MASS(J,I)
          END DO
          END DO
!*******PRINT THE INPUT MATRICES*****************************
          WRITE(2,*)' STIFFNESS MATRIX'
          DO I = 1,N
          WRITE(2,'(6(4X,F10.4))')(STIF(I,J), J = 1,N)
          END DO
          WRITE(2,*)' MASS MATRIX'
```

```
          DO I = 1,N
          WRITE(2,'(6(4X,F10.4))')(MASS(I,J), J = 1,N)
          END DO
!*******READ RAYLEIGH DAMPING CONSTANTS*******************
!       C = ALPHA*M + BETA*K
          READ(1,*)ALPHA,BETA
          WRITE(2,'(A,E14.6,A,E14.6)')'ALPHA = ',ALPHA,'   BETA = ',BETA
!*******READ NEWMARK CONSTANTS ALPHA AND DELTA *******
          READ(1,*)NALPHA,NDELTA
          WRITE(2,'(A,F10.4,A,F10.4,A,E14.6)')'NALPHA = ',NALPHA,&
          &'NDELTA = ',NDELTA,
!*******READ TIME INCREMENT DELTA T***************************
          READ(1,*)DELTAT
          WRITE(2,'(A,F10.4)')'DELTA T = ',DELTAT
          DELTAT2=DELTAT**2
          C1=1/(NALPHA*DELTAT2)
          C2=NDELTA/(NALPHA*DELTAT)
          C3=C1+C2*ALPHA
          C4=1.0+C2*BETA
          STIFBAR=C3*MASS+C4*STIF
          WRITE(2,*)' STIFBAR MATRIX'
          DO I = 1,N
          WRITE(2,'(6(4X,E14.6))')(STIFBAR(I,J), J = 1,N)
          END DO
          C5=0.5/NALPHA-1
          C6=0.5*DELTAT*(NDELTA/NALPHA-2)
        MBAR=(C5+C6*ALPHA)*MASS+C6*BETA*STIF
          WRITE(2,*)' MBAR MATRIX'
          DO I = 1,N
          WRITE(2,'(6(4X,E14.6))')(MBAR(I,J), J = 1,N)
          END DO
          C7=1/(DELTAT*NALPHA)
          C8=1-NDELTA/NALPHA
          CBAR=(C7+C8*ALPHA)*MASS+C8*BETA*STIF
          WRITE(2,*)' CBAR MATRIX'
          DO I = 1,N
          WRITE(2,'(6(4X,E14.6))')(CBAR(I,J), J = 1,N)
          END DO
          KBAR=C3*MASS+(C4-1)*STIF
          WRITE(2,*)' KBAR MATRIX'
          DO I = 1,N
          WRITE(2,'(6(4X,E14.6))')(KBAR(I,J), J = 1,N)
          END DO
          C9=-C6*DELTAT
!*******CHOLESKY FACTORISATION OF STIFBAR************
**********************************************************
*       THIS SECTION IS SAME AS IN PROGRAM WILSON        *
*                                                        *
**********************************************************
!******SET INITIAL DISPLACEMENT AND VELOCITY VECTORS**
!       AS ZERO********************************************
          VEL0=0.0
          DIS0=0.0
!*******INITIAL ACCELERATION OF ALL MASSES AS EQUAL*****
!       TO BASE ACCELERATION*******************************
```

```
        ACLR0=ACLR(1)
!*******PRINT DISPLACEMENT, VELOCITY AND ACCELERATION**
        WRITE(2,*)'TIME'
        WRITE(2,*)'DISPLACEMENTS  VELOCITY   ACCLERATION'
        WRITE(2,'(F10.4)')TIME(1)
        DO I = 1,N
        WRITE(2,'(3(4X,E14.6))')DIS0(I),VEL0(I),ACLR0(I)
        END DO
!*******START THE ITERATIONS*******************************
        NMAXM2=NMAX-2
        DO I1 = 2,NMAXM2
        I1M1=I1-1
!*******CALCULATE THE MASS TIMES ACCELERATION AT EACH *
!       LEVEL**********************************************
        DO J= 1,N
        F(J)=-ACLR(I1)*MASS(J,J)
        END DO
!*******CALCULATE THE NET FORCE****************************
        F=F+MATMUL(KBAR,DIS0)+MATMUL(CBAR,VEL0)&
        &+MATMUL(MBAR,ACLR0)
        WRITE(2,'(6(4X,E14.6))')(F(J), J = 1,N)

!******USE FORWARD SUBSTITUTION TO SOLVE *********
!      U TRANSFORM ×Y = F****************************
        Y(1)=F(1)/U(1,1)
        DO I = 2,N
        SUM = 0
        IM1=I-1
        DO J= 1,IM1
        SUM = SUM + U(J,I)*Y(J)
        END DO
        Y(I)= (F(I)-SUM)/U(I,I)
        END DO
!*******USE BACK SUBSTITUTION TO SOLVE U × X = Y*******
!********X STORES THE DISPLACEMENT AT TIME (T+DT)*****
        X(N)=Y(N)/U(N,N)
        NM1=N-1
        DO I= NM1,1,-1
        SUM = 0
        IP1=I+1
        DO J= N,IP1,-1
        SUM=SUM + U(I,J)*X(J)
        END DO
        X(I)=(Y(I)-SUM)/U(I,I)
        END DO
        DIS1=X
        ACLR1=C1*(X-DIS0)-C7*VEL0-C5*ACLR0
        VEL1=VEL0+DELTAT*((1-NDELTA)*ACLR0+NDELTA*ACLR1)
        WRITE(2,'(F10.4)')TIME(I1)
        DO J = 1,N
        WRITE(2,'(3(4X,E14.6))')DIS1(J),VEL1(J),ACLR1(J)
        END DO
        ACLR0=ACLR1
        VEL0=VEL1
        DIS0=DIS1
```

```
        END DO
        END PROGRAM NEWMARK
```

11.7.3 Program CENTRAL

```
!************************************************************
! DIRECT INTEGRATION BY CENTRAL DIFFERENCES METHOD
! LUMPED MASS MATRIX ASSUMED TO BE DIAGONAL
! STRUCTURE SUBJECTED TO BASE ACCELERATION
! SOLUTION OF SIMULTANEOUS EQUATIONS USING CHOLESKY FACTORISATION
!*****************************************************************
        PROGRAM CENTRAL
!*******VARIABLE DECLARATION*******************************
        IMPLICIT NONE
        REAL:: SUM,FACTOR,STEP,T,C1,C2,C3,C4,C5,C6,&
        &ALPHA,BETA,DELTAT,DELTAT2,TMAX,T1,T2,A1,A2
        INTEGER::I,J,IM1,IP1,K,N,NM1,JP1,NDPTS,NMAX,NMAXM2,&
        &I1,I1M1
        REAL,ALLOCATABLE:: STIF(:,:),MASS(:,:),STIFM1(:,:),&
        &STIF0(:,:),STIF1(:,:),ACLR0(:),VEL0(:),DIS0(:),DIS1(:),&
        &U(:,:), F(:),Y(:),X(:),DISM1(:)
         REAL,ALLOCATABLE:: AINPUT(:),TINPUT(:),ACLR(:),TIME(:)
!*******FILE MANAGEMENT***********************************
        OPEN(UNIT=1, FILE='D:\USER\FORTRAN\CENTREIN.TXT',&
        &STATUS='OLD',ACTION='READ')
        OPEN(UNIT=2, FILE ='D:\USER\FORTRAN\CENTREOUT.TXT',&
        &STATUS='REPLACE',ACTION='WRITE')
!*******INPUT SIZE OF MATRICES*******************************
        READ(1,*) N
        ALLOCATE(STIF(N,N),MASS(N,N),STIFM1(N,N),&
        &STIF0(N,N),STIF1(N,N),ACLR0(N),VEL0(N),DIS0(N),DIS1(N),&
        &U(N,N), F(N),Y(N),X(N),DISM1(N))
!******INPUT UPPER TRIANGLE OF STIFFNESS AND MASS MATRICES*
!******INPUT THE DATA ROW BY ROW****************************
        READ(1,*) ((STIF(I,J),J=I,N),I=1,N)
        READ(1,*) ((MASS(I,J),J=I,N),I=1,N)
!*******FILL UP THE LOWER SYMMETRIC HALF*****************
        DO I = 2,N
        DO J = 1,(I-1)
        STIF(I,J)=STIF(J,I)
        MASS(I,J)=MASS(J,I)
        END DO
        END DO
!******PRINT THE INPUT MATRICES*****************************
        WRITE(2,*)' STIFFNESS MATRIX'
        DO I = 1,N
        WRITE(2,'(6(4X,F10.4))')(STIF(I,J), J = 1,N)
        END DO
        WRITE(2,*)' MASS MATRIX'
        DO I = 1,N
        WRITE(2,'(6(4X,F10.4))')(MASS(I,J), J = 1,N)
        END DO
!*******READ RAYLEIGH DAMPING CONSTANTS*********
!       C = ALPHA*M + BETA*K**************************
```

```
          READ(1,*)ALPHA,BETA
          WRITE(2,'(A,E14.6,A,E14.6)')'ALPHA = ',ALPHA,'   BETA = ',BETA
!*******READ TIME INCREMENT DELTA T*************************
          READ(1,*)DELTAT
          WRITE(2,'(A,E14.6)')'DELTA T = ',DELTAT
          DELTAT2=DELTAT**2
          C1=1/DELTAT2
          C2=1/DELTAT
          C3=C1+C2*ALPHA
          C4=C2*BETA
          STIF1=C3*MASS+C4*STIF
          WRITE(2,*)' STIF1 MATRIX'
          DO I = 1,N
          WRITE(2,'(6(4X,E14.6))')(STIF1(I,J), J = 1,N)
          END DO
          C5=-C1+C2*ALPHA
          STIFM1=C5*MASS+C4*STIF
          WRITE(2,*)' STIFM1 MATRIX'
          DO I = 1,N
          WRITE(2,'(6(4X,E14.6))')(STIFM1(I,J), J = 1,N)
          END DO
          C6=2*C1 +ALPHA
          STIF0=C6*MASS - STIF
          WRITE(2,*)' STIF0 MATRIX'
          DO I = 1,N
          WRITE(2,'(6(4X,E14.6))')(STIF0(I,J), J = 1,N)
          END DO
!*******CHOLESKY FACTORISATION OF STIFBAR************
**********************************************************
*       THIS SECTION IS SAME AS IN PROGRAM WILSON          *
*                                                        *
**********************************************************
!*******SET INITIAL DISPLACEMENT AND VELOCITY VECTORS
!       AS ZERO*******************************************
          VEL0=0.0
          DIS0=0.0
!******INITIAL ACCELERATION OF ALL MASSES AS EQUAL
!       TO BASE ACCELERATION****************************
          ACLR0=ACLR(1)
!******CALCULATE THE DISPACEMENT AT TIME = -DELATA T**
          DISM1=DIS0-DELTAT*VEL0+0.5*DELTAT2*ACLR0
          WRITE(2,*)'TIME'
          WRITE(2,*)'DISPLACEMENTS  VELOCITY   ACCLERATION'
          WRITE(2,'(F10.4)')TIME(1)
          DO I = 1,N
          WRITE(2,'(4(4X,E14.6))')DISM1(I),DIS0(I),VEL0(I),ACLR0(I)
          END DO
!*******START THE ITERATIONS*****************************
          NMAXM2=NMAX-2
          DO I1 = 2,NMAXM2
          I1M1=I1-1
!*******CALCULATE THE MASS TIMES ACCELERATION AT
!       EACH LEVEL******************************************
          DO J= 1,N
          F(J)=-ACLR(I1M1)*MASS(J,J)
```

```
          END DO
!*******CALCULATE THE NET FORCE****************************
          F=F+MATMUL(STIF0,DIS0)+MATMUL(STIFM1,DISM1)
          WRITE(2,*)' NET FORCE'
          WRITE(2,'(6(4X,E14.6))')(F(J), J = 1,N)
!*******USE FORWARD SUBSTITUTION TO SOLVE
!         U TRANSFORM ×Y = F***************
          Y(1)=F(1)/U(1,1)
          DO I = 2,N
          SUM = 0
          IM1=I-1
          DO J= 1,IM1
          SUM = SUM + U(J,I)*Y(J)
          END DO
          Y(I)= (F(I)-SUM)/U(I,I)
          END DO
!*******USE BACK SUBSTITUTION TO SOLVE U ×X = Y*********
!*******VECTOR X HAS  DISPLACEMENT AT TIME (T+DT)******
          X(N)=Y(N)/U(N,N)
          NM1=N-1
          DO I= NM1,1,-1
          SUM = 0
          IP1=I+1
          DO J= N,IP1,-1
          SUM=SUM + U(I,J)*X(J)
          END DO
          X(I)=(Y(I)-SUM)/U(I,I)
          END DO
          DIS1=X
          VEL0=(0.5/DELTAT)*(X-DISM1)
          ACLR0=(1.0/DELTAT2)*(DISM1-2*DIS0+DIS1)
          WRITE(2,'(F10.4)')TIME(I1)
          DO J = 1,N
          WRITE(2,'(3(4X,E14.6))')DIS0(J),VEL0(J),ACLR0(J)
          END DO
!*******RESET THE INITIAL VALUES**************************
          DISM1=DIS0
          DIS0=DIS1
          END DO
          END PROGRAM CENTRAL
```

CHAPTER 12

SUBROUTINES AND PROGRAMS

12.1 INTRODUCTION

In this chapter, a number of subroutines and programs will be described. The logic of each subroutine and the meaning of variables will be described to aid understanding of the subroutines. Note that when a subroutine is used in a main program, any variable not declared in the subroutine should be declared in the main program.

12.2 TERMINOLOGY USED IN THE PROGRAMS

In the programs given, the following terminology is used. The reader should become familiar with the terminology, as it significantly eases the understanding of the programs.

NDOFN: This defines the number of degrees of freedom allowable at a joint. For example, in the case of a two-dimensional bar element or in an in-plane quadratic element, since there are only two translations at each joint (node), therefore NDOFN = 2. However in the case of two-dimensional rigid-jointed frame element NDOFN = 3, because at each node there are two translations and a rotation. Similarly in the case of a plane grid element or in a thin plate-bending element NDOFN = 3, as at each joint there is a translation along the z axis and rotations about x and y axes. In a flat shell element NDOFN = 6, because at each node there are three translations and three rotations. Variables IDOFN or JDOFN vary from 1 to NDOFN.

NNODE: This refers to the number of nodes defining an element. In the case of all line elements and in the case of finite strips with no auxiliary lines, only two joints (nodal lines) define the element. Therefore NNODE = 2. In the case of an isoparametric quadratic element, NNODE = 8. Variables INODE or JNODE vary from 1 to NNODE.

NEVAB: This simply defines the size of the element stiffness matrix and is given by NEVAB = NNODE × NDOFN. Variables IEVAB or JEVAB vary from 1 to NEVAB.

ESTIF: This matrix refers to the element stiffness matrix of size NEVAB × NEVAB.

NPOIN: Total number of nodes in a structure. Variables IPOIN or JPOIN vary from 1 to NPOIN.

NELEM: Total number of elements in a structure. Variables IELEM or JELEM vary from 1 to NELEM.

COORD: This array stores the coordinates of all the nodes of the structure.
LNODS: This array stores the node numbers of all elements in the structure.
NODFRE: This array stores the node freedoms of all the nodes of the structure.
MEMDIS: This vector of length NEVAB stores the node freedoms of all the nodes of an element.

12.3 SUBROUTINES FOR DATA INPUT

Three subroutines have been used specifically to assist data input. These will be described in detail.

12.3.1 Subroutine CORDGEN

Given the coordinates of the node at the beginning and end of a line and the number of equally spaced nodes in-between, this subroutine generates the node numbers and the coordinates of the intermediate nodes.

The variables are: NODSTR= NODe number at the STaRt of a line, NODEND= NODe number at the END of a line, NODINC= NODe number INCrement for successive nodes, NSTEP = (NODEND-NODSTR)/NODINC, CORINC(IDIME) = CoORdinate INCrement between successive nodes, IDIME = 1 for x and 2 for y coordinate.

Calculations start by inputting a node number at the beginning of a line, NSTEP =1 and the coordinates of that node. This node becomes the NODSTR. After that input NODEND, the node number at the end of a set of equally spaced nodes, NSTEP and the coordinates of NODEND. This is continued until NODEND = last node in the mesh.

```
!*****************************************************************************
        SUBROUTINE CORDGEN
!***VARIABLE DECLARATION******************************************************
        IMPLICIT NONE
        INTEGER::NODEND,NODSTR,NSTEP,NODDIF,NODINC,NODE1,NODE2,NODE3
        REAL:: CORINC(3)
!***START OF AN ENDLESS DO LOOP***********************************************
        DO
!***READ THE LAST NODE NUMBER IN A LINE AND THE NUMBER OF STEPS BETWEEN**
!*** THE PREVIOUSLY READ NODE AND THE NODE AT !   THE END OF THE LINE********
        READ(1,*) NODEND,NSTEP
!***READ THE COORDINATES OF THE NODE AT THE END OF THE LINE
        READ(1,*)(COORD(NODEND,IDIME),IDIME=1,NDIME)
        IF(NSTEP<0)THEN
        WRITE(2,*)' NSTEP IS NEGATIVE'
        NFAIL=1
        EXIT
        ELSE IF(NSTEP > 1) THEN
        NODDIF = NODEND - NODSTR
        NODINC = NODDIF/NSTEP
        DO IDIME = 1, NDIME
```

```
CORINC(IDIME) = (COORD(NODEND,IDIME)-&
&COORD(NODSTR,IDIME))/FLOAT(NSTEP)
END DO
NODE1 = NODSTR + NODINC
NODE2 = NODEND - NODINC
DO INODE = NODE1, NODE2, NODINC
NODE3 = INODE - NODINC
DO IDIME = 1, NDIME
COORD(INODE,IDIME) = COORD(NODE3,IDIME) + &
&CORINC(IDIME)
END DO
END DO
IF(NODEND = =NPOIN) EXIT
END IF
NODSTR = NODEND
END DO
RETURN
END SUBROUTINE CORDGEN
```

12.3.2 Subroutine INTGEN

This subroutine is identical to the subroutine CORDGEN, except that the matrices it operates on are all integer matrices. This subroutine can be used to generate data for the following three matrices.

LNODS: i^{th} row of this array stores the node numbers corresponding to i^{th} element in the mesh. If the node number of a set of elements are increasing in equal increments, the data for the nodes corresponding to intermediate elements can be generated.

MATNO: This array is really a vector. Each row of this array stores the material numbers corresponding to that element in the mesh. If the material numbers are remaining the same for a set of elements, the data for intermediate elements can be generated.

NODFRE: This array stores the node freedom numbers of the nodes. Subroutine INTGEN is used to generate this array only when some node freedoms of some of the nodes in a structure are constrained to be identical.

```
!*************************************************************************************
SUBROUTINE INTGEN(IA,NVAR,NMAX)
!***VARIABLE DECLARATION*************************************************************
IMPLICIT NONE
INTEGER::IEND,ISTR,NSTEP,IVAR,JVAR,NVAR,NMAX,IDIF,INC,I1,I2,I3,I
INTEGER :: IA(NMAX,NVAR),INCR(NVAR)
!***START OF AN ENDLESS DO LOOP******************************************************
DO
READ(1,*)IEND,NSTEP
READ(1,*)(IA(IEND,JVAR),JVAR=1,NVAR)
IF(NSTEP<0)THEN
WRITE(2,*)'ERROR IN DATA.  NSTEP IS NEGATIVE'
NFAIL=1
EXIT
ELSE IF(NSTEP > 1) THEN
```

```
        IDIF=IEND-ISTR
        INC=IDIF/NSTEP
        DO IVAR=1,NVAR
        INCR(IVAR)=(IA(IEND,IVAR) - IA(ISTR,IVAR))/NSTEP
        END DO
        I1=ISTR+INC
        I2=IEND-INC
        DO I=I1,I2,INC
        I3=I-INC
        DO IVAR=1,NVAR
        IA(I,IVAR)=IA(I3,IVAR)+INCR(IVAR)
        END DO
        END DO
        END IF
        IF(IEND = =NMAX) EXIT
        ISTR = IEND
        END DO
        RETURN
        END SUBROUTINE INTGEN
```

12.3.3 Subroutine NFGEN

This subroutine generates the NFGEN array from the data in two arrays. NRESND = number of nodes where one or more degrees of freedom are restrained. Array NDFFIX(NRESND) stores the node numbers corresponding to NRESND nodes. The array IFPRE(NRESND,NDOFN) stores the data about which freedoms are constrained.

For example, if NRESND = 4, the constrained node numbers are 45, 56, 72, 109, then

NDFFIX(1) = 45, NDFFIX(2) = 56, NDFFIX(3) = 72, NDFFIX(4) = 109.

IFPRE(1,1) = 0, IFPRE(1,1) = 1, IFPRE(1,1) = 0. This indicates that at node 45, only the second freedom is restrained.

IFPRE(2,1) = 1, IFPRE(1,1) = 1, IFPRE(1,1) = 0. This indicates that at node 56, the first two freedoms are restrained.

IFPRE(3,1) = 1, IFPRE(1,1) = 0, IFPRE(1,1) = 1. This indicates that at node 72, the first and third freedoms are restrained.

IFPRE(1,1) = 1, IFPRE(1,1) = 1, IFPRE(1,1) = 1. This indicates that at node 109, all freedoms are restrained.

```
!*****************************************************************************************
        SUBROUTINE NFGEN
        KOUNT =0
!***CYCLE ON ALL NPOIN POINTS**************************************
        DO IPOIN=1,NPOIN
        JRESND=0
!***CHECK THROUGH NRESND NODES TO SEE IF IPOIN IS A RESTRAINED NODE****
        DO IRESND=1,NRESND
        IF(IPOIN = = NDFFIX(IRESND)) THEN
        JRESND=IRESND
!***IF IPOIN IS A RESTRAINED NODE STOP SEARCHING*************************
        EXIT
```

```
          END IF
          END DO
!***IF IT IS NOT A RESTRAINED NODE, THEN CALCULATE THE NODE  FREEDOM **
!****BY INCREASING EACH FREEDOM BY 1 **************************************
          IF(JRESND = = 0) THEN
          DO  IDOFN = 1, NDOFN
          KOUNT=KOUNT+1
          NODFRE(IPOIN,IDOFN)=KOUNT
          END DO
!***CHECK THROUGH NDOFN FREEDOMS TO SEE WHICH FREEDOMS*************
!*** ARE RESTRAINED AND MODIFY THE NODE FREEDOMS **********************
          ELSE
          DO IDOFN = 1, NDOFN
          JDOFN=IFPRE(JRESND,IDOFN)
          IF(JDOFN  = =1) THEN
! ***IT IS RESTRAINED FREEDOM, THEN THE CORRESPONDING NODE*************
!**** FREEDOM IS ZERO*********************************************************
          NODFRE(IPOIN,IDOFN)=0
          ELSE
!***IT IS NOT A RESTRAINED FREEDOM, THEN THE CORRESPONDING *************
!***NODE FREEDOM IS  INCREASED BY ONE*************************************
          KOUNT=KOUNT+1
          NODFRE(IPOIN,IDOFN)=KOUNT
          END IF
          END DO
          END IF
          END DO
          RETURN
          END SUBROUTINE NFGEN
```

12.4 SUBROUTINES TO DETERMINE THE SIZE OF STIFFNESS MATRIX

This section describes a series of subroutines which generate data about the size, bandwidth, etc. of global stiffness and mass matrices.

12.4.1 Subroutine MAXEQN

This subroutine calculates the maximum number MAXEQN of degrees of freedom of the entire structure. It scans through the array NODFRE(NPOIN,NDOFN) through all IPON = 1, NPOIN and IDOF = 1, NDOFN and determines the largest value. This determines the size of the global stiffness and mass matrices.

```
!*****************************************************************************
          SUBROUTINE MAXEQN
          NEQNS=0
          DO IPOIN=1, NPOIN
          DO JDOFN=1,NDOFN
          IF(NODFRE(IPOIN,JDOFN)  > NEQNS) NEQNS=NODFRE(IPOIN,JDOFN)
          END DO
          END DO
```

```
        RETURN
        END SUBROUTINE MAXEQN
```

12.4.2 Subroutine BANDWD

This subroutine calculates the maximum half bandwidth. It scans through all the IELEM = 1, NELEM elements and for each element through all its INODE = 1, NNODE nodes and for each node through all the IDOFN = 1, NDOFN degrees of freedom and determines the maximum difference between the node freedoms.

```
!*******************************************************************************
        SUBROUTINE BANDWD
!***VARIABLE DECLARATION*********************************************
        IMPLICIT NONE
        INTEGER::LBAND, INODE, KEVAB
        NBAND=0
!***CYCLE ON ALL THE ELEMENTS***************************************
        ELEMENT:DO IELEM=1,NELEM
!***CYCLE ON ALL NODES OF THE ELEMENT
        NODE:DO INODE = 1,NNODE
!***DETERMINE THE NODE FREEDOMS CORRESPONDING TO THE NODES OF THE
!    ELEMENT AND STORE IT IN ARRAY MEMDIS(NEVAB).
        DOFN:DO IDOFN=1,NDOFN
        KEVAB= (INODE-1)*NDOFN+IDOFN
        MEMDIS(KEVAB)=NODFRE(LNODS(IELEM,INODE),IDOFN)
        END DO DOFN
        END DO NODE
!***CALCULATE THE HALF BANDWIDTH LBAND FOR THE ELEMENT BY FINDING THE
!    MAXIMUM DIFFERENCE BETWEEN THE FREEDOM AT THE NODES
        DO IEVAB=1,NEVAB
        IF(MEMDIS(IEVAB) > 0) THEN
        DO JEVAB=1,NEVAB
        IF(MEMDIS(JEVAB) > 0) THEN
        LBAND = ABS(MEMDIS(IEVAB) - MEMDIS(JEVAB))
        IF(LBAND>NBAND) NBAND=LBAND
        END IF
        END DO
        END IF
        END DO
        END DO ELEMENT
        RETURN
        END SUBROUTINE BANDWD
```

12.4.3 Subroutine NSKYLINE

This subroutine calculates the position of a diagonal element and the number of elements which will remain non-zero in the column corresponding to each diagonal element of structural stiffness and mass matrices when stored as a vector. The basic idea is to choose IEQNS = 1, NEQNS and then check through IELEM = 1, NELEM elements to see if the freedom number at INODE = 1, NNODE and for each node for IDOFN = 1, NDOFN if any node freedom is equal to the chosen equation number. If it is, then calculate the maximum difference between the freedom number less than the chosen equation and the equation plus 1. The maximum difference thus calculated is the maximum height of the column NCOLM(IEQNS) corresponding to the chosen equation. The position of any diagonal NDIA(IEQNS) is equal to the position of the previous diagonal + the height above the previous diagonal. MAXCOL = total length of the vector = NDIA(NEQNS) + NCOLM(NEQNS) - 1.

As explained in Section 4.7.3, in vector format, the position of $(i,j)^{th}$ element in the square format is in position $DIA(j) + j - i$, $j > i$ and $(j - i + 1) \leq COL(j)$.

```
!***************************************************************************
        SUBROUTINE NSKYLINE
!***SET THE FIRST DIAGONAL ELEMENT IN POSITION 1 AND NUMBER OF***
!*** ELEMENT IN  COLUMN 1 AS 1****************************************
!***VARIABLE DECLARATION*********************************************
        IMPLICIT NONE
        INTEGER::IEQNS,IBIG,JFRDOM,IDIF,IEQNS1
        NCOLM(1) = 1
        NDIA(1) = 1
!***CHOOSE THE EQUATION NUMBER***************************************
        DO  IEQNS=2,NEQNS
!***THE MINIMUM NUMBER OF ELEMENTS IN A COLUMN IS 1****************
        IBIG=1
!***CYCLE ON ALL ELEMENTS********************************************
        DO  IELEM=1,NELEM
        KOUNT=0
!***CHECK IF ANY OF THE FREEDOMS AT THE NODE CORRESPOND TO********
!*** THE CHOSEN  EQUATION********************************************
        DO  INODE=1,NNODE
        DO  IDOFN=1,NDOFN
        JFRDOM=NODFRE(LNODS(IELEM,INODE),IDOFN)
        IF(JFRDOM = = IEQNS) THEN
        KOUNT=1
        EXIT
        END IF
        END DO
        IF(JFRDOM = = IEQNS) THEN
        KOUNT=1
        EXIT
        END IF
        END DO
        IF(KOUNT = =1)  THEN
!***FIND THE DIFFERENCE BETWEEN THE CHOSEN EQUATION AND THE ******
!***APPROPRIATE FREEDOMS AT THE NODES*******************************
        DO  INODE=1,NNODE
        DO  IDOFN=1,NDOFN
        JFRDOM=NODFRE(LNODS(IELEM,INODE),IDOFN)
        IF(JFRDOM > 0  .AND. JFRDOM  <IEQNS) THEN
        IDIF=IEQNS - JFRDOM + 1
        END IF
        IF(IDIF  >IBIG)IBIG=IDIF
        END DO
        END DO
```

```
        END IF
        END DO
        NCOLM(IEQNS)=IBIG
        END DO
!***CALCULATE THE POSITION OF DIAGONALS*************************
        DO  IEQNS=2,NEQNS
        IEQNS1=IEQNS-1
        NDIA(IEQNS)=NDIA(IEQNS1)+NCOLM(IEQNS1)
        END DO
!***CALCULATE THE TOTAL LENGTH OF THE VECTOR******************
        MAXCOL=NDIA(NEQNS)+NCOLM(NEQNS)-1
        RETURN
        END  SUBROUTINE NSKYLINE
```

12.5 SUBROUTINES FOR CHOLESKY FACTORISATION

The following subroutines are associated with the Cholesky factorisation and the forward and back substitution required to solve simultaneous equations.

12.5.1 Subroutine CHOLESKY

This subroutine calculates the Cholesky decomposition of GSTIF = $U^T * U$ explained in Section 4.4.3. The stiffness matrix GSTIF and the factor matrix U are both stored in vector format. As explained in Section 4.7.3, in vector format, the position of $(i,j)^{th}$ element in the square format is in position $(DIA(j) + j - i)$, $j > i$ and $(j - i + 1) \leq COL(j)$.

```
!*****************************************************************************
        SUBROUTINE CHOLESKY(U)
!***VARIABLE DECLARATION*****************************************************
        IMPLICIT NONE
        REAL::FACTOR, SUM
        INTEGER::NBAND1,IJ,JMAX,I,J,L,K,KI,KJ,IK1,JK1
        REAL:: U(MAXCOL)
        NBAND1 = NBAND+1
        U(1) = SQRT(GSTIF(1))
        FACTOR = 1.0/U(1)
!***START WITH THE FIRST ROW************************************************
        DO J = 2,NBAND1
        IF(J <=NCOLM(J))THEN
        IJ=NDIA(J)+J-1
        U(IJ) = GSTIF(IJ)*FACTOR
        END IF
        END DO
!***DO THE ROWS 2 TO N*******************************************************
        DO I = 2,NEQNS
!***CHECK HOW FAR TO THE RIGHT OF DIAGONAL ONE NEEDS TO GO*********
        JMAX = I + NBAND
        IF(JMAX > NEQNS) JMAX = NEQNS
        DO  J = I,JMAX
        IF((J-I+1) <= NCOLM(J)) THEN
```

```
!***SUM OVER K= 1 TO (I-1) THE PRODUCT OF U(K,I) ×U(K,J)********************
        SUM = 0
        L = I-1
        DO K = 1,L
!*****CALCULATE THE POSITIONS OF THE ELEMENTS IN THE VECTOR FORMAT**
        KI=NDIA(I) + I-K
        IK1=I-K+1 - NCOLM(I)
        IF(IK1 <= 0)  THEN
        KJ =NDIA(J)+ J-K
        JK1=J-K + 1 - NCOLM(J)
        IF(JK1 <= 0) THEN
        SUM = SUM - U(KI)*U(KJ)
        END IF
        END IF
        END DO
        IJ=NDIA(J) +J-I
        SUM = SUM + GSTIF(IJ)
        IF(J = = I) THEN
        U(IJ) = SQRT(SUM)
        ELSE
        U(IJ) = SUM/U(NDIA(I))
        END IF
        END IF
        END DO
        END DO
        RETURN
        END SUBROUTINE CHOLESKY
```

12.5.2 Subroutine BACKSUB(A,B,C)

This subroutine carries out the back substitution. The matrices involved are A×C = B. A is an upper triangular matrix stored in vector format. B is given and it is required to determine C by back substitution. This subroutine will be involved in solving the simultaneous equation by the Cholesky method.

```
!*************************************************************************
        SUBROUTINE BACKSUB(A,B,C)
!***VARIABLE DECLARATION*****************************************
        IMPLICIT NONE
        INTEGER::IVECTR,IVARBL,JVARBL,IJ,JCOLMS,JCOLM1,NEQNS1
        REAL::SUM
        REAL:: A(MAXCOL),B(NEQNS,NVECTR),C(NEQNS,NVECTR)
        NEQNS1 = NEQNS-1
!***CYCLE ON ALL NVECTOR*****************************************
        VECTOR:DO  IVECTR = 1,NVECTR
!***CALCULATE THE VALUE OF THE LAST VARIABLE******************
        C(NEQNS,IVECTR)=B(NEQNS,IVECTR)/A(NDIA(NEQNS))
!***CALCULATE THE VALUES OF OTHER VARIABLES*****************
        VARIABLE: DO  IVARBL=1,NEQNS1
        JVARBL=NEQNS-IVARBL
        SUM=0.0
        JCOLM1 = JVARBL+1
        DO JCOLMS = JCOLM1,NEQNS
        IJ=NDIA(JCOLMS) +JCOLMS -JVARBL
```

```
IF( (JCOLMS - JVARBL +1) <= NCOLM(JCOLMS) )THEN
SUM=SUM+A(IJ)*C(JCOLMS,IVECTR)
END IF
END DO
C(JVARBL,IVECTR)=(B(JVARBL,IVECTR)-SUM)/A(NDIA(JVARBL))
END DO  VARIABLE
END DO  VECTOR
RETURN
END SUBROUTINE BACKSUB
```

12.5.3 Subroutine FORWSUB(A,B,C)

This subroutine carries out the forward substitution. The matrices involved are $A^T \times C = B$. A is an upper triangular matrix stored in vector format. B is given and it is required to determine C by forward substitution. This subroutine will be involved in solving the simultaneous equation by the Cholesky method.

```
!****************************************************************
        SUBROUTINE FORWSUB(A,B,C)
!***VARIABLE DECLARATION************************************
        IMPLICIT NONE
        INTEGER::IVECTR,IEQNS,IEQNS1,IJ,JEQNS
        REAL::SUM
        REAL:: A(MAXCOL),B(NEQNS,NVECTR),C(NEQNS,NVECTR)
!***CYCLE ON ALL NVECTOR
        VECTOR: DO  IVECTR = 1,NVECTR
!***CALCULATE THE FIRST VARIABLE******************************
        C(1,IVECTR)=B(1,IVECTR)/A(1)
!***CALCULATE THE REST OF VARIABLES*************************
        EQUATIONS: DO  IEQNS=2,NEQNS
        IEQNS1 = IEQNS-1
        SUM=0.0
        DO  JEQNS=1,IEQNS1
        IF( (IEQNS-JEQNS+1) <= NCOLM(IEQNS) ) THEN
        IJ=NDIA(IEQNS)+IEQNS-JEQNS
        SUM=SUM+A(IJ)*C(JEQNS,IVECTR)
        END IF
        END DO
        C(IEQNS,IVECTR)=(B(IEQNS,IVECTR)-SUM)/A(NDIA(IEQNS))
        END DO EQUATIONS
        END DO VECTOR
        RETURN
        END SUBROUTINE FORWSUB
```

12.6 STIFFNESS MATRICES OF LINE ELEMENTS

Two subroutines are used to compute the stiffness matrix of line elements. The expressions for element stiffness matrices in global coordinates developed in Section 10.5 are coded.

12.6.1 Subroutine STIFRJ

This subroutine calculates the element stiffness matrix of a two-dimensional rigid-jointed frame element and assembles the structural stiffness matrix in vector format.

```
!*****************************************************************************
        SUBROUTINE STIFRJ
!***VARIABLE DECLARATION****************************************************
        IMPLICIT NONE
         INTEGER::IEND,JEND, IEVAB,JEVAB,IEVAB1,IJ,ICOLM
        REAL::YOUNG, SHEARM, YINERA, AREA, ZASF, XPROJ,YPROJ, SPAN,XL, &
        &YM,EI,EIBYL,EIBYL2,EIBYL3,AEBYL,BETA,BETA1,S11,S12,Q11,T11
!*** CYCLE ON ALL ELEMENTS  AND CALCULATE ELEMENT STIFFNESS MATRIX***
!***  AND ADD TO THE STRUCTURAL STIFFNESS MATRIX**************************
         ELEMENT: DO  IELEM=1,NELEM
         LPROP=MATNO(IELEM,1)
!*** RETRIEVE THE MATERIAL AND CROSS-SECTIONAL PROPERTIES OF ELEMENTS*
        YOUNG=PROPS(LPROP,1)
        SHEARM=PROPS(LPROP,2)
        YINERA=PROPS(LPROP,3)
        AREA=PROPS(LPROP,4)
        ZASF=PROPS(LPROP,5)
!***RETRIEVE THE NODES CORRESPONDING TO THE TWO ENDS OF THE ELEMENT**
        IEND=LNODS(IELEM,1)
        JEND=LNODS(IELEM,2)
!***CALCULATE THE ELEMENT LENGTH, DIRECTION COSINES ETC***************
        XPROJ=COORD(JEND,1)-COORD(IEND,1)
        YPROJ=COORD(JEND,2)-COORD(IEND,2)
        SPAN=SQRT(XPROJ*XPROJ + YPROJ*YPROJ)
        XL=XPROJ/SPAN
        YM=YPROJ/SPAN
        EI=YOUNG*YINERA
        GA=SHEARM*AREA
        EIBYL=EI/SPAN
        EIBYL2=EIBYL/SPAN
        EIBYL3=EIBYL2/SPAN
        GJBYL=TINERA*SHEARM/SPAN
        BETA=12.0*EIBYL2*ZASF/GA
        BETA1=1.0/(1.0+BETA)
!***LOCAL AXIS BENDING STIFFNESS COEFFICIENTS  FOR A BEAM ELEMENT******
        S11= (4.0+BETA)*BETA1
        S12= (2.0 - BETA)*BETA1
        Q11=S11+S12
        T11=2.0*Q11
        T11=T11*EIBYL3
        S11=S11*EIBYL
        S12=S12*EIBYL
        Q11=Q11*EIBYL2
!***CALCULATE THE GLOBAL AXIS UPPER TRIANGLE OF ELEMENT STIFFNESS***
!**** MATRIX*****************************************************************
        ESTIF(1,1)= T11*YM*YM + AEBYL*XL*XL
        ESTIF(1,2)= (-T11 + AEBYL)*XL*YM
        ESTIF(1,3)= Q11*YM
        ESTIF(1,4)= -ESTIF(1,1)
        ESTIF(1,5)= -ESTIF(1,2)
```

```
        ESTIF(1,6)= ESTIF(1,3)
        ESTIF(2,2)= T11*XL*XL + AEBYL*YM*YM
        ESTIF(2,3)= -Q11*XL
        ESTIF(2,4)= ESTIF(1,5)
        ESTIF(2,5)= -ESTIF(2,2)
        ESTIF(2,6)= ESTIF(2,3)
        ESTIF(3,3)= S11
        ESTIF(3,4)= -ESTIF(1,3)
        ESTIF(3,5)= -ESTIF(2,3)
        ESTIF(3,6)= S12
        ESTIF(4,4)= ESTIF(1,1)
        ESTIF(4,5)= ESTIF(1,2)
        ESTIF(4,6)= -ESTIF(1,3)
        ESTIF(5,5)= ESTIF(2,2)
        ESTIF(5,6)= ESTIF(3,5)
        ESTIF(6,6)= S11
!***FILL UP THE SYMMETRICAL LOWER HALF OF THE ESTIF MATRIX*********
        DO IEVAB=2,NEVAB
        IEVAB1=IEVAB-1
        DO JEVAB=1,IEVAB1
        ESTIF(IEVAB,JEVAB)=ESTIF(JEVAB,IEVAB)
        END DO
        END DO
!***RETRIEVE THE NODE FREEDOMS CORRESPONDING TO THE NODES AT THE**
!*** ENDS OF THE ELEMENT AND STORE IN ARRAY MEMDIS(NDOFN)*************
        DO  IDOFN=1,NDOFN
        MEMDIS(IDOFN)=NODFRE(IEND,IDOFN)
        MEMDIS(IDOFN+NDOFN)=NODFRE(JEND,IDOFN)
        END DO
!***ASSEMBLE THE STRUCTURAL STIFFNESS MATRIX IN THE VECTOR FORM****
        DO  IEVAB=1,NEVAB
        IF(MEMDIS(IEVAB) > 0) THEN
        DO  JEVAB=1,NEVAB
        IF(MEMDIS(JEVAB) > 0  .AND. MEMDIS(JEVAB) >= MEMDIS(IEVAB) ) THEN
!***CALCULATE THE ELEMENT POSITION IN THE VECTOR FORMAT***************
        IJ=NDIA(MEMDIS(JEVAB)) + MEMDIS(JEVAB)-MEMDIS(IEVAB)
        GSTIF(IJ)=GSTIF(IJ)+ESTIF(IEVAB,JEVAB)
        END IF
        END DO
        END IF
        END DO
        END DO ELEMENT
        RETURN
        END SUBROUTINE STIFRJ
```

12.6.2 Subroutine STIFGR

This subroutine calculates the element stiffness matrix of plane grid element and assembles the structural stiffness matrix in vector format. This is similar to subroutine STIFRJ.

```
!*************************************************************************
    SUBROUTINE STIFGR
!***VARIABLE DECLARATION****************************************************
```

```
      IMPLICIT NONE
      INTEGER::IEND,JEND, IEVAB,JEVAB,IEVAB1,IJ,ICOLM
      REAL::YOUNG, SHEARM, YINERA, TINERA, AREA, ZASF,XPROJ,YPROJ,SPAN,&
      &XL,YM,EI,GA,EIBYL,EIBYL2,EIBYL3,GJBYL,AEBYL,BETA,BETA1,S11,S12,Q11,&
      &T11
!***CYCLE ON ALL THE ELEMENTS, CALCULATE THE ELEMENT STIFFNESS********
!*** MATRIX AND ADD AS APPROPRIATE TO THE STRUCTURAL STIFFNESS MATRIX*
      DO IELEM=1,NELEM
      LPROP=MATNO(IELEM,1)
!***RETRIEVE THE GEOMETRIC AND MATERIAL PROPERTIES OF THE ELEMENT*****
      YOUNG=PROPS(LPROP,1)
      SHEARM=PROPS(LPROP,2)
      YINERA=PROPS(LPROP,3)
      TINERA=PROPS(LPROP,4)
      AREA=PROPS(LPROP,5)
      ZASF=PROPS(LPROP,6)
!***RETRIEVE THE NODES CORRESPONDING TO THE TWO ENDS OF THE ELEMENT***
      IEND=LNODS(IELEM,1)
      JEND=LNODS(IELEM,2)
!***CALCULATE THE ELEMENT LENGTH, DIRECTION COSINES ETC.*****************
      XPROJ=COORD(JEND,1)-COORD(IEND,1)
      YPROJ=COORD(JEND,2)-COORD(IEND,2)
      SPAN=SQRT(XPROJ*XPROJ + YPROJ*YPROJ)
      XL=XPROJ/SPAN
      YM=YPROJ/SPAN
      EI=YOUNG*YINERA
      GA=SHEARM*AREA
      EIBYL=EI/SPAN
      EIBYL2=EIBYL/SPAN
      EIBYL3=EIBYL2/SPAN
      GJBYL=TINERA*SHEARM/SPAN
      BETA=12.0*EIBYL2*ZASF/GA
      BETA1=1.0/(1.0+BETA)
!***LOCAL AXIS BENDING STIFFNESS COEFFICIENTS FOR A BEAM ELEMENT****
      S11= (4.0 + BETA)*BETA1
      S12= (2.0 - BETA)*BETA1
      Q11=S11+S12
      T11=2.0*Q11
      T11=T11*EIBYL3
      S11=S11*EIBYL
      S12=S12*EIBYL
      Q11=Q11*EIBYL2
!***CALCULATE THE  GLOBAL AXIS UPPER TRIANGLE OF ELEMENT STIFFNESS**
!*** MATRIX*****************************************************************
      ESTIF(1,1)= T11
      ESTIF(1,2)= Q11*YM
      ESTIF(1,3)= -Q11*XL
      ESTIF(1,4)= -T11
      ESTIF(1,5)= ESTIF(1,2)
      ESTIF(1,6)= ESTIF(1,3)
      ESTIF(2,2)= S11*YM*YM + GJBYL*XL*XL
      ESTIF(2,3)= (-S11 + GJBYL)*XL*YM
      ESTIF(2,4)= -ESTIF(1,2)
      ESTIF(2,5)= S12*YM*YM - GJBYL*XL*XL
      ESTIF(2,6)= (-S12 - GJBYL)*XL*YM
```

```
      ESTIF(3,3)= S11*XL*XL + GJBYL*YM*YM
      ESTIF(3,4)= -ESTIF(1,3)
      ESTIF(3,5)= -(S11 + GJBYL)*XL*YM
      ESTIF(3,6)= S12*XL*XL - GJBYL*YM*YM
      ESTIF(4,4)= T11
      ESTIF(4,5)= ESTIF(2,4)
      ESTIF(4,6)= ESTIF(3,4)
      ESTIF(5,5)= ESTIF(2,2)
      ESTIF(5,6)= ESTIF(2,3)
      ESTIF(6,6)= ESTIF(3,3)
!***FILL UP THE SYMMETRICAL LOWER HALF OF THE ESTIF MATRIX********
      DO IEVAB=2,NEVAB
      IEVAB1=IEVAB-1
      DO  JEVAB=1,IEVAB1
      ESTIF(IEVAB,JEVAB)=ESTIF(JEVAB,IEVAB)
      END DO
      END DO
!***RETRIEVE THE NODE FREEDOMS CORRESPONDING TO THE NODES AT****
!*** THE ENDS OF THE ELEMENT.  STORE IN THE ARRAY MEMDIS(NDOFN)****
      DO  IDOFN=1,NDOFN
      MEMDIS(IDOFN)=NODFRE(IEND,IDOFN)
      MEMDIS(IDOFN+NDOFN)=NODFRE(JEND,IDOFN)
      END DO
!***ASSEMBLE THE STRUCTURAL STIFFNESS MATRIX IN VECTOR FORM******
      DO IEVAB=1,NEVAB
      IF(MEMDIS(IEVAB) > 0) THEN
      DO JEVAB=1,NEVAB
      IF(MEMDIS(JEVAB) > 0  .AND. MEMDIS(JEVAB) >= MEMDIS(IEVAB) ) THEN
!***CALCULATE THE ELEMENT POSITION IN THE VECTOR FORMAT*************
      IJ=NDIA(MEMDIS(JEVAB)) + MEMDIS(JEVAB)-MEMDIS(IEVAB)
      GSTIF(IJ)=GSTIF(IJ)+ESTIF(IEVAB,JEVAB)
      END IF
      END DO
      END IF
      END DO
      END DO
      RETURN
      END SUBROUTINE STIFGR
```

12.7 CONSISTENT MASS MATRICES OF LINE ELEMENTS

One subroutine is used to read the mass at joints and on elements and two other subroutines are used to calculate the mass matrix for line elements. The expressions for consistent mass matrix is as explained in Sections 7.12.3 and 7.12.5. The mass matrix due to concentrated mass at the joints is assumed not to contribute to rotational degrees of freedom.

12.7.1 Subroutine READMASFRAME

This subroutine is used to read and store data on mass at joints and on elements.

```
!*********************************************************************************
      SUBROUTINE READMASFRAME
!***NLODND = NUMBER OF NODES WHERE CONCENTRATED MASS ACT**********
      IF(NLODND > 0) THEN
      WRITE(2,'(2/,A,I4,//)')' NUMBER OF LOADED NODES = ',NLODND
      WRITE(2,*)' NODE      CONC. MASS'
!***ARRAY MASPT AND XMASND STORE RESPECTIVELY, THE NODE NUMBER***
!*** AND THE MASS AT THE JOINT***************************************************
      DO ILODND=1,NLODND
      READ(1,*)MASPT(ILODND)
      READ(1,*)XMASND(ILODND)
      WRITE(2,'(I3,5X,E14.6)')MASPT(ILODND),XMASND(ILODND)
      END DO
      END IF
!***NLODEL = NUMBER OF ELEMENTS CARRYING DISTRIBUTED MASS**********
      IF(NLODEL > 0)  THEN
      WRITE(2,'(2/,A,I5)')' NO. OF ELEMENTS CARRYING DISTRIBUTED MASS   =&
      & ',NLODEL
      WRITE(2,'(2/,A)')' ELEMENT    UDL MASS'
!***ARRAY MASEL AND XMASEL STORE RESPECTIVELY,  THE ELEMENT NUMBER**
!*** AND THE MASS ON THE ELEMENT************************************************
      DO  ILODEL=1,NLODEL
      READ(1,*)MASEL(ILODEL)
      READ(1,*)XMASEL(ILODEL)
      WRITE(2,'(I5,E14.6)')MASEL(ILODEL),XMASEL(ILODEL)
      END DO
      END IF
      RETURN
      END SUBROUTINE READMASFRAME
```

12.7.2 Subroutine SMASRJ

This subroutine computes the element mass matrix (consistent + concentrated) and assembles the structural mass matrix in vector format for a two-dimensional rigid-jointed structure.

```
!*********************************************************************************
      SUBROUTINE SMASRJ
!***VARIABLE DECLARATION*********************************************************
      IMPLICIT NONE
      INTEGER::IEND,JEND, IEVAB,JEVAB,IEVAB1,IJ,I,ILODEL,JLODEL,ILODND,ICOLM
      REAL::YOUNG, SHEARM, YINERA, AREA, ZASF, XPROJ,YPROJ,SPAN,SPAN2,&
      &XL,YM,BETA,BETA1,BETA2,UDLMASS,CONST,CONST1,CONST2,S11,S12,Q11,&
      &Q12,T11,T12,A1,A2
!***DEAL WITH CONSISTENT MASS MATRIX FIRST************************************
      IF(NLODEL >0) THEN
!***CYCLE ON ALL ELEMENTS WITH DISTRIBUTED MASS**************************
      ELEMENTS: DO  ILODEL=1,NLODEL
!***RETRIEVE THE NUMBER OF ELEMENT AND  THE UDL MASS ON THE ELEMENTS**
      IELEM=MASEL(ILODEL)
      UDLMASS=XMASEL(ILODEL)
!***RETRIEVE THE RELEVANT MATERIAL PROPERTIES****************************
      LPROP=MATNO(IELEM,1)
```

```
        YOUNG=PROPS(LPROP,1)
        SHEARM=PROPS(LPROP,2)
        YINERA=PROPS(LPROP,3)
        AREA=PROPS(LPROP,4)
        ZASF=PROPS(LPROP,5)
!*** RETRIEVE THE NODES CORRESPONDING TO THE TWO ENDS OF THE ELEMENT**
! ***CALCULATE THE ELEMENT LENGTH AND THE DIRECTION COSINES***********
        IEND=LNODS(IELEM,1)
        JEND=LNODS(IELEM,2)
        XPROJ=COORD(JEND,1)-COORD(IEND,1)
        YPROJ=COORD(JEND,2)-COORD(IEND,2)
        SPAN=SQRT(XPROJ*XPROJ + YPROJ*YPROJ)
        XL=XPROJ/SPAN
        YM=YPROJ/SPAN
        SPAN2=SPAN*SPAN
        BETA=12.0*YOUNG*YINERA*ZASF/(SHEARM*AREA*SPAN2)
        BETA1=BETA+1.0
        BETA2=BETA*BETA
!***CALCULATE THE MEMBER AXIS COEFFICIENTS OF MASS MATRIX DUE TO**
!*** UDL MASS. BENDING AND SHEAR DEFORMATION CONTRIBUTION*********
        CONST = UDLMASS*SPAN/(BETA1*BETA1)
        CONST1 = CONST*YINERA/(AREA*SPAN2)
        S11= (1.0/105.0 + BETA/60.0 + BETA2/120.0)*SPAN2*CONST&
        &+ (2.0/15.0 +BETA/6.0  + BETA2/3.0)*SPAN2*CONST1
        S12= -(1.0/140.0 + BETA/60.0 + BETA2/120.0)*SPAN2*CONST&
        &-(1.0/30.0 + BETA/6.0 - BETA2/6.0)*SPAN2*CONST1
         Q11= (11.0/210.0 + 11.0*BETA/120.0 + BETA2/24.0)*SPAN*CONST&
        & + (0.5*BETA -0.1)*SPAN*CONST1
        Q12= -(13.0/420.0 + 3.0*BETA/40.0 + BETA2/24.0)*SPAN*CONST&
        & + (0.5*BETA - 0.1)*SPAN*CONST1
        T11= (13.0/35.0 + 0.7*BETA + BETA2/3.0)*CONST + 1.2*CONST1
        T12= -(54.0/420.0 + 0.3*BETA + BETA2/6.0)*CONST + 1.2*CONST1
!***AXIAL DEFORMATION CONTRIBUTION*************************
        CONST2 = UDLMASS*SPAN/6.0
        A1 = 2.0*CONST2
        A2 = CONST2
!***CALCULATE THE GLOBAL AXIS ELEMENT MASS MATRIX*********
        EMASS(1,1)=T11*YM*YM +A1*XL*XL
        EMASS(1,2)=(-T11 + A1)*XL*YM
        EMASS(1,3)=Q11*YM
        EMASS(1,4)=-T12*YM*YM+A2*XL*XL
        EMASS(1,5)=(T12 + A2)*XL*YM
        EMASS(1,6)=Q12*YM
        EMASS(2,2)=T11*XL*XL + A1*YM*YM
        EMASS(2,3)=-Q11*XL
        EMASS(2,4)=EMASS(1,5)
        EMASS(2,5)=-T12*XL*XL+ A2*YM*YM
        EMASS(2,6)=-Q12*XL
        EMASS(3,3)=S11
        EMASS(3,4)=-EMASS(1,6)
        EMASS(3,5)=-EMASS(2,6)
        EMASS(3,6)=S12
        EMASS(4,4)=EMASS(1,1)
        EMASS(4,5)=EMASS(1,2)
        EMASS(4,6)=-EMASS(1,3)
```

```
          EMASS(5,5)=EMASS(2,2)
          EMASS(5,6)=-EMASS(2,3)
          EMASS(6,6)=S11
!***FILL UP THE SYMMETRICAL LOWER HALF OF THE EMASS MATRIX**
          DO IEVAB=2,NEVAB
          IEVAB1=IEVAB-1
          DO JEVAB=1,IEVAB1
          EMASS(IEVAB,JEVAB)=EMASS(JEVAB,IEVAB)
          END DO
          END DO
!***RETRIEVE THE NODE FREEDOMS CORRESPONDING TO THE NODES AT **
!***THE ENDS OF THE ELEMENT. THIS IS STORED IN ARRAY MEMDIS(NDOFN)*
          DO IDOFN=1,NDOFN
          MEMDIS(IDOFN)=NODFRE(IEND,IDOFN)
          MEMDIS(IDOFN+NDOFN)=NODFRE(JEND,IDOFN)
          END DO
!***ASSEMBLE THE STRUCTURAL MASS MATRIX IN VECTOR FORM************
          DO IEVAB=1,NEVAB
          IF(MEMDIS(IEVAB) > 0) THEN
          DO JEVAB=1,NEVAB
          IF(MEMDIS(JEVAB) > 0  .AND. MEMDIS(JEVAB) >= MEMDIS(IEVAB) ) THEN
!***CALCULATE THE ELEMENT POSITION IN THE VECTOR FORMAT**************
          IJ=NDIA(MEMDIS(JEVAB)) + MEMDIS(JEVAB)-MEMDIS(IEVAB)
          GMASS(IJ)=GMASS(IJ)+EMASS(IEVAB,JEVAB)
          END IF
          END DO
          END IF
          END DO
          END DO ELEMENTS
!***END OF ELEMENT LOOP******************************************************
          END IF
!***MODIFY THE MASS MATRIX FOR THE EFFECT OF CONCENTRATED MASSES**
!****IT IS ASSUMED THAT THE CONCENTRATED MASS EXERTS INERTIAL FORCE**
!*** WITH RESPECT TO TRANSLATIONS CORRESPONDING TOIDOFN = 1 AND 2 ONLY*
          IF(NLODND > 0) THEN
          DO ILODND=1, NLODND
          DO IDOFN = 1,2
          I=NODFRE(MASPT(ILODND),IDOFN)
          IF(I > 0) THEN
          IJ=NDIA(I)
          GMASS(IJ)=GMASS(IJ) + XMASND(ILODND)
          END IF
          END DO
          END DO
          END IF
          RETURN
          END SUBROUTINE SMASRJ
```

12.7.3 Subroutine SMASGR

This subroutine computes the element mass matrix (consistent + concentrated) and assembles the structural mass matrix in vector format for a plane grid structure.

```
!*******************************************************************************
```

```
      SUBROUTINE SMASGR
!***VARIABLE DECLARATION*************************************************
      IMPLICIT NONE
      INTEGER::IEND,JEND, IEVAB,JEVAB,IEVAB1,IJ,I,ILODEL,JLODEL,ILODND
      REAL::YOUNG, SHEARM, YINERA, AREA, ZASF, POLAR,XPROJ,YPROJ,&
      &SPAN,SPAN2,XL,YM,BETA,BETA1,BETA2,UDLMASS,CONST,CONST1,&
      &CONST2,S11,S12,Q11,Q12,T11,T12,A1,A2
      IF(NLODEL >0) THEN
!***CONSISTENT MASS CALCULATION******************************************
      ELEMENTS: DO  ILODEL=1,NLODEL
!***RETRIEVE ELEMENT AND MASS ON ELEMENT*********************************
      IELEM=MASEL(ILODEL)
      UDLMASS=XMASEL(ILODEL)
!***RETRIEVE THE PROPERTIES OF THE ELEMENT*******************************
      LPROP=MATNO(IELEM,1)
      YOUNG=PROPS(LPROP,1)
      SHEARM=PROPS(LPROP,2)
      YINERA=PROPS(LPROP,3)
      AREA=PROPS(LPROP,5)
      ZASF=PROPS(LPROP,6)
      POLAR = PROPS(LPROP,7)
!***RETRIEVE THE NODES CORRESPONDING TO THE TWO ENDS OF THE ELEMENT**
       IEND=LNODS(IELEM,1)
      JEND=LNODS(IELEM,2)
!***CALCULATE THE ELEMENT LENGTH AND THE DIRECTION COSINES************
      XPROJ=COORD(JEND,1)-COORD(IEND,1)
      YPROJ=COORD(JEND,2)-COORD(IEND,2)
      SPAN=SQRT(XPROJ*XPROJ + YPROJ*YPROJ)
      XL=XPROJ/SPAN
      YM=YPROJ/SPAN
      SPAN2=SPAN*SPAN
      BETA=12.0*YOUNG*YINERA*ZASF/(SHEARM*AREA*SPAN2)
      BETA2=BETA*BETA
      BETA1=BETA+1.0
!***CALCULATE THE MEMBER AXIS COEFFICIENTS OF MASS MATRIX DUE*****
!***TO BENDING AND SHEAR DEFORMATION************************************
      CONST = UDLMASS*SPAN/BETA1/BETA1
      CONST1 = CONST*YINERA/AREA/SPAN2
      S11= (1.0/105.0 + BETA/60.0 + BETA2/120.0)*SPAN2*CONST&
      & + (2.0/15.0 + BETA/6.0 + BETA2/3.0)*SPAN2*CONST1
      S12 =-(1.0/140.0 + BETA/60.0 + BETA2/120.0)*SPAN2*CONST&
      &- (1.0/30.0 + BETA/6.0 - BETA2/6.0)*SPAN2*CONST1
      Q11= (11.0/210.0 + 11.0*BETA/120.0 + BETA2/24.0)*SPAN*CONST&
      &+ (0.5*BETA - 0.1)*SPAN*CONST1
      Q12= -(13.0/420.0 + 3.0*BETA/40.0 + BETA2/24.0)*SPAN*CONST&
      &-(0.1 - 0.5*BETA)*SPAN*CONST1
      T11= (13.0/35.0 + 0.7*BETA + BETA2/3.0)*CONST + 1.2*CONST1
      T12= -(9.0/70.0 + 0.3*BETA + BETA2/6.0)*CONST + 1.2*CONST1
!***CALCULATE THE COEFFICIENTS OF MASS MATRIX DUE TO TORSIONAL **
!**DEFORMATION**********************************************************
      CONST2 = UDLMASS*SPAN*POLAR/AREA/6.0
      A1 = 2.0*CONST2
      A2 = CONST2
!***CALCULATE THE GLOBAL AXIS ELEMENT MASS MATRIX************************
      EMASS(1,1)= T11
```

```
        EMASS(1,2)= Q11*YM
        EMASS(1,3)= -Q11*XL
        EMASS(1,4)= -T12
        EMASS(1,5)= Q12*YM
        EMASS(1,6)= -Q12*XL
        EMASS(2,2)= S11*YM*YM + A1*XL*XL
        EMASS(2,3)= -(S11 - A1)*XL*YM
        EMASS(2,4)= -EMASS(1,5)
        EMASS(2,5)= S12*YM*YM + A2*XL*XL
        EMASS(2,6)= -(S12 - A2)*XL*YM
        EMASS(3,3)= S11*XL*XL + A1*YM*YM
        EMASS(3,4)= -EMASS(1,6)
        EMASS(3,5)= EMASS(2,6)
        EMASS(3,6)= S12*XL*XL + A2*YM*YM
        EMASS(4,4)= EMASS(1,1)
        EMASS(4,5)= -EMASS(1,2)
        EMASS(4,6)= EMASS(1,3)
        EMASS(5,5)= EMASS(2,2)
        EMASS(5,6)= EMASS(2,3)
        EMASS(6,6)= EMASS(3,3)
!***FILL UP THE SYMMETRICAL LOWER HALF OF THE EMASS MATRIX****
        DO  IEVAB=2,NEVAB
        IEVAB1=IEVAB-1
        DO  JEVAB=1,IEVAB1
        EMASS(IEVAB,JEVAB)=EMASS(JEVAB,IEVAB)
        END DO
        END DO
!***RETRIEVE THE NODE FREEDOMS CORRESPONDING TO THE NODES AT****
!*** THE ENDS OF THE ELEMENT AND  STORE IN ARRAY MEMDIS(NDOFN)***
        DO  IDOFN=1,NDOFN
        MEMDIS(IDOFN)=NODFRE(IEND,IDOFN)
        MEMDIS(IDOFN+NDOFN)=NODFRE(JEND,IDOFN)
        END DO
!***ASSEMBLE THE STRUCTURAL MASS MATRIX IN VECTOR FORM******
        DO  IEVAB=1,NEVAB
         IF(MEMDIS(IEVAB) > 0) THEN
        DO  JEVAB=1,NEVAB
        IF(MEMDIS(JEVAB) > 0  .AND. MEMDIS(JEVAB) >= MEMDIS(IEVAB) ) THEN
!***CALCULATE THE ELEMENT POSITION IN THE VECTOR FORMAT**************
        IJ=NDIA(MEMDIS(JEVAB)) + MEMDIS(JEVAB)-MEMDIS(IEVAB)
        GMASS(IJ)=GMASS(IJ)+EMASS(IEVAB,JEVAB)
        END IF
        END DO
        END IF
        END DO
        END DO ELEMENTS
        END IF
!***MODIFY THE MASS MATRIX FOR THE EFFECT OF CONCENTRATED MASSES.**
!***IT IS ASSUMED THAT THE CONCENTRATED MASS EXERTS INERTIAL FORCE**
!*** WITH RESPECT TO TRANSLATION CORRESPONDING TO  IDOFN=1 ONLY******
        IF(NLODND  > 0) THEN
        DO  ILODND=1, NLODND
        I=NODFRE(MASPT(ILODND),1)
        IF(I /= 0) THEN
        IJ=NDIA(I)
```

```
        GMASS(IJ)=GMASS(IJ) + XMASND(ILODND)
        END IF
        END DO
        END IF
        RETURN
        END SUBROUTINE SMASGR
```

12.8 SIMULTANEOUS ITERATION

Simultaneous iteration was described in Section 6.2. In this section, the subroutines that accomplish this task are described.

12.8.1 Subroutine SIMIT

This subroutine carries out simultaneous iteration. The steps involved in simultaneous iteration were described in Section 6.2.2. This subroutine basically consists of a series of other subroutines which perform the individual steps of the process. The subroutines called are CHOLESKY, GENVEC1, ORTHO, BACKSUB, MULT3, FORWSUB, FORMW and MULT5. Subroutines CHOLESKY, BACKSUB and FORWSUB have been described in Section 12.5. Other subroutines will be described in the next section.

```
!***************************************************************************
        SUBROUTINE SIMIT
!***VARIABLE DECLARATION****************************************************
        IMPLICIT NONE
        REAL::BIG,C,TEMP,OMEGA
        INTEGER::IVECTR,JVECTR,KVECTR,IFRDOM,ICYCLE,IEQNS,I,IPRINT
        REAL::U(MAXCOL),Y(NEQNS,NVECTR),Z(NEQNS,NVECTR),&
        &V(NEQNS,NVECTR),W(NEQNS,NVECTR),BM(NVECTR),DIS(3)
!*****CHOLESKY DECOMPOSITION OF  GSTIF ********************************
        CALL CHOLESKY(U)
!***GENERATE TRIAL EIGEN-VECTORS*****************************************
        CALL GENVEC1
!***ITERATE ON TRIAL VECTORS*********************************************
        CYCLES: DO  ICYCLE = 1,30
!***ORTHOGONALISE THE EIGEN-VECTORS************************************
        CALL ORTHO
!***START OF INVERSE ITERATION********************************************
!***DETERMINE Y FROM U ×Y =  VECTOR.  USE  BACK SUBSTITUTION *****
        CALL BACKSUB(U,VECTOR,Y)
!***MULTIPLY GMASS × Y = Z************************************************
        CALL MULT3(GMASS,Y,Z)
!***DETERMINE U ×  V = Z.  USE FORWARD SUBSTITUTION*****************
        CALL FORWSUB(U,Z,V)
!***CALCULATE W.  W = V × C.  C IS CALCULATED FROM B = VECTORᵀ × V**
!**** MATRICES B AND C ARE NOT EXPLICITLY FORMED*******************
        CALL FORMW(W,V,BM)
!***REORDER W ACCORDING TO DOMINANT VALUES OF DIAGONALS OF B*
        DO  IVECTR=1,NVECTR-1
        BIG=0
```

```
        DO JVECTR=IVECTR,NVECTR
        C = BM(JVECTR)
        IF(C > BIG) THEN
        BIG=C
        KVECTR=JVECTR
        END IF
        END DO
        BM(KVECTR)=BM(IVECTR)
        BM(IVECTR)=BIG
        DO IEQNS=1,NEQNS
        TEMP = W(IEQNS,IVECTR)
        W(IEQNS,IVECTR)=W(IEQNS,KVECTR)
        W(IEQNS,KVECTR)=TEMP
        END DO
        END DO
!***REPLACE VECTOR BY W******************************************
        VECTOR =W
        END DO CYCLES
!***CALCULATE THE  EIGEN-VECTORS Y OF GENERAL EIGEN-VALUE ***
!***PROBLEM.  U×Y=VECTOR.  USE BACK SUBSTITUTION TO FIND Y**
        CALL BACKSUB(U,VECTOR,Y)
!***MAKE THE EIGEN-VECTORS M ORTHOGONAL**********************
        CALL MULT5(GMASS,Y)
!***PRINT THE VALUE OF EIGEN-VALUE AND EIGEN-VECTORS************
        DO  IVECTR=1,NVECTR
        FREQ2(IVECTR) = 1.0/BM(IVECTR)
        OMEGA=SQRT(FREQ2(IVECTR))
        WRITE(2,'(//,A,I3,A,E14.6)')'EIGEN-VALUE NO. = ',IVECTR,'   VALUE = ',OMEGA
        WRITE(2,*)'EIGEN-VECTOR S'
        DO  IPOIN = 1,NPOIN
        DO  IDOFN = 1,NDOFN
        DIS(IDOFN) = 0.0
        IFRDOM = NODFRE(IPOIN,IDOFN)
        IF(IFRDOM .NE. 0) DIS(IDOFN) = Y(IFRDOM,IVECTR)
        END DO
        WRITE(2,'(I6,5E14.6)')IPOIN,(DIS(IDOFN),IDOFN = 1,NDOFN)
        END DO
        END DO
        RETURN
        END SUBROUTINE SIMIT
```

12.8.1.1 Subroutine GENVEC1

This subroutine generates initial starting eigen-vector for all the NVECTR eigen-values. The vector used has a value of 2.0 in the 'diagonal' position and unity elsewhere.

```
!*************************************************************
        SUBROUTINE GENVEC1
!***VARIABLE DECLARATION*************************************
        INTEGER::IEQNS,JVECTR
        DO IEQNS = 1,NEQNS
        DO JVECTR = 1, NVECTR
        VECTOR(IEQNS,JVECTR) = 1.0
```

```
        IF(IEQNS = = JVECTR) VECTOR(IEQNS,JVECTR) = 2.0
        END DO
        END DO
        RETURN
        END SUBROUTINE GENVEC1
```

12.8.1.2 Subroutine ORTHO

This subroutine uses modified Gram-Schmidt orthogonalisation to convert columns of array VECTOR into a set of orthogonal vectors.

```
!***********************************************************************
        SUBROUTINE ORTHO
!***VARIABLE DECLARATION**********************************************
        IMPLICIT NONE
        INTEGER::IVECTR, IVECTR1, JEQNS, KVECTR
         REAL::SUM, DENOM, ALPHA
!***MODIFIED GRAM-SCHMIDT ORTHOGONALISATION PROCEDURE*********
        DO  IVECTR = 2,NVECTR
        IVECTR1 = IVECTR -1
        SUM = 0.0
        DO  JEQNS = 1,NEQNS
        SUM = SUM + VECTOR(JEQNS,IVECTR1)*VECTOR(JEQNS,IVECTR1)
        END DO
        DENOM = 1.0/SUM
        DO  KVECTR = IVECTR,NVECTR
        SUM = 0.0
        DO  JEQNS = 1,NEQNS
        SUM = SUM + VECTOR(JEQNS,IVECTR1)*VECTOR(JEQNS,KVECTR)
        END DO
        ALPHA = SUM*DENOM
        DO JEQNS = 1,NEQNS
        VECTOR(JEQNS,KVECTR) = VECTOR(JEQNS,KVECTR)-&
        &ALPHA*VECTOR(JEQNS,IVECTR1)
        END DO
        END DO
        END DO
!***MAKE THE VECTORS OF UNIT LENGTH*********************************
        CALL UNIT
        RETURN
        END SUBROUTINE ORTHO
```

12.8.1.3 Subroutine UNIT

This subroutine converts the vectors to unit length.

```
!***********************************************************
        SUBROUTINE UNIT
!***VARIABLE DECLARATION*********************************
        IMPLICIT NONE
        INTEGER::IVECTR,IEQNS
        REAL::SUM
        DO  IVECTR = 1,NVECTR
        SUM = 0
```

```
        DO IEQNS = 1,NEQNS
        SUM = SUM + VECTOR(IEQNS,IVECTR)*VECTOR(IEQNS,IVECTR)
        END DO
        SUM = SQRT(1.0/SUM)
        DO  IEQNS = 1,NEQNS
        VECTOR(IEQNS,IVECTR)=VECTOR(IEQNS,IVECTR)*SUM
        END DO
        END DO
        RETURN
        END SUBROUTINE UNIT
```

12.8.1.4 Subroutine MULT3

This subroutine multiples matrix A held in vector format with a full matrix B and stores it in matrix C as a full matrix.

```
!*********************************************************************
        SUBROUTINE MULT3(A,B,C)
!***VARIABLE DECLARATION*********************************************
        INTEGER::IEQNS,JVECTR,KEQNS,IK,KOUNT
        REAL:: A(MAXCOL),B(NEQNS,NVECTR),C(NEQNS,NVECTR)
        DO IEQNS = 1,NEQNS
        DO JVECTR = 1,NVECTR
        C(IEQNS,JVECTR)=0.0
        DO KEQNS=1,NEQNS
!***CONSIDER ELEMENTS TO THE RIGHT OF DIAGONAL AND DETERMINE ***
!****ITS POSITION IN VECTOR FORMAT**********************************
        IF(KEQNS >= IEQNS) THEN
        IK=NDIA(KEQNS) + KEQNS-IEQNS
        KOUNT = KEQNS-IEQNS+1 - NCOLM(KEQNS)
        END IF
!***CONSIDER ELEMENTS TO THE LEFT OF THE DIAGONAL AND DETERMINE**
!*** ITS POSITION IN VECTOR FORMAT**********************************
        IF(KEQNS .LT. IEQNS) THEN
         K=NDIA(IEQNS) +IEQNS -KEQNS
        KOUNT=IEQNS-KEQNS+1 - NCOLM(IEQNS)
        END IF
        IF(KOUNT <= 0) C(IEQNS,JVECTR) = C(IEQNS,JVECTR) + A(IK)*B(KEQNS,JVECTR)
        END DO
        END DO
        END DO
        RETURN
        END  SUBROUTINE MULT3
```

12.8.1.5 Subroutine FORMW

This subroutine calculates the matrix product $W = V \times T$, where T is a function of B. $B = VECTOR^T \times V$. Matrices B and T are not stored but the necessary parts are computed as required. Details of calculations are given in Section 6.2.2.1.

```
!*********************************************************************
    SUBROUTINE FORMW(W,V,BM)
!***VARIABLE DECLARATION*********************************************
        IMPLICIT NONE
```

```
        INTEGER::IVECTR,IVECTR1,JVECTR,KEQNS,IEQNS
        REAL::BII,BJJ,BIJ,DIFF,S,TIJ,SUM
        REAL:: W(NEQNS,NVECTR),V(NEQNS,NVECTR),BM(NVECTR)
!***COPY V INTO W**********************************************************
        W=V
!***CALCULATE THE NECESSARY COEFFICIENTS OF B*************************
        IVECTR: DO  IVECTR = 1,NVECTR
        IVECTR1=IVECTR+1
        IF(IVECTR1 <= NVECTR) THEN
        JVECTR: DO  JVECTR = IVECTR1, NVECTR
        BII=0
        BJJ=0
        BIJ=0
        KEQNS: DO KEQNS=1,NEQNS
        BIJ=BIJ+VECTOR(KEQNS,IVECTR)*V(KEQNS,JVECTR)
        BII=BII+VECTOR(KEQNS,IVECTR)*V(KEQNS,IVECTR)
        BJJ=BJJ+VECTOR(KEQNS,JVECTR)*V(KEQNS,JVECTR)
        END DO KEQNS
!***FORM THE COEFFICIENT OF T(I,J)******************************
        DIFF = BII-BJJ
        IF(DIFF >=0)THEN
        S = +1.0
        ELSE
        S= -1.0
        END IF
        TIJ = -2.0*BIJ/(DIFF+S*SQRT(DIFF*DIFF+16.0*BIJ*BIJ))
!***FORM W****************************************************
        DO IEQNS=1,NEQNS
        W(IEQNS,IVECTR)=W(IEQNS,IVECTR)-V(IEQNS,JVECTR)*TIJ
        W(IEQNS,JVECTR)=W(IEQNS,JVECTR)+V(IEQNS,IVECTR)*TIJ
        END DO
        END DO JVECTR
!***STORE DIAGONAL ELEMENTS OF B MATRIX FOR REARRANGING ****
!***THE VECTORS****************************************************
        BM(IVECTR)=BII
        END IF
        END DO IVECTR
!***CALCULATE BM(NVECTR)AS IT HAS NOT BEEN DONE BEFORE******
        SUM = 0
        DO KEQNS = 1,NEQNS
        SUM = SUM + VECTOR(KEQNS,NVECTR)*V(KEQNS,NVECTR)
        END DO
        BM(NVECTR) = SUM
        RETURN
        END   SUBROUTINE FORMW
```

12.8.1.6 Subroutine MULT5

This subroutine is used to M normalise the eigen-vectors.

```
!*****************************************************************
         SUBROUTINE MULT5(A,B)
!***THIS SUBROUTINE MULTIPLES A MATRIX HELD IN VECTOR FORMAT***
!***WITH A FULL MATRIX B AND STORES IT IN C AS A FULL MATRIX A B = C*
!****FINALLY D = B TRANSFORM C
```

```
!***VARIABLE DECLARATION**************************************************
        IMPLICIT NONE
        INTEGER::IEQNS,KOUNT,KEQNS,IK,JVECTR,IVECTR
        REAL::D
        REAL:: A(MAXCOL),B(NEQNS,NVECTR),C(NEQNS,NVECTR)
        CALL MULT3(A,B,C)
!***M NORMALISE THE EIGEN-VECTORS*****************************************
!***D PRODUCT OF THE COLUMNS OF B TRANSFORM BY ROWS OF C **
! **MODIFY COLUMNS OF B BY MULTIPLYING THEM BY SQUARE ROOT ***
!***OF D TO ENSURE MATRIX B IS M ORTHOGONAL***********************
        DO IVECTR = 1, NVECTR
        D = 0.0
        DO KEQNS = 1, NEQNS
        D = D + B(KEQNS,IVECTR) * C(KEQNS,IVECTR)
        END DO
        D = 1.0/SQRT(D)
        DO IEQNS = 1, NEQNS
        B(IEQNS,IVECTR) = B(IEQNS,IVECTR)*D
        END DO
        END DO
        RETURN
        END SUBROUTINE MULT5
```

12.9 STURM SEQUENCE CHECK

Eigen-values are calculated using the subroutine SIMIT. However to ensure that no eigen-value has been missed, it is necessary to use the Sturm sequence and corresponding sign counts. This is explained in Section 5.9.2. Two subroutines are used for this purpose. Subroutine GAUSRED forms the matrix $A = K - \omega^2 M$ for a given value of ω^2 and reduces the matrix A to upper triangular form so that the determinants of the sub-matrices of A can be calculated. Then the subroutine STRUM is used to do the sign agreement counts.

12.9.1 Subroutine STURM

Subroutine STURM calculates, by Sturm sequence check on the sign agreements, the number of eigen-values between 0.95 and 1.05 of the eigen-value calculated. It is important that the difference in the sign counts should be one. This check is repeated for all the NVECTR values.

```
!*****************************************************************************
        SUBROUTINE STURM
!***VARIABLE DECLARATION*******************************************************
        IMPLICIT NONE
        REAL::OMEGAS,OMEGAS1,OMEGAS2,FREQ
        INTEGER::IFREQ,IVECTR,IMAXCOL,ICHANGE,ISIGN,ICHANG1,ICHANG2,&
        &I1,I2,IEQNS
        REAL:: A(MAXCOL)
        WRITE(2,*)'FREQUENCY NO.  OMEGA SQ.  AGGRE1     AGREE2'
!****START THE ITERATION ON ALL THE NVECTR EIGEN-VALUES *******************
        VECTORS:  DO IVECTR = 1, NVECTR
```

```
        OMEGAS = FREQ2(IVECTR)
        OMEGAS1 = 0.95*OMEGAS
        OMEGAS2 = 1.05*OMEGAS
!***THE SIGN AGREEMENT WILL BE CHECKED ONLY FOR 0.95 AND 1.05 OF THE**
!*** CHOSEN EIGEN-VALUE ******************************************************
        FREQUENCY:DO  IFREQ = 1, 2
        FREQ = OMEGAS1
        IF(IFREQ = = 2) FREQ = OMEGAS2
!***CALCULATE A = (K - OMEGA SQUARED × M)**************************
        DO  IMAXCOL = 1, MAXCOL
        A(IMAXCOL) = GSTIF(IMAXCOL)- FREQ*GMASS(IMAXCOL)
        END DO
!***REDUCE (K - OMEGA SQ. × M) TO UPPER TRIANGULAR FORM BY ***********
!*** GAUSSIAN ELIMINATION**************************************************
        CALL GAUSRED(A)
        IF(NFAIL = = 1) THEN
        WRITE(2,*)'ZERO PIVOT IN GAUSRED'
        EXIT
        END IF
!***SIGN AGREEMENT COUNTS CALCULATION*********************************
        ICHANGE = 0
        I1 = 1
        EQUATIONS: DO  IEQNS = 1, NEQNS
        ISIGN = -1
        IF( A(NDIA(IEQNS)) > 0.0) ISIGN = 1
        I2 = ISIGN*I1
        IF(I2 /= I1) ICHANGE = ICHANGE + 1
        I1 = I2
        END DO EQUATIONS
        IF(IFREQ = = 1) ICHANG1 = ICHANGE
        IF(IFREQ = = 2) ICHANG2 = ICHANGE
        END DO FREQUENCY
        IF (NFAIL /=1) THEN
        WRITE(2,'(I6,5X,E14.6,6X,I5,5X,I5)')IVECTR,OMEGAS, ICHANG1,ICHANG2
        END IF
        END DO VECTORS
        RETURN
        END  SUBROUTINE STURM
```

12.9.2 Subroutine GAUSRED

Determinants of sub-matrices are calculated by first of all carrying out Gaussian elimination on the matrix A = GSTIF - Lamda × GMASS. The matrix A is stored in vector format. Gaussian elimination method is described in Section 4.5.

```
!*****************************************************************************
        SUBROUTINE GAUSRED(A)
!***VARIABLE DECLARATION*****************************************************
        IMPLICIT NONE
I       NTEGER::KEQNS,NEQNS1,NBAND1,IBAND,JBAND,JCOLMS,IROWS,KI,KJ,IJ
        REAL::PIVOT,QUOT
        REAL:: A(MAXCOL)
        NBAND1 = NBAND + 1
        NEQNS1 = NEQNS -1
```

```
!***CHOOSE THE KEQNS TH UNKNOWN TO BE ELIMINATED FROM ALL THE *******
!***SUCCEDING EQUATIONS. START WITH UNKNOWN 1 DOWN TO (NEQNS -1)****
        DO KEQNS = 1, NEQNS1
        PIVOT = A(NDIA(KEQNS))
        IF(ABS(PIVOT) .LT. 1.0E-30) THEN
        NFAIL=1
        EXIT
        ELSE
        PIVOT = 1.0/PIVOT
!***SINCE THE VARIABLE TO BE ELIMINATED WILL APPEAR ONLY IN THE NBAND**
!*** EQUATIONS AFTER THE PIVOT EQUATION, COMPUTATIONS NEED TO BE********
!*** RESTRICTED TO ONLY THOSE EQUATIONS************************************
        DO IBAND = 1, NBAND
        IROWS = KEQNS + IBAND
        IF(IROWS > NEQNS) EXIT
!***THE COEFFICIENT OF THE UNKNOWN TO BE ELIMINATED FROM THE IROWS***
!*** IS GIVEN BY THE ELEMENT IN THE UPPER TRIANGLE IN POSITION (KEQNS,IROWS).
!***SINCE I = KEQNS, J = IROWS = KEQNS + IBAND, IN THE VECTOR FORMAT***********
!***THIS ELEMENT IS IN POSITION*************************************************
!***KPOSN = NDIA(IROW) + IROWS – KEQNS = NDIA(IROWS) + IBAND******************
!*** THE ELEMENT EXISTS ONLY IF ( J-I +1) = (IBAND +1 ) ≤ NCOLM(J)*****************
        IF((IBAND + 1) <= NCOLM(IROWS) ) THEN
        KI = NDIA(IROWS) + IBAND
!***IF THE ELEMENT IN (KEQNS,IROWS) =0, THEN NO NEED TO ALTER THE ROW**
        IF(A(KI) /= 0.0) THEN
!***CALCULATE THE MULTIPLYING FACTOR***********************************
        QUOT = A(KI)*PIVOT
!***MODIFY THE ELEMENTS OF THE IROWS. CONSIDER ONLY ELEMENTS IN UPPER**
!***TRIANGLE. IN ANY ROW ONE NEED TO CONSIDER ELEMENTS FROM THE********
!*** DIAGONAL EXTENDING UP TO NBAND ONLY*********************************
        DO JBAND = 1, NBAND1
        JCOLMS = IROWS + JBAND -1
        IF (JCOLMS <= NEQNS) THEN
!***THE ELEMENT × QUOT WHICH WILL MODIFY THE (IROWS,JCOLMS) ELEMENT IS ***
!***IN POSITION(KEQNS, JCOLMS). THE ELEMENT IS ZERO IF**********************
!*** (JCOLMS-KEQNS + 1) > NCOLMS(JCOLMS)*************************************
        IF((JCOLMS - KEQNS +1) <= NCOLM(JCOLMS)) THEN
        KJ = NDIA(JCOLMS) + JCOLMS - KEQNS
!***THE ELEMENT WHICH WILL BE MODIFIED IS IN POSITION (IROWS, JCOLMS)*****
!***THE ELEMENT IS ZERO IF (JCOLMS-IROWS + 1) > NCOLMS(JCOLMS)*************
        IF((JCOLMS - IROWS +1) <= NCOLM(JCOLMS)) THEN
        IJ = NDIA(JCOLMS) + JCOLMS - IROWS
        A(IJ) = A(IJ) - QUOT*A(KJ)
        END IF
        END IF
        END IF
        END DO
        END IF
        END IF
        END DO
        ENDIF
        END DO
        RETURN
        END SUBROUTINE GAUSRED
```

12.10 Program FRASIM.F90

Program FRASIM.F90 calculates by simultaneous iteration, the first N natural frequencies in ascending order and the corresponding mode shapes of both two-dimensional rigid-jointed structures and plane grid structures.

```
          PROGRAM FRASIM
!*********VARIABLE DECLARATION*************************************************
         IMPLICIT NONE
         INTEGER::NPROB,IPROB,NFAIL,NNODE,JNODE,INODE,NDOFN,IDOFN,JDOFN,&
         &NDIME,IDIME,JDIME,NEVAB,IEVAB,JEVAB,ISTRUCT,NPROP,NPOIN,IPOIN,&
         &IELEM,NELEM,NRESND,IRESND,JRESND,NOPTN,NMATS,IMATS,IPROP,&
         &LPROP,KOUNT,NLODEL,JLODEL,ILODEL,NEQNS,NBAND,NLODND,ILODND,&
         &NVECTR,MAXCOL
!**********DEFINE INTEGER ARRAYS WHICH ARE ALLOCATABLE*******************
         INTEGER,ALLOCATABLE::NDFFIX(:),IFPRE(:,:),NODFRE(:,:),LNODS(:,:),&
         &MEMDIS(:),MATNO(:,:),MASEL(:),MASPT(:),NCOLM(:),NDIA(:)
!**********DEFINE REAL ARRAYS WHICH ARE ALLOCATABLE********************
         REAL,ALLOCATABLE::COORD(:,:),PROPS(:,:),ESTIF(:,:),EMASS(:,:),XMASEL(:),&
         &XMASND(:),GSTIF(:),GMASS(:),FREQ2(:),VECTOR(:,:)
!*********DATA AND OUTPUT TRANSMISSION FROM AND TO FILES****************
         OPEN(UNIT=1,FILE=NFILEIN,STATUS='OLD',ACTION='READ')
         OPEN(UNIT=2,FILE=NFILEOUT,STATUS='REPLACE',ACTION='WRITE')
!***START READING THE NUMBER OF PROBLEMS TO BE ANALYSED*******************
         READ(1,*)NPROB
         WRITE(2,'(2/,A,I8)')' TOTAL NUMBER OF PROBLEMS = ',NPROB
!***START THE CYCLE ON THE NUMBER OF PROBLEMS********************************
         PROBLEMS:DO  IPROB = 1, NPROB
         READ(1,*)TITLE
         WRITE(2,'(4/,A,I3)')'PROBLEM NUMBER = ',IPROB
         WRITE(2,'(A)')TITLE
!***SET THE BASIC PARAMETERS OF THE STRUCTURE*****************************
         NFAIL=0
         NNODE=2
         NDOFN=3
         NDIME=2
         NEVAB=6
         READ(1,*)ISTRUCT
         IF(ISTRUCT = = 1)THEN
         WRITE(2,*)' THIS IS A TWO-DIMENSIONAL RIGID-JOINTED STRUCTURE'
         NPROP=5
         ELSE IF(ISTRUCT = =2) THEN
         WRITE(2,*)' THIS A PLANE GRID STRUCTURE'
         NPROP=7
         END IF
!***READ AND PRINT THE BASIC PARAMETERS OF THE PARTICULAR PROBLEM**
         READ(1,*)NPOIN,NELEM,NRESND,NMATS
         WRITE(2,'(A,I5)')'NUMBER OF NODES = ',NPOIN
           WRITE(2,'(A,I5)')'NUMBER OF MEMBERS = ',NELEM
         WRITE(2,'(A,I5)')'NUMBER OF RESTRAINED NODES = ',NRESND
         WRITE(2,'(A,I5)')'NUMBER OF DIFFERENT TYPES OF MATERIALS = ',NMATS
!******** ALLOCATE ARRAYS****************************************************
         ALLOCATE(COORD(NPOIN,NDIME),LNODS(NELEM,NNODE),&
         &MATNO(NELEM,1),PROPS(NMATS,7),NODFRE(NPOIN,NDOFN),&
```

```
        &NDFFIX(NRESND),IFPRE(NRESND,NDOFN),ESTIF(NEVAB,NEVAB),&
        &MEMDIS(NEVAB),EMASS(NEVAB,NEVAB))
!***READ THE COORDINATES OF SOME OF THE JOINTS, GENERATE THE*****
!*** OMITTED DATA AND PRINT THE COORDINATES OF ALL THE NODES***
!****NFAIL=1 MEANS THAT ERROR IN DATA ******************************
        CALL CORDGEN
        IF(NFAIL= =1) EXIT
        WRITE(2,'(2/,A)')' COORDINATES OF NODES'
        WRITE(2,'(2/,A)')'   NODE    X-CORD      Y-CORD'
        DO  IPOIN=1,NPOIN
        WRITE(2,'(1X,I5,3(5X,F10.3))')IPOIN,(COORD(IPOIN,JDIME),JDIME=1,NDIME)
        END DO
!***READ THE NODE NUMBERS DEFINING THE ENDS OF SOME OF THE *******
!***ELEMENTS AND GENERATE THE OMITTED DATA************************
        CALL INTGEN(LNODS,NNODE,NELEM)
        IF(NFAIL= =1) EXIT
!***READ THE MATERIAL NUMBERS OF SOME OF THE ELEMENTS AND******
!**** GENERATE THE OMITTED DATA*************************************
        CALL INTGEN(MATNO,1,NELEM)
        IF(NFAIL= =1) EXIT
!***PRINT THE MATERIAL NUMBERS AND THE NODE NUMBERS AT THE ENDS**
!*** FOR ALL THE ELEMENTS*********************************************
        WRITE(2,'(3/,A)')' ELEMENT NO.  MATERIAL NO.   NODE-1   NODE-2'
        DO  IELEM = 1, NELEM
        WRITE(2,'(1X,I5,11X,I5,9X,I5,5X,I5,6X,I5)')IELEM,MATNO(IELEM,1),&
        &(LNODS(IELEM,JNODE),JNODE=1,NNODE)
        END DO
!***READ NOPTN. THIS TELLS WHETHER THE NODE FREEDOM ARRAY IS****
!***INPUT AS DATA (FOR NOPTN=1) OR TO BE GENERATED FROM THE DATA ABOUT**
!*** THE NODES WHERE ONE MORE DISPLACEMENTS ARE ZERO(NOPTN=0)*********
        READ(1,*)NOPTN
!***READ NODE FREEDOMS OF SOME OF THE NODES AND GENERATE THE OMITTED**
!***DATA*****************************************************************
        IF(NOPTN= =1) THEN
        CALL INTGEN(NODFRE,NDOFN,NPOIN)
        IF(NFAIL= =1)  EXIT
        ELSE
!***READ AND ECHO DATA ABOUT THE NODE NUMBERS AND FIXITY STATE AT NODES*
!***WHERE ONE OR MORE FREEDOMS ARE EQUAL TO ZERO****************************
        DO  IRESND=1,NRESND
        READ(1,*)NDFFIX(IRESND),(IFPRE(IRESND,JDOFN),JDOFN=1,NDOFN)
        END DO
        WRITE(2,'(3/,A)')' DEGREES OF RESTRAINT AT RESTRAINED NODES'
        WRITE(2,'(3/,A)')' 1 = FIXED AND 0 = FREE'
        IF(ISTRUCT= =2)WRITE(2,'(3/,A)')'NODE NO.  Z-FRDOM    THETAX-FRDOM&
        &      THETAY-FREEDOM'
        IF(ISTRUCT= =1) WRITE(2,'(3/,A)')'NODE NO.  X-FRDOM     Y-FRDOM      &
        &THETAZ-FRDOM'
        DO IRESND=1,NRESND
        WRITE(2,'(I5,10X,I5,8X,I5,16X,I5)')NDFFIX(IRESND),(IFPRE(IRESND,JDOFN),&
        &JDOFN=1,NDOFN)
        END DO
!***CALL THE SUBROUTINE TO GENERATE THE NODE FREEDOM ARRAY FROM****
!****THE DATA ABOUT THE RESTRAINED NODES**********************************
        CALL NFGEN
```

```
        END IF
        WRITE(2,'(2/,A)')' DEGREES OF FREEDOM AT NODES'
        IF(ISTRUCT= =1) WRITE(2,'(2/,A)')' NODE NUMBER     X-FRDOM.   Y-FRDOM&
        &.  THETAZ-FRDOM.'
        IF(ISTRUCT= =2) WRITE(2,'(2/,A)') 'NODE NUMBER     Z-FRDOM.  THETAX-
        &FRDOM.  THETAY-FRDOM.'
        DO IPOIN=1,NPOIN
        WRITE(2,'(I3,5X,I5,10X,I5,9X,I5)')IPOIN,(NODFRE(IPOIN,IDOFN),IDOFN=1,NDOFN)
        END DO
!***READ AND PRINT THE MATERIAL PROPERTIES *******************************
        DO IMATS=1,NMATS
        READ(1,*)(PROPS(IMATS,IPROP),IPROP=1,NPROP)
        END DO
        WRITE(2,'(2/,A)')' PROPERTIES OF DIFFERENT MATERIALS'
        IF(ISTRUCT= =1) WRITE(2,'(2/,A)')' MATERIAL NO.  YOUNGS MOD.  SHEAR&
        & MOD.  INERTIA       AREA  AREA SHEAR FACTOR'
         IF(ISTRUCT= =2) WRITE(2,'(2/,A)')' MATERIAL NO.  YOUNGS MOD.  SHEAR&
        &MOD.  INERTIA      TOR INERTIA       AREA       ASF.  &
        &POLAR INERTIA'
        DO IMATS=1,NMATS
        WRITE(2,'(2X,I5,8X,7(E14.6,2X))')IMATS,(PROPS(IMATS,IPROP),IPROP=1,NPROP)
        END DO
!***CALCULATE THE SIZE OF GSTIF*************************************************
        CALL MAXEQN
        WRITE(2,'(A,I5)')'NUMBER OF EQUATIONS =  ',NEQNS
!******  ALLOCATE ARRAYS*********************************************************
        ALLOCATE(NCOLM(NEQNS),NDIA(NEQNS))
!***CALCULATE THE HALF BANDWIDTH OF GSTIF***************************************
        CALL BANDWD
        WRITE(2,'(A,I5)')'HALF BAND WIDTH =  ',NBAND
!***CALL THE SUBROUTINE TO DETERMINE THE DETAILS OF SKYLINE FORMAT******
!***AND TOTAL LENGTH OF THE COLUMN VECTOR*********************************
        CALL NSKYLINE
!*********  ALLOCATE ARRAYS**************************************************
        ALLOCATE(GSTIF(MAXCOL),GMASS(MAXCOL))
!***READ THE CONCENTRATED MASSES AT THE NODES AND UDL MASS ON THE****
!***ELEMENTS AND STORE IN THE ARRAY MASPT(NLODND),XMASND(NLODND) ***
!***AND MASEL(NLODEL),XMASEL(NLODEL)**************************************
        READ(1,*)NLODND,NLODEL
        WRITE(2,'(/,A,I5)')'NUMBER OF NODES WITH CONCENTRATED MASS&
        &=',NLODND
        WRITE(2,'(/,A,I5)')'NUMBER OF ELEMENTS WITH DISTRIBUTED MASS&
        &=',NLODEL
!****** **ALLOCATE ARRAYS**************************************************
        IF (NLODND >0)  ALLOCATE(MASPT(NLODND), XMASND(NLODND))
        IF (NLODEL >0)  ALLOCATE(MASEL(NLODEL), XMASEL(NLODEL))
        CALL READMASFRAME
!***ZERO AND THEN FORM THE STRUCTURAL STIFFNESS AND MASS MATRICES**
        GSTIF=0.0
        GMASS=0.0
!***FORM GLOBAL STIFFNESS AND MASS MATRICES***************************
        IF(ISTRUCT= =1) THEN
        CALL STIFRJ
        CALL SMASRJ
        ELSE
```

```
        CALL STIFGR
        CALL SMASGR
        END IF
!***READ THE NUMBER OF EIGEN-PAIRS REQUIRED INCLUDING GUARD VECTORS*
        READ(1,*)NVECTR
        WRITE(2,'(/,A,I5)')'NUMBER OF EIGEN VALUES REQUIRED = ',NVECTR
!****** **ALLOCATE ARRAYS**********************************************
        ALLOCATE(VECTOR(NEQNS,NVECTR),FREQ2(NVECTR))
!*****CALL THE SIMULTANEOUS ITERATION SUBROUTINE***********
        CALL SIMIT
!*********CALL STURM SEQUENCE CHECK SUBROUTINE*************
        CALL STURM
        END DO PROBLEMS
        CONTAINS
!*****************************************************************
! ****INSERT ALL SUBROUTINES USED ONE AFTER ANOTHER********
!*****************************************************************
        END PROGRAM FRASIM
```

12.11 FINITE ELEMENT ANALYSIS: PLATE BENDING

Finite element analysis of Mindlin plates proceeds in a manner very similar to the frame analysis program described in Section 12.10. The same subroutines will be used, except for the calculation of element stiffness and consistent mass matrices. These are described in the next section.

12.11.1 Subroutine DSTIFPB

The basic steps in the calculation of stiffness and consistent mass matrix for a quadratic element are described in Section 8.8. The subroutine DSTIFPB follows those steps closely. It calls a series of other subroutines which are described later.

```
!*****************************************************************
        SUBROUTINE DSTIFPB
!*****VARIABLE DECLARATION*****************************************
        IMPLICIT NONE
        INTEGER::NGAUS1,LNODE,KGASP,IGAUS,JGAUS,K,K1,K2,IPOSN,IPOSN1,I&
        &POSN2,JPOSN,JPOSN1,JPOSN2,KSTRESS,IEVAB1,KFRDOM,IJ
        REAL::SUBGR,ELMASS,THICK,XINERIA,YOUNG,POISS,EXISP,ETASP,DAREA,&
        &DETJACOB,CONST,CONST1,CONST2
        REAL:: BTDB(NEVAB,NEVAB),BFLEX(3,NEVAB),BSHER(2,NEVAB),DFLEX(3,3)
!***FOR REDUCED INTEGRATION FOR SHEAR PART ONLY THE ORDER OF**********
!***INTEGRATION IS ONE LESS THAN FULL INTEGRATION*************************
        NGAUS1 = NGAUS -1
!*** LOOP OVER ALL ELEMENTS*******************************************
        DO IELEM=1,NELEM
!***RETRIEVE THE MATERIAL PROPERTIES OF THE ELEMENT*******************
        LPROP=MATNO(IELEM,1)
        YOUNG=PROPS(LPROP,1)
        POISS=PROPS(LPROP,2)
        THICK = PROPS(LPROP,3)
```

```
        ELMASS = PROPS(LPROP,4)
        SUBGR=PROPS(LPROP,5)
!*** RETRIEVE THE COORDINATES OF THE ELEMENT NODAL POINTS*********
        DO  INODE=1,NNODE
        LNODE=LNODS(IELEM,INODE)
        DO  IDIME=1,NDIME
        ELCOD(IDIME,INODE)=COORD(LNODE,IDIME)
        END DO
        END DO
!*** INITIALISE THE ELEMENT STIFFNESS AND MASS MATRICES*******
        EMASS = 0.0
        ESTIF=0.0
!***DO THE FLEXURE PART ONLY***************************************
!*** SET UP GAUSSIAN INTEGRATION CONSTANTS FOR 3 X 3 INTEGRATION**
        CALL GAUSINT (NGAUS)
!*** EVALUATE THE D-MATRIX FOR FLEXURE PART***********************
        CALL DFLEXR(DFLEX)
!*** ENTER LOOPS FOR AREA NUMERICAL INTEGRATION*******************
        KGASP = 0
        DO  IGAUS=1,NGAUS
        DO  JGAUS=1,NGAUS
        KGASP=KGASP+1
        EXISP=POSGP(IGAUS)
        ETASP=POSGP(JGAUS)
!*** CALCULATE THE SHAPEF FUNCTIONS, ELEMENTAL AREA, ETC************
        CALL SHAPEF8NODE(EXISP,ETASP)
        CALL JACOBIAN(DETJACOB,KGASP)
        DAREA=DETJACOB*WEIGP(IGAUS)*WEIGP(JGAUS)
!*** CALCULATE THE B AND BTDB MATRICES*****************************
        CALL BFLEXR(BFLEX)
        BTDB=MATMUL(TRANSPOSE(BFLEX),MATMUL(DFLEX,BFLEX))
!*** CALCULATE THE ELEMENT STIFFNESSES MATRIX**********************
        ESTIF = ESTIF + DAREA* BTDB
!***ADD THE STIFFNESS DUE TO SUBGRADE MODULUS TO THE ELEMENT********
!*** STIFFNESS MATRIX. ALSO FORMULATE THE ELEMENT MASS MATRIX*******
        DO INODE = 1, NNODE
        IPOSN = (INODE-1)*NDOFN + 1
        IPOSN1 = IPOSN + 1
        IPOSN2 = IPOSN + 2
        DO  JNODE = INODE , NNODE
        CONST = SHAPEFUN(INODE)*SHAPEFUN(JNODE)*DAREA
        CONST1 = CONST*ELMASS
        CONST2 = CONST1*XINERIA
        JPOSN = (JNODE-1)*NDOFN + 1
        JPOSN1 = JPOSN + 1
        JPOSN2 = JPOSN + 2
        EMASS(IPOSN,JPOSN) = EMASS(IPOSN,JPOSN) + CONST1
        EMASS(IPOSN1,JPOSN1) = EMASS(IPOSN1,JPOSN1) + CONST2
        EMASS(IPOSN2,JPOSN2) = EMASS(IPOSN2,JPOSN2) + CONST2
        ESTIF(IPOSN,JPOSN)=ESTIF(IPOSN,JPOSN)+SUBGR*CONST
        END DO
        END DO
        END DO
        END DO
!***THE ABOVE CALCULATIONS COMPLETE NGAUS × NGAUS INTEGRATION**
```

```
!***FOR FLEXURAL PART ONLY*********************************************
!***CALCULATE THE SHEAR CONTRIBUTION TO THE STIFFNESS MATRIX***
!***USE REDUCED INTEGRATION*******************************************
!***SET UP NUMERICAL INTEGRATION VALUES FOR REDUCED INTEGRATION*
        CALL GAUSINT(NGAUS1)
!***CALCULATE SHEAR STIFFNESS = G×AREA IN SHEAR*********************
        CONST = YOUNG*THICK/(2.4*(1.0+POISS))
!*** ENTER LOOPS FOR AREA NUMERICAL INTEGRATION******************
        KGASP = 0
        DO IGAUS=1,NGAUS1
        DO  JGAUS=1,NGAUS1
        KGASP=KGASP+1
        EXISP=POSGP(IGAUS)
        ETASP=POSGP(JGAUS)
!*** CALCULATE THE SHAPE FUNCTIONS, ELEMENTAL AREA, ETC*********
        CALL SHAPEF8NODE(EXISP,ETASP)
        CALL JACOBIAN(DETJACOB,KGASP)
        DAREA=DETJACOB*WEIGP(IGAUS)*WEIGP(JGAUS)
!*** CALCULATE THE B AND BTDB MATRICES*****************************
        CALL BSHEAR(BSHER)
!***CALCULATE B TRANSFORM × B × G ×T/1.2 ********************************
        ESTIF=ESTIF+CONST*DAREA*MATMUL(TRANSPOSE(BSHER),BSHER)
        END DO
        END DO
!***THIS COMPLETES THE REDUCED INTEGRATION FOR SHEAR PART**********
!***CONSTRUCT THE LOWER TRIANGLE OF THE STIFFNESS AND MASS MATRICES*
        DO IEVAB=2,NEVAB
        IEVAB1 = IEVAB -1
        DO  JEVAB=1,IEVAB1
        EMASS(IEVAB,JEVAB) = EMASS(JEVAB,IEVAB)
        ESTIF(IEVAB,JEVAB)=ESTIF(JEVAB,IEVAB)
        END DO
        END DO
!***CALCULATE THE NODE FREEDOMS OF THE NODES OF THE ELEMENT AND ****
!***STORE  IN AN ARRAY MEMDIS*************************************************
        DO  INODE = 1,NNODE
        DO  JDOFN = 1,NDOFN
        KFRDOM = (INODE-1)*NDOFN + JDOFN
        MEMDIS(KFRDOM) = NODFRE(LNODS(IELEM,INODE),JDOFN)
        END DO
        END DO
!***ADD THE ELEMENT STIFFNESS AND MASS MATRICES TO STRUCTURAL ***
!*** AND MASS MATRICES**************************************************
        DO IEVAB = 1,NEVAB
        K1=MEMDIS(IEVAB)
        IF( K1> 0)  THEN
        DO  JEVAB = 1,NEVAB
        K2 = MEMDIS(JEVAB)
        IF (K2 > 0 .AND. K2 >= MEMDIS(IEVAB)) THEN
        IJ = NDIA(K2) + K2 - K1
        GSTIF(IJ) = GSTIF(IJ) + ESTIF(IEVAB,JEVAB)
        GMASS(IJ) = GMASS(IJ) + EMASS(IEVAB,JEVAB)
        END IF
        END DO
        END IF
```

```
        END DO
        END DO
!***ADD THE CONCENTRATED MASSES TO THE APPROPRIATE PLACES IN****
!***MASS MATRIX FOR ACCELERATION IN TRANSLATION DIRECTION ONLY**
        IF(NLODND > 0) THEN
        DO  ILODND = 1,NLODND
        K =NDIA(NODFRE(MASPT(ILODND),1))
        GMASS(K) = GMASS(K) + XMASND(ILODND)
        END DO
        END IF
        RETURN
        END SUBROUTINE DSTIFPB
```

12.11.1.1 Subroutine GAUSINT(NORDER)

This subroutine simply sets up the weights and abscissas for Gaussian integration depending on the order of integration used.

```
!**********************************************************************************
   SUBROUTINE GAUSINT(NORDER)
!***VARIABLE DECLARATION*********************************************************
        IMPLICIT NONE
         INTEGER::NORDER
        IF(NORDER.EQ.2)THEN
        POSGP(1)=-0.577350269189626
        WEIGP(1)=1.0
        POSGP(2)=0.577350269189626
        WEIGP(2)=1.0
        END IF
        IF(NORDER.EQ.3) THEN
        POSGP(1)=-0.774596669241483
        POSGP(2)=0.0
        POSGP(3)=0.774596669241483
        WEIGP(1)=0.555555555555556
        WEIGP(2)=0.888888888888889
        WEIGP(3)=0.555555555555556
        END IF
        RETURN
        END SUBROUTINE GAUSINT
```

12.11.1.2 Subroutine SHAPEF8NODE

Given the values of the Gaussian coordinates (ξ,η), this subroutine calculates the value of the eight shape functions $n_i(\xi,\eta)$, and their derivatives.

```
!**********************************************************************************
   SUBROUTINE SHAPEF8NODE(S,T)
!***VARIABLE DECLARATION*********************************************************
        IMPLICIT NONE
        REAL::S,T,S2,T2,SS,TT,ST,SST,STT,ST2
        S2=S*2.0
        T2=T*2.0
        SS=S*S
        TT=T*T
```

```
        ST=S*T
        SST=S*S*T
        STT=S*T*T
        ST2=S*T*2.0
!*** SHAPE FUNCTIONS**********************************************************
        SHAPEFUN(1)=(-1.0+ST+SS+TT-SST-STT)/4.0
        SHAPEFUN(2)=(1.0-T-SS+SST)/2.0
        SHAPEFUN(3)=(-1.0-ST+SS+TT-SST+STT)/4.0
        SHAPEFUN(4)=(1.0+S-TT-STT)/2.0
        SHAPEFUN(5)=(-1.0+ST+SS+TT+SST+STT)/4.0
        SHAPEFUN(6)=(1.0+T-SS-SST)/2.0
        SHAPEFUN(7)=(-1.0-ST+SS+TT+SST-STT)/4.0
        SHAPEFUN(8)=(1.0-S-TT+STT)/2.0
!*** SHAPE FUNCTION DERIVITIVES WITH RESPECT TO S******************
        DERIV(1,1)=(T+S2-ST2-TT)/4.0
        DERIV(1,2)=-S+ST
        DERIV(1,3)=(-T+S2-ST2+TT)/4.0
        DERIV(1,4)=(1.0-TT)/2.0
        DERIV(1,5)=(T+S2+ST2+TT)/4.0
        DERIV(1,6)=-S-ST
        DERIV(1,7)=(-T+S2+ST2-TT)/4.0
        DERIV(1,8)=(-1.0+TT)/2.0
!*** SHAPE FUNCTION DERIVITIVES WITH RESPECT TO T********************
        DERIV(2,1)=(S+T2-SS-ST2)/4.0
        DERIV(2,2)=(-1.0+SS)/2.0
        DERIV(2,3)=(-S+T2-SS+ST2)/4.0
        DERIV(2,4)=-T-ST
        DERIV(2,5)=(S+T2+SS+ST2)/4.0
        DERIV(2,6)=(1.0-SS)/2.0
        DERIV(2,7)=(-S+T2+SS-ST2)/4.0
        DERIV(2,8)=-T+ST
        RETURN
        END SUBROUTINE SHAPEF8NODE
```

12.11.1.3 Subroutine JACOBIAN

This subroutine calculates the Cartesian coordinates of the Gauss points and the Jacobian matrix, its determinant and inverse for two-dimensional elements.

```
!***************************************************************************
        SUBROUTINE JACOBIAN(DETJACOB,KGASP)
!***VARIABLE DECLARATION*******************************************************
        IMPLICIT NONE
        INTEGER::KGASP
        REAL::DETJACOB
        REAL:: XJACM(NDIME,NDIME),XJACI(NDIME,NDIME)
!*** CALCULATE (X,Y) COORDINATES OF SAMPLING POINT****************************
        DO IDIME=1,NDIME
        GPCOD(IDIME,KGASP)=0.0
        DO INODE=1,NNODE
        GPCOD(IDIME,KGASP)=GPCOD(IDIME,KGASP)+&
        &ELCOD(IDIME,INODE)*SHAPEFUN(INODE)
        END DO
        END DO
```

```
!*** CREATE JACOBIAN MATRIX XJACM*****************************************
        DO IDIME=1,NDIME
        DO JDIME=1,NDIME
        XJACM(IDIME,JDIME)=0.0
        DO INODE=1,NNODE
        XJACM(IDIME,JDIME)=XJACM(IDIME,JDIME)+&
        &DERIV(IDIME,INODE)*ELCOD(JDIME,INODE)
        END DO
        END DO
        END DO
!*** CALCULATE DETERMINANT AND INVERSE OF JACOBIAN MATRIX*********
        DETJACOB=XJACM(1,1)*XJACM(2,2)-XJACM(1,2)*XJACM(2,1)
        IF(DETJACOB >0.0) THEN
        DETJACOB=1.0/DETJACOB
        XJACI(1,1)=XJACM(2,2)*DETJACOB
        XJACI(2,2)=XJACM(1,1)*DETJACOB
        XJACI(1,2)=-XJACM(1,2)*DETJACOB
        XJACI(2,1)=-XJACM(2,1)*DETJACOB
!*** CALCULATE CARTESIAN DERIVITIVES*****************************************8
        DO IDIME=1,NDIME
        DO INODE=1,NNODE
        CARTD(IDIME,INODE)=0.0
        DO JDIME=1,NDIME
        CARTD(IDIME,INODE)=CARTD(IDIME,INODE)+&
        &XJACI(IDIME,JDIME)*DERIV(JDIME,INODE)
        END DO
        END DO
        END DO
        ELSE
        NFAIL=1
        WRITE(2,'(A,I5)')' PROGRAM HALTED IN JACOBIAN BECAUSE OF ZERO OR &
        &NEGATIVE AREA IN ELEMENT NUMBER = ',IELEM
        END IF
        RETURN
        END SUBROUTINE JACOBIAN
```

12.11.1.4 Subroutine DFLEXR

This subroutine calculates the flexural part of D_f matrix of elastic rigidities for plate-bending element as given in equation (8.105).

```
!*****************************************************************
        SUBROUTINE DFLEXR(DFLEX)
!***VARIABLE DECLARATION*******************************************************
        IMPLICIT NONE
        REAL::YOUNG,POISS,THICK
        REAL:: DFLEX(3,3)
!***ZERO THE ELASTIC RIGIDITY MATRIX****************************************
        DFLEX=0.0
        YOUNG=PROPS(LPROP,1)
        POISS=PROPS(LPROP,2)
        THICK=PROPS(LPROP,3)
        DFLEX(1,1)=YOUNG*THICK*THICK*THICK/(12.0*(1.0-POISS*POISS))
        DFLEX(1,2)=POISS*DFLEX(1,1)
```

```
        DFLEX(2,2)=DFLEX(1,1)
        DFLEX(2,1)=DFLEX(1,2)
        DFLEX(3,3)=(1.0-POISS)*DFLEX(1,1)/2.0
        RETURN
        END SUBROUTINE DFLEXR
```

12.11.1.5 Subroutine BFLEXR

This subroutine calculates the flexural part of strain matrix B_f for plate-bending element as given in equation (8.113).

```
!**********************************************************************
        SUBROUTINE BFLEXR(BFLEX)
!****VARIABLE DECLARATION*******************************************
        IMPLICIT NONE
        INTEGER:: ICURVAT,J2,J3
        REAL:: BFLEX(3,NEVAB)
!***ZERO B MATRIX*********************************************************
        BFLEX=0.0
        DO  INODE= 1,NNODE
        J2 = 2+(INODE-1)*NDOFN
        J3 = J2+1
        BFLEX(1,J2)=-CARTD(1,INODE)
        BFLEX(2,J3)=-CARTD(2,INODE)
        BFLEX(3,J2)=-CARTD(2,INODE)
        BFLEX(3,J3)=-CARTD(1,INODE)
        END DO
        RETURN
        END  SUBROUTINE BFLEXR
```

12.11.1.6 Subroutine BSHEAR

This subroutine calculates the shear part of strain matrix B_s for plate-bending element as given in equation (8.115).

```
!**********************************************************************
   SUBROUTINE BSHEAR(BSHER)
!***VARIABLE DECLARATION*******************************************
        IMPLICIT NONE
        INTEGER:: ISHEAR,J1,J2,J3
        REAL:: BSHER(2,NEVAB)
!***ZERO B MATRIX*********************************************************
        BSHER=0.0
        DO  INODE= 1,NNODE
        J1 = 1 +(INODE-1)*NDOFN
        J2 = J1 + 1
        J3 = J2+1
        BSHER(1,J1)=CARTD(1,INODE)
        BSHER(1,J2) =-SHAPEFUN(INODE)
        BSHER(2,J1)=CARTD(2,INODE)
        BSHER(2,J3) = -SHAPEFUN(INODE)
        END DO
        RETURN
        END SUBROUTINE BSHEAR
```

12.11.1.7 Subroutine READMAS

This subroutine reads and stores the data about node numbers and the corresponding masses at the nodes. This subroutine is similar to subroutine **READMASFRAME** in Section 12.6.1. The only difference is that uniformly distributed mass is not read in this subroutine, because its value is treated as a 'material property'.

```
!*****************************************************************
        SUBROUTINE READMAS
        WRITE(2,'(A)')' NODE      CONC. MASS'
         DO ILODND=1,NLODND
        READ(1,*)MASPT(ILODND)
        READ(1,*)XMASND(ILODND)
        WRITE(2,'(I3,5X,E14.6)')MASPT(ILODND),XMASND(ILODND)
        END DO
        RETURN
        END SUBROUTINE READMAS
```

12.12 PROGRAM DPLSIM.F90

This program calculates in ascending order the first N natural frequencies and the corresponding eigen-vectors of elastic Mindlin plates. The solution is by simultaneous iteration method.

```
!*****************************************************************
        PROGRAM DPLSIM
!*** VARIABLE DECLARATION*********************************************
        IMPLICIT NONE
        INTEGER::NPROB,IPROB,NFAIL,NNODE,JNODE,INODE,NDOFN,IDOFN,JDOFN,&
        &NDIME,IDIME,JDIME,NEVAB,IEVAB,JEVAB,NPROP,NPOIN,IPOIN,IELEM,&
        &NELEM,NRESND,IRESND,JRESND,NOPTN,NMATS,IMATS,IPROP,LPROP,&
        &KOUNT,NLODEL,JLODEL,ILODEL,NEQNS,NBAND,NLODND,ILODND,NVECTR,&
        &MAXCOL,NGAUS,NSTRE
!********* DEFINE INTEGER ARRAYS WHICH ARE ALLOCATABLE******************
        INTEGER,ALLOCATABLE::NDFFIX(:),IFPRE(:,:),NODFRE(:,:),LNODS(:,:),&
        &MEMDIS(:),MATNO(:,:),MASEL(:),MASPT(:),NCOLM(:),NDIA(:)
!**********DEFINE REAL ARRAYS WHICH ARE ALLOCATABLE**********************
        REAL,ALLOCATABLE::COORD(:,:),PROPS(:,:),ESTIF(:,:),EMASS(:,:),XMASEL(:),&
        &XMASND(:),GSTIF(:),GMASS(:),FREQ2(:),VECTOR(:,:),ELCOD(:,:),SHAPEFUN(:),&
        &DERIV(:,:),CARTD(:,:),POSGP(:),WEIGP(:),GPCOD(:,:)
! ****DATA AND OUTPUT TRANSMISSION FROM AND TO FILES******************
        OPEN(UNIT=1,FILE=NFILEIN,STATUS='OLD',ACTION='READ')
        OPEN(UNIT=2,FILE=NFILEOUT,STATUS='REPLACE',ACTION='WRITE')
!***START READING THE NUMBER OF PROBLEMS TO BE ANALYSED***************
        READ(1,*)NPROB
        WRITE(2,'(2/,A,I8)')' TOTAL NUMBER OF PROBLEMS = ',NPROB
!***START THE CYCLE ON THE NUMBER OF PROBLEMS***************************
        PROBLEMS:DO IPROB = 1, NPROB
        READ(1,*)TITLE
        WRITE(2,'(4/,A,I3)')'PROBLEM NUMBER = ',IPROB
        WRITE(2,'(A)')TITLE
```

```
!***SET THE BASIC PARAMETERS OF THE QUADRATIC ISOPARAMETRIC ELEMENT**
        NFAIL=0
        NNODE=8
        NDOFN=3
        NDIME=2
        NEVAB=24
        NPROP=5
        NSTRE=5
        NGAUS=3
!**READ AND PRINT THE BASIC PARAMETERS OF THE PARTICULAR PROBLEM**
        READ(1,*)NPOIN,NELEM,NRESND,NMATS
        WRITE(2,'(A,I5)')'NUMBER OF NODES = ',NPOIN
        WRITE(2,'(A,I5)')'NUMBER OF ELEMENTS = ',NELEM
        WRITE(2,'(A,I5)')'NUMBER OF RESTRAINED NODES = ',NRESND
        WRITE(2,'(A,I5)')'NUMBER OF DIFFERENT TYPES OF MATERIALS = ',NMATS
!****ALLOCATE ARRAYS*********************************************************
        ALLOCATE(COORD(NPOIN,NDIME),LNODS(NELEM,NNODE),&
        &MATNO(NELEM,1),PROPS(NMATS,7),NODFRE(NPOIN,NDOFN),&
        &NDFFIX(NRESND),IFPRE(NRESND,NDOFN),ESTIF(NEVAB,NEVAB),&
        &MEMDIS(NEVAB),EMASS(NEVAB,NEVAB),ELCOD(NDIME,NNODE),&
        &SHAPEFUN(NNODE),DERIV(NDIME,NNODE),CARTD(NDIME,NNODE),&
        &POSGP(NGAUS),WEIGP(NGAUS),GPCOD(NDIME,9))
!***NOTE NFAIL=1 MEANS THAT ERROR IN DATA*****************************
!***READ THE COORDINATES OF SOME OF THE NODES, GENERATE THE*********
!****OMITTED DATA AND PRINT THE COORDINATES OF ALL THE NODES********
        CALL CORDGEN
        IF(NFAIL= =1) EXIT
        WRITE(2,'(2/,A)')' COORDINATES OF NODES'
        WRITE(2,'(2/,A)')'   NODE      X-CORD       Y-CORD'
        DO IPOIN=1,NPOIN
        WRITE(2,'(1X,I5,3(5X,F10.3))')IPOIN,(COORD(IPOIN,JDIME),JDIME=1,NDIME)
        END DO
!***READ THE NODE NUMBERS CORRESPONDING TO SOME OF THE ELEMENTS***
!***AND GENERATE THE OMITTED DATA****************************************
        CALL INTGEN(LNODS,NNODE,NELEM)
        IF(NFAIL= =1) EXIT
!***READ THE MATERIAL NUMBERS OF SOME OF THE ELEMENTS AND GENERATE**
!***THE OMITTED DATA*******************************************************
        CALL INTGEN(MATNO,1,NELEM)
        IF(NFAIL= =1) EXIT
!***PRINT THE MATERIAL NUMBERS AND THE NODE NUMBERS OF THE ELEMENTS**
        WRITE(2,'(3/,A)')' ELEMENT NO.  MATERIAL NO.  NODE-1  NODE-2 &
        &NODE-3 NODE-4 NODE-5  NODE-6 NODE-7 NODE-8'
        DO IELEM = 1, NELEM
        WRITE(2,'(1X,I5,8X,I5,8X,8(I3,2X))')IELEM,MATNO(IELEM,1),&
        &(LNODS(IELEM,JNODE),JNODE=1,NNODE)
        END DO
!***READ NOPTN. THIS TELLS WHETHER THE NODE FREEDOM ARRAY IS FED**
!***AS DATA (FOR NOPTN=1) OR TO BE GENERATED FROM THE DATA ABOUT THE**
!***NODES WHERE ONE OR MORE DISPLACEMENTS ARE ZERO(NOPTN=0)************
        READ(1,*)NOPTN
!***READ NODE FREEDOMS OF SOME OF THE NODES AND GENERATE OMITTED DATA*
        IF(NOPTN= =1) THEN
        CALL INTGEN(NODFRE,NDOFN,NPOIN)
        IF(NFAIL= =1) EXIT
```

```
        ELSE
!***READ AND ECHO DATA ABOUT THE NODE NUMBERS AND FIXITY STATE AT NODES*
!***WHERE ONE OR MORE FREEDOMS ARE EQUAL TO ZERO**************************
        DO  IRESND=1,NRESND
        READ(1,*)NDFFIX(IRESND),(IFPRE(IRESND,JDOFN),JDOFN=1,NDOFN)
        END DO
        WRITE(2,'(3/,A)')' DEGREES OF RESTRAINT AT RESTRAINED NODES'
        WRITE(2,'(3/,A)')' 1 = FIXED AND 0 = FREE'
        WRITE(2,'(3/,A)')'NODE NO.   W      SLOPE= DW/DX        SLOPE=DW/DY'
        DO  IRESND=1,NRESND
        WRITE(2,'(I5,10X,I5,8X,I5,16X,I5)')NDFFIX(IRESND),(IFPRE(IRESND,JDOFN),&
        &JDOFN=1,NDOFN)
        END DO
!***CALL THE SUBROUTINE TO GENERATE THE NODE FREEDOM ARRAY FROM***
!***THE DATA ABOUT THE RESTRAINED NODES*********************************
        CALL NFGEN
        END IF
        WRITE(2,'(2/,A)')' DEGREES OF FREEDOM AT NODES'
        WRITE(2,'(2/,A)') 'NODE NUMBER     W    SLOPE=DW/DX   SLOPE=DW/DY'
        DO IPOIN=1,NPOIN
        WRITE(2,'(I3,5X,I5,10X,I5,9X,I5)')IPOIN,(NODFRE(IPOIN,IDOFN),IDOFN=1,NDOFN)
        END DO
!***READ AND PRINT THE MATERIAL PROPERTIES OF DIFFERENT MATERIALS*******
        DO  IMATS=1,NMATS
        READ(1,*)(PROPS(IMATS,IPROP),IPROP=1,NPROP)
        END DO
        WRITE(2,'(2/,A)')' PROPERTIES OF DIFFERENT MATERIALS'
        WRITE(2,'(2/,A)')' MATERIAL NO.   YOUNGS MOD.   POISS.RATIO&
         &    THICKNESS       UDL. MASS    SUB GRADE. MOD '
        DO  IMATS=1,NMATS
        WRITE(2,'(2X,I5,8X,5(E14.6,2X))')IMATS,(PROPS(IMATS,IPROP),IPROP=1,NPROP)
        END DO
!***CALCULATE THE SIZE OF GSTIF AND GMASS MATRICES*************************
        CALL MAXEQN
! ***ALLOCATE ARRAYS**********************************************************
        ALLOCATE(NCOLM(NEQNS),NDIA(NEQNS))
!***CALCULATE THE HALF BANDWIDTH*****************************************
        CALL BANDWD
!***CALL THE SUBROUTINE TO DETERMINE THE  DETAILS OF VECTOR STORAGE **
!***OF GSTIF AND GMASS*******************************************************
        CALL NSKYLINE
        WRITE(2,'(A,I5)')'TOTAL COLUMN LENGTH =  ',MAXCOL
! ***ALLOCATE ARRAYS**********************************************************
        ALLOCATE(GSTIF(MAXCOL),GMASS(MAXCOL))
!***READ THE CONCENTRATED MASSES AT THE NODES AND STORE IN THE ****
!***ARRAY MASPT(NLODND),XMASND(NLODND)*********************************
        READ(1,*)NLODND
        WRITE(2,'(/,A,I5)')'NUMBER OF NODES WITH CONCENTRATED MASS =',NLODND
        IF (NLODND >0)  THEN
! ***ALLOCATE ARRAYS**********************************************************
        ALLOCATE(MASPT(NLODND), XMASND(NLODND))
        CALL READMAS
        END IF
!***ZERO AND THEN FORM THE STRUCTURAL STIFFNESS AND MASS MATRICES***
        GSTIF=0.0
```

```
        GMASS=0.0
        CALL DSTIFPB
!***READ THE NUMBER OF EIGEN-PAIRS REQUIRED. INCLUDE GUARD VECTORS**
        READ(1,*)NVECTR
        WRITE(2,'(/,A,I5)')'NUMBER OF EIGENVALUES REQUIRED = ',NVECTR
! ***ALLOCATE ARRAYS*********************************************************
        ALLOCATE(VECTOR(NEQNS,NVECTR),FREQ2(NVECTR))
!***CALL THE SUBROUTINE TO CARRY OUT SIMULTANEOUS ITERATION**
        CALL SIMIT
!**CALL SUBROUTINE TO DO STURM CHECK**************************
        CALL STURM
        END DO PROBLEMS
        CONTAINS
!*****************************************************************
! ****INSERT ALL SUBROUTINES USED ONE AFTER ANOTHER********
!*****************************************************************
        END PROGRAM DPLSIM
```

12.13 SUBROUTINES FOR DIRECT INTEGRATION

Five subroutines including the subroutine GAUSRED described in Section 12.9.2 are used to carry out the direct integration of the differential equations. The four subroutines are described below.

12.13.1 Subroutine MULT6(A,B,C)

This subroutine is similar to the subroutine MULT3 described in Section 12.8.1.4, except for the fact that it multiples matrix A held in vector format with a vector B and stores the product in a vector C.

```
!* ***********************************************************
        SUBROUTINE MULT6(A,B,C)
!***VARIABLE DECLARATION************************************
        IMPLICIT NONE
        INTEGER::IEQNS,KEQNS,IK,KOUNT
        REAL:: A(MAXCOL),B(NEQNS),C(NEQNS)
        DO IEQNS = 1,NEQNS
        C(IEQNS)=0.0
        DO KEQNS=1,NEQNS
!* **CONSIDER DIAGONAL AND ELEMENTS TO THE RIGHT OF DIAGONAL**
        IF(KEQNS >= IEQNS) THEN
        IK=NDIA(KEQNS) + KEQNS-IEQNS
        KOUNT = KEQNS-IEQNS +1- NCOLM(KEQNS)
        END IF
!* **CONSIDER ELEMENTS TO THE LEFT OF THE DIAGONAL*************
        IF(KEQNS < IEQNS) THEN
        IK=NDIA(IEQNS) +IEQNS -KEQNS
        KOUNT=IEQNS-KEQNS+1 - NCOLM(IEQNS)
        END IF
        IF(KOUNT <=0) C(IEQNS)=C(IEQNS) + A(IK)*B(KEQNS)
        END DO
        END DO
```

```
        RETURN
        END SUBROUTINE MULT6
```

12.13.2 Subroutine BACKS1(A,C,B)

This subroutine is similar to the subroutine BACKSUB described in Section 12.5.2 except that in the first section, it reduces the 'force' vector B by Gaussian elimination method and in the second section, the 'displacement' vector C is determined by back substitution. It is assumed that the main matrix A has already been reduced to upper triangular form by Gaussian elimination method and is stored in skyline format.

```
!*******************************************************************
    SUBROUTINE BACKS1(A,C,B)
!***VARIABLE DECLARATION*********************************************
        IMPLICIT NONE
        INTEGER::NEQNS1,KEQNS,IBAND,IROWS,KI,IVARBL,JVARBL,JCOLM1,&
        &JCOLMS,IJ
        REAL::PIVOT,QUOT,SUM
         REAL:: A(MAXCOL),B(NEQNS),C(NEQNS)
        NEQNS1 = NEQNS-1
!***REDUCE THE LOAD VECTOR*******************************************
        DO KEQNS = 1, NEQNS1
        PIVOT = 1.0/A(NDIA(KEQNS))
        DO IBAND = 1, NBAND
        IROWS = KEQNS + IBAND
        IF(IROWS > NEQNS) EXIT
        IF((IBAND +1) <= NCOLM(IROWS)) THEN
        KI = NDIA(IROWS) + IBAND
        IF(A(KI) /= 0.0 ) THEN
        QUOT = A(KI)*PIVOT
        B(IROWS) = B(IROWS) - QUOT*B(KEQNS)
        END IF
        END IF
        END DO
        END DO
 !***SOLVE FOR DISPLACEMENTS: BACK SUBSTITUTION PHASE***********
        C(NEQNS)=B(NEQNS)/A(NDIA(NEQNS))
         DO IVARBL=1,NEQNS1
        JVARBL=NEQNS-IVARBL
        SUM=0.0
        JCOLM1 = JVARBL+1
        DO JCOLMS = JCOLM1,NEQNS
        IJ=NDIA(JCOLMS) +JCOLMS -JVARBL
        IF( (JCOLMS - JVARBL +1) <= NCOLM(JCOLMS))THEN
        SUM=SUM+A(IJ)*C(JCOLMS)
        END IF
        END DO
        C(JVARBL)=(B(JVARBL)-SUM)/A(NDIA(JVARBL))
        END DO
        RETURN
        END SUBROUTINE BACKS1
```

12.13.3 Subroutine FORSCAL

The input force-time relationship is stored in three vectors. The variable NDATAPN = number of force-time data points describing the force-time input. The variable NLODPTS is the number of nodes at which force is applied. For each of the NLODPTS nodes, the force–time relationship is stored in the vectors FORCE(NDATAPN,NLODPTS) and TIME(NDATAPN,NLODPTS). The data is read in the subroutine STEP. The subroutine FORSCAL calculates the force at any time T.

```
!*****************************************************************************
        SUBROUTINE FORSCAL(FORCE,TIME,T,FORSVEC,LOADPT,NDATAPN,&
        & NLODPTS)
!***VARIABLE DECLARATION********************************************
        IMPLICIT NONE
        INTEGER::ILODPTS,NLODPTS,IFRDOM,IDATAPN,NDATAPN,I1,I2
        REAL::T1,T2,T,F1,F2
        REAL:: TIME(20,20),FORCE(20,20),FORSVEC(NEQNS),LOADPT(20)
!****CALCULATE THE FORCE AT ALL THE NODES WHERE IT IS APPLIED****
        NODE: DO ILODPTS = 1, NLODPTS
        INODE = LOADPT(ILODPTS)
!***FORCE IS ASSUMED TO ACT ONLY IN THE HORIZONTAL DIRECTION**
        IFRDOM = NODFRE(INODE,1)
!***INTERPOLATE FROM THE INPUT FORCE-TIME RELATIONSHIP********
        DATAPOINT: DO IDATAPN = 2, NDATAPN
        I1 = IDATAPN -1
        I2 = IDATAPN
        T1 = TIME(I1,ILODPTS)
        T2 = TIME(I2,ILODPTS)
        IF(T <= T2) THEN
        F1 = FORCE(I1,ILODPTS)
        F2 = FORCE(I2,ILODPTS)
        IF(T = =T1) THEN
        FORSVEC(IFRDOM) = F1
        ELSE IF(T = =T2) THEN
        FORSVEC(IFRDOM) = F2
        ELSE IF(T > T1 .AND. T < T2) THEN
        FORSVEC(IFRDOM)=F1+(F2-F1)*(T-T1)/(T2-T1)
        END IF
        END IF
        END DO DATAPOINT
        END DO NODE
        RETURN
        END SUBROUTINE FORSCAL
```

12.13.4 Subroutine ACELCAL

This subroutine is similar to the subroutine FORSCAL. However it is used to calculate at any time T, the acceleration at the base interpolated from the input acceleration at base-time relationship stored in two vectors. The variable

NPOINTS = number of sets of acceleration-time data points. For each of the NPOINTS, the data is stored in the matrix ACELBAS(NPOINTS,2), the first column contains time data and the second column the corresponding acceleration data. The data is read in the subroutine STEP. The subroutine ACELCAL calculates the acceleration at the base at any time T.

```
!******************************************************************
        SUBROUTINE ACELCAL(ACELBAS,T,FORSVEC,NPOINTS)
!***VARIABLE DECLARATION*****************************************
        IMPLICIT NONE
        INTEGER::IFRDOM,IPOINTS,NPOINTS,I1,I2
        REAL::XMASS,T1,T2,T,A1,A2,A3
        REAL:: ACELBAS(20,20),FORSVEC(NEQNS)
!***FORCE IN THE HORIZONTAL DIRECTION IS EQUAL TO THE BASE ACCELERATION **
!     TIMES THE CONCENTRATED MASS AT THE NODE*******************************
        DO IPOINTS = 2, NPOINTS
        I1 = IPOINTS-1
        I2 = IPOINTS
        T1 = ACELBAS(I1,1)
        T2 = ACELBAS(I2,1)
        IF(T < = T2) THEN
        A1 = ACELBAS(I1,2)
        A2 = ACELBAS(I2,2)
        IF(T = =T1) THEN
        A3=A1
        ELSE IF(T = = T2) THEN
        A3=A2
        ELSE IF(T > T1 .AND. T < T2)THEN
        A3 = A1 + (A2-A1)*(T-T1)/(T2-T1)
        END IF
        END IF
        END DO
!****CALCULATE THE APPLIED FORCE AS = -MASS TIMES THE ACCELERATION*
!***ONLY IN THE HORIZONTAL DIRECTION**********************************
        DO IPOIN = 1, NPOIN
        IFRDOM = NODFRE(IPOIN,1)
        IF(IFRDOM /= 0) FORSVEC(IFRDOM) =  -A3*GMASS(NDIA(IFRDOM))
        END DO
        RETURN
    END SUBROUTINE ACELCAL
```

12.13.5 Subroutine STEP

This subroutine carries out time step integration using Wilson's theta method. It calls the five subroutines described earlier.

```
!******************************************************************
        SUBROUTINE STEP
!***VARIABLE DECLARATION**********************************************
        IMPLICIT NONE
        INTEGER::IFORS,IQUAKE,NPOINTS,IPOINTS,I,NDATAPN,NLODPTS,ILODPTS,&
        &IDATAPN,IFRDOM,IEQNS
        REAL::DT,THETA,TDT,CONST,CONST2,THETA1,C0,C1,C2,C3,C4,C5,C6,C7,C8,&
        &TMAX,ALPHA,BETA,T0,T1,T
```

```
        REAL:: GBAR(MAXCOL),DAMP(MAXCOL),U0(NEQNS),V0(NEQNS),A0(NEQNS),&
        &U1(NEQNS),V1(NEQNS),A1(NEQNS),FORCE(20,20),TIME(20,20),LOADPT(20),&
        &FORSVEC(NEQNS),U0K(NEQNS),V0C(NEQNS),GREDM(MAXCOL),&
        &ACELBAS(20,20),D1(NEQNS),D2(NEQNS),FORST(NEQNS),FORST1(NEQNS),&
        D3(NEQNS),D4(NEQNS),UTDT(NEQNS)
!***READ THE VALUE OF TIME STEP AND CALCULATE ADDITIONAL PARAMETERS**
!*** OF TIME STEP.  TAKE Θ = 1.4**************************************************
        READ(1, *)DT
        WRITE(2,' (A,F10.4)' )' TIME STEP = ',DT
        THETA = 1.4
        TDT = THETA*DT
        CONST = 1.0/(TDT)
        CONST2 = CONST*CONST
        THETA1 = 1.0/THETA
        C0 = 6.0*CONST2
        C1 = 3.0*CONST
        C2 = 2.0*C1
        C3 = 0.5/CONST
        C4 = C0*THETA1
        C5 = -C2*THETA1
        C6 = 1.0 - 3.0*THETA1
        C7 = 0.5*DT
        C8 = 0.166666667*DT*DT
!***READ THE MAXIMUM TIME UP TO WHICH RESPONSE IS DESIRED*******
        READ(1, *)TMAX
        WRITE(2,'(//,(A,F10.4))' )'MAXIMUM TIME UP TO WHICH RESPONSE IS DESIRED &
        &= ',TMAX
!***ASSUME RAYLEIGH DAMPING IE. C = ALPHA × K + BETA × M*****************
        READ(1,*)ALPHA, BETA
        WRITE(2,' (/,A)' )' RAYLEIGH DAMPING ASSUMED  C = ALPHA*K + BETA*M'
        WRITE(2,'(//,2(A,F10.4,6X))' )'ALPHA = ',ALPHA,'BETA = ',BETA
!****CALCULATE THE DAMPING MATRIX****************************************
        DAMP=ALPHA*GSTIF+BETA*GMASS
!***FORM  GBAR = K + C0 × M + C1 × C*******************************************
        GBAR = GSTIF + C0*GMASS + C1*DAMP
!***REDUCE GBAR BY GAUSSIAN ELIMINATION*****************************
        CALL GAUSRED(GBAR)
!***INITIALISE DISPLACEMENT, VELOCITY, AND FORCE VECTORS*********
        U0 = 0.0
        V0 = 0.0
!***FORST = FORCE AT TIME T AND FORST1 = FORCE AT TIME T+DT.
        FORST = 0.0
        FORST1 = 0.0
!***FORSVEC = NET FORCE AT TIME T + THETA × DT********************************
        FORSVEC = 0.0
!***READ MARKERS TO IDENTIFY IF BASE ACCELERATION DUE TO EARTHQUAKE**
!*** IS PRESCRIBED, OR FORCES AT THE NODES ARE PRESCRIBED*****************
        READ(1,*)IQUAKE,IFORS
!***DEAL WITH BASE ACCELERATION FIRST*****************************************
        IF (IQUAKE = = 1) THEN
!***READ BASE ACCELERATION-TIME DATA AND STORE IT IN AN ARRAY ACELBAS**
        WRITE(2,'(//,A)')'BASE ACCELERATION DATA'
        WRITE(2,' (//,A)' )'TIME              ACCELERATION'
        READ(1, *)NPOINTS
        DO  IPOINTS = 1, NPOINTS
```

```
        READ(1,*)ACELBAS(IPOINTS,1), ACELBAS(IPOINTS,2)
        WRITE(2,' (E14.6,10X,E14.6)' )ACELBAS(IPOINTS,1),ACELBAS(IPOINTS,2)
        END DO
!***CALCULATE THE FORCE GENERATED BY BASE ACCELERATION AT TIME T = 0*
!**FORCE = - MASS CORRESPONDING TO A TRANSLATION IN LINE WITH BASE****
!***MOVEMENT FOR WHICH ACCELERATION IS GIVEN * BASE ACELERATION***
!***ONLY HORIZONTAL ACCELERATION IS CONSIDERED************************
        DO IPOIN = 1, NPOIN
        I = NODFRE(IPOIN,1)
        IF(I/=0)FORSVEC(I) = -GMASS(NDIA(I))*ACELBAS(1,2)
        END DO
!***CALCULATE THE INITIAL ACCELERATION A0**************************
! ***SINCE GMASS IS REQUIRED LATER, STORE GMASS  IN GREDM*************
        GREDM = GMASS
!***CARRY OUT GAUSSIAN REDUCTION OF GMASS ***********************
        CALL GAUSRED(GREDM)
        CALL BACKS1(GREDM,A0,FORSVEC)
        ELSE
!****DEAL WITH PRESCRIBED FORCES AT SOME NODES**********************
!* **READ THE NUMBER OF NODES WHERE LOADS ARE APPLIED AND READ ***
!   THE DATA FOR THOSE FORCES.  NOTE THAT FORCES ARE APPLIED IN THE ***
!   DIRECTION OF HORIZONTAL TRANSLATION IN A FRAME AND TRANSLATION **
!   NORMAL TO THE PLANE GRID***********************************************
        WRITE(2,'(A)')'FORCES APPLIED AT NODES'
        WRITE(2,'(A)')'NOTE: LOAD IS APPLIED IN THE DIRECTION OF A &
        &TRANSLATION ONLY'
        READ(1,*)NDATAPN,NLODPTS
        WRITE(2,'(//,2(A,I4,6X))' )'NUMBER OF DATA POINTS = ',NDATAPN,&
        &'NUMBER OF NODES AT WHICH LOAD IS APPLIED = ',NLODPTS
        WRITE(2,' (//,3(A,6X))' )'NODE NUMBER   FORCE          TIME'
        DO ILODPTS = 1, NLODPTS
        READ(1,  *)LOADPT(ILODPTS)
        WRITE(2,'(/,A,I5)') 'LOAD HISTORY AT NODE = ',LOADPT(ILODPTS)
        DO  IDATAPN = 1, NDATAPN
        READ(1,  *)FORCE(IDATAPN,ILODPTS),TIME(IDATAPN,ILODPTS)
        WRITE(2,' (E14.6, 6X,E14.6)' )FORCE(IDATAPN,ILODPTS),TIME(IDATAPN.ILODPTS)
        END DO
        END DO
!* **CALCULATE THE FORCE VALUES AT TIME ZERO******************************
        DO ILODPTS = 1, NLODPTS
        INODE = LOADPT(ILODPTS)
        IFRDOM = NODFRE(INODE,1)
        IF(IFRDOM /= 0)FORSVEC(IFRDOM) = FORCE(1,ILODPTS)
        END DO
!* **CALCULATE THE INITIAL ACCELERATION A0*********************************
!****SINCE GMASS IS REQUIRED LATER, STORE GMASS  IN GREDM*************
        GREDM = GMASS
!***CARRY OUT GAUSSIAN REDUCTION OF GMASS ***********************
         CALL GAUSRED(GREDM)
        CALL BACKS1(GREDM,A0,FORSVEC)
        END IF
!***AS WE NO LONGER NEED THE GAUSSIAN REDUCED MASS MATRIX, DEALLOCATE**
        DEALLOCATE(GREDM)
!***THIS COMPLETES ALL THE PRELIMINARIES.  WE CAN NOW ENTER TIME STEP**
!***INTEGRATION.  START TIME STEPPING******************************************
```

```
        T0 = 0.0
        WRITE(2,' (A,F10.4)' ) 'TIME = ', T0
        WRITE(2,' (A)' ) 'FREEDOM NO. DISPLACEMENT      VELOCITY&
        &        ACCELERATION'
        DO IEQNS = 1, NEQNS
        WRITE(2,'(I6,4X,3(E14.6,6X))')IEQNS,U0(IEQNS), V0(IEQNS), A0(IEQNS)
        END DO
!***START OF AN ENDLESS DO LOOP****************************************
        DO
        IF(T > TMAX) EXIT
        T = T0 + TDT
        T1 = T0 + DT
!* **CALCULATE VECTORS D1 AND D2****************************************
        D1 = C0*U0 + C2*V0 + 2.0*A0
        D2 = C1*U0 + 2.0*V0 + C3*A0
!* **MULTIPLY D3 = M × D1 AND D4 = C × D2************************************
        CALL MULT6(GMASS,D1,D3)
        CALL MULT6(DAMP,D2,D4)
        IF(IQUAKE = = 1) THEN
        CALL ACELCAL(ACELBAS,T0,FORST,NPOINTS)
        CALL ACELCAL(ACELBAS,T1,FORST1,NPOINTS)
        ELSE IF(IFORS = = 1) THEN
        CALL FORSCAL(FORCE,TIME,T0,FORST,LOADPT,NDATAPN,NLODPTS)
        CALL FORSCAL(FORCE,TIME,T1,FORST1,LOADPT,NDATAPN,NLODPTS)
        END IF
!* **CALCULATE THE NET FORCE VECTOR*************************************
        FORSVEC = (1.0 - THETA)*FORST + THETA*FORST1 + D3 + D4
!* **SOLVE FOR DISPLACEMENTS AT T0 + THETA × DT***************************
        CALL BACKS1(GBAR,UTDT,FORSVEC)
!* **CALCULATE THE VALUES OF DISPLACEMENTS ETC. AT (T + DT)***********
        WRITE(2,'(A,F10.4)') 'TIME =  ',T1
        WRITE(2,' (A)' ) 'FREEDOM NO. DISPLACEMENT      VELOCITY&
        &        ACCELERATION'
        A1 = C4*(UTDT - U0) + C5*V0 + C6*A0
        V1 = C7*(A1 + A0) + V0
        U1 = U0 + DT*V0 + C8*(A1 + 2*A0)
        DO IEQNS=1,NEQNS
        WRITE(2,'(I5,4X,3(E14.6,6X))' )IEQNS,U1(IEQNS),V1(IEQNS),A1(IEQNS)
        END DO
!***SET THE INITIAL VALUES FOR NEXT CYCLE*****************************
        U0 = U1
        V0 = V1
        A0 = A1
        T0 = T1
        END DO
        RETURN
        END SUBROUTINE STEP
```

12.13.6 PROGRAM FRASTEP.F90

This program is used for the analysis of elastic frames by direct integration by the Wilson theta method. It is very similar to the program FRASIM.F90, except that instead of the subroutine SIMIT and the associated subroutines, the subroutine

STEP and the associated subroutines is used. In the variable declaration, the variable NVECTR is not needed.

```
!*****************************************************************
        PROGRAM FRASTEP
!***VARIABLE DECLARATION***********************************************
     FROM HERE ON EXACTLY AS FOR FRASIM.F90.
        CALL SMASGR
        END IF
        TILL HERE
!*******************************************************************
!******CALL THE SUBROUTINE TO SOLVE BY TIME STEPPING*****************
        CALL STEP
        END DO PROBLEMS
        CONTAINS
!****************************************************************
! ****INSERT ALL RELEVANT SUBROUTINES USED ONE AFTER ANOTHER*
!*******************************************************************
        END PROGRAM FRASTEP
```

12.14 PROGRAM DPLSTEP.F90

This program is used for analysing thick elastic plates by direct integration. It is very similar to the program DPLSIM.F90, except that instead of the subroutine SIMIT and the associated subroutines, the subroutine STEP and the associated subroutines is used. In the variable declaration, the variable NVECTR is not needed and the changes made in adopting FRASIM to FRASTEP are the same steps that are needed in making the conversion in this case also.

12.15 FINITE STRIP METHOD: PLATE BENDING

In Chapter 9, the finite strip method was described. The procedure is very similar to the plate analysis program using finite elements, except for the fact that the finite strip rather than finite elements are used.

12.15.1 Subroutine DSTIFSTP

This subroutine calculates the element stiffness and mass matrices for Reissner-Mindlin plate-bending finite strip with single point integration. The stiffness and mass coefficients are given in Section 9.5.

```
!*********************************************************************
        SUBROUTINE DSTIFSTP
!****VARIABLE DECLARATION**************************************
        IMPLICIT NONE
        INTEGER::NODE1,NODE2,LNODE,IEVAB1,KFRDOM,IJ,K1,K2,K
        REAL::EX,EY,G,POISSX,POISSY,DX,D1,DY,DXY,GAS,ELMASS,THICK,XINERTIA,&
        &CONST,B,B2,B3,ALPHAM,ALPHAM2,CONST1,CONST2,CONST3
```

```
!*** LOOP OVER EACH ELEMENT*****************************************************
        ELEMENT: DO IELEM=1,NELEM
        LPROP=MATNO(IELEM,1)
        EX=PROPS(LPROP,1)
        POISSX=PROPS(LPROP,2)
        EY=PROPS(LPROP,3)
        POISSY=EX*POISSX/EY
        G=PROPS(LPROP,4)
        ELMASS = PROPS(LPROP,5)
        THICK = PROPS(LPROP,6)
        XINERTIA = (THICK**3)/12.0
        DX=XINERTIA*EX/(1-POISSX**2)
        DY=XINERTIA*EY/(1-POISSY**2)
        D1=DX*POISSX
        DXY= G*XINERTIA
        GAS=THICK*G/1.2
!*** EVALUATE THE COORDINATES OF THE ELEMENT NODAL POINTS******
        NODE1=LNODS(IELEM,1)
        NODE2=LNODS(IELEM,2)
        B=COORD(NODE2,1)-COORD(NODE1,1)
!*** INITIALISE THE ELEMENT STIFFNESS AND MASS MATRICES**************
        EMASS = 0.0
        ESTIF=0.0
!***CALCULATE THE ELEMENT STIFFNESS COEFFICIENTS********************
        ALPHAM= IHARM*3.1415926/SPAN
        B2=B*B
        ALPHAM2=ALPHAM**2
        CONST1=SPAN/(8*B)
        CONST2=4/(B*B)
        CONST3=SPAN*B/8
        ESTIF(1,1)=GAS*CONST1*(4+ALPHAM2*B2)
        ESTIF(1,2)=GAS*SPAN/4
        ESTIF(1,3)=-GAS*ALPHAM*CONST3
        ESTIF(1,4)=GAS*CONST1*(-4+ALPHAM2*B2)
        ESTIF(1,5)=ESTIF(1,2)
        ESTIF(1,6)=ESTIF(1,3)
        ESTIF(2,2)=CONST3*(DX*CONST2+DXY*ALPHAM2+GAS)
        ESTIF(2,3)=ALPHAM*(SPAN/4)*(D1-DXY)
        ESTIF(2,4)=-ESTIF(1,2)
        ESTIF(2,5)=CONST3*(-DX*CONST2+DXY*ALPHAM2+GAS)
        ESTIF(2,6)=ALPHAM*(SPAN/4)*(D1+DXY)
        ESTIF(3,3)=CONST3*(DX*ALPHAM2+ CONST2*DXY+GAS)
        ESTIF(3,4)=-GAS*ALPHAM*SPAN/8
        ESTIF(3,5)=-ESTIF(2,6)
        ESTIF(3,6)=CONST3*(DX*ALPHAM2- CONST2*DXY+GAS)
        ESTIF(4,4)=ESTIF(1,1)
        ESTIF(4,5)=-ESTIF(1,2)
        ESTIF(4,6)=ESTIF(1,3)
        ESTIF(5,5)=ESTIF(2,2)
        ESTIF(5,6)=-ESTIF(2,3)
        ESTIF(6,6)=ESTIF(3,3)
!***CALCULATE THE ELEMENT MASS MATRIX COEFFICIENTS***********
        CONST=ELMASS*B*SPAN/6.0
        EMASS(1,1)=CONST
        EMASS(1,4)=0.5*CONST
```

```
        EMASS(4,4)=EMASS(1,1)
!***CONSTRUCT THE LOWER TRIANGLE OF THE STIFFNESS AND MASS MATRICES*
        DO IEVAB=2,NEVAB
        IEVAB1 = IEVAB -1
        DO  JEVAB=1,IEVAB1
        EMASS(IEVAB,JEVAB) = EMASS(JEVAB,IEVAB)
        ESTIF(IEVAB,JEVAB)=ESTIF(JEVAB,IEVAB)
        END DO
        END DO
!***CALCULATE THE NODE FREEDOMS OF THE NODES OF THE ELEMENT AND ***
!***STORE IN AN ARRAY MEMDIS*******************************************
        DO  INODE = 1,NNODE
        DO  JDOFN = 1,NDOFN
        KFRDOM = (INODE-1)*NDOFN + JDOFN
        MEMDIS(KFRDOM) = NODFRE(LNODS(IELEM,INODE),JDOFN)
        END DO
        END DO
!***ADD THE ELEMENT STIFFNESS AND MASS MATRICES TO STRUCTURAL ******
!***STIFFNESS AND MASS MATRICES****************************************
        DO  IEVAB = 1,NEVAB
        K1=MEMDIS(IEVAB)
        IF( K1> 0)  THEN
        DO JEVAB = 1,NEVAB
        K2 = MEMDIS(JEVAB)
        IF (K2 > 0 .AND. K2 >= MEMDIS(IEVAB)) THEN
        IJ = NDIA(K2) + K2 - K1
        GSTIF(IJ) = GSTIF(IJ) + ESTIF(IEVAB,JEVAB)
        GMASS(IJ) = GMASS(IJ) + EMASS(IEVAB,JEVAB)
        END IF
        END DO
        END IF
        END DO
        END DO ELEMENT
!***ADD THE CONCENTRATED MASSES TO THE APPROPRIATE PLACES IN****
!***MASS MATRIX.  ONLY ACCELERATION IN THE VERTICAL DIRECTION IS***
!***CONSIDERED*********************************************************
        IF(NLODND > 0) THEN
        DO  ILODND = 1,NLODND
        K =NDIA(NODFRE(MASPT(ILODND),1))
        GMASS(K) = GMASS(K) + XMASND(ILODND)
        END DO
        END IF
        RETURN
        END   SUBROUTINE DSTIFSTP
```

12.15.2 PROGRAM STRIPSIM.F90

This program uses the finite strip method to calculate the stiffness and mass matrix and then solve the eigen-value problem by simultaneous iteration method. It follows the same steps as the program DPLSIM.F90. Apart from minor changes which are highlighted in bold, the main change is the replacement of the subroutine DSTIFPB by the subroutine DSTIFSTP.

```
!*************************************************************************************
```

```
        PROGRAM STPSIM
!*** VARIABLE DECLARATION*************************************************
        IMPLICIT NONE
        INTEGER::NPROB,IPROB,NFAIL,NNODE,JNODE,INODE,NDOFN,IDOFN,JDOFN,&
        &NDIME,IDIME,JDIME,NEVAB,IEVAB,JEVAB,NPROP,NPOIN,IPOIN,IELEM,&
        &NELEM,NRESND,IRESND,JRESND,NMATS,IMATS,IPROP,LPROP,KOUNT,&
        &NLODEL,JLODEL,ILODEL,NEQNS,NBAND,NLODND,ILODND,NVECTR,&
        &MAXCOL,NLINE,NHARM,IHARM
        REAL::SPAN
!***DEFINE INTEGER ARRAYS WHICH ARE ALLOCATABLE*********************
         INTEGER,ALLOCATABLE::NDFFIX(:),IFPRE(:,:),NODFRE(:,:),LNODS(:,:),&
        &MEMDIS(:),MATNO(:,:),MASEL(:),MASPT(:),NCOLM(:),NDIA(:)
!***DEFINE REAL ARRAYS WHICH ARE ALLOCATABLE**************************
        REAL,ALLOCATABLE::COORD(:,:),PROPS(:,:),ESTIF(:,:),EMASS(:,:),XMASEL(:),&
        &XMASND(:),GSTIF(:),GMASS(:),FREQ2(:),VECTOR(:,:)
!***DATA AND OUTPUT TRANSMISSION FROM AND TO FILES************************
        CHARACTER(LEN=40):: NFILEIN, NFILEOUT,TITLE
        PRINT*,'INPUT NAME OF DATA FILE'
        READ*,NFILEIN
        PRINT*,'INPUT NAME OF OUTPUT FILE'
        READ*,NFILEOUT
        OPEN(UNIT=1,FILE=NFILEIN,STATUS='OLD',ACTION='READ')
        OPEN(UNIT=2,FILE=NFILEOUT,STATUS='REPLACE',ACTION='WRITE')
!***START READING THE NUMBER OF PROBLEMS TO BE ANALYSED**************
        READ(1,*)NPROB
        WRITE(2,'(2/,A,I8)')' TOTAL NUMBER OF PROBLEMS = ',NPROB
!***START THE CYCLE ON THE NUMBER OF PROBLEMS*****************************
        PROBLEMS:DO  IPROB = 1, NPROB
        READ(1,*)TITLE
        WRITE(2,'(4/,A,I3)')'PROBLEM NUMBER = ',IPROB
        WRITE(2,'(A)')TITLE
!***SET THE BASIC PARAMETERS ***********************************************
        NFAIL=0
        NNODE=2
        NDOFN=3
        NDIME=1
        NEVAB=6
        NPROP=6
!**READ AND PRINT THE BASIC PARAMETERS OF THE PARTICULAR PROBLEM****
        READ(1,*)NLINE,NRESND,NMATS,NHARM
        NELEM= NLINE-1
        NPOIN=NLINE
        WRITE(2,'(A,I5)')'NUMBER OF NODAL LINES = ',NLINE
        WRITE(2,'(A,I5)')'NUMBER OF STRIPS = ',NELEM
        WRITE(2,'(A,I5)')'NUMBER OF RESTRAINED NODES = ',NRESND
        WRITE(2,'(A,I5)')'NUMBER OF DIFFERENT TYPES OF MATERIALS = ',NMATS
        WRITE(2,'(A,I5)')'NUMBER OF HARMONICS = ',NHARM
!********* ALLOCATE ARRAYS****************************************
        ALLOCATE(COORD(NPOIN,NDIME),LNODS(NELEM,NNODE),&
        &MATNO(NELEM,1),PROPS(NMATS,NPROP),NODFRE(NPOIN,NDOFN),&
        &NDFFIX(NRESND),IFPRE(NRESND,NDOFN),ESTIF(NEVAB,NEVAB),&
        &MEMDIS(NEVAB),EMASS(NEVAB,NEVAB))
!***NFAIL =1 MEANS ERROR IN DATA******************************************
!***READ THE COORDINATES OF SOME OF THE NODAL LINES, GENERATE THE***
! ***  OMITTED DATA AND PRINT THE COORDINATES OF ALL THE NODES******
```

```
        READ(1,*)SPAN
        WRITE(2,'(A,F10.4)')'SIMPLY SUPPORTED SPAN OF PLATE = ',SPAN
        CALL CORDGEN
        IF(NFAIL==1) EXIT
        WRITE(2,'(2/,A)')' COORDINATES OF NODAL LINES'
        WRITE(2,'(2/,A)')'   NODE      X-CORD'
        DO IPOIN=1,NPOIN
        WRITE(2,'(1X,I5,F10.3)')IPOIN,COORD(IPOIN,1)
        END DO
!***READ THE NODE NUMBERS CORRESPONDING TO SOME OF THE ELEMENTS**
!***AND GENERATE THE OMITTED DATA*****************************************
        CALL INTGEN(LNODS,NNODE,NELEM)
        IF(NFAIL==1) EXIT
!***READ THE MATERIAL NUMBERS OF SOME OF THE ELEMENTS AND ********
!***GENERATE THE OMITTED DATA****************************************
        CALL INTGEN(MATNO,1,NELEM)
        IF(NFAIL==1) EXIT
!***PRINT THE MATERIAL NUMBERS AND THE NODE NUMBERS OF ELEMENTS**
        WRITE(2,'(3/,A)')' ELEMENT NO.  MATERIAL NO.        NODES'
        DO IELEM = 1, NELEM
        WRITE(2,'(1X,I5,8X,I5,8X,8(I3,2X))')IELEM,MATNO(IELEM,1),&
        &(LNODS(IELEM,JNODE),JNODE=1,NNODE)
        END DO
        IF(NRESND > 0) THEN
!***READ AND ECHO DATA ABOUT THE NODE NUMBERS AND FIXITY STATE AT ***
!***NODES WHERE ONE OR MORE FREEDOMS ARE EQUAL TO ZERO****************
        DO IRESND=1,NRESND
        READ(1,*)NDFFIX(IRESND),(IFPRE(IRESND,JDOFN),JDOFN=1,NDOFN)
        END DO
        WRITE(2,'(3/,A)')' DEGREES OF RESTRAINT AT RESTRAINED NODES'
        WRITE(2,'(3/,A)')' 1 = FIXED AND 0 = FREE'
        WRITE(2,'(3/,A)')'NODE      W      PHIX      PHIY'
        DO IRESND=1,NRESND
        WRITE(2,'(I5,4X,I5,8X,I5,8X,I5)')NDFFIX(IRESND),(IFPRE(IRESND,JDOFN),&
        &JDOFN=1,NDOFN)
        END DO
        END IF
!***CALL THE SUBROUTINE TO GENERATE THE NODE FREEDOM ARRAY FROM***
!***THE DATA ABOUT THE RESTRAINED NODES**********************************
        CALL NFGEN
        WRITE(2,'(2/,A)')' DEGREES OF FREEDOM AT NODES'
        WRITE(2,'(2/,A)') 'NODE     W       PHIX       PHIY'
        DO IPOIN=1,NPOIN
        WRITE(2,'(I3,5X,I5,10X,I5,9X,I5)')IPOIN,(NODFRE(IPOIN,IDOFN),IDOFN=1,NDOFN)
        END DO
!***READ THE MATERIAL PROPERTIES OF DIFFERENT MATERIALS*****************
        DO IMATS=1,NMATS
        READ(1,*)(PROPS(IMATS,IPROP),IPROP=1,NPROP)
        END DO
        WRITE(2,'(2/,A)')' PROPERTIES OF DIFFERENT MATERIALS'
        WRITE(2,'(2/,A)')' MATERIAL NO.  EX   NUX  EY   G &
         &   UDL. MASS      THICKNESS'
        DO IMATS=1,NMATS
        WRITE(2,'(2X,I5,8X,5(E14.6,2X))')IMATS,(PROPS(IMATS,IPROP),IPROP=1,NPROP)
        END DO
```

```
        FROM HERE ON EXACTLY AS FOR PROGRAM DPLSIM.F90
        ALLOCATE(VECTOR(NEQNS,NVECTR),FREQ2(NVECTR))
        TILL HERE
!*****START CYCLING ON HARMONICS***********************************
        HARMONICS:DO IHARM=1,NHARM
!***FIRST ZERO AND THEN FORM THE STRUCTURAL STIFFNESS AND MASS MATRICES
        GSTIF=0.0
        GMASS=0.0
        CALL DSTIFSTP
        CALL SIMIT
        CALL STURM
        END DO HARMONICS
        END DO PROBLEMS
!*************************************************************
        CONTAINS
!*************************************************************
!*******PUT ALL THE NECESSARY SUBROUTINES HERE***************
!*******************************************************************
        END PROGRAM STP3SIM
```

12.16 SPLINE FINITE STRIP

The use of cubic splines as an interpolation function was described in Section 9.6. The main difference from the program DPLSIM.F90 will be the reading of data for spline strips and of course the calculation of stiffness and mass matrix. These changes are described in the following sections.

12.16.1 Subroutine BANDWDSPLN

This subroutine is similar to the subroutine BANDWD except for minor changes because of the spline strip property.

```
!*********************************************************************
        SUBROUTINE BANDWDSPLN
!***VARIBLE DECLARATION**********************************************
        IMPLICIT NONE
        INTEGER::LBAND,INODE,KEVAB,IKNOTS
        NBAND=0
!***CYCLE ON ALL THE ELEMENTS*****************************************
        DO IELEM=1,NELEM
        DO IKNOTS=1,NKNOTS
!***DETERMINE THE NODES OF THE ELEMENT*******************************
!***DETERMINE THE NODE FREEDOMS CORRESPONDING TO THE NODES OF THE***
!*** ELEMENT AND STORE IT IN ARRAY MEMDIS(NDOFN)*********************
        DO IDOFN=1,NDOFN
        KEVAB=(IKNOTS-1)*NDOFN + IDOFN
        MEMDIS(KEVAB)=0
        IF(LKNOTS(IELEM,IKNOTS) >0)MEMDIS(KEVAB)=&
        &NODFRE(LKNOTS(IELEM,IKNOTS),IDOFN)
        END DO
        END DO
```

```
!***CALCULATE THE HALF BANDWIDTH LBAND FOR THE ELEMENT**********
        DO IEVAB=1,NEVAB
        IF(MEMDIS(IEVAB) > 0) THEN
        DO  JEVAB=1,NEVAB
        IF(MEMDIS(JEVAB) > 0) THEN
        LBAND = ABS(MEMDIS(IEVAB) - MEMDIS(JEVAB))
        IF(LBAND>NBAND) NBAND=LBAND
        END IF
        END DO
        END IF
        END DO
        END DO
        RETURN
        END SUBROUTINE BANDWDSPLN
```

12.16.2 Subroutine NSKYLINESPLN

This subroutine is similar to the subroutine NSKYLINE except for minor changes because of the spline strip property.

```
!******************************************************************
   SUBROUTINE NSKYLINESPLN
!***VARIABLE DECLARATION*********************************************
        IMPLICIT NONE
        INTEGER::IEQNS,IBIG,JFRDOM,IDIF,IEQNS1, IKNOTS
        NCOLM(1) = 1
        NDIA(1) = 1
        DO  IEQNS=2,NEQNS
        IBIG=1
        DO IELEM=1,NELEM
        KOUNT=0
        DO IKNOTS=1,NKNOTS
        IF(LKNOTS(IELEM,IKNOTS) > 0) THEN
        DO  IDOFN=1,NDOFN
        JFRDOM=NODFRE(LKNOTS(IELEM,IKNOTS),IDOFN)
        IF(JFRDOM > 0 .AND. JFRDOM = = IEQNS) THEN
        KOUNT=1
        EXIT
        END IF
        END DO
        IF(JFRDOM > 0 .AND. JFRDOM = = IEQNS) THEN
        KOUNT=1
        EXIT
        END IF
        END IF
        END DO
        IF(KOUNT = =1)  THEN
        DO  IKNOTS=1,NKNOTS
        IF(LKNOTS(IELEM,IKNOTS) > 0) THEN
        DO IDOFN=1,NDOFN
        JFRDOM=NODFRE(LKNOTS(IELEM,IKNOTS),IDOFN)
        IF(JFRDOM > 0  .AND. JFRDOM  <IEQNS) THEN
        IDIF=IEQNS - JFRDOM + 1
        END IF
```

```
IF(IDIF > IBIG) IBIG=IDIF
END DO
END IF
END DO
END IF
END DO
NCOLM(IEQNS)=IBIG
END DO
DO IEQNS=2,NEQNS
IEQNS1=IEQNS-1
NDIA(IEQNS)=NDIA(IEQNS1)+NCOLM(IEQNS1)
END DO
MAXCOL=NDIA(NEQNS)+NCOLM(NEQNS)-1
RETURN
END SUBROUTINE NSKYLINESPLN
```

12.16.3 Subroutine DSPLNPB

This subroutine calculates the stiffness and mass matrices of cubic spline plate bending finite strip.

```
!*****************************************************************
        SUBROUTINE DSPLNPB
!***VARIABLE DECLARATION**************************************
        IMPLICIT NONE
        INTEGER::NSECTN1,NSECTN2,ISTRIP,IS1,IS2,ISECTN, I,J,J1,J2,I1,I2,I3,I4,&
        &KFRDOM,JNODE,INODE,IGAUS,J3,J4,K,IJ,IKNOTS,JKNOTS
        REAL::D11,D12,D22 ,D33,B,H, XI,W,B2,BH,BH2,BH3,BH4, CONST,ELMASS,&
        &CONST2, CONST3,CONST4 ,SPAN
        REAL:: F(4),DF(4),D2F(4),PHI(5,4),DPHI(5,4),D2PHI(5,4),FF(4,4),&
        &D2FD2F(4,4),DFDF(4,4),FD2F(4,4)
!***ABSCISSA AND WEIGHT FACTORS FOR FOUR POINT GAUSSIAN INTEGRATION**
        NGAUS = 4
        POSGP(1) = -0.861136312
        POSGP(2) = -0.33998104
        POSGP(3) = 0.33998104
        POSGP(4) = 0.861136312
        WEIGHP(1) = 0.347854845
        WEIGHP(2) = 0.65214516
        WEIGHP(3) = 0.65214516
        WEIGHP(4) = 0.347854845
!***NUMBER OF KNOTS IN A LINE IS EQUAL TO THE NUMBER OF SECTIONS + 1**
        NSECTN1 = NSECTN - 1
        NSECTN2 = NSECTN -2
!***CYCLE ON THE NUMBER OF STRIPS*******************************************
        STRIP: DO ISTRIP = 1, NSTRIP
!***START CALCULATING THE ELEMENT STIFFNESS MATRICES FOR THE NSECTIONS**
!***IN THE STRIP**************************************************************
        IS1 = NTYPE(ISTRIP,1)
        IS2 = NTYPE(ISTRIP,2)
        SECTION: DO ISECTN = 1,NSECTN
        IELEM = (ISTRIP -1)*NSECTN + ISECTN
!***DETERMINE THE PROPERTIES OF THE PARTICULAR ELEMENT******************
        LPROP = MATNO(IELEM,1)
```

```
        D11 = PROPS(LPROP,1)
        D12 = PROPS(LPROP,2)
        D22 = PROPS(LPROP,3)
        D33 = PROPS(LPROP,4)
        B = COORD(LNODS(IELEM,3),1) - COORD(LNODS(IELEM,4),1)
        H =  COORD(LNODS(IELEM,3),2) - COORD(LNODS(IELEM,1),2)
!***ZERO THE ELEMENT STIFFNESS AND MASS MATRICES**************
        EMASS = 0.0
        ESTIF = 0.0
!***FOR THE SECTION UNDER CONSIDERATION, CALCULATE THE NECESSARY***
!***INTEGRALS OF SPLINE FUNCTIONS AND THEIR DERIVATIVES.  ZERO THEM**
!*** TO START WITH***********************************************************
        FF = 0.0
        DFDF = 0.0
        D2FD2F = 0.0
        FD2F = 0.0
!***GAUSSIAN INTEGRATION**************************************************
        GAUSS:   DO  IGAUS = 1,NGAUS
        XI = POSGP(IGAUS)
        W = WEIGHP(IGAUS)
        CALL SPLINE(PHI,DPHI,D2PHI,XI,IS1,IS2)
        F(4) = PHI(3,4)
        F(3) = PHI(3,3)
        F(2) = PHI(3,2)
        F(1) = PHI(3,1)
        DF(4) = DPHI(3,4)
        DF(3) = DPHI(3,3)
        DF(2) = DPHI(3,2)
        DF(1) = DPHI(3,1)
        D2F(4) = D2PHI(3,4)
        D2F(3) = D2PHI(3,3)
        D2F(2) = D2PHI(3,2)
        D2F(1) = D2PHI(3,1)
!*** CONSIDER THE FIRST SECTION******************************************
        IF(ISECTN = = 1) THEN
        F(4) = 0.0
        F(3) = PHI(1,3)
        F(2) = PHI(2,2)
        F(1) = PHI(3,1)
        DF(4) = 0.0
        DF(3) = DPHI(1,3)
        DF(2) = DPHI(2,2)
        DF(1) = DPHI(3,1)
        D2F(4) = 0.0
        D2F(3) = D2PHI(1,3)
        D2F(2) = D2PHI(2,2)
        D2F(1) = D2PHI(3,1)
        END IF
!***NEXT CONSIDER THE SECOND SECTION**********************
        IF(ISECTN = = 2 ) THEN
        IF(NSECTN = = 4) THEN
        F(1)= PHI(4,1)
        F(2)= PHI(3,2)
        F(3)= PHI(2,3)
        F(4)= PHI(1,4)
```

```
            DF(1)= DPHI(4,1)
            DF(2)= DPHI(3,2)
            DF(3)= DPHI(2,3)
            DF(4)= DPHI(1,4)
            D2F(1)= D2PHI(4,1)
            D2F(2)= D2PHI(3,2)
            D2F(3)= D2PHI(2,3)
            D2F(4)= D2PHI(1,4)
            ELSE  IF(NSECTN > 4) THEN
            F(4) = PHI(1,4)
            F(3) = PHI(2,3)
            F(2) = PHI(3,2)
            F(1) = PHI(3,1)
            DF(4) = DPHI(1,4)
            DF(3) = DPHI(2,3)
            DF(2) = DPHI(3,2)
            DF(1) = DPHI(3,1)
            D2F(4) = D2PHI(1,4)
            D2F(3) = D2PHI(2,3)
            D2F(2) = D2PHI(3,2)
            D2F(1) = D2PHI(3,1)
            END IF
            END IF
!***CONSIDER THE THIRD SECTION*****************************
            IF(ISECTN = = 3) THEN
            IF(NSECTN = = 4)THEN
            F(1)=PHI(5,1)
            F(2)=PHI(4,2)
            F(3)=PHI(3,3)
            F(4)=PHI(2,4)
            DF(1) = DPHI(5,1)
            DF(2) = DPHI(4,2)
            DF(3) = DPHI(3,3)
            DF(4) = DPHI(2,4)
            D2F(1) = D2PHI(5,1)
            D2F(2) = D2PHI(4,2)
            D2F(3) = D2PHI(3,3)
            D2F(4) = D2PHI(2,4)
            ELSE IF(NSECTN = = 5) THEN
            F(1)=PHI(4,1)
             F(2)=PHI(3,2)
             F(3)=PHI(3,3)
             F(4)=PHI(2,4)
            DF(1) = DPHI(4,1)
            DF(2) = DPHI(3,2)
            DF(3) = DPHI(3,3)
             DF(4) = DPHI(2,4)
            D2F(1) = D2PHI(4,1)
            D2F(2) = D2PHI(3,2)
            D2F(3) = D2PHI(3,3)
            D2F(4) = D2PHI(2,4)
            ELSE IF(NSECTN > 5) THEN
            F(4) = PHI(2,4)
             F(3) = PHI(3,3)
            F(2) = PHI(3,2)
```

```
          F(1) = PHI(3,1)
          DF(4) = DPHI(2,4)
          DF(3) = DPHI(3,3)
          DF(2) = DPHI(3,2)
          DF(1) = DPHI(3,1)
          D2F(4) = D2PHI(2,4)
          D2F(3) = D2PHI(3,3)
          D2F(2) = D2PHI(3,2)
          D2F(1) = D2PHI(3,1)
           END IF
           END IF
!***CONSIDER THE LAST BUT TWO SECTION********************
          IF(ISECTN = = NSECTN2  .AND. NSECTN > 5) THEN
          F(4) = PHI(3,4)
          F(3) = PHI(3,3)
          F(2) = PHI(3,2)
          F(1) = PHI(4,1)
          DF(4) = DPHI(3,4)
          DF(3) = DPHI(3,3)
          DF(2) = DPHI(3,2)
          DF(1) = DPHI(4,1)
          D2F(4) = D2PHI(3,4)
          D2F(3) = D2PHI(3,3)
          D2F(2) = D2PHI(3,2)
          D2F(1) = D2PHI(4,1)
           END IF
!***CONSIDER THE LAST BUT ONE SECTION*************************
          IF(ISECTN = = NSECTN1  .AND. NSECTN  > 4) THEN
          F(4) = PHI(3,4)
           F(3) = PHI(3,3)
          F(2) = PHI(4,2)
          F(1) = PHI(5,1)
          DF(4) = DPHI(3,4)
          DF(3) = DPHI(3,3)
          DF(2) = DPHI(4,2)
          DF(2) = DPHI(5,1)
          D2F(4) = D2PHI(3,4)
          D2F(3) = D2PHI(3,3)
          D2F(2) = D2PHI(4,2)
          D2F(1) = D2PHI(5,1)
           END IF
!***CONSIDER THE LAST SECTION**********************************
          IF(ISECTN = = NSECTN) THEN
          F(4) = PHI(3,4)
          F(3) = PHI(4,3)
          F(2) = PHI(5,2)
          F(1) = 0.0
          DF(4) = DPHI(3,4)
          DF(3) = DPHI(4,3)
          DF(2) = DPHI(5,2)
          DF(1) = 0.0
          D2F(4) = D2PHI(3,4)
          D2F(3) = D2PHI(4,3)
          D2F(2) = D2PHI(5,2)
          D2F(1) = 0.0
```

```
          END IF
!***CALCULATE THE INTEGRALS ASSOCIATED WITH THE SPLINES************
!***CORRESPONDING TO THAT SECTION. FOUR POINT INTEGRATION RULE IS**
!***USED BECAUSE IN THE PRODUCT OF SPLINE FUNCTIONS THE HIGHEST ***
!***DEGREE OF POLYNOMIAL IS SIX*********************************************
          DO I= 1,4
          I1=5-I
          DO J = 1,4
          J1=5-J
          FF(I,J) = FF(I,J) + F(I1)*F(J1)*W*0.5*H
          DFDF(I,J) = DFDF(I,J) + DF(I1)*DF(J1)*W*0.5*H
          D2FD2F(I,J) = D2FD2F(I,J) + D2F(I1)*D2F(J1)*W*0.5*H
          FD2F(I,J) = FD2F(I,J) + F(I1)*D2F(J1)*W*0.5*H
          END DO
          END DO
          END DO   GAUSS
!***************************************************************************
!***THIS COMPLETES THE NUMERICAL INTEGRATION TO CALCULATE THE***
!***INTEGRALS ASSOCAITED WITH SPLINE FUNCTIONS. INTEGRALS ASSOCIATED**
!*** WITH THE 'BEAM FUNCTIONS' IN THE X-DIRECTION HAVE BEEN CALCULATED **
!***BY ANALYTICAL INTEGRATION************************************************
!***CALCULATE THE ELEMENT STIFFNESS MATRIX FOR THE SECTION**************
          B2 = B*B
          BH = B/H
          BH2 = BH*BH
          BH3 = BH*BH2
          BH4 = BH2*BH2
          CONST = 1.0/(420.0*B2*B)
          D11 = D11*CONST
          D12 = D12*CONST
          D22 = D22*CONST
          D33 = D33*CONST
          DO I = 1,4
          I1 = (I-1)*4 + 1
          I2 = I1+1
          I3 = I2+1
          I4 = I3+1
          DO J = 1,4
          J1 = (J-1)*4 + 1
          J2 = J1 + 1
          J3 = J2 + 1
          J4 = J3 + 1
          ESTIF(I1,J1) = D11*5040.0*FF(I,J) - D12*504.0*BH2*(FD2F(I,J) +&
          &FD2F(J,I)) + D22*156.0*BH4*D2FD2F(I,J) +&
          & D33*2016.0*BH2*DFDF(I,J)
          ESTIF(I1,J2) = B*(D11*2520*FF(I,J) - D12*42.0*BH2*(FD2F(I,J)&
          &+ 11.0*FD2F(J,I)) + D22*22.0*BH4*D2FD2F(I,J) +&
          & D33*168.0*BH2*DFDF(I,J))
          ESTIF(I1,J3) = -D11*5040.0*FF(I,J) + D12*504.0*BH2*(FD2F(I,J) +&
          &FD2F(J,I)) + D22*54.0*BH4*D2FD2F(I,J) -&
          & D33*2016.0*BH2*DFDF(I,J)
          ESTIF(I1,J4) = B*(D11*2520*FF(I,J) - D12*42.0*BH2*(FD2F(I,J)&
          &+ FD2F(J,I))- D22*13.0*BH4*D2FD2F(I,J) +&
          & D33*168.0*BH2*DFDF(I,J))
          ESTIF(I2,J1) = B*(D11*2520.0*FF(I,J) - D12*42.0*BH2*(11.0*&
```

```
        &FD2F(I,J)+ FD2F(J,I)) + D22*22.0*BH4*D2FD2F(I,J) +&
        & D33*168.0*BH2*DFDF(I,J))
        ESTIF(I2,J2) = B2*(D11*1680*FF(I,J)- D12*56.0*BH2*(FD2F(I,J)&
        &+ FD2F(J,I)) + D22*4.0*BH4*D2FD2F(I,J) +&
        & D33*224.0*BH2*DFDF(I,J))
        ESTIF(I2,J3) = B*(-D11*2520.0*FF(I,J) + D12*42.0*BH2*(FD2F(I,J)&
        &+ FD2F(J,I)) + D22*13.0*BH4*D2FD2F(I,J) -&
        & D33*168.0*BH2*DFDF(I,J))
        ESTIF(I2,J4) = B2*(D11*840*FF(I,J) + D12*14.0*BH2*(FD2F(I,J)&
        &+ FD2F(J,I))- D22*3.0*BH4*D2FD2F(I,J) - D33*56.0*BH2*DFDF(I,J))
        ESTIF(I3,J1) = -D11*5040.0*FF(I,J) + D12*504.0*BH2*(FD2F(I,J) +&
        &FD2F(J,I)) + D22*54.0*BH4*D2FD2F(I,J) -&
        & D33*2016.0*BH2*DFDF(I,J)
        ESTIF(I3,J2) = B*(-D11*2520*FF(I,J) + D12*42.0*BH2*(FD2F(I,J)&
        &+ FD2F(J,I)) + D22*13.0*BH4*D2FD2F(I,J) -&
        & D33*168.0*BH2*DFDF(I,J))
        ESTIF(I3,J3) = D11*5040.0*FF(I,J) - D12*504.0*BH2*(FD2F(I,J) +&
        &FD2F(J,I)) + D22*156.0*BH4*D2FD2F(I,J) +&
        & D33*2016.0*BH2*DFDF(I,J)
        ESTIF(I3,J4) = B*(-D11*2520*FF(I,J) + D12*42.0*BH2*(&
        &FD2F(I,J) + 11.0*FD2F(J,I))- D22*22.0*BH4*D2FD2F(I,J) -&
        & D33*168.0*BH2*DFDF(I,J))
        ESTIF(I4,J1) = B*(D11*2520.0*FF(I,J) - D12*42.0*BH2*(FD2F(I,J)+&
        &FD2F(J,I)) - D22*13.0*BH4*D2FD2F(I,J) +&
        & D33*168.0*BH2*DFDF(I,J))
        ESTIF(I4,J2) = B2*(D11*840*FF(I,J)+ D12*14.0*BH2*(FD2F(I,J)&
        &+ FD2F(J,I)) - D22*3.0*BH4*D2FD2F(I,J) -&
        & D33*56.0*BH2*DFDF(I,J))
        ESTIF(I4,J3) = B*(-D11*2520.0*FF(I,J) + D12*42.0*BH2*(11.0*&
        &FD2F(I,J) + FD2F(J,I)) - D22*22.0*BH4*D2FD2F(I,J) -&
        & D33*168.0*BH2*DFDF(I,J))
        ESTIF(I4,J4) = B2*(D11*1680*FF(I,J) - D12*56.0*BH2*(FD2F(I,J)&
        &+ FD2F(J,I)) + D22*4.0*BH4*D2FD2F(I,J) +&
        & D33*224.0*BH2*DFDF(I,J))
!***CALCULATE ELEMENT MASS MATRIX*********************************
        ELMASS = PROPS(LPROP,5)
        CONST2 = ELMASS*B/420.0*FF(I,J)
        CONST3 = CONST2*B
        CONST4 = CONST3*B
        EMASS(I1,J1) = CONST2*156.0
        EMASS(I1,J2) = CONST3*22.0
        EMASS(I1,J3) = CONST2*54.0
        EMASS(I1,J4) = -CONST3*13.0
        EMASS(I2,J1) =  CONST3*22.0
        EMASS(I2,J2) = CONST4*4
        EMASS(I2,J3) = CONST3*13.0
        EMASS(I2,J4) = -CONST4*3.0
        EMASS(I3,J1) = CONST2*54.0
        EMASS(I3,J2) = CONST3*13.0
        EMASS(I3,J3) = CONST2*156.0
        EMASS(I3,J4) = -CONST3*22.0
        EMASS(I4,J1) = -CONST3*13.0
        EMASS(I4,J2) = -CONST4*3.0
        EMASS(I4,J3) = -CONST3*22.0
        EMASS(I4,J4) =  CONST4*4.0
```

```
        END DO
        END DO
!***CALCULATE THE NODE FREEDOMS OF THE NODES OF THE ELEMENT AND ***
!***STORE IN AN ARRAY MEMDIS. NOTE THAT THE ACTUAL AREA OF ****
!***INTEGRATION EXTENDS IN THE Y-DIRECTION OVER ONLY TWO KNOTS, ***
!***THE INTEGRATION INVOLVES FOUR SPLINES FOR EACH OF THE VARIABLES**
        DO IKNOTS = 1,NKNOTS
        JKNOTS = LKNOTS(IELEM,IKNOTS)
        DO JDOFN = 1, NDOFN
        KFRDOM = (IKNOTS-1)*NDOFN + JDOFN
        MEMDIS(KFRDOM) = 0
        IF(JKNOTS > 0) MEMDIS(KFRDOM) = NODFRE(LKNOTS(IELEM,IKNOTS),JDOFN)
        END DO
        END DO
!***ADD THE ELEMENT STIFFNESS AND MASS MATRICES TO STRUCTURAL STIFFNESS**
!***AND MASS MATRICES****************************************************
        DO IEVAB = 1,NEVAB
        IF(MEMDIS(IEVAB) > 0) THEN
        DO JEVAB = 1,NEVAB
        K = MEMDIS(JEVAB)
        IF (K > 0 .AND. K >= MEMDIS(IEVAB)) THEN
        IJ = NDIA(MEMDIS(JEVAB)) + MEMDIS(JEVAB) - MEMDIS(IEVAB)
        GSTIF(IJ) = GSTIF(IJ) + ESTIF(IEVAB,JEVAB)
        GMASS(IJ) = GMASS(IJ) + EMASS(IEVAB,JEVAB)
        END IF
        END DO
        END IF
        END DO
!***THIS COMPLETES ALL CALCULATIONS ON ALL SECTIONS IN A STRIP****
        END DO SECTION
!***THIS COMPLETES ALL CALCULATIONS ON ALL STRIPS*******************
        END DO STRIP
!***ADD THE LINE MASSES TO THE APPROPRIATE PLACES IN MASS MATRIX***
!*** ONLY ACCELERATION IN THE VERTICAL DIRECTION IS CONSIDERED*****
        IF(NLODND > 0) THEN
        DO ILODND = 1,NLODND
        K =NDIA(NODFRE(MASPT(ILODND),1))
        GMASS(K) = GMASS(K) + XMASND(ILODND)*0.5*SPAN
        END DO
        END IF
        RETURN
        END SUBROUTINE DSPLNPB
```

12.16.4 Subroutine SPLINE

This subroutine calculates the value of the spline functions and their derivatives. As shown in Fig. 9.5, in each element four splines contribute to the displacement. Therefore we need to calculate the value of the spline and its derivatives at the selected Gauss points. A further complication arises from that fact as discussed in Section 9.6.3, depending on the boundary conditions, the two end splines are modified. The subroutine SPLINE takes account of all these factors.

```
!****************************************************************
        SUBROUTINE SPLINE(PHI,DPHI,D2PHI,XI,IS1,IS2)
!***VARIABLE DECLARATION*************************************
        IMPLICIT NONE
        REAL:: F(4),DF(4),D2F(4),PHI(5,4),DPHI(5,4),D2PHI(5,4)
        REAL::Y,Y2,Y3,XI
        INTEGER::IS1,IS2
        Y = 0.5*(XI + 1.0)
        Y2 = Y*Y
        Y3 = Y2*Y
!***DETERMINE THE BASIC SPLINE FUNCTION AND THEIR DERIVATIVES*****
!***FOR THE FOUR SEGMENTS*********************************************
        F(1) = Y3/6.0
        F(2) = (1.0 + 3.0*Y + 3.0*Y2 - 3.0*Y3)/6.0
        F(3) = (4.0 - 6.0*Y2 + 3.0*Y3)/6.0
        F(4) = (1.0 - 3.0*Y + 3.0*Y2 - Y3)/6.0
        DF(1) = 0.5*Y2
        DF(2) = 0.5*(1.0 + 2.0*Y - 3.0*Y2)
        DF(3) = 0.5*(-4.0*Y + 3.0*Y2)
        DF(4) = 0.5*(-1.0 + 2.0*Y - Y2)
        D2F(1) = Y
        D2F(2) = 1.0 - 3.0*Y
        D2F(3) = -2.0 + 3.0*Y
        D2F(4) = 1.0 - Y
!***NOTE THAT PHI(3,1) TO PHI(3,4) STORE FOR THE FOUR SEGMENTS, THE ***
!***VALUES OF BASIC SPLINE**************************************************
        PHI(3,1) = F(1)
        PHI(3,2) = F(2)
        PHI(3,3) = F(3)
        PHI(3,4) = F(4)
!***CORRESPONDING FIRST DERIVATIVES*************************************
        DPHI(3,1) = DF(1)
        DPHI(3,2) = DF(2)
        DPHI(3,3) = DF(3)
        DPHI(3,4) = DF(4)
!***CORRESPONDING SECOND DERIVATIVES**********************************
        D2PHI(3,1) = D2F(1)
        D2PHI(3,2) = D2F(2)
        D2PHI(3,3) = D2F(3)
        D2PHI(3,4) = D2F(4)
!***CALCULATE THE DIFFERENT TYPES OF SPLINES REQUIRED FOR *****
!***DIFFERENT TYPES OF  END CONDITIONS*****************************
!***FIRST OF ALL CONSIDER THE  FIRST END****************************
        IF(IS1 = = 1 ) THEN
! ***WE ARE CONSIDERING A SIMPLY SUPPORTED FIRST END****************
! ***PHI(1,1) TO PHI(1,4) STORE FOR THE FOUR SEGMENTS THE VALUES******
! ***OF MODIFIED VALUES OF FIRST SPLINE AT THE SIMPLY SUPPORTED END**
!***NOTE THAT PHIBAR1 = PHI1 - 4 × PHI0***************************************
        PHI(1,1) = F(1) - 4.0*F(2)
        PHI(1,2) = F(2) - 4.0*F(3)
        PHI(1,3) = F(3) - 4.0*F(4)
        PHI(1,4) = F(4)
!***CORRESPONDING FIRST DERIVATIVES*********************************
        DPHI(1,1) = DF(1) - 4.0*DF(2)
```

```
        DPHI(1,2) = DF(2) - 4.0*DF(3)
        DPHI(1,3) = DF(3) - 4.0*DF(4)
        DPHI(1,4) = DF(4)
!***CORRESPONDING SECOND DERIVATIVES****************************
        D2PHI(1,1) = D2F(1) - 4.0*D2F(2)
        D2PHI(1,2) = D2F(2) - 4.0*D2F(3)
        D2PHI(1,3) = D2F(3) - 4.0*D2F(4)
        D2PHI(1,4) = D2F(4)
!***PHI(2,1) TO PHI(2,4) STORE FOR THE FOUR SEGMENTS THE VALUES**********
!***OF MODIFIED VALUES OF SECOND SPLINE AT THE SIMPLY SUPPORTED END**
!***NOTE THAT PHIBAR2 = PHI2 - PHI0**********************************************
        PHI(2,1) = F(1) - F(3)
        PHI(2,2) = F(2) - F(4)
        PHI(2,3) = F(3)
        PHI(2,4) = F(4)
!***CORRESPONDING FIRST DERIVATIVES************************
        DPHI(2,1) = DF(1) - DF(3)
        DPHI(2,2) = DF(2) - DF(4)
        DPHI(2,3) = DF(3)
        DPHI(2,4) = DF(4)
!***CORRESPONDING SECOND DERIVATIVES********************
        D2PHI(2,1) = D2F(1) - D2F(3)
        D2PHI(2,2) = D2F(2) - D2F(4)
        D2PHI(2,3) = D2F(3)
        D2PHI(2,4) = D2F(4)
!*****************************************************************
        ELSE IF(IS1 = = 2) THEN
!***WE ARE CONSIDERING A CLAMPED NEAR END*********************
!***PHI(1,1) TO PHI(1,4) STORE FOR THE FOUR SEGMENTS THE VALUES***
!***OF MODIFIED FIRST SPLINE AT THE CLAMPED NEAR END.************
!***FIRST SPLINE IS ELIMINATED****************************************
        PHI(1,1) = 0.0
        PHI(1,2) = 0.0
        PHI(1,3) = 0.0
        PHI(1,4) = 0.0
!***CORRESPONDING FIRST DERIVATIVES*******************************
        DPHI(1,1) = 0.0
        DPHI(1,2) = 0.0
        DPHI(1,3) = 0.0
        DPHI(1,4) = 0.0
!***CORRESPONDING SECOND DERIVATIVES****************************
        D2PHI(1,1) = 0.0
        D2PHI(1,2) = 0.0
        D2PHI(1,3) = 0.0
        D2PHI(1,4) = 0.0
!*********************************************************************
!***PHI(2,1) TO PHI(2,4) STORE FOR THE FOUR SEGMENTS THE VALUES*****
!***OF MODIFIED SECOND SPLINE AT THE CLAMPED NEAR END.************
!***PHI2BAR = PHI0-0.5 × PHI1+PHI2*****************************************
        PHI(2,1) = F(1) - 0.5*F(2) + F(3)
        PHI(2,2) = F(2) - 0.5*F(3) + F(4)
        PHI(2,3) = F(3) - 0.5*F(4)
        PHI(2,4) = F(4)
!***CORRESPONDING FIRST DERIVATIVES**********************************
        DPHI(2,1) = DF(1) - 0.5*DF(2) + DF(3)
```

```
        DPHI(2,2) = DF(2) - 0.5*DF(3) + DF(4)
        DPHI(2,3) = DF(3) - 0.5*DF(4)
        DPHI(2,4) = DF(4)
!***CORRESPONDING SECOND DERIVATIVES*********************************
        D2PHI(2,1) = D2F(1) - 0.5*D2F(2) + D2F(3)
        D2PHI(2,2) = D2F(2) - 0.5*D2F(3) + D2F(4)
        D2PHI(2,3) = D2F(3) - 0.5*D2F(4)
        D2PHI(2,4) = D2F(4)
        END IF
!***THIS COMPLETES THE CALCULATION OF MODIFIED SPLINES AT THE ******
!***FIRST END, WHETHER SIMPLY SUPPORTED OR CLAMPED*****************
!***NOW CONSIDER THE FAR END********************************************
        IF (IS2 = =1) THEN
!***WE ARE CONSIDERING A SIMPLY SUPPORTED END*******************
!***PHI(4,1) TO PHI(4,4) STORE FOR THE FOUR SEGMENTS THE VALUES****
!***OF MODIFIED MTH SPLINE AT THE SIMPLY SUPPORTED END***********
!***PHIBAR M = PHI M - PHI M+2*******************************************
        PHI(4,1) = F(1)
        PHI(4,2) = F(2)
        PHI(4,3) = F(3) - F(1)
        PHI(4,4) = F(4) - F(2)
!***CORRESPONDING FIRST DERIVATIVES**********************************
        DPHI(4,1) = DF(1)
        DPHI(4,2) = DF(2)
        DPHI(4,3) = DF(3) - DF(1)
        DPHI(4,4) = DF(4) - DF(2)
!***CORRESPONDING SECOND DERIVATIVES*******************************
        D2PHI(4,1) = D2F(1)
         D2PHI(4,2) = D2F(2)
        D2PHI(4,3) = D2F(3) - D2F(1)
        D2PHI(4,4) = D2F(4) - D2F(2)
!***PHI(5,1) TO PHI(5,4) STORE FOR THE FOUR SEGMENTS THE VALUES****
!***OF MODIFIED (M + 1)TH SPLINE AT THE SIMPLY SUPPORTED END******
!***PHIBARM+1 = PHIM+1 - 4 PHIM+2******************************************
        PHI(5,1) = F(1)
        PHI(5,2) = F(2) - 4.0*F(1)
        PHI(5,3) = F(3) - 4.0*F(2)
        PHI(5,4) = F(4) - 4.0*F(3)
!***CORRESPONDING FIRST DERIVATIVES**********************************
        DPHI(5,1) = DF(1)
        DPHI(5,2) = DF(2) - 4.0*DF(1)
        DPHI(5,3) = DF(3) - 4.0*DF(2)
        DPHI(5,4) = DF(4) - 4.0*DF(3)
!***CORRESPONDING SECOND DERIVATIVES*****************************
        D2PHI(5,1) = D2F(1)
        D2PHI(5,2) = D2F(2) - 4.0*D2F(1)
        D2PHI(5,3) = D2F(3) - 4.0*D2F(2)
        D2PHI(5,4) = D2F(4) - 4.0*D2F(3)
!***WE ARE CONSIDERING THE FIXED END*******************************
        ELSE IF(IS2= =2) THEN
!***PHI(4,1) TO PHI(4,4) STORE FOR THE FOUR SEGMENTS THE VALUES****
!***OF MODIFIED MTH SPLINE AT THE CLAMPED END*********************
!***PHIBARM = PHIM - 0.5 × PHI(M+1) + PHI(M+2)********************************
        PHI(4,1) = F(1)
        PHI(4,2) = F(2) - 0.5*F(1)
```

```
        PHI(4,3) = F(3) - 0.5*F(2) + F(1)
        PHI(4,4) = F(4) - 0.5*F(3) + F(2)
!***CORRESPONDING FIRST DERIVATIVES**********************************
        DPHI(4,1) = DF(1)
        DPHI(4,2) = DF(2) - 0.5*DF(1)
        DPHI(4,3) = DF(3) - 0.5*DF(2) + DF(1)
        DPHI(4,4) = DF(4) - 0.5*DF(3) + DF(2)
!***CORRESPONDING SECOND DERIVATIVES*****************************
        D2PHI(4,1) = D2F(1)
        D2PHI(4,2) = D2F(2) - 0.5*D2F(1)
        D2PHI(4,3) = D2F(3) - 0.5*D2F(2) + D2F(1)
        D2PHI(4,4) = D2F(4) - 0.5*D2F(3) + D2F(2)
!***PHI(5,1) TO PHI(5,4) STORE FOR THE FOUR SEGMENTS THE VALUES*****
!***OF MODIFIED (M + 1)TH SPLINE AT THE CLAMPED END************
!***(M+1)TH SPLINE IS ELIMINATED**********************************
        PHI(5,1) = 0.0
        PHI(5,2) = 0.0
        PHI(5,3) = 0.0
        PHI(5,4) = 0.0
!***CORRESPONDING FIRST DERIVATIVES*****************************
        DPHI(5,1) = 0.0
        DPHI(5,2) = 0.0
        DPHI(5,3) = 0.0
        DPHI(5,4) = 0.0
!***CORRESPONDING SECOND DERIVATIVES**************************
        D2PHI(5,1) = 0.0
        D2PHI(5,2) = 0.0
        D2PHI(5,3) = 0.0
        D2PHI(5,4) = 0.0
        END IF
        RETURN
        END SUBROUTINE SPLINE
```

12.17 PROGRAM SPLNSIM.F90

This program is similar to program DPLSIM.F90. The only difference is that the calculations use cubic spline strips. The main differences are highlighted in bold. In the y direction, for an element contributions come from in all eight splines, four splines from each edge parallel to the y axis and centred on four nodes. These eight nodes are called KNOTS. For an element, one whose edges lie on a supported edge, the splines are assumed to be centred on nodes whose values are zero, because the subroutine SPLINE uses the modified splines. The other variables used are

NSTRIP = number of strips in the y direction.

NSECTN= this is the number of elements to which the strip is divided.

```
*******************************************************************
        PROGRAM SPLNSIM
!***VARIABLE DECLARATION*******************************************
        IMPLICIT NONE
        INTEGER::NPROB,IPROB,NFAIL,NNODE,JNODE,INODE,NDOFN,IDOFN,JDOFN,&
        &NDIME,IDIME,JDIME,NEVAB,IEVAB,JEVAB,NPROP,NPOIN,IPOIN,IELEM,&
        &NELEM,NRESND,IRESND,JRESND,NOPTN,NMATS,IMATS,IPROP,LPROP,&
```

```
       &KOUNT,NLODEL,JLODEL,ILODEL,NEQNS,NBAND,NLODND,ILODND,NVECTR,&
       &MAXCOL,NGAUS,NSTRE,NSTRIP,NSECTN,ISTRIP,NKNOTS,JKNOTS
!***DEFINE INTEGER ARRAYS WHICH ARE ALLOCATABLE***************************
       INTEGER,ALLOCATABLE::NDFFIX(:),IFPRE(:,:),NODFRE(:,:),LNODS(:,:),&
       &MEMDIS(:),MATNO(:,:),MASEL(:),MASPT(:),NCOLM(:),NDIA(:),NTYPE(:,:),&
       &LKNOTS(:,:)
!***DEFINE REAL ARRAYS WHICH ARE ALLOCATABLE*****************************
       REAL,ALLOCATABLE::COORD(:,:),PROPS(:,:),ESTIF(:,:),EMASS(:,:),XMASEL(:),&
       &XMASND(:),GSTIF(:),GMASS(:),FREQ2(:),VECTOR(:,:),ELCOD(:,:),SHAPEFUN(:),&
       &DERIV(:,:),CARTD(:,:),POSGP(:),WEIGHP(:),GPCOD(:,:)
!***DATA AND OUTPUT TRANSMISSION FROM AND TO FILES************************
       CHARACTER(LEN=70):: NFILEIN, NFILEOUT,TITLE
       PRINT*,'INPUT NAME OF DATA FILE'
       READ*,NFILEIN
       PRINT*,'INPUT NAME OF OUTPUT FILE'
       READ*,NFILEOUT
       OPEN(UNIT=1,FILE=NFILEIN,STATUS='OLD',ACTION='READ')
       OPEN(UNIT=2,FILE=NFILEOUT,STATUS='REPLACE',ACTION='WRITE')
!***START READING THE NUMBER OF PROBLEMS TO BE ANALYSED************
       READ(1,*)NPROB
       WRITE(2,'(2/,A,I8)')' TOTAL NUMBER OF PROBLEMS = ',NPROB
!***START THE CYCLE ON THE NUMBER OF PROBLEMS*************************
       PROBLEMS:DO  IPROB = 1, NPROB
       READ(1,*)TITLE
       WRITE(2,'(//,A,I3)')'PROBLEM NUMBER = ',IPROB
       WRITE(2,'(A)')TITLE
!***SET THE BASIC PARAMETERS OF THE SPLINE STRUCTURE****************
!***NEVAB = 16 BECAUSE IN EACH ELEMENT CONTRIBUTIONS FROM EIGHT****
!*** NODES AT EACH OF WHICH THERE ARE TWO DEGREES OF FREEDOM*******
       NFAIL=0
       NNODE=4
       NDOFN=2
       NDIME=2
       NEVAB=16
       NPROP=5
       NGAUS=4
       NKNOTS=8
!***READ AND PRINT THE BASIC PARAMETERS OF THE PARTICULAR PROBLEM*****
       READ(1,*)NSTRIP,NSECTN,NPOIN,NRESND,NMATS
       NELEM=NSTRIP*NSECTN
       WRITE(2,'(/,A,I5)')'NUMBER OF STRIPS = ',NSTRIP
       WRITE(2,'(/,A,I5)')'NUMBER OF SECTIONS = ',NSECTN
       WRITE(2,'(/,A,I5)')'NUMBER OF MEMBERS = ',NELEM
       WRITE(2,'(/,A,I5)')'NUMBER OF NODES = ',NPOIN
       WRITE(2,'(/,A,I5)')'NUMBER OF RESTRAINED NODES = ',NRESND
       WRITE(2,'(/,A,I5)')'NUMBER OF DIFFERENT TYPES OF MATERIALS = ',NMATS
       IF(NSECTN <4) THEN
       WRITE(2,'(A)')'PROGRAM CANNOT HANDLE LESS THAN NSECTN =4'
       NFAIL=1
       EXIT
       END IF
!***ALLOCATE ARRAYS**************************************************
       ALLOCATE(COORD(NPOIN,NDIME),LNODS(NELEM,NNODE),&
       &MATNO(NELEM,1),PROPS(NMATS,5),NODFRE(NPOIN,NDOFN),&
       &NDFFIX(NRESND),IFPRE(NRESND,NDOFN),ESTIF(NEVAB,NEVAB),&
```

```
        &MEMDIS(NEVAB),EMASS(NEVAB,NEVAB),ELCOD(NDIME,NNODE),&
        &SHAPEFUN(NNODE),DERIV(NDIME,NNODE),CARTD(NDIME,NNODE),&
        &POSGP(NGAUS),WEIGHP(NGAUS),GPCOD(NDIME,9),NTYPE(NSTRIP,2),&
        &LKNOTS(NELEM,8))
!***READ THE TYPE OF EDGE FOR THE STRIPS. SIMPLY SUPPORTED = 1, ******
!***CLAMPED = 2, FREE = 0. READ NEAR END FIRST***************************
        WRITE(2,'(//,A)')'TYPE OF EDGE FOR THE STRIPS  SIMPLE SUPPORT = 1  &
        &CLAMPED SUPPORT = 2 AND UNRESTRAINED = 0'
         WRITE(2,'(//,A)')' STRIP NO.  EDGE 1     EDGE 2'
        CALL INTGEN(NTYPE,2,NSTRIP)
        DO  ISTRIP = 1, NSTRIP
        WRITE(2,'(3(I6,6X))')ISTRIP,NTYPE(ISTRIP,1),NTYPE(ISTRIP,2)
        END  DO
!***READ THE COORDINATES OF SOME OF THE JOINTS, GENERATE THE***********
!***OMITTED DATA AND PRINT THE COORDINATES OF ALL THE NODES***********
        CALL CORDGEN
        F(NFAIL= =1) EXIT
        WRITE(2,'(/,A)')' COORDINATES OF NODES'
        WRITE(2,'(/,A)')'    NODE      X-CORD         Y-CORD'
        DO  IPOIN=1,NPOIN
        WRITE(2,'(1X,I5,3(5X,F10.3))')IPOIN,(COORD(IPOIN,JDIME),JDIME=1,NDIME)
        END DO
!***READ THE NODE NUMBERS CORRESPONDING TO SOME OF THE ********
!***ELEMENTS AND GENERATE THE OMITTED DATA**********************
        CALL INTGEN(LNODS,NNODE,NELEM)
    IF(NFAIL= =1) EXIT
!***READ THE MATERIAL NUMBERS OF SOME OF THE ELEMENTS AND******
!*** GENERATE THE OMITTED DATA**************************************
        CALL INTGEN(MATNO,1,NELEM)
        IF(NFAIL= =1) EXIT
!***PRINT THE MATERIAL NUMBERS AND THE NODE NUMBERS OF  ELEMENTS****
        WRITE(2,'(3/,A)')' ELEMENT NO.   MATERIAL NO.   NODE-1   NODE-2  &
        &  NODE-3   NODE-4'
        DO  IELEM = 1, NELEM
        WRITE(2,'(1X,I5,11X,I5,9X,8(I5,5X))')IELEM,MATNO(IELEM,1),&
        &(LNODS(IELEM,JNODE),JNODE=1,NNODE)
        END DO
!***READ NOPTN. THIS TELLS WHETHER THE NODE FREEDOM ARRAY IS **********
!**INPUT AS DATA (FOR NOPTN=1) OR TO BE GENERATED FROM THE DATA********
!*** ABOUT THE NODES WHERE ONE OR MORE DISPLACEMENTS ARE ZERO(NOPTN=0)*
        READ(1,*)NOPTN
!***READ NODE FREEDOMS OF SOME OF THE NODES AND GENERATE THE ******
!***OMITTED DATA*******************************************************
        IF(NOPTN= =1) THEN
        CALL INTGEN(NODFRE,NDOFN,NPOIN)
        IF(NFAIL= =1)  EXIT
        ELSE
!***READ AND ECHO DATA ABOUT THE NODE NUMBERS AND FIXITY STATE AT*****
!*** NODES WHERE ONE OR MORE FREEDOMS ARE EQUAL TO ZERO***************
        DO  IRESND=1,NRESND
        READ(1,*)NDFFIX(IRESND),(IFPRE(IRESND,JDOFN),JDOFN=1,NDOFN)
        END DO
        WRITE(2,'(3/,A)')' DEGREES OF RESTRAINT AT RESTRAINED NODES'
        WRITE(2,'(/,A)')' 1 = FIXED AND 0 = FREE'
        WRITE(2,'(//,A)')'NODE NO.   Z-FRDOM.  THETAY-FRDOM'
```

```
        DO IRESND=1,NRESND
        WRITE(2,'(I5,10X,I5,8X,I5,16X,I5)')NDFFIX(IRESND),(IFPRE(IRESND,JDOFN),&
        &JDOFN=1,NDOFN)
        END DO
!***CALL THE SUBROUTINE TO GENERATE THE NODE FREEDOM ARRAY********
!*** FROM THE DATA ABOUT THE RESTRAINED NODES************************
        CALL NFGEN
        END IF
        WRITE(2,'(2/,A)')' DEGREES OF FREEDOM AT NODES'
        WRITE(2,'(2/,A)') 'NODE NO.   Z-FRDOM.   THETAY-FRDOM.'
        DO  IPOIN=1,NPOIN
        WRITE(2,'(I3,5X,I5,10X,I5,9X,I5)')IPOIN,(NODFRE(IPOIN,IDOFN),IDOFN=1,NDOFN)
        END DO
!***READ AND PRINT THE PROPERTIES OF DIFFERENT MATERIALS******************
        DO IMATS=1,NMATS
        READ(1,*)(PROPS(IMATS,IPROP),IPROP=1,NPROP)
         END DO
        WRITE(2,'(2/,A)')' PROPERTIES OF DIFFERENT MATERIALS'
        WRITE(2,'(2/,A)')' MATERIAL NO.   D11           D12           D22&
        &            D33              UDL. MASS '
        DO  IMATS=1,NMATS
        WRITE(2,'(2X,I5,8X,5(E14.6,2X))')IMATS,(PROPS(IMATS,IPROP),IPROP=1,NPROP)
        END DO
!***READ THE KNOTS AFFECTING THE ELEMENT*********************************
        CALL INTGEN(LKNOTS,NKNOTS,NELEM)
        IF(NFAIL= =1) EXIT
!***PRINT THE KNOT NUMBERS RELEVANT FOR ALL THE ELEMENTS*********
        WRITE(2,'(3/,A)')' ELEMENT    KNOT-1     KNOT-2   KNOT-3&
        &   KNOT-4    KNOT-5    KNOT-6     KNOT-7     KNOT-8'
        DO  IELEM = 1, NELEM
        WRITE(2,'(1X,I5,3X,8(I5,5X))')IELEM,(LKNOTS(IELEM,JKNOTS),&
        &JKNOTS=1,NKNOTS)
        END DO
!***CALCULATE THE MAXIMUM NUMBER OF EQUATIONS*************
        CALL MAXEQN
!***CALCULATE THE HALF BANDWIDTH*****************************
        CALL BANDWDSPLN
!***ALLOCATE ARRAYS*********************************************
        ALLOCATE(NCOLM(NEQNS),NDIA(NEQNS))
!***CALL THE SUBROUTINE TO DETERMINE THE  PARAMETERS OF VECTOR ***
!***FORMAT STORAGE**************************************************
        CALL NSKYLINESPLN
!***ALLOCATE ARRAYS*********************************************
        ALLOCATE(GSTIF(MAXCOL),GMASS(MAXCOL))
!***READ THE CONCENTRATED MASSES AT THE NODES AND STORE IN THE ***
!***ARRAY MASPT(NLODND),XMASND(NLODND)****************************
        READ(1,*)NLODND
        WRITE(2,'(/,A,I5)')'NUMBER OF NODES WITH CONCENTRATED MASS&
        &=',NLODND
!***ALLOCATE ARRAYS*********************************************
        IF (NLODND >0)  THEN
        ALLOCATE(MASPT(NLODND), XMASND(NLODND))
        WRITE(2,'(A)')' NODE       LINE MASS'
        CALL READMAS
        END IF
```

```
!***FIRST ZERO AND THEN FORM GLOBAL STIFFNESS AND MASS MATRICES***
        GSTIF=0.0
        GMASS=0.0
        CALL DSPLNPB
!***READ AND ECHO THE NUMBER OF EIGEN-PAIRS REQUIRED. INCLUDE ***
!***GUARD VECTORS*****************************************************
        READ(1,*)NVECTR
        WRITE(2,'(/,A,I5)')'NUMBER OF EIGEN VALUES REQUIRED = ',NVECTR
!***ALLOCATE ARRAYS*******************************************
        ALLOCATE(VECTOR(NEQNS,NVECTR),FREQ2(NVECTR))
        CALL SIMIT
        CALL STURM
        END DO PROBLEMS
        CONTAINS
!**********************************************************
!***INCLUDE ALL RELEVANT SUBROUTINES*****************
!**********************************************************
        END PROGRAM SPLNSIM
```

12.18 DATA PREPARATION FOR PROGRAMS

This section summarises the data preparation for three programs. Others can be prepared in a similar manner.

12.18.1 DATA INPUT TO PROGRAM FRASIM.F90

Initially the program will ask for the following data to be input on the screen.

- Name of input data file name.
- Name of output result file name.

Once this is done, the following data will be read from a file whose name has been given above.

- Number of Problems to be analysed (NPROB).

Now follows NPROB sets of data

- Title of Problem.
- ISTRUCT.

(ISTRUCT = 1, two-dimensional rigid-jointed structure, ISTRUCT = 2, plane grid).

- NPOIN, NELEM, NRESND, NMATS.
- Input coordinates of nodes. The subroutine CORDGEN can generate missing data.

Input the NODEND(node number = 1), NSTEP (=1)
X AND Y COORDINATES OF NODE 1

After this only input data to generate intermediate data if there are equally spaced nodes. In this case then just input
NODEND (= last node in the set of equally spaced nodes), NSTEP (= number of nodes from the previously read node to the NODEND)

X AND Y COORDINATE OF last node in the line namely NODEND
Continue until all the necessary information about coordinates data is input
[Finally if the last NODEND data does not correspond to the node NPOIN, input it as
NODEND (Last number = NPOIN), NSTEP (= 1)
X AND Y COORDINATE OF NPOIN]

- Input node number at the ends of element. The subroutine INTGEN can generate missing data.

Input the
IEND (Element number = 1), NSTEP (= 1)
First and Second Nodes at the ends of Element 1

After this, input data to generate intermediate data if there are equally incremented node numbers. In this case then just input
IEND (= last element in the set of equally incremented node numbers), NSTEP (= number of elements from the previously read element number to the IEND)
First and Second Nodes at the ends of Element IEND

Continue until all the necessary information about nodes at the ends of elements data is input.
[Finally if the last IEND data does not correspond to the node NELEM, input it as
IEND (Last number = NELEM), NSTEP (= 1)
First and Second Nodes at the ends of Element NELEM]

- Input material number of elements. The subroutine INTGEN can generate missing data.

Input the
IEND (Element number = 1), NSTEP (= 1)
Material Number of Element 1

After this only input data to generate intermediate data if there are equally incremented element numbers with the same material number In this case then just input
IEND (= last element in the set of same material numbers), NSTEP (= number of elements from the previously read element number to the IEND)
Material Number of Element IEND

Continue until all the necessary information about material numbers of elements data is input
[Finally if the last IEND data does not correspond to the node NELEM, input it as
IEND (Last number = NELEM), NSTEP (= 1)
Material number of NELEM

- Input NOPTN.

If NOPTN = 0, then
Input fixity data at NRESND number of nodes
There are thus NRESND sets of data
Node number restrained, Fixity code (1 = fixed, 0 = free) for the three freedoms at the node
....

....

If NOPTN = 1, then

- Input the node freedoms at all the NPOIN nodes. The subroutine INTGEN can generate missing data.

Input the
IEND(Node number = 1), NSTEP (= 1)
The three node freedom numbers corresponding to the node = 1

After this only input data to generate intermediate data if there are equally incremented node numbers and freedom numbers. In this case then just input
IEND (= last Node in the set of equally incremented node numbers with equally incremented freedom numbers), NSTEP (= number of nodes from the previously read node number to the IEND)
The three freedom numbers corresponding node IEND

Continue until all the necessary information about nodes and corresponding node freedom data is input
[Finally if the last IEND data does not correspond to the node NPOIN, input it as
IEND (Last number = NPOIN), NSTEP (= 1)
The three freedom numbers corresponding to node NPOIN.

- Material properties corresponding to NMATS number of materials.

If it is a two-dimensional rigid-jointed structure, then there are thus NMATS sets of data.
Material data for the first material: E, G, I, A, Area shear factor

....

....

If it is a plane grid structure, then there are thus NMATS sets of data
Material data for first material: E, G, I, J, A, Area shear factor, I Polar

....

- Data about mass on elements and mass at nodes.

NLODND = number of nodes with concentrated mass, NLODEL = number of elements with uniformly distributed mass.

- Data about concentrated mass at nodes.

NLODND sets of data to follow. If NLODND = 0, Skip this section
Node number with a concentrated mass
Value of concentrated mass

....

....

- Data about uniformly distributed mass on elements.

NLODEL sets of data to follow. If NLODEL = 0, Skip this section
Element number with uniformly distributed mass
Value of uniformly distributed mass

....

....

- Input NVECTR, the number of eigen-pairs required.

12.18.2 DATA INPUT TO PROGRAM DPLSIM.F90

Initially the program will ask for the following data to be input on the screen.

- Name of input data file name.
- Name of output result file name.

Once this is done, the following data will be read from a file whose name has been given above.

- Number of Problems to be analysed (NPROB).

Now follows NPROB sets of data

- Title of Problem.
- NPOIN, NELEM, NRESND, NMATS.
- Input coordinates of nodes.

This section is similar to program FRASIM.F90

- Input node numbers of element in the *anticlockwise* direction. Start at bottom-left corner node.

This section is similar to program FRASIM.F90

- Input material number of elements.

This section is similar to program FRASIM.F90

- Input NOPTN.

This section is similar to program FRASIM.F90

If NOPTN = 0, then input fixity data at NRESND number of nodes. Fixity code (1= fixed, 0 = free) for the three freedoms at the node in the order w, dw/dx and dw/dy

....

....

If NOPTN = 1, then input the node freedoms at all the NPOIN nodes. The three node freedom numbers (in the order w, dw/dx and dw/dy) corresponding to the node = 1. This section is similar to program FRASIM.F90

- Material properties corresponding to NMATS number of materials.

Material data is: E, Poisson's ratio, thickness, uniformly distributed Mass, subgrade modulus.

....

- Data about concentrated mass.

This section similar to program FRASIM.F90, except that no data need be input on distributed mass.

- Input NVECTR, the number of eigen-pairs required.

12.18.3 DATA INPUT TO PROGRAM SPLNSIM.F90

Initially the program will ask for the following data to be input on the screen.

- Name of input data file name.
- Name of output result file name.

Once this is done, the following data will be read from a file whose name has been given above.

- Number of problems to be analysed (NPROB).

Now follows NPROB sets of data

- Title of Problem.
- NSTRIP, NSECTN, NPOIN, NRESND, NMATS.

NSTRIP refers to the number of finite strips in the y direction into which the structure is divided.
NSECTN refers to the number of elements into which the strip is divided.
NPOIN = number of nodes in the structure.
NRESND refers to restrained nodes along the edges parallel to the y direction. Can only handle where all the nodes along an edge are restrained in displacement on z direction only or in the rotation about the y axis only or, of course, fully restrained in both displacement as well as rotation.
NMATS refers to different elastic rigidities D11, D12, D22, D33 and uniformly distributed mass on the element.

- Input types of edge conditions on the edges parallel to x axis.

The subroutine INTGEN is used to generate missing data
IEND (Strip number = 1), NSTEP (= 1)
End condition at edge 1, End condition at Edge 2 for strip 1
After this only input data to generate intermediate data if there are equally incremented node numbers. In this case then just input
IEND (= last strip in the set of equally incremented strip numbers), NSTEP (= number of strips from the previously read strip number to the IEND)
End condition at edge 1, End condition at Edge 2 for strip IEND
Continue until all the necessary information about the end conditions at the ends of strips data is input
[Finally if the last IEND data does not correspond to the node NSTRIP, input it as
IEND (Last number = NSTRIP), NSTEP (=1)
End condition at edge 1, End condition at Edge 2 for strip NSTRIP

- Input coordinates of nodes.
- Input node numbers of element

This section is similar to program FRASIM.F90.
It is important to note that the element number is calculated as IELEM = (ISTRIP - 1) × NSECTN + ISECTN, ISTRIP = 1 to NSTRIP and ISECTN = 1 to NSECTN
Therefore the first element is the first section in the first strip. The numbering is continued until the end section of that strip. The next element starts at the first section of the second strip and so on.

- Input material number of elements.
- Input NOPTN.
- Material properties corresponding to NMATS number of materials. Material data D11, D12, D22, D33, uniformly distributed mass.

This section is similar to program FRASIM.F90.

Fixity code (1 = fixed, 0 = free) for the two freedoms at the node in the order w and dw/dy

- Data about the KNOTS affecting the element.

For each element splines from eight knots affect the displacement. Four splines centred at knots at left-hand edge and four other splines centred at knots on the right-hand edge parallel to the y axis.
Input the data in the following order: knot 1 left, knot 1 right, knot 2 left, knot 2 right, knot 3 left, knot 3 right, knot 4 left, knot 4 right.
Note that for the first element in a strip, knot 1 left, knot 1 right do not exist. Input them as zeros.
Similarly for the last element in a strip, knot 4 left, knot 4 right do not exist. Input them as zeros.

- Data about concentrated mass at nodes.
- Input NVECTR, the number of eigen-pairs needed.

This section is similar to program DPLSIM.F90.

12.18.4 DATA INPUT TO PROGRAM FRASTEP.F90

From the start, data input is the same as for program FRASIM.F90 except that the last data NVECTR should not be input. The new data starts here.
Input time increment: Δt.
Input maximum time up to which response required: TMAX.
Input Rayleigh damping ($C = \alpha K + \beta M$) constants: α , β
Input :IQUAKE, IFORS
IQUAKE = 1 for earthquake base acceleration-time history input in the horizontal direction.
IFORS = 1 for force–time history at selected nodes in x and y directions for two-dimensional rigid-jointed structure or in z direction for plane grid.
If IQUAKE = 1, then
Input how many points of acceleration-time data are given: NPOINTS
Input NPOINTS sets of data of: Time, Acceleration.
If IFORS = 1, then
Input how many points of force-time data are given: NDATAPN
Input at how many nodes these forces are applied: NLODPTS
Input NLODPTS sets of data of: Force, Time.

12.18.5 DATA INPUT TO PROGRAM DPLSTEP.F90

From the start, data input is same as for program DPLSIM.F90 except that the last data NVECTR should not be input. The new data is as for FRASTEP.F90.

SELECTED REFERENCES

General Mathematics

Jeffrey, A., 1989, *Mathematics for engineers and scientists.* (Van Nostrand Reinhold), 4th Edition.

Strang, G., 1988, *Linear algebra and its applications.* (Harcourt, Brace, Jovanovich), 3rd Edition.

Analysis of Eigen-value Problems

Golub, G.H., and Van Loan, C.F., 1996, *Matrix Computations.* (The Johns Hopkins University), 3rd Edition.

Jennings, A., and McKeown, J.J., 1992, *Matrix Computation.* (Wiley), 2nd Edition.

Parlett, B.N., 1980, *The symmetric eigenvalue problem.* (Prentice-Hall).

Rajendran, S., and Narasimhan, M.V., 1994, An accelerated subspace iteration method, *International Journal for Numerical Methods in Engineering*, Vol. 37, pp.141-153.

Scraton, R.E., 1987, *Further Numerical Methods in BASIC.* (Arnold).

Sehmi, N.S., 1989, *Large order structural eigenanalysis techniques.* (Ellis Horwood).

Simpson, A., 1984, On the solution of S(ω)x = 0 by a Newtonian procedure, *Journal of Sound and Vibration*, Vol. 97, no. 1, pp. 153-164.

Wilkinson, J.H., 1988, *The algebaic eigenvalue problem.* (Oxford University Press).

Williams, F.W., and Kennedy, D., 1988, Reliable use of determinants to solve non-linear structural eigenvalue problems efficiently, *International Journal for Numerical Methods in Engineering*, Vol. 26, pp.1825-1841.

Programming Aspects

Bhatt, P., 1986, *Programming the matrix analysis of skeletal structures.* (Ellis Horwood).

Hinton, E., 1988, *Numerical methods and software for dynamic analysis of plates and shells*, (Pineridge Press).

Hinton, E., and Owen, D.R.J., 1977, *Finite element programming.* (Academic Press).

Hinton, E., and Owen, D.R.J., 1980, *Finite element in plasticity.* (Pineridge Press).

Hinton, E., and Owen, D.R.J., 1984, *Finite element software for plates and shells.* (Pineridge Press).

Hou-Cheng Huang, 1989, *Static and dynamic analysis of plates and shells.* (Springer-Verlag).

Smith, I.M., and Griffiths, D.V., 1998, *Programming the finite element method.* (Wiley), 3rd Edition.

Fortran 90

Hahn, B.D., 1994, *Fortran 90 for scientists and engineers.* (Arnold).
Smith, I.M., 1995, *Programming in Fortran-90.* (Wiley).

Galerkin's Weighted Residual Method

Finlayson, B.A., 1972, *The method of weighted residuals and variational principles.* (Academic Press).
Ottosen, N., and Peterson, H., 1992, *Finite element method.* (Prentice-Hall).
Zienkiewicz, O.C., and Morgan, K., 1983, *Finite elements and approximation.* (Wiley).

Structural Dynamics

Blevins, R.D., 1979, *Formulas for natural frequency and mode shape.* (Van Nostrand).
Chopra, A.K., 1982, *Dynamics of structures: A primer.* (Earthquake Engineering Research Institute).
Chopra, A.K., 1995, *Dynamics of structures.* (Prentice Hall).
Clough, R.W., and Penzien, J., 1993, *Dynamics of structures.* (McGraw-Hill), 2nd Edition.
Fertis, D.G., 1995, *Mechanical and structural vibration.* (Wiley).
Humar, J.L., 1990, *Dynamics of structures.* (Prentice Hall).
Tedesco, J.W., McDougal, W.G., and Allen Ross, C., 1999, *Structural dynamics.* (Addison-Wesley).

Finite Elements

Bathe, K.J., 1996, *Finite element procedures.* (Prentice-Hall).
Crisfield, M.A., 1986, *Finite elements and solution procedures for structural analysis, Vol. 1, Linear analysis.* (Pineridge Press).
Crisfield, M.A., 1991, *Non-linear finite element analysis of solids and structures, Vol. 1, Essentials.* 1991 and *Vol. 2 Advanced topics.* 1997, (Wiley).
Hitchins, D. (Ed.), 1992, *A finite element dynamics primer, NAFEMS.*
Hughes, T.J.R., 1987, *The finite element method.* (Prentice-Hall).
Peyt, M., 1990, *Introduction to finite element vibration analysis.* (Wiley).
Weaver, W., and Johnston, P.R., 1987, *Structural dynamics by finite elements.* (Prentice Hall).
Zienkiewicz, O.C., and Taylor, R.L., 2000, *The finite element method, Volumes 1 and 2.* (Butterworth-Heinemann).

Finite Strip Method

Cheung, M.S., Li, W., and Chidiac, S.E., 1996, *Finite strip analysis of bridges.* (Spon).

Earthquake Engineering

Gupta, A.K., 1990, *Response spectrum method.* (Blackwell Scientific).

Housner, G., and Jennings, P.C., 1982, *Earthquake design criteria.* (Earthquake Engineering Research Institute).

Newmark, N.M., and Hall, W.J., 1982, *Earthquake spectra and design.* (Earthquake Engineering Research Institute).

Design of Earthquake Resistant Structures, Use of Codes, etc.

Booth, E. (Ed.), 1994, *Concrete structures in earthquake regions.* (Longman Scientifica and Technical).

Green, N.B., 1987, *Earthquake resitant building design and construction.* (Elsevier), 3rd Edition.

Key, D., 1988, *Earthquake design practice for buildings.* (Thomas Telford).

Penelis, G.G., and Kappos, A.J., 1997, *Earthquake-resistant concrete structures.* (Spon).

Soil-Structure Interaction

Wolf, J.P., 1985, *Dynamic soil-structure interaction.* (Prentice-Hall).

Practical Guidelines

Bachman, H., and Ammann, W., 1994, *Vibration problems in structures: practical guidelines.* (Birkhauser Verlag).

Jeary, A., 1997, *Designer's guide to the dynamic response of structures.* (Spon).

Theory of Elasticity and Plates

Timoshenko, S.P., and Goodier, J.N., 1970, *Theory of elasticity.* (McGraw-Hill), 3rd Edition.

Timoshenko, S.P., and Woinowsky-Krieger, S., 1959, *Theory of plates and shells.* (McGraw-Hill), 2nd Edition.

Skeletal Structures

Bhatt, P., 1981, *Problems in structural analysis by matrix methods.* (Construction Press).

Offshore Structures

Barltrop, N.D.P., and Adams, A.J., 1991, *Dynamics of fixed marine structures.* (Butterworth-Heinemann), 3rd Edition.

Brebbia, C.A., and Walker, S., 1979, *Dynamic analysis of offshore structures.* (Newnes-Butterworths).

INDEX